D1207051

# HEALTH EFFECTS OF
# EXPOSURE TO
# RADON

## BEIR VI

Committee on Health Risks of Exposure to Radon (BEIR VI)

Board on Radiation Effects Research

Commission on Life Sciences

National Research Council

NATIONAL ACADEMY PRESS
Washington, D.C. 1999

**NATIONAL ACADEMY PRESS • 2101 Constitution Avenue, NW • Washington, D.C. 20418**

NOTICE: The project that is the subject of this report was approved by the Governing Board of the National Research Council, whose members are drawn from the councils of the National Academy of Sciences, the National Academy of Engineering, and the Institute of Medicine. The members of the committee responsible for the report were chosen for their special competences and with regard to appropriate balance.

This report was prepared under Grant No. X820576-01-0 between the National Academy of Sciences and the Environmental Protection Agency.

*Health Risks of Exposure to Radon: BEIR VI* is available for sale from the National Academy Press, 2101 Constitution Avenue, N.W., Lock Box 285, Washington, DC, 20055. Call 800-624-6242 or 202-334-3938 (Washington Metropolitan Area); Internet, http://www.nap.edu

## COMMITTEE ON HEALTH RISKS
## OF EXPOSURE TO RADON  (BEIR VI)

JONATHAN M. SAMET (*Chairman*), Department of Epidemiology, The
Johns Hopkins University, Baltimore, Maryland

DAVID BRENNER, College of Physicians and Surgeons, Columbia
University, New York

ANTONE L. BROOKS, Washington State University at Tri-Cities, Richland,
Washington

WILLIAM H ELLETT, National Research Council (ret.), Crofton, Maryland

ETHEL S. GILBERT, Radiation Epidemiology Branch, National Cancer
Institute, Bethesda, Maryland

DUDLEY T. GOODHEAD, Medical Research Council, Oxfordshire, England

ERIC J. HALL, College of Physicians and Surgeons, Columbia University,
New York

PHILIP K. HOPKE, Department of Chemistry, Clarkson University, Potsdam,
New York

DANIEL KREWSKI, Faculty of Medicine, University of Ottawa, and Health
Protection Branch, Health Canada, Ottawa, Canada

JAY H. LUBIN, Biostatistics Branch, National Cancer Institute, Bethesda,
Maryland

ROGER O. McCLELLAN, Chemical Industry Institute of Toxicology,
Research Triangle Park, North Carolina

PAUL L. ZIEMER, School of Health Sciences, Purdue University, West
Lafayette, Indiana

## NATIONAL RESEARCH COUNCIL STAFF

EVAN B. DOUPLE, Study Director; Director, Board on Radiation Effects
Research (as of April 1997)

JOHN D. ZIMBRICK, Director, Board on Radiation Effects Research (through
December 1996)

AMY NOEL O'HARA, Project Assistant (through November 10, 1997)

CATHERINE S. BERKLEY, Administrative Associate

DORIS E. TAYLOR, Administrative Assistant

NORMAN GROSSBLATT, Editor

## SPONSOR'S PROJECT OFFICER

SUSAN CONRATH, U.S. Environmental Protection Agency

ANITA SCHMIDT, U.S. Environmental Protection Agency

The National Academy of Sciences is a private, nonprofit, self-perpetuating society of distinguished scholars engaged in scientific and engineering research, dedicated to the furtherance of science and technology and to their use for the general welfare. Upon the authority of the charter granted to it by the Congress in 1863, the Academy has a mandate that requires it to advise the federal government on scientific and technical matters. Dr. Bruce M. Alberts is president of the National Academy of Sciences.

The National Academy of Engineering was established in 1964, under the charter of the National Academy of Sciences, as a parallel organization of outstanding engineers. It is autonomous in its administration and in the selection of its members, sharing with the National Academy of Sciences the responsibility for advising the federal government. The National Academy of Engineering also sponsors engineering programs aimed at meeting national needs, encourages education and research, and recognizes the superior achievements of engineers. Dr. William A. Wulf is the president of the National Academy of Engineering.

The Institute of Medicine was established in 1970 by the National Academy of Sciences to secure the services of eminent members of appropriate professions in the examination of policy matters pertaining to the health of the public. The Institute acts under the responsibility given to the National Academy of Sciences by its congressional charter to be an adviser to the federal government and, upon its own initiative, to identify issues of medical care, research, and education. Dr. Kenneth I. Shine is president of the Institute of Medicine.

The National Research Council was organized by the National Academy of Sciences in 1916 to associate the broad community of science and technology with the Academy's purposes of furthering knowledge and advising the federal government. Functioning in accordance with general policies determined by the Academy, the Council has become the principal operating agency of both the National Academy of Sciences and the National Academy of Engineering in providing services to the government, the public, and the scientific and engineering communities. The Council is administered jointly by both Academies and the Institute of Medicine. Dr. Bruce M. Alberts and Dr. William A. Wulf are chairman and vice chairman, respectively, of the National Research Council.

# Preface

Nearly a decade has passed since the fourth in a series of studies called Biological Effects of Ionizing Radiation (BEIR) assessed the risks posed by exposure to radon and other alpha emitters in 1988. Radon, a gas emitted into homes from the soil, from water and from building materials, becomes trapped in homes. Its radioactive daughters, the progeny of radioactive decay, are inhaled into human lungs, where further decay results in the exposure of lung cells to densely ionizing alpha particles. On the basis of considerable experience gained by studying health effects in uranium and other miners who worked in radon-rich environments, the radioactive radon progeny were identified as a cause of lung cancer. It has not been clear whether radon poses a similar risk of causing lung cancer in men, women, and children exposed at generally lower levels found in homes, but homeowners are concerned about this potential risk, and the Environmental Protection Agency (EPA) suggests "action levels," concentrations to which citizens are encouraged to reduce their levels of exposure.

Motivated by the ubiquitous exposure of the general population to radon and the continued concern about the risks of exposure to this natural radioactive carcinogen, EPA sought the advice of the National Academy of Sciences and the National Research Council in re-examining the approaches to assessing the lung-cancer risk associated with radon concentrations in the domestic environment. As requested by EPA, the National Research Council first conducted a scoping study; in its 1994 report *The Health Effects of Exposure to Radon: Time for Reassessment?,* the BEIR VI Phase I committee concluded that sufficient and appropriate new evidence had become available since the 1989 BEIR IV report to justify a new National Research Council BEIR VI study. Consequently, the

BEIR VI committee, consisting of 13 scientists with the expertise required to explore EPA's charge, was formed in 1994 to reexamine the risk of health effects posed by exposure to radon in homes.

The report that follows includes 4 chapters and 7 appendixes. The chapters provide the committee's principal findings; supporting analyses and other related evidence are presented in the appendixes. After an introduction to the radon problem in chapter 1, the biologic evidence on mechanisms of radon-related lung cancer is reviewed in chapter 2, which summarizes information on issues identified by the committee as critical to assessing lung-cancer risk from radon exposure. Chapter 3 presents the committee's risk models and 3 major risk projections, provides the rationale for the committee's modeling decisions, describes the committee's preferred risk models and the projections of lung-cancer risk resulting from their use, and addresses the issues related to uncertainty in the risk projections. Chapter 4 reviews the evidence on health effects other than lung cancer that result from exposure to radon progeny.

The appendixes support the committee's findings and should be consulted by readers who want in-depth coverage of specific issues. Appendix A is devoted to previously used risk models, the committee's methods and approach to risk modeling, and the details of the uncertainty considerations and analyses. Radon dosimetry is discussed in appendix B, which updates a 1991 National Research Council report on this subject. The dominant cause of lung cancer in the United States and many other countries is tobacco-smoking; appendix C provides additional information on the risks of lung cancer in relation to smoking and reviews the available data on the combined effects of smoking and exposure to radon and its progeny.

Appendixes D, E, and F address the information available from epidemiologic studies on underground miners. Appendix D summarizes the characteristics and designs of the miner studies. Appendix E and its annexes review the details of the exposures in individual miner studies and provide the proceedings of a workshop on exposure estimation. Appendix F addresses exposures to agents other than radon—such as diesel exhaust, silica, and arsenic—which might be relevant to estimating lung-cancer risk posed by radon. The findings of studies—principally ecologic or case-control epidemiologic studies conducted to estimate directly the risk posed by radon exposure in homes—are described in appendix G. The findings of several individual studies and combined analyses of the case-control studies are summarized. Finally, a compilation of the literature cited in the report and a glossary of technical terms used in the report follows appendix G.

## ACKNOWLEDGMENTS

In the course of doing its work, the committee held several meetings, including a meeting in which the scientific community and other interested groups were invited to present suggestions regarding the problem. Other information-gathering meetings were held in which scientists with specific expertise were invited to discuss the results of their work. Several workshops were held in which the committee focused on particular aspects of the radon problem. The committee is grateful to participants in a January 1996 workshop, in which over 30 scientists discussed findings of work relevant to the shape of the dose-response curve at low doses of radiation with densely deposited energy applicable to the alpha particles emitted by radon and its progeny. The results of that workshop, which was sponsored by the Department of Energy's Office of Health and Environmental Research, were useful to the committee in selecting a risk-assessment model.

The committee is appreciative to the many scientists who contributed directly or indirectly to the work of the BEIR VI committee. Seven persons and groups deserve special recognition and thanks. First, the committee is indebted to the principal investigators (PIs) and other scientists who directed and analyzed the various cohort studies of miners and the case-control studies of lung cancer and radiation in the general population. Those scientists willingly and graciously provided their data to the committee and enabled the committee to apply analytical techniques to original data in order to derive its models and its risk estimates. For the miner studies, the scientists included Shu-Xiang Yao, Xiang-Zhen Xuan, and Jay H. Lubin (Chinese tin miners), Emil Kunz and Ladislav Tomášek (Czechoslovakian, now the Czech Republic, Uranium miners), Geoffrey Howe (Beaverlodge, Canada and Port Radium, Canada uranium miners), Howard I. Morrison (Newfoundland, Canada fluorspar miners), Robert A. Kusiak and Jan Muller (Ontario, Canada uranium miners), Margot Tirmarche (French uranium miners), Alistair Woodward (South Australian uranium miners), Edward P. Radford and Christer Edling (Swedish iron miners), Richard W. Hornung (Colorado, U.S.A. uranium miners), and Jonathan M. Samet (New Mexico, U.S.A. uranium miners).

Second, the committee sponsored a workshop on exposures of miners to radon progeny. The following persons, knowledgeable about the history of the mining industry with experience in geology or mine and ventilation engineering, were invited to the workshop and made contributions that were much appreciated: William Chenoweth, James Cleveland, Andreas George, and Douglas Chambers. During the workshop, Daniel Stram and Duncan Thomas provided advice to the committee regarding quantification of the measurement errors associated with the mine exposures.

Third, the committee expresses its thanks to Anthony James, who provided computational analyses of doses to lung cells resulting from radon and its progeny.

Fourth, the committee thanks Susan Rose of the Department of Energy for inviting the BEIR VI chair and study director to participate in meetings of the PIs of the case-control studies in Europe and North America in which plans for pooling of data and joint analyses were discussed.

Fifth, three scientists from Health Canada provided valuable computational analyses and assisted committee members in the development and application of the risk assessment: Shesh N. Rai, Yong Wang, and Jan M. Zielinski. Their contributions were important in the development of chapter 3 and appendix A.

Sixth, the committee is especially grateful to Rosalyn Yalow, an orginal committee member whose health prohibited her from serving as a final author of the study. Dr. Yalow was especially effective in insuring that the committee not overlook the influence of cigarette-smoking on the lung-cancer problem.

Seventh, this report has been reviewed by individuals chosen for their diverse perspectives and technical expertise, in accordance with procedures approved by the National Research Council's Report Review Committee. The purpose of this independent review is to provide candid and critical comments that will assist the authors and the National Research Council in making the published report as sound as possible and to ensure that the report meets institutional standards for objectivity, evidence, and responsiveness to the study charge. The content of the review comments and draft manuscript remain confidential to protect the integrity of the deliberative process. We wish to thank the following individuals for their participation in the review of this report: Gregg Claycamp (University of Pittsburgh School of Public Health), Merril Eisenbud (deceased), Robert Forster (University of Pennsylvania School of Medicine), Naomi Harley (New York University Medical Center), Werner Hofmann (University of Salzburg), Maureen Henderson (University of Washington), Richard Hornung (University of Cincinnati), Donald Mattison (University of Pittsburgh School of Public Health), Suresh Moolgavkar (University of Washington), Lincoln Moses (Stanford University), Louise Ryan (Harvard University School of Public Health), and Duncan Thomas (University of Southern California). Although the individuals listed above have provided many constructive comments and suggestions, responsibility for the final content of this report rests solely with the authoring committee and the National Research Council.

The committee appreciates the assistance of the Board on Radiation Effects Research (BRER) staff who supported the committee's work, especially Doris Taylor for her assistance with meeting and travel arrangements and Amy Noel O'Hara and Dennis Gleeson, Jr., for their attention to the details of manuscript preparation.

The BEIR VI committee has faced many challenges in preparing this report. Committee members are hopeful that the report will meet the needs of the EPA as it considers risk management strategies for indoor radon and also will be informative to the public as homeowners make decisions about testing for radon and lowering radon concentrations in their homes.

Jonathan Samet, M.D.
Chair, Committee on Health Effects of
Exposure to Radon, (BEIR VI)

# Contents

# Public Summary:
# The Health Effects of
# Exposure to Indoor Radon

Radon is a naturally occurring gas that seeps out of rocks and soil. Radon comes from uranium that has been in the ground since the time the earth was formed, and the rate of radon seepage is variable, partly because the amounts of uranium in the soil vary considerably. Radon flows from the soil into outdoor air and also into the air in homes from the movement of gases in the soil beneath homes. Outside air typically contains very low levels of radon, but it builds up to higher concentrations indoors when it is unable to disperse. Some underground mines, especially uranium mines, contain much higher levels of radon.

Although radon is chemically inert and electrically uncharged, it is radioactive, which means that radon atoms in the air can spontaneously decay, or change to other atoms. When the resulting atoms, called radon progeny, are formed, they are electrically charged and can attach themselves to tiny dust particles in indoor air. These dust particles can easily be inhaled into the lung and can adhere to the lining of the lung. The deposited atoms decay, or change, by emitting a type of radiation called alpha radiation, which has the potential to damage cells in the lung. Alpha radiations can disrupt DNA of these lung cells. This DNA damage has the potential to be one step in a chain of events that can lead to cancer. Alpha radiations travel only extremely short distances in the body. Thus, alpha radiations from decay of radon progeny in the lungs cannot reach cells in any other organs, so it is likely that lung cancer is the only potentially important cancer hazard posed by radon in indoor air.

For a century, it has been known that some underground miners suffered from higher rates of lung cancer than the general population. In recent decades, a growing body of evidence has causally linked their lung cancers to exposure to high levels of radon and also to cigarette-smoking. The connection between radon and lung cancer in miners has raised concern that radon in homes might be causing lung cancer in the general population, although the radon levels in most homes are much lower than in most mines. The National Research Council study, which has been carried out by the sixth Committee on Biological Effects of Ionizing Radiation (BEIR VI), has used the most recent information available to estimate the risks posed by exposure to radon in homes.

The most direct way to assess the risks posed by radon in homes is to measure radon exposures among people who have lung cancer and compare them with exposures among people who have not developed lung cancer. Several such studies have been completed, and several are under way. The studies have not produced a definitive answer, primarily because the risk is likely to be very small at the low exposure encountered from most homes and because it is difficult to estimate radon exposures that people have received over their lifetimes. In addition, it is clear that far more lung cancers are caused by smoking than are caused by radon.

Since a valid risk estimate could not be derived only from the results of studies in homes, the BEIR VI committee chose to use the lung-cancer information from studies of miners, who are more heavily exposed to radon, to estimate the risks posed by radon exposures in homes. In particular, the committee has drawn on 11 major studies of underground miners, which together involved about 68,000 men, of whom 2,700 have died from lung cancer. The committee statistically analyzed the data to describe how risk of death from lung cancer depended on exposure. In this way, the committee derived two models for lung-cancer risk from radon exposure.

In converting radon risks from mines to homes, the committee was faced with several problems. First, most miners received radon exposures that were, on the average, many times larger than those of people in most homes; people in a few homes actually receive radon exposures similar to those of some miners. It was necessary for the committee to estimate the risks posed by exposures to radon in homes on the basis of observed lung-cancer deaths caused by higher exposures in mines. The committee agreed with several earlier groups of experts that the risk of developing lung cancer increases linearly as the exposure increases; for example, doubling the exposure doubles the risk, and halving the exposure halves the risk. Furthermore, the existing biologic evidence suggests that any exposure, even very low, to radon might pose some risk. However, from the evidence now available, a threshold exposure, that is, a level of exposure below which there is no effect of radon, cannot be excluded.

The second problem is that the majority of miners in the studies are smokers and all inhale dust and other pollutants in mines. Because radon and cigarette smoke both cause lung cancer, it is complicated to disentangle the effects of the 2 kinds of exposure. That makes it especially difficult to estimate radon risks for nonsmokers in homes using the evidence from miners. A final problem is that the miners were almost all men, whereas the population exposed to radon in homes includes men, women, and children.

The committee used the information from miners and supplemented it with information from laboratory studies of how radon causes lung cancer. Then, with facts about the U.S. population, including measurements of radon levels in homes, it estimated the number of lung-cancer deaths due to radon in homes. In 1995, about 157,400 people died of lung cancer (from all causes including smoking and radon exposure) in the United States. Of the 95,400 men who died of lung cancer, about 95% were probably ever-smokers; of the 62,000 women, about 90% were probably ever-smokers. Approximately 11,000 lung-cancer deaths are estimated to have occurred in never-smokers in 1995.

The BEIR VI committee's preferred central estimates, depending on which one of the two models are used, are that about 1 in 10 or 1 in 7 of all lung-cancer deaths—amounting to central estimates of about 15,400 or 21,800 per year in the United States—can be attributed to radon among ever-smokers and never-smokers together. Although 15,400 or 21,800 total radon-related lung-cancer deaths per year are the committee's central estimates, uncertainties are involved in these estimates. The committee's preferred estimate of the uncertainties was obtained by using a simplified analysis of a constant relative risk model based on observations closest to residential exposure levels. The number of radon-related lung-cancer deaths resulting from that analysis could be as low as 3,000 or as high as 33,000 each year. Most of the radon-related lung cancers occur among ever-smokers, and because of synergism between smoking and radon, many of the cancers in ever-smokers could be prevented by either tobacco control or reduction of radon exposure. The committee's best estimate is that among the 11,000 lung-cancer deaths each year in never-smokers, 2,100 or 2,900, depending on the model used, are radon-related lung cancers.

Radon, being naturally occurring, cannot be entirely eliminated from our homes. Of the deaths that the committee attributes to radon (both independently and through joint action with smoking), perhaps one-third could be avoided by reducing radon in homes where it is above the "action guideline level" of 148 $Bqm^{-3}$ (4 $pCiL^{-1}$) to below the action levels recommended by the Environmental Protection Agency.

The risk of lung cancer caused by smoking is much higher than the risk of lung cancer caused by indoor radon. Most of the radon-related deaths among smokers would not have occurred if the victims had not smoked. Furthermore, there is evidence for a synergistic interaction between smoking and radon. In other words, the number of cancers induced in ever-smokers by radon is greater than one would expect from the additive effects of smoking alone and radon alone. Nevertheless, the estimated 15,400 or 21,800 deaths attributed to radon in combination with cigarette-smoking and radon alone in never-smokers constitute a public-health problem.

# Executive Summary

## INTRODUCTION

This National Research Council's report of the sixth Committee on Biological Effects of Ionizing Radiations (BEIR VI) addresses the risk of lung cancer associated with exposure to radon and its radioactive progeny. Radon, a naturally occurring gas formed from the decay of uranium in the earth, has been conclusively shown in epidemiologic studies of underground miners to cause lung cancer. There is supporting evidence from experimental studies of animals that confirm radon as a cause of lung cancer and from molecular and cellular studies that provide an understanding of the mechanisms by which radon causes lung cancer.

In addition to being present at high concentrations in many types of underground mines, radon is found in homes and is also present outdoors. Extensive measurements of radon concentrations in homes show that although concentrations vary widely, radon is universally present, raising concerns that radon in homes increases lung-cancer risk for the general population, especially those who spend a majority of their time indoors at home. For the purpose of developing public policy to manage the risk associated with indoor radon, there is a need to characterize the possible risks across the range of exposures received by the population. The higher end of that range of exposures is comparable to those exposures that caused lung cancer in underground miners. The lower end of that range includes exposures received from an average indoor lifetime exposure which is at least one order of magnitude lower.

Risk models, which mathematically represent the relationship between exposure and risk, have been developed and used to assess the lung-cancer risks

*4*

associated with indoor radon. For example, the precursor to this committee, the BEIR IV committee, developed one such model on the basis of statistical analysis of data from 4 epidemiologic studies of underground miners. The BEIR IV model has been widely used to estimate the risk posed by indoor radon. Since the 1988 publication of the BEIR IV report, substantial new evidence on radon has become available: new epidemiologic studies of miners have been completed, existing studies have been extended, and analysis of the pooled data from 11 principal epidemiologic studies of underground miners has been conducted involving a total of 68,000 miners and to date, 2,700 deaths from lung cancer. Other lines of scientific evidence relevant to assessing radon risks have also advanced, including findings on the molecular and cellular basis of carcinogenesis by alpha particles. Radon itself does not directly cause lung cancer but alpha particles from radon progeny directly damage target lung cells to cause cancer. There is additional information for calculating the dose of alpha particles received by the lung from inhaled radon progeny, the topic of a 1991 follow-up report to the BEIR IV report, the report of the National Research Council's Panel on Dosimetric Assumptions. Finally, during the last decade, a number of epidemiologic case-control studies that estimated the risk associated with indoor radon directly have also been implemented.

The BEIR VI committee faced the task of estimating the risks associated with indoor radon across the full range of exposures and providing an indication of the uncertainty to be attached to risk estimates across this range. In preparing this report, the BEIR VI committee, in response to its charge, reviewed the entire body of data on radon and lung cancer, integrating findings from epidemiologic studies with evidence from animal experiments and other lines of laboratory investigation. The committee also considered the substantial evidence on smoking and cancer and the more limited evidence on the combined effect of smoking and radon. The report's elements include comprehensive reviews of the cellular and molecular basis of radon carcinogenesis and of the dosimetry of radon in the respiratory tract, of the epidemiologic studies of miners and the general population, and of the combined effects of radon and other occupational carcinogens with tobacco-smoking. The committee describes its preferred risk models, applies the models to estimate the risk posed by indoor radon, and characterizes uncertainties associated with the risk estimates.

## THE MECHANISTIC BASIS OF RADON-INDUCED LUNG CANCER

Information on radon carcinogenesis comes from molecular, cellular, animal, and human (or epidemiologic) studies. Radiation carcinogenesis, in common with any other form of cancer induction, is likely to be a complex multistep process that can be influenced by other agents and genetic factors at each step. Since our current state of knowledge precludes a systematic quantitative description of all steps from early subcellular lesions to observed malignancy, the committee used epidemiologic data to develop and quantify an empirical model of the exposure-risk relationship for lung cancer. The committee did draw extensively, however, on findings from molecular, cellular, and animal studies in developing its risk assessment for the general population.

The committee's review of the cellular and molecular evidence was central to the specification of the risk model. This review led to the selection of a linear nonthreshold relation between lung-cancer risk and radon exposure. However, the committee acknowledged that other relationships, including threshold and curvilinear relationships, cannot be excluded with complete confidence, particularly at the lowest levels of exposure. At low radon exposures, typical of those in homes, a lung epithelial cell would rarely be traversed by more than one alpha particle per human lifespan. As exposure decreases, the insult to cell nuclei that are traversed by alpha particles remains the same as at higher exposures, but the number of traversed nuclei decreases proportionally. There is good evidence that a single alpha particle can cause major genomic changes in a cell, including mutation and transformation. Even allowing for a substantial degree of repair, the passage of a single alpha particle has the potential to cause irreparable damage in cells that are not killed. In addition, there is convincing evidence that most cancers are of monoclonal origin, that is, they originate from damage to a single cell. These observations provide a mechanistic basis for a linear relationship between alpha-particle dose and cancer risk at exposure levels at which the probability of the traversal of a cell by more than one alpha particle is very small, that is, at exposure levels at which most cells are never traversed by even one alpha particle. On the basis of these mechanistic considerations, and in the absence of credible evidence to the contrary, the committee adopted a linear-nonthreshold model for the relationship between radon exposure and lung-cancer risk. However, the committee recognized that it could not exclude the possibility of a threshold relationship between exposure and lung cancer risk at very low levels of radon exposure.

Extrapolation from higher to lower radon exposures is also influenced by the inverse dose-rate effect, an increasing effect of a given total exposure as the rate of exposure is decreased, as demonstrated by experiments in vivo and in vitro for high-LET radiation, including alpha particles, and in miner data. This dose-rate effect, whatever its underlying mechanism, is likely to occur at exposure levels at which multiple particle traversals per cell nucleus occur. Mechanistic, experimental, and epidemiologic considerations support the disappearance of the effect at low exposure corresponding to an average of much less than one traversal per cell location, as in most indoor exposures. Extrapolating radon risk from the full range of miner exposures to low indoor exposures involves extrapolating from a situation in which multiple alpha-particle traversals of target nuclei occur to one in which they are rare; such an extrapolation would be from circumstances in which the inverse dose-rate effect might be important to one in which it is likely to be nonexistent. These considerations indicated a need to assess risks of radon in homes on the basis of miner data corresponding to as low an exposure as possible, or to use a risk model that accounts for the diminution of an inverse exposure-rate effect with decreasing exposure.

The committee also reviewed other evidence relevant to the biologic basis of its risk assessment approach. For the combined effect of smoking and radon, animal studies provided conflicting evidence on synergism, and there is uncertainty as to the relevance of the animal experiments to the patterns of smoking by people. Early attempts to identify a molecular "signature" of prior alpha-particle damage through the identification of unusual point mutations in specific genes have not yet proven useful, although approaches based on specific chromosomal aberrations show some promise, and all the principal histologic types of lung cancer can be associated with radon exposure. Available evidence, albeit limited, supports the likelihood that a typical human population would have a broad spectrum of susceptibility to alpha-particle-induced carcinogenesis.

## THE BEIR VI RISK MODELS

For estimating the risk imposed by exposure to indoor radon, the committee chose an empirical approach based on analysis of data from radon-exposed miners. Other approaches that the committee considered but did not use included a "dosimetric" approach, and use of "biologically-motivated" risk models. A dosimetric approach, in which radon risks are estimated by applying risk estimates from A-bomb survivor studies to estimates of radiation doses delivered to the lung, was not pursued because of the major differences in the type of radiation and exposure patterns compared with radon-progeny exposure. A biologic-based approach to modeling with a description of the various processes leading to radon-induced cancer was not followed primarily because of the present incomplete state of knowledge of many of these processes.

The committee turned to the empirical analysis of epidemiologic data as the basis for developing its risk model. Two sources of information were available: data from the epidemiologic studies of underground miners and data from the case-control studies of indoor radon and lung cancer in the general population. Both groups include ever-smokers and never-smokers. Although the case-control studies provide direct estimates of indoor radon risk, the estimates obtained from these studies are very imprecise, particularly if estimated for never-smokers or ever-smokers separately, because the excess lung-cancer risk is likely to be small. Other weaknesses of the case-control studies are errors in estimating exposure and the limited potential for studying modifying factors, particularly cigarette smoking. Nonetheless, the committee considered the findings of a meta-analysis of the 8 completed studies.

In developing its risk models, the committee started with the recently reported analyses by Lubin and colleagues of data from 11 studies of underground miners—uranium miners in Colorado, New Mexico, France, Australia, the Czech

Republic, and in Port Radium, Beaverlodge, and Ontario in Canada; metal miners in Sweden; tin miners in China; and fluorspar miners in Canada. The data for 4 studies were updated with new information. These 11 studies offered a substantially greater data resource than had been available to the BEIR IV committee. The 11 epidemiologic studies covered a range of mining environments, times, and countries, and their methods of data collection differed in some respects.

The committee analyzed the data with a relative-risk model in which radon exposure has a multiplicative effect on the background rate of lung cancer. In particular, the committee modeled the excess relative risk (ERR), which represents the multiplicative increment to the excess disease risk beyond background resulting from exposure. The model represents the ERR as a linear function of past exposure to radon. This model allows the effect of exposure to vary flexibly with the length of time that has passed since the exposure, with the exposure rate, and with the attained age. The mathematical form of the model for ERR is:

$$ERR = \beta(w_{5-14} + \theta_{15-24}\, w_{15-24} + \theta_{25+}\, w_{25+})\phi_{age}\gamma_z$$

The parameter $\beta$ represents the slope of the exposure-risk relationship for the assumed reference categories of the modifying factors. Exposure at any particular age has 4 components: exposure in the last 5 years—excluded as not biologically relevant to cancer risk—and exposures in 3 windows of past time, namely 5-14, 15-24, and 25 or more years previously. Those exposures are labeled $w_{5-14}$, $w_{15-24}$, and $w_{25+}$, respectively, and each is allowed to have its own relative level of effect, $\theta_{5-14}$ (set equal to unity), $\theta_{15-24}$, and $\theta_{25+}$, respectively. With this weighting system, total exposure can be calculated as $w^* = w_{5-14} + \theta_{15-24}\, w_{15-24} + \theta_{25+}\, w_{25+}$. The rate of exposure also affects risk through the parameter $\gamma_z$; thus, the effect of a particular level of exposure increases with decreasing exposure rate, as indexed either by the duration of exposure or the average concentration at which exposure was received. The ERR also declines with increasing age, as described by the parameter $\phi_{age}$.

Based on this analysis, the committee developed two preferred risk models referred to as the exposure-age-concentration model and the exposure-age-duration model. These two models differ only with respect to the parameter $\gamma_z$, which represents either duration of exposure or the average concentration over the time of the exposure. The models were equally preferred by the committee. The new models are similar in form to the BEIR IV model, but have an additional term for exposure rate and more-detailed categories for the time-since-exposure windows and for attained age.

## RISK ASSESSMENT

The committee's risk models can be used to project the lung-cancer risk associated with radon exposure, both for individuals and for the entire US population. To extend the models that were developed from miner data to the general population, the committee needed to make a set of assumptions on the following key issues.

## Lung Dosimetry of Radon Progeny

Physical and biologic differences between the circumstances of exposures of male miners working underground and of men, women, and children in their homes could lead to differing doses at the same exposures. The committee estimated the value of a dimensionless parameter, termed the "K factor" in prior reports, that characterizes the comparative doses to lung cells in homes and mines for the same exposure. Using a model to estimate the dose to the cells in the lung, and incorporating new information on the input parameters of the model, the committee found that the doses per unit exposure in mines and homes were essentially the same. Thus, K is calculated to be about 1 for men, women and children (age 10 years), and slightly above K = 1 for infants (age 1). Consequently, a value of 1 was used in making the risk projections.

## Extrapolation of Risks at Higher Exposures to Lower Exposures

Average exposures received by the miners in the epidemiologic studies are about one order of magnitude higher than average indoor exposures, although the lowest exposures of some miners overlap with some of the highest indoor exposures. To estimate risks of indoor radon exposures, it is thus necessary to make an assumption about the shape of the exposure-risk relationship across the lower range of the distribution of radon exposures.

The committee selected a linear-nonthreshold relationship relating exposure to risk for the relatively low exposures at issue for indoor radon. This assumption has significant implications for risk projections. Support for this assumption came primarily from the committee's review of the mechanistic information on alpha-particle-induced carcinogenesis. Corroborating information included evidence for linearity in the miner studies at the lower range of exposures, and the linearity and magnitude of risk observed in the meta-analysis of the case-control studies, which was fully consistent with extrapolation of the miner data. Although a linear-nonthreshold model was selected, the committee recognized that a threshold—that is, a level of exposure with no added risk—could exist and not be identifiable from the available epidemiologic data.

## Exposure Rate

At higher exposures, the committee found evidence in the miner data of an inverse exposure-rate effect. Theoretical considerations suggested that the inverse exposure-rate effect found in the miner data should not modify risks for typical indoor exposures. Consequently, the exposure-rate effect in the lowest range of miner exposure rates was applied for relevant indoor exposures without further adjustment.

## Combined Effect of Smoking and Radon

Apart from the results of very limited in vitro and animal experiments, the only source of evidence on the combined effect of the 2 carcinogens (cigarette smoke and radon) was the data from 6 of the miner studies. Analysis of those data indicated a synergistic effect of the two exposures acting together, which was characterized as submultiplicative, i.e., less than the anticipated effect if the joint effect were the product of the risks from the two agents individually, but more than if the joint effect were the sum of the individual risks. The committee applied a full multiplicative relation of the joint effect of smoking and exposure to radon, as done by the BEIR IV committee, and also a submultiplicative relationship. Although the committee could not precisely characterize the joint effect of smoking and radon exposure, the submultiplicative relation was preferred by the committee because it was found to be more consistent with the available data.

## Risks for Women

The risk model is based on epidemiologic studies of male miners. The effect of radon exposure on lung-cancer risk in women might be different from that in men because of differing lung dosimetry or other factors related to gender. The K factor was calculated separately for women and men, but did not differ by gender. The committee also could not identify strong evidence indicative of differing susceptibility to lung carcinogens by sex. Consequently, the model was extended directly to women, with the assumption that the excess risk imposed by radon progeny estimated from the male miners multiplies the background lung cancer rates for women, which are presently substantially lower than for men.

## Risks Associated with Exposures in Childhood

Evidence was available from only one study of miners on whether risk was different for exposures received during childhood, during adolescence, and during adulthood. There was not a clear indication of the effect of age at exposure. The committee made no specific adjustment for exposures received at earlier ages. The K factor for children aged 10 was calculated as 1 and the value for infants was only slightly higher (about 1.08).

## Characterization of Radon Risks

In making its calculations, the committee used the latest data on lung cancer mortality for 1985-1989 and for smoking prevalence for the U.S. in 1993. To characterize the lung-cancer risk posed to the population by indoor radon, the two models for the exposure-risk relationship were applied to the distribution of exposures received by the population to estimate the burden of lung cancer sustained by the population as a result of indoor-radon exposure. To characterize risks to the population, we have used the population attributable risk (AR), which indicates how much of the lung-cancer burden could, in theory, be prevented if all exposures to radon were reduced to the background level of radon in outdoor air. The AR estimates include cases in ever-smokers and never-smokers. To characterize the risk to specific individuals, the committee calculated the lifetime relative risk (LRR), which describes the relative increment in lung-cancer risk resulting from exposure to indoor radon beyond that from exposure to outdoor-background concentrations of radon.

## Radon-Attributable Risks

LRRs were computed using the committee's risk models. Estimates were computed for exposure scenarios which reflect concentrations of indoor radon of interest. Table ES-1 shows the estimated LRRs for lifetime exposures at various constant radon concentrations. The LRR values are quite similar for the preferred 2 models: exposure-age-concentration and exposure-age-duration. The LRR values estimated by the BEIR VI models and the BEIR IV model are also similar, in spite of the addition of exposure rate to the new models. As anticipated, LRR values increase with exposure. Women have a somewhat steeper increment in LRR with increasing exposure because of differing mortality patterns.

Attributable risks for lung cancer from indoor radon in the US population were computed with the committee's 2 preferred models and compared with the BEIR IV results. Based on the National Residential Radon Survey, the committee assumed a log-normal distribution for residential radon concentration, with a median of 24.3 Bqm$^{-3}$ (0.67 pCiL$^{-1}$) and a geometric standard deviation of 3.1 (Marcinowski 1994). The AR was calculated for the entire US population and for males and females and ever-smokers and never-smokers under the preferred submultiplicative model (Table ES-2). For the entire population, the ARs calculated with the new models ranged from about 10% to 14% and were higher than estimates based on the BEIR IV model. Under the submultiplicative assumption which was described on page ES-9, the attributable risk estimates for ever-smokers tended to be lower than estimates for never-smokers, although the numbers of cases are far greater in ever-smokers than in never-smokers.

These AR estimates for the general population are further broken down with respect to the distribution of indoor concentrations in Table ES-3. This analysis provides a picture of the potential consequences of alternative mitigation strategies that might be used for risk-management purposes. The findings were the same for the committee's 2 models. The radon concentration distribution is highly skewed, with homes with higher radon concentrations contributing dispro-

TABLE ES-1  Estimated lifetime relative risk (LRR) of lung cancer for lifetime indoor exposure to radon[a]

| Exposure[b] | | | | | Exposure-age-concentration model | | | | Exposure-age-duration model | | | |
|---|---|---|---|---|---|---|---|---|---|---|---|---|
| | | | | | Male | | Female | | Male | | Female | |
| WLM/y | Jhm⁻³/y | Bqm⁻³ | pCiL⁻¹ | WL | Ever-smoker | Never-smoker | Ever-smoker | Never-smoker | Ever-smoker | Never-smoker | Ever-smoker | Never-smoker |
| 0.10 | 0.00035 | 25 | 0.7 | 0.003 | 1.081 | 1.194 | 1.089 | 1.206 | 1.054 | 1.130 | 1.059 | 1.137 |
| 0.19 | 0.00067 | 50 | 1.4 | 0.005 | 1.161 | 1.388 | 1.177 | 1.411 | 1.108 | 1.259 | 1.118 | 1.274 |
| 0.39 | 0.00137 | 100 | 2.7 | 0.011 | 1.318 | 1.775 | 1.352 | 1.821 | 1.214 | 1.518 | 1.235 | 1.547 |
| 0.58 | 0.00203 | 150 | 4.1 | 0.016 | 1.471 | 2.159 | 1.525 | 2.229 | 1.318 | 1.776 | 1.352 | 1.819 |
| 0.78 | 0.00273 | 200 | 5.4 | 0.022 | 1.619 | 2.542 | 1.694 | 2.637 | 1.420 | 2.033 | 1.466 | 2.091 |
| 1.56 | 0.00546 | 400 | 10.8 | 0.043 | 2.174 | 4.057 | 2.349 | 4.255 | 1.809 | 3.053 | 1.915 | 3.174 |
| 3.12 | 0.01092 | 800 | 21.6 | 0.086 | 3.120 | 7.008 | 3.549 | 7.440 | 2.507 | 5.058 | 2.760 | 5.317 |

[a]Based on a submultiplicative relationship between tobacco-smoking and radon.

[b]Exposures are represented by concentrations in bequerels per cubic meter (Bqm⁻³), picocuries per liter (pCiL⁻¹), or Working Levels (WL), assumed to be constant for home occupancy at the 70% level and 40% equilibrium between radon and its progeny, and also by joules-hours per cubic meter per year (Jhm⁻³/y) and Working Level Months per year (WLM/y).  For definitions of these terms, see the Glossary at the end of this report.

TABLE ES-2   Estimated attributable risk (AR$^a$) for lung-cancer death from domestic exposure to radon using 1985-89 U.S. population mortality rates based on selected risk models

| Model | Population | Ever-smokers$^b$ | Never-smokers$^b$ |
|---|---|---|---|
| **Males** | | | |
| Committee's preferred models | | | |
|   Exposure-age-concentration | 0.141 | 0.125 | 0.258 |
|   Exposure-age-duration | 0.099 | 0.087 | 0.189 |
| Other Models | | | |
|   CRR$^c$ (<0.175 Jhm$^{-3}$; <50 WLM) | 0.109 | 0.096 | 0.209 |
|   BEIR IV | 0.082 | 0.071 | 0.158 |
| **Females** | | | |
| Committee's preferred models | | | |
|   Exposure-age-concentration | 0.153 | 0.137 | 0.269 |
|   Exposure-age-duration | 0.108 | 0.096 | 0.197 |
| Other Models | | | |
|   CRR$^c$ (<0.175 Jhm$^{-3}$; <50 WLM) | 0.114 | 0.101 | 0.209 |
|   BEIR IV | 0.087 | 0.077 | 0.163 |

$^a$AR = the risk of lung cancer death attributed to radon in populations exposed to radon divided by the total risk of lung cancer death in a population.
$^b$Based on a submultiplicative relationship between tobacco-smoking and radon.
$^c$CRR = constant relative risk.

portionately to AR. Only 13% of the calculated AR is estimated to be contributed by the 50% of homes below the median concentration of about 25 Bqm$^{-3}$ (0.7 pCiL$^{-1}$) and about 30% by homes below the mean of about 46 Bqm$^{-3}$ (1.25 pCiL$^{-1}$). Homes above 148 Bqm$^{-3}$ (4 pCiL$^{-1}$), the current action level established by the Environmental Protection Agency, contribute about 30% percent of the AR. This contribution to the total AR is indicative of the potential magnitude of avoidable deaths with a risk management program based on the current action guideline. While 10-15 percent of all lung cancers are estimated to be attributable to indoor radon, eliminating exposures in excess of 148 Bqm$^{-3}$ (4 pCiL$^{-1}$) would prevent about 3 to 4 percent of all lung cancers, or, about one-third of the radon-attributable lung cancers.

The ARs were reestimated with assumption of thresholds, levels below which cancer risk is not increased, at 37, 74, or 148 Bqm$^{-3}$ (1, 2, or 4 pCiL$^{-1}$). Even though the committee assumed that risk was most likely linear with exposure at lower levels, this analysis was conducted to illustrate the impact of assuming a threshold on risk-management decisions. Assuming an action level of 148 Bqm$^{-3}$ (4 pCiL$^{-1}$) for mitigation, postulating a threshold reduces the total number of lung-cancer deaths that are attributable to indoor radon and also the number of lung-cancer deaths that can be prevented by reducing levels in homes to zero. For assumed thresholds below 148 Bqm$^{-3}$ (4 pCiL$^{-1}$), there is little impact on the

TABLE ES-3  Distribution of attributable risks for U.S. males from indoor residential radon exposure under BEIR VI models

| Exposure Range (Bqm⁻³) | % of Homes in Range | Exposure-age-concentration model Contribution to AR | | | Exposure-age-duration model Contribution to AR | | |
|---|---|---|---|---|---|---|---|
| | | Actual | % | Cumulative % | Actual | % | Cumulative % |
| 0- 25 | 49.9 | 0.018 | 12.8 | 12.8 | 0.013 | 12.8 | 12.8 |
| 26- 50 | 23.4 | 0.026 | 18.5 | 31.3 | 0.018 | 18.4 | 31.2 |
| 51- 75 | 10.4 | 0.020 | 14.2 | 45.5 | 0.014 | 14.2 | 45.4 |
| 76-100 | 5.4 | 0.015 | 10.5 | 56.0 | 0.010 | 10.5 | 55.9 |
| 101-150 | 5.2 | 0.020 | 13.9 | 69.9 | 0.004 | 13.9 | 69.8 |
| 151-200 | 2.4 | 0.013 | 9.2 | 79.1 | 0.009 | 9.2 | 79.0 |
| 201-300 | 1.8 | 0.014 | 9.6 | 88.7 | 0.010 | 9.7 | 88.7 |
| 301-400 | 0.7 | 0.007 | 5.2 | 93.9 | 0.005 | 5.3 | 94.0 |
| 401-600 | 0.4 | 0.006 | 4.5 | 98.4 | 0.005 | 4.6 | 98.6 |
| 601+ | 0.4 | 0.002 | 1.5 | 99.9 | 0.001 | 1.6 | 100.2 |
| Total | 100.0 | 0.141 | 100.0 | | 0.099 | 100.0 | |

estimated numbers of preventable lung cancers by mitigation of homes with radon concentrations above 148 Bqm$^{-3}$ (4 pCiL$^{-1}$).

These AR estimates can be translated into numbers of lung-cancer deaths (Table ES-4). In 1995, there were approximately 157,400 lung-cancer deaths—95,400 in men and 62,000 in women—in the United States. Most occurred in smokers and it is estimated that 95% of cases occurred in men and 90% in women. Table ES-4 shows the estimated lung-cancer deaths in the United States attributable to indoor radon progeny exposure under the BEIR VI models. A review of the data presented in Table ES-4 reveals some differences in the calculated radon-attributable lung-cancer deaths using the exposure-age-concentration model and the exposure-age-duration model. Further variability is evident for both models depending on the approach used to estimate the influence of cigarette-smoking on lung-cancer risk. The use of the two models with two approaches to dealing with smoking yields an array of estimates of lung-cancer risk attributable to radon exposure, and provides an indication of the influence of the model and of incorporating the effects of tobacco-smoking on the projections of population risk. The range of calculated values, however, is not a complete

TABLE ES-4　Estimated number of lung cancer deaths for the U.S. for 1995 attributable to indoor residential radon progeny exposure

| Population | Number of lung-cancer deaths | Lung-cancer deaths attributable to Rn progeny exposure (No.) | |
| --- | --- | --- | --- |
| | | Exposure-age-concentration model | Exposure-age-duration model |
| **Males**[a] | | | |
| Total | 95,400 | 12,500[b] | 8,800[b] |
| Ever-smokers | 90,600 | 11,300 | 7,900 |
| Never-smokers | 4,800 | 1,200 | 900 |
| **Females**[a] | | | |
| Total | 62,000 | 9,300 | 6,600 |
| Ever-smokers | 55,800 | 7,600 | 5,400 |
| Never-smokers | 6,200 | 1,700 | 1,200 |
| **Males and Females** | | | |
| Total | 157,400 | 21,800 | 15,400 |
| Ever-smokers | 146,400 | 18,900 | 13,300 |
| Never-smokers | 11,000 | 2,900 | 2,100 |

[a]Assuming 95% of all lung cancers among males occurs among ever-smokers; 90% of lung cancers among females occurs among ever-smokers.

[b]Estimates based on applying a smoking adjustment to the risk models, multiplying the baseline estimated attributable risk per exposure by 0.9 for ever-smokers and by 2.0 for never-smokers, implying a submultiplicative relationship between radon-progeny exposure and smoking.

reflection of the uncertainty in estimating the lung-cancer risks of radon exposures and especially for never-smokers at low levels of radon exposure.

## Uncertainty Considerations

Quantitative estimates of the lung cancer risk imposed by radon are subject to uncertainties—uncertainties that need to be understood in using the risk projections as a basis for making risk-management decisions (see Table ES-5). Broad categories of uncertainties can be identified, including uncertainties arising from the miner data used to derive the lung-cancer risk models and the models themselves, from the representation of the relationship between exposure and dose, from the exposure-distribution data, from the demographic and lung-cancer mortality data, and from the assumptions made in extending the committee's models from the exposures received by the miners to those received by the general population. The committee addressed those sources of uncertainty qualitatively and, to a certain extent, quantitatively.

TABLE ES-5   Sources of uncertainty in estimates of lifetime risk of lung-cancer mortality resulting from exposure to radon in homes

I  Sources of uncertainty arising from the model relating lung-cancer risk to exposure
  A  Uncertainties in parameter estimates derived from miner data
    1  Sampling variation in the underground miner data;
    2  Errors and limitations in the underground miner data;
      a)  Errors in health-effects data including vital status and information on cause of death;
      b)  Errors in data on exposure to radon and radon progeny including estimated cumulative exposures, exposure rates and durations;
      c)  Limitations in data on other exposures including data on smoking and on other exposures such as arsenic.
  B  Uncertainties in application of the lung-cancer exposure-response model and in its application to residential exposure to the general U.S. population
    1  Shape of the exposure/exposure rate response function for estimates at varying exposures and exposure rates;
    2  Temporal expression of risks;
    3  Dependence of risks on sex;
    4  Dependence of risks on age at exposure;
    5  Dependence risks on smoking status.
II  Sources of uncertainty arising from differences in radon progeny dosimetry in mines and in homes
III  Sources of uncertainty arising from estimating the exposure distribution for the U.S. population exposure distribution model
    1  Estimate of the average radon concentration;
    2  Estimate of the average equilibrium fraction;
    3  Estimate of the average occupancy factor.
IV  Sources of uncertainty in the demographic data used to calculate lifetime risk

The committee's models of lung-cancer risk were based on analyses of data from epidemiologic studies of miners. There are undoubtedly errors in the estimates of exposures to radon progeny for the miners, and information was limited on other key exposures including cigarette smoking and arsenic. The committee could not identify any overall systematic bias in the exposure estimates for radon progeny, but random errors might have led to an underestimation of the slope of the exposure-risk relationship. Although 6 of 11 study cohorts had some smoking information, sparse information on smoking limited the committee's characterization of the combined effects of smoking and radon-progeny exposure and precluded precise estimation of the risk of radon-progeny exposure in never-smokers.

The committee's models may not correctly specify the true relationship between radon exposure and lung-cancer risk. The models assume a linear-multiplicative relationship without threshold between radon exposure and risk. While the miner data provide evidence of linearity across the range of exposures received in the mines, the assumption of linearity down to the lowest exposures was based on mechanistic considerations that could not be validated against observational data. Alternative exposure-risk relations, including relations with a threshold, may be operative at the lowest exposures. However, the committee's analysis showed that assumption of a threshold up to exposures at 148 Bqm$^{-3}$ (4 pCiL$^{-1}$) had little impact on the numbers of lung-cancer deaths theoretically preventable by mitigation of exposures above that level.

Additional sources of uncertainty in the risk projections reflect the approach used to evaluate possibly differing lung dosimetry for miners and for the general population, the limited information on cigarette-smoking, and the lack of data on risks of exposures of children and women.

The committee applied new quantitative methods for uncertainty analysis to evaluate the impact of variability and uncertainty in the model parameters on the attributable risk. Since not all sources of uncertainty could be characterized, this analysis was intended to be illustrative and not to replace the committee's more comprehensive qualitative analysis.

The quantitative analysis conducted by the committee provided limits within which the AR was considered to lie with 95% certainty. For the exposure-age-concentration model, the uncertainty interval around the central estimate of AR (14%) for the entire population ranged from about 9 to 25%. This range reflects a substantial degree of uncertainty in the AR estimate, although the shape of the uncertainty distributions indicated that values near the central estimates were much more likely than values near the upper and lower limits. For the exposure-age-duration model, the uncertainty interval ranged from 7 to 17% and was centered at about 10%. The committee's preferred uncertainty limits were obtained using a simple constant-relative-risk model fitted to the miner data below 0.175 Jhm$^{-3}$ (50 WLM), which is based on observations at exposures closest to residential exposure levels. The latter analysis, which minimizes the degree of extrapolation outside the range of the miner data, led to uncertainty limits of 2 to 21%, with a central estimate of about 11%.

## Effects of Radon Exposure Other Than Lung Cancer

Health effects of exposure to radon progeny other than lung cancer have been of concern, including other malignancies and non-malignant respiratory diseases in miners. The findings of several ecologic studies in the general population have indicated a possible effect of radon exposure in increasing risk for several types of non-lung cancers and leukemias. A pooled analysis of 11 miner studies, differing in one study from the data used by the committee, showed no evidence of excess risk for cancers other than the lung. The committee concluded that the findings in the miners could be reasonably extended to the general population and that there is no basis for considering that effects would be observed in the range of typical exposures of the general population that would not be observed in the underground miners exposed at generally much higher levels.

The committee reviewed new studies of non-malignant respiratory disease in uranium miners. A case series of uranium miners with pulmonary fibrosis supported the possibility that exposures to radon progeny may cause fibrosis of the pulmonary interstitium, but the case series is insufficient to establish a causal link to radon progeny specifically.

## CONCLUSIONS

Radon is one of the most extensively investigated human carcinogens. The carcinogenicity of radon is convincingly documented through epidemiologic studies of underground miners, all showing a markedly increased risk of lung cancer. The exposure-response relationship has been well characterized by analyses of the epidemiologic data from the miner studies, and a number of modifiers of the exposure-response relationship have been identified, including exposure rate, age, and smoking. For residences in the United States, a large national survey provides information on typical exposures and on the range of exposures.

On the basis of the epidemiologic evidence from miners and understanding of the genomic damage caused by alpha particles, the committee concluded that exposure to radon in homes is expected to be a cause of lung cancer in the general population. According to the committee's two preferred risk models, the number of lung-cancer cases due to residential radon exposure in the United States was projected to be 15,400 (exposure-age-duration model) or 21,800 (exposure-age-concentration model). Although these represent the best estimates that can be made at this time, the committee's uncertainty analyses using the constant relative risk model suggested that the number of cases could range from about 3,000 to 33,000[1]. Nonetheless, this indicates a public-health problem and makes indoor radon the second leading cause of lung cancer after cigarette-smoking.

The full number of attributed deaths can be prevented through radon mitigation only by eliminating radon in homes, a theoretical scenario that cannot be reasonably achieved. Nonetheless, the burden of lung-cancer deaths attributed to the upper end of the exposure distribution is expected to be reduced by lowering radon concentrations. Perhaps one-third of the radon-attributed cases (about 4% of the total lung-cancer deaths) would be avoided if all homes had concentrations below the Environmental Protection Agency's action guideline of 148 $Bqm^{-3}$ (4 $pCiL^{-1}$); of these, about 87% would be in ever-smokers. It can be noted that the deaths from radon-attributable lung cancer in smokers could most efficiently be reduced through tobacco-control measures, in that most of the radon-related deaths among smokers would not have occurred if the victims had not smoked.

The committee's model and general approach to assessing lung-cancer risks posed by indoor radon and cigarette-smoking are subject to considerable uncertainty because of gaps in our scientific knowledge of effects at low levels of exposure. This uncertainty should be reduced as an improved understanding develops of molecular and cellular events in the induction of lung cancer at low levels of exposure to radon and other toxicants and of the role of various factors influencing susceptibility to lung cancer. The long-term follow-up of miner populations is strongly encouraged, as is completion of the case-control studies of residential exposures now in progress. The committee encourages further meta-analysis and pooling of case-control data. However, the committee recommends that new case-control studies not be initiated until those in progress are completed, data are analyzed and synthesized, and judgments rendered as to the likely value of further residential studies.

Despite the limitations of existing data, the committee found key observational and experimental data that, along with theoretical considerations in radiobiology and carcinogenesis, provided a basis for the models developed and used to estimate radon-attributable lung-cancer risks. The major shortcomings in the existing data relate to estimating lung-cancer risks near 148 $Bqm^{-3}$ (4 $pCiL^{-1}$) and down to the average indoor level of 46 $Bqm^{-3}$ (1.24 $pCiL^{-1}$), especially the risks to never-smokers. The qualitative and quantitative uncertainty analyses indicated the actual number of radon-attributable lung-cancer deaths could be either greater or lower than the committee's central estimates. This uncertainty did not change the committee's view that indoor radon should be considered as a cause of lung cancer in the general population that is amenable to reduction. However, the attributable risk for smoking, the leading cause of lung cancer, is far greater than for radon, the second leading cause. Lung cancer in the general population and in miners is related to both risk factors and is amenable to prevention.

---

[1]The 95% upper confidence limit for the exposure-age-concentration model was approximately 38,600, but such an upper limit was highly unlikely based on the committee's review and analyses.

# 1

# Introduction

## RADON AND LUNG CANCER: AN OVERVIEW

Radon-222, a noble gas resulting from the decay of naturally occurring uranium-238, was the first occupational respiratory carcinogen to be identified. As early as the 1500s, Agricola chronicled unusually high mortality from respiratory disease among underground metal miners in the Erz Mountains of eastern Europe (Hoover and Hoover 1950). In 1879, Harting and Hesse (1879) described autopsy findings that documented pulmonary malignancy in miners in that region and by early in the 20th century the malignancy was shown to be primary carcinoma of the lung (Arnstein 1913). The finding of high levels of radon in mines in the region and in the nearby mines of Joachimsthal in Czechoslovakia, where miners also had high lung-cancer rates, led to the hypothesis that radon was the cause of the lung cancers (Ludwig and Lorenser 1924; Pirchan and Sikl 1932; also, see Jacobi 1994 and Proctor 1995 for a historical review). That hypothesis was not uniformly accepted until the findings of epidemiologic studies of other groups of radon-exposed underground miners were reported during the 1950s and 1960s (see Lorenz 1944 for an early view of the evidence and Proctor 1995 for a review of the controversy; NRC 1988 summarizes the epidemiologic evidence). Radon has now been classified as a human carcinogen by the International Agency for Research on Cancer (IARC 1988).

Radon is an alpha-particle emitter that decays with a half-life of 3.8 d into a short-lived series of progeny that have been referred to historically as radon daughters but are now more often termed radon-decay products or radon progeny (Figure 1-1). Radon itself was initially considered to be the direct cause of the lung cancer in the miners. However, Harley proposed in his doctoral thesis at

MASS
NUMBER

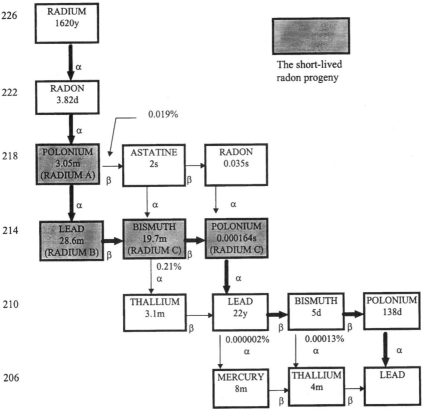

FIGURE 1-1    The radon-decay chain.  An arrow pointing downward indicates decay by alpha-particle emission; an arrow pointing to the right indicates decay by beta-particle emission.  The historical symbols for the nuclides are in parentheses below the modern symbols.  Most decay takes place along the unbranched chain marked with thick arrows. The negligible percentage of decay along the thin arrows is shown at critical points.  The end of the chain, lead-206, is stable, not radioactive.  Half-lives of each isotope are shown as seconds (s), minutes (m), days (d), or years (y).  Modified from NRC (1988).

Rensselear Polytechnic Institute that it was the decay products of radon, and not radon, that delivered the pertinent dose to lung cells (Harley 1952, 1953, 1980). Bale learned of this thesis when he visited the Health and Safety Laboratory of the U.S. Atomic Energy Commission, where Harley had done his work, and he confirmed the calculations of lung dose from radon and thoron.  Alpha particles released by 2 radioisotopes in the radon-decay chain, polonium-218 and polonium-214, deliver to target cells in the respiratory epithelium the energy that is considered to cause radon-associated lung cancer (NRC 1991).

Evidence on radon and lung cancer is now available from about 20 epidemiologic studies of underground miners, including 11 studies that provided quantitative information on the exposure-response relationship between radon and lung-cancer risk (Lubin and others 1995). Those studies and several epidemiologic findings before them, continue to support the implementation of regulatory programs to reduce exposures of underground miners to radon and to provide compensation for occupational lung cancer (Samet 1992).

Although the progeny of radon are now a well-recognized cause of lung cancer, radon itself has again become a topic of controversy and public-health concern because it has been found to be a ubiquitous indoor air pollutant to which all persons are exposed (Cole 1993 and Proctor 1995 review the controversy). Radon was found to be present in indoor air as early as the 1950s, but the potential health implications received little attention until the late 1970s (Proctor 1995). In Scandinavia, housing surveys in the 1970s documented the presence of relatively high radon concentrations in homes built with materials containing medium-rich alum shale. Sparse data from the United States provided a similar indication of contamination of indoor air with radon. In 1984, a man triggered the radiation detector on entering the nuclear power plant where he worked. His home was found to have radon concentrations well beyond those permitted in underground mines. Other homes with high concentrations were identified later, and an enlarging database on indoor radon concentrations showed that the problem was widespread (Nero 1986).

The evidence on radon and lung cancer is now extensive. Initially, research was driven by the need to characterize the risks faced by underground miners so that exposure limits that would keep risks to an acceptable level could be set. The work emphasized epidemiologic studies of the uranium and other underground miners exposed to radon, but animal studies were also conducted to address the modifying effects of such factors as the presence of ore dust and diesel exhaust, cigarette-smoking, and dose rate. Models of the respiratory tract were developed to characterize the relationship between exposure to radon progeny and dose of alpha energy delivered to target cells in the respiratory epithelium. In the last decade, research has reflected the need to improve the understanding of the risks posed to the general population by indoor radon. Epidemiologic studies have been conducted to assess the general population's risk of lung cancer associated with indoor radon and complementary animal and laboratory studies have been carried out to address uncertainties in assessment of the risks associated with indoor radon. As a result, we have gained a rich body of evidence on radon and lung cancer that addresses all facets of the problem within the framework of exposure, dose, and response (Figure 1-2). The new techniques of cellular and molecular biology also have brought new insights into how alpha particles injure the genetic material of cells and cause cancer (NRC 1994).

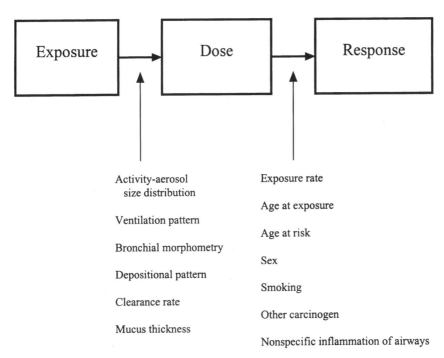

FIGURE 1-2   Factors influencing the relationship between radon exposure and lung-cancer risk.  Modified from NRC (1991).

During the past 3 decades, a series of studies have been conducted within the operation of the National Academy of Sciences' National Research Council to evaluate the risks to human health following exposure to ionizing radiations. This series of risk assessments by committees on the Biological Effects of Ionizing Radiations, or BEIR, included a BEIR IV report in 1988 (NRC 1988) which focused on health effects of radon and other alpha-particle emitters.

The context for the work of the 6th Committee on the Biological Effects of Ionizing Radiations (BEIR VI), in the National Research Council's Board on Radiation Effects Research (BRER), has been set by concern about the risk posed to the public by indoor radon. As a basis for developing public policy on indoor radon, the risks associated with indoor radon across the range of residential exposures received by the population must be characterized. That range extends from exposures at the higher end comparable to those received by miners found to be at increased risk for lung cancer down to exposures at an average indoor concentration of about 50 becquerels/cubic meter, or 50 Bqm$^{-3}$ (1.4 picocuries/liter, or 1.4 pCiL$^{-1}$), which are much lower than the exposures of most of the miners included in the epidemiologic studies. The degree of uncertainty in the risk estimates increases from the high end of the exposure range where risks are directly measured, to the low end where risks must be extrapolated; the degree of uncertainty at the low end of the range has contributed to persistent questioning of the appropriateness of directing risk-management strategies uniformly across the full range of population exposures. In the United States, for example, the Environmental Protection Agency (EPA) has called for the voluntary measurement of indoor radon concentration in single-family residences and mitigation of radon in homes with average annual concentrations above 148 Bqm$^{-3}$ (4 pCiL$^{-1}$) (USEPA 1992c).

The BEIR VI committee thus faced the task of characterizing the risks associated with indoor radon across the full range of exposures and providing an indication of the uncertainty to be attached to risk estimates across that range. At the higher end of the range, exposures of the general population overlap those received by miners, and the extensive findings on risks in miners provide a reasonably accurate picture of the risks likely to be sustained by the population. At the lower end of the range, risks are estimated by extrapolating from the miner data and are consistent with the results of a meta-analysis of domestic-exposure studies (Lubin and Boice 1997). The extrapolation requires assumptions about the relationship between exposure to radon and lung-cancer risk and a careful exploration of the comparative doses from alpha particles delivered to the lung in the mining and indoor environments. The extrapolation also requires assumptions on the potential modifying effects of cigarette-smoking, age at exposure, and sex. These assumptions are a source of uncertainty in estimates of the risk of indoor radon, but choices can be supported by epidemiologic and experimental data. In preparing this report, the BEIR VI committee considered the entire body of evidence on radon and lung cancer, integrating findings from epidemiologic studies with evidence from animal experiments and other lines of laboratory investigation.

Radon, of course, is only one of the causes of lung cancer (Table 1-1). In fact, the epidemic of lung cancer in the United States and many other countries largely reflects trends in cigarette-smoking, the dominant cause of lung cancer (USDHHS 1989; Burns 1994; Mason 1994). Of the approximately 170,000 lung cancer cases in 1996, most will be in cigarette smokers and thus avoidable in principle through smoking prevention and cessation. Synergism between radon and tobacco smoking implies heightened risks from radon in ever-smokers but cases caused by the joint effect of smoking and radon can be prevented by avoidance of smoking. Thus, risk projections of the number of lung-cancer cases attributable to radon should be interpreted with acknowledgment that most lung-cancer cases and deaths can be prevented by eliminating smoking.

## PRIOR REPORTS ON THE RISK ASSOCIATED WITH RADON

There have been numerous assessments of the risks posed by radon, in both mining and indoor environments. Some have been generated by previous BEIR

TABLE 1-1    Risk factors for lung cancer

- Active cigarette-smoking
- Passive cigarette-smoking
- Radon
- Occupational carcinogens:
  Arsenic
  Asbestos
  Chromates
  Chloromethyl ethers
  Nickel
- Polycyclic aromatic hydrocarbons
- Family history
- Fibrotic lung disorders
- Ambient air pollution

committees and others by committees of the National Council on Radiation Protection and Measurements (NCRP) and the International Commission on Radiological Protection (ICRP). The BEIR IV report (NRC 1988) reviewed the principal risk models published through the 1980s. Those models were based on either the dosimetric or the epidemiologic approach. In the dosimetric approach, a model is used to estimate the dose from alpha-particles delivered to the lung; the lung-cancer risk is then estimated with a risk coefficient based on risks in populations exposed to low linear-energy-transfer (low-LET) radiation, such as the Japanese atomic-bomb survivors, and adjusted for the greater potency of alpha radiation by using a quality factor for the relative biologic effectiveness (RBE) of alpha radiation in causing cancer, historically assumed to be the same as the ICRP quality factor of 20. The epidemiologic approach employs a risk model based on the studies of underground miners. The BEIR IV committee analyzed data from 4 cohorts of underground miners, using regression methods to develop a model that described the relationship between excess relative risk and exposure to radon progeny during 3 temporal periods of time since exposure. The model also incorporated an age dependence of the effect of radon on lung-cancer risk.

After the BEIR IV (NRC 1988) report, 2 new risk models were reported, both based on the epidemiologic approach. ICRP addressed exposures at home and at work in its *Publication 65: Protection Against Radon-222 at Home and at Work* (ICRP 1994a). That report used a model for excess relative risk that incorporated time dependence of the effect of radon; the risk coefficient was that reported by the BEIR IV committee. The 2nd model was based on an extensive pooled analysis of data on 11 cohorts of underground miners (Lubin and others 1995). The general analytic approach followed that of the BEIR IV committee. The models for excess relative risk incorporated exposure during windows defined by time since exposure and included age-dependence of the effect of radon. The rate at which exposure was received also had an important effect on lung-cancer risk: risk increased as exposure rate decreased. Two models were described, one describing the exposure-rate effect by variables for the duration of exposure and the other by variables for exposure rate.

Early in its deliberations, the committee critically assessed the various possible approaches to risk estimation from exposure to low levels of radon. The possible methodologies for generating such risk estimates are summarized in Figure 1-3, and are discussed in chapter 3. In addition to the previously used dosimetric and epidemiologic approaches, statistical models can also be applied that are based on a biologic construct of the effect of radon.

The dosimetric approach focuses on computing the doses deposited in lung target cells by alpha particles emitted from radon progeny deposited in various compartments of the lung. The approach relies on applying risk estimates from

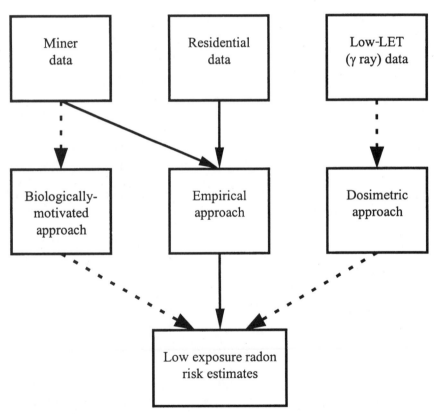

FIGURE 1-3    Possible methodologies for generating risk estimates for low exposures to radon.

populations exposed to other radiations, such as from the atomic-bomb survivors in Hiroshima and Nagasaki who were exposed to primarily gamma rays and neutrons, to the doses estimated in the lung cells which are believed to become lung cancers.

The biologically-motivated approach would use data from molecular, cellular, and animal studies to generate parameters which could be used to refine models of carcinogenesis. The miner data could be used in conjunction with the models to see if effects observed in nonhuman systems might be altered by responses in the intact human.

Finally, the empirical approach would use the state-of-the-art statistical methodologies to analyze epidemiologic data from miner and residential studies. In brief, of the three data sources (miner studies, domestic radon exposure studies, and populations exposed to γ rays), use of data from γ-ray exposed groups (particularly A-bomb survivors) was rejected due to the many assumptions needed to generate risk estimates for prolonged localized exposure to densely-ionizing radiation, based on risks for acute whole-body exposures to γ rays. Use of data from studies of residential radon exposure was rejected for the primary risk estimation due to the very limited statistical power available in these studies. However, the residential data were used as validation and support for the low-dose risk estimates, in that the results were compared with, and shown to be consistent with, risk estimates extrapolated from analyses of the miner data.

Of the three modeling approaches possible for analyzing the epidemiologic data (dosimetric approach, biologically-motivated models, empirical approach), the dosimetric approach was rejected for the same reasons as was the use of A-bomb survivor data—the uncertainties involved in extrapolating from the effects of acute whole-body γ-ray exposures to prolonged localized exposure to densely-ionizing radiation. The use of biologically-motivated models—such as the two-stage clonal expansion model—was rejected due to our limited knowledge of the complete mechanisms involved in radon-induced carcinogenesis. As described in chapters 2 and 3, however, mechanistic information was used to guide specific parts of the empirical modeling, whenever possible.

In summary, the committee chose to follow the overall methodology both of BEIR IV and the pooled analysis of Lubin and others (1994a), and to base risk estimates on an empirically-based analysis of miner data. This approach provided the committee with the invaluable starting point of existing data bases and methodologies, some well-characterized and others with deficiencies and limitations as described in chapter 3.

## POPULATION EXPOSURE TO RADON

The principal place of exposure to radon is the home; the predominant contribution of exposure in the home reflects the amount of time spent there and the general pattern of radon concentrations in buildings (Harley 1991; NCRP 1991;

Samet and Spengler 1991).  The major source of the radon in a home is the soil gas that enters from beneath and around the structure (NCRP 1984).  Radon-contaminated water and radium-rich building materials can also contribute radon.  Under some circumstances, water can contribute a substantial amount of the radon in air, but water is a relatively minor source of population exposure to radon progeny.

Surveys of indoor radon concentrations in the United States and many other countries have shown that radon is ubiquitous indoors, typically at a concentration only one hundredth to one-tenth that found in the underground mines and shown to be associated with lung cancer.  For the survey conducted by the U.S. Environmental Protection Agency, the distribution of concentrations is roughly lognormal; the arithmetic mean concentration is about 46.25 Bqm$^{-3}$ (1.25 pCiL$^{-1}$) (Figure 1-4).  On the basis of the survey information, mean exposure of the general population is about 0.0007 Jhm$^{-3}$ (0.2 WLM) per year or about 0.049 Jhm$^{-3}$ (14 WLM) in a lifetime (NCRP 1984, 1991).

## THE COMMITTEE'S APPROACH

The BEIR VI committee was constituted after a BEIR VI phase 1 committee determined that sufficient evidence had become available since the 1988 publication of the BEIR IV report to justify a new study (NRC 1994).  The charge extended to the BEIR VI committee was broad (Table 1-2).  Its principal goals were to examine evidence of effects of low-level exposure to radon progeny and to develop a mathematical model for the lung-cancer risk associated with radon.  In developing this model, the committee was to address key uncertainties, including the combined effect of smoking and radon, the effect of exposure rate, the effects of exposures to agents other than radon in the mines, and the consequences of errors in exposure estimates.  The committee was also asked to review relevant evidence from radiobiologic studies, to reassess exposure-dose relations, and to evaluate the potential utility of the case-control studies of indoor radon.

The multidisciplinary BEIR VI phase 2 committee worked in topic-oriented groups that addressed exposure and dosimetry; molecular and cellular aspects of radon carcinogenesis, including the findings of in vitro approaches and animal studies; epidemiologic studies of miners;  case-control studies in the general population; and risk modeling.  When the committee began its work, it was recognized that the data on the 11 cohorts of miners would be essential for the development of a new risk model; the committee obtained the cooperation of the principal investigators for the individual cohort studies so that additional analyses of the data could be undertaken to develop a risk model.  During the course of deliberations by the committee, it became apparent that a new meta-analysis of exposure in homes was powerful enough to yield useful risk estimates associated with domestic exposures.

The committee also recognized that its work could not be artificially sepa-

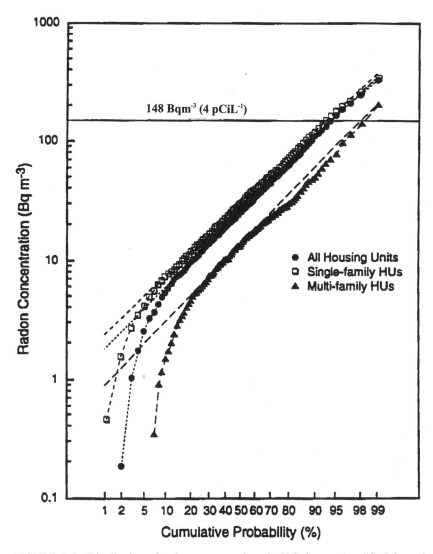

FIGURE 1-4    Distribution of radon concentrations in U.S. homes (modified from the National Residential Radon Survey; USEPA 1994).

rated from the context set by controversy and debate over the management of the risk posed by indoor radon. In the United States, the EPA has pursued a broad national strategy to reduce radon risks through voluntary testing of homes and mitigation of radon in those with an annual average above 148 $Bqm^{-3}$ (4 $pCiL^{-1}$). EPA finds a basis for risk management in its projections of 7,000-30,000 lung-cancer deaths per year attributable to residential radon exposure (Page 1993).

TABLE 1-2   Charge to the BEIR VI committee

**Radiation Biology and Carcinogenesis**
- Evaluate experimental evidence on inverse dose-rate relations and its implications.
- Attempt to correlate risks of radon-induced lung tumors with those of tumors induced by external exposure to low-LET radiation.
- Examine in more detail the induction and repair of molecular changes after exposure to alpha particles.

**Exposure-Dose Relations**
- Gain access to additional information that is relevant to radon dosimetry from the use of noninvasive methods for monitoring ventilation and from an improved model.
- Use activity-weighted size-distribution data that have become available on residences.
- Use the recent data on growth of particles from various sources typically found in homes.
- Use biologic dosimetry to reduce the uncertainties associated with the exposure-dose relationship and apply the reduced uncertainties in various aspects of lung dosimetry to risk calculations.

**Studies of Miners**
- Critique the recently completed  analyses of pooled data on 11 cohorts and whatever new risk models emerge from them and use the database to suggest additional analyses.
- Model the pooled data with emphasis on biologically-driven modeling.
- Formally evaluate sources of exposure error in the miner cohorts and the consequences of those errors for risk estimation and risk assessment.

**Studies of Lung Cancer in the General Population**
- Evaluate and interpret the results of completed case-control and ecologic studies, including evaluation of their limitations and uncertainties.
- Determine the appropriate role of case-control studies in developing a risk model of residential exposure to radon.
- Make general recommendations regarding the potential role of case-control studies that are not yet complete.
- Estimate the statistical power of the completed studies for various end points as a group and identify the expected upper and lower confidence limits according to the assumptions of the recommended model and alternative models.
- Estimate the potential of the completed studies for providing information on the modifying effects of smoking and other factors, taking into account uncertainties in exposure estimates.
- Make recommendations on the desirability of initiating new case-control or ecologic studies of residential radon exposure.

**Modifications to the Phase 1 Recommendations**
- Assess the validity and scientific reliability of analytic approaches used by various investigators.
- Examine in detail the interaction between radon exposure and cigarette-smoking on the basis of data on US and other miners.
- Reexamine the effect of the rate of exposure to radon on the incidence of lung cancer.
- Critically examine exposure estimates for miner cohorts and reassess the consequences of the exposure rates.
- Propose, in light of the phase 1 findings, a mathematical model of risk based on more complete and up-to-date U.S. and international miner data.

## TABLE 1-2   Continued

- Assess the role of arsenic, silica, and other contaminants in mines on the consequences of exposure.
- Examine the uncertainties associated with the miner studies.

**Analysis of Data from Studies of Residential Exposure**
- Critically review studies and comment on their strengths and weaknesses and their current and future roles in risk assessment.
- If EPA and BRER agree that is feasible, update the assumptions and estimates in the Research Council report that compared miner and home dosimetry, focusing on recent data on such physical and biologic factors as aerosol size distribution, ultrafine fraction, equilibrium fraction, and hygroscopicity and deposition of radon daughters in the respiratory tract.
- Test and possibly revise models in light of available residential data.
- Consider the contribution of radon-220 to risk in mines and homes.
- Examine the effects of age, sex, and smoking on radon-associated risk.
- Incorporate concepts from cellular and molecular biology into models for risk assessment.

Most of those projected deaths are attributable to the lower range of exposures, well below the exposures and exposure rates sustained by the miners, so there is substantial uncertainty as to the total burden of lung cancer resulting from indoor radon. The agency's action guideline of 148 Bqm$^{-3}$ (4 pCiL$^{-1}$) is also in this range, as are the action levels set by other countries. The BEIR VI committee therefore characterized the certainty that could be attached to risk estimates made with its model, moving from a high level of confidence in estimates at exposures comparable with those received by the underground miners to a substantially lower degree of confidence in estimates related to typical indoor exposures.

## CRITICAL ISSUES

The BEIR VI committee identified 6 issues deemed critical in characterizing the risks associated with indoor radon. Each deals with a point of uncertainty in formulating a risk model and estimating the risk posed to a population by indoor radon. In the review of the evidence, those critical issues served as points of synthesis for identifying the limits of current evidence for risk assessment. The issues are briefly described here.

### Extrapolation from Higher to Lower Exposures

Early in its work, the committee recognized that the evidence on quantitative risks from studies of underground miners would remain the principal basis for estimating the risk associated with indoor radon. Most of the evidence from the miners is based on cumulative exposures at least 10 times the estimates of typical lifetime residential exposure, that is, about 0.049 Jhm$^{-3}$ (14 WLM) (NCRP 1984). Consequently, extension of risk models based on data from miners to the general population rests on extrapolation below the exposure range of most of the miner data. A linear-nonthreshold model has been assumed for this purpose by other groups developing radon risk models, including the BEIR IV committee (NRC 1988). Risk projections are sensitive to the extrapolation model chosen. For example, a model that incorporated a threshold at lower levels of exposure, such as, below the mean of about 0.0035 Jhm$^{-3}$ (1 WLM), would project far fewer radon-caused lung cancers than a nonthreshold model, because of the distribution of population exposures (Figure 1-4).

## Extrapolation from Higher to Lower Exposure Rates

The underground miners included in the epidemiologic studies typically received their occupational exposures to radon over several years (Lubin and others 1994). However, exposure to indoor radon takes place across the full lifetime, so exposure rates are generally far lower for the radon exposures received by the general population than for the exposures received by the miners in the epidemiologic studies. The analysis of pooled data on 11 cohorts of underground miners indicated an inverse exposure-rate effect; that is, the effect of exposure increased as the rate at which the exposure was received diminished (Lubin and others 1994). Laboratory systems have shown a parallel inverse dose-rate effect for some end points (NRC 1994). The BEIR VI phase 1 committee recognized that the full BEIR VI committee would need to address the extrapolation of exposure-rate effects for exposures received by the miners to those typically received by the general population.

## Interactions of Radon Progeny with Other Agents

Lung cancer can be caused by a number of inhaled environmental agents, including tobacco smoke (Table 1-1) (Lubin and others 1995). Agents in the mining environment other than radon progeny have been associated with lung cancer, including diesel exhaust, arsenic, and silica. Those agents might potentially confound the relationship between radon progeny and lung cancer or modify the risk associated with radon progeny. The majority of the miners included in the epidemiologic studies also were probably cigarette smokers, although the information on smoking in the various cohorts is incomplete (Lubin and others 1994). Exposure to radon progeny is an established cause of lung cancer in those who have never smoked (NRC 1988), but the risks to ever-smokers and never-smokers have not been separately characterized with great precision. Lack of information on the combined effect of smoking and radon introduces uncertainty in extending a risk model based on data from miners to the general population.

Interpretation of risk projections for the general population is further clouded by synergism between smoking and radon progeny: some lung-cancer cases reflect the joint effect of the two agents and are in principle preventable by removing either agent.

## Susceptibility

The development of lung cancer is considered to be a multistep process involving a sequence of genetic changes that ultimately results in malignant transformation. A rapidly growing body of literature describes some of these changes, for example, in tumor-suppressor genes and oncogenes (Economou and others 1994). There are many points in this sequence, from carcinogen metabolism to DNA repair, at which genetic susceptibility might influence lung-cancer risk. In fact, there is descriptive evidence from epidemiologic studies that a family history of lung cancer is associated with increased risk (Economou and others 1994). The evidence has not addressed radon progeny specifically, but it was deemed sufficiently compelling by the BEIR VI committee to warrant its being considered as one of the critical issues. In particular, the committee considered the implications of identifying population groups at risk for lung cancer generally or for lung cancer caused by radon progeny specifically.

## Links Between Biologic Evidence and Risk Models

The BEIR VI phase 1 committee recognized that key sources of uncertainty listed above as the first 4 critical issues would be best addressed by risk models that directly reflect current biologic understanding directly and are not simply empiric and based on analysis of observational data. The assumptions to be made with regard to the relationship between exposure and risk and with regard to the inverse exposure-rate effect were judged to be particularly critical. Consequently, the phase 2 committee was constituted to include the full range of scientific disciplines concerned with radon carcinogenesis, and a subcommittee, the Molecular and Cellular Working Group, synthesized the evidence with a goal of providing guidance to a second subcommittee, the Risk Modeling Working Group.

## Signatures of Radon Effects

Lung cancer constitutes a clinically and histologically heterogeneous group of malignancies. There are 4 principal histologic types: squamous cell carcinoma, adenocarcinoma, small cell carcinoma, and large cell carcinoma. Each has somewhat different clinical characteristics. In addition, substantial heterogeneity at the molecular level has been demonstrated with markers of genetic change. This heterogeneity, initially at the light-microscopy level and now at the genetic level, has motivated investigations to determine whether specific lung cancers can be linked to radon; that is, is there a "signature" that identifies a particular cancer as caused by radon? Demonstration of a signature would add specificity to risk assessments and even permit the designation of lung cancers in individuals as caused by radon or some other factor. The committee reviewed the literature for evidence of signatures of radon-related lung cancer

## OVERVIEW OF COMMITTEE RISK ASSESSMENT

The committee's approach to assessing the risks associated with indoor radon had multiple components. Analysis of the miner data supplied a model describing the relationship between cumulative exposure to radon and lung-cancer risk; that model development is described in chapter 3 and appendix A. The greatest proportion of the population's exposure to radon is at the lower end of the distribution of residential concentrations (Figure 1-4), a distribution which results in exposures across the lifespan substantially lower than the cumulative exposures received by most of the miners in the epidemiologic studies.

As a basis for selecting the preferred exposure-response model, the committee reviewed relevant biologic information on mechanisms of DNA damage by alpha particles and the dosimetry of alpha particles at exposures relevant to environmental exposures. This review, described in chapter 2, led to the conclusion that, at typical indoor exposure levels and over a typical lifetime, target-cell locations in the respiratory tract would be traversed by no more than a single alpha particle. Even allowing for DNA repair and cellular lethality, such a single alpha-particle traversal is known to produce, with significant probability, large-scale damage to the DNA in surviving traversed cells. Thus, at low exposure levels, a further decrease in the exposure would result in a proportionate decrease in the number of target cells traversed by single alpha particles, and thus suffering large scale DNA damage. Each of these heavily damaged traversed cells would still be expected to show the same probability of initiating the sequence of events that ultimately lead to carcinogenesis. Taken together, these observations provide a rationale for the committee's assumption that the dose-response relationship for radon-induced lung cancer is likely to be best described by a linear model with no threshold in dose.

The committee noted, as discussed elsewhere in this report, that statistical considerations preclude a direct investigation of low-dose thresholds for radon-induced carcinogenesis, at least in human populations. However, the view of the committee was that the mechanistic and experimental considerations were sufficiently compelling to support assumption of a linear/nonthreshold dose-response relationship at low radon exposures. It is also important to note that the linear/nonthreshold approach which the committee has adopted is based on mechanistic considerations relating to the nature of alpha-particle induced energy deposition and damage; whether these considerations also hold for other carcinogens, such as x rays, was not an issue that was addressed by the committee.

In developing its risk model, the committee conducted analyses of the full miner data set and additional analyses limited to the exposure range of < 0.35 Jhm$^{-3}$ (less than 100 WLM) and focused on < 0.175 Jhm$^{-3}$ (less than 50 WLM) which is an exposure range involving less than one alpha-particle traversal per cell nucleus location depending on the assumed target cell. An exposure of 0.35 Jhm$^{-3}$ (100 WLM) results in about 0.84 traversals for basal cell nuclei, about 1.88 traversals for bronchiolar secretory-cell nuclei, and about 3.25 traversals per bronchial secretory-cell nuclei. Additional details regarding alpha-particle traversals of basal- and secretory-cell nuclei can be found in Harley and others (1996) who used a large data base of bronchial-cell nuclei morphometry. The committee's preferred risk model describes lung-cancer risk with a linear non-threshold relationship between exposure and risk. The model describes risk as time-dependent and age-dependent.

When the risk model is applied to the general population, possible differences between miners and the general population need consideration; that is, similar exposures to radon progeny in homes and miners may not result in the same doses of alpha energy for target cells. Following the approach of the BEIR IV committee, the BEIR VI committee used a mathematical model of the lung to calculate doses to target cells received by miners and by the general population for a given exposure. Those doses were used to determine whether an adjustment was needed in extending risks from miners to the general population.

The committee's model describes the increase in the relative risk of lung cancer associated with exposure to radon progeny. To project the lung-cancer risks associated with exposure to indoor radon, the committee needed to assume a background rate of lung-cancer mortality. Assumptions will also be needed to extend the model to the full life span, to women, and to ever-smokers and never-smokers. The bases for these assumptions and associated uncertainties are described in the report.

# 2

# The Mechanistic Basis of Radon-Induced Lung Cancer

## INTRODUCTION

This chapter summarizes the current state of knowledge of the various processes that are presently considered to be involved in the induction of cancer by radon progeny. Inclusion of the chapter was motivated by the desirability of providing, where possible, a biological and mechanistic framework for epidemiologic analysis of risk models. Various components of the process of radon-induced carcinogenesis are understood to some degree, but we do not yet have a complete mechanism-based understanding of the entire process. In particular, although our understanding of the various radiation-related components (such as the effects of dose and dose rate) is at least semiquantitative, our knowledge of the various steps in the carcinogenic process (particularly at the genetic level) is at best qualitative in spite of important research findings since the publication of the report of the 4th Committee on the Biological Effects of Ionizing Radiations, BEIR IV (NRC 1988).

Consequently, systematic quantitative mechanism-based (biophysical) modeling of the entire process of lung-cancer induction by radon progeny is beyond present capabilities. However, some elements, such as dose-rate effects, can be modeled on the basis of specific assumptions and used to guide epidemiologic analyses and risk modeling.

In broad terms, the types of information available on radon carcinogenesis can be characterized as molecular, cellular, animal, and human. All contribute to our current understanding of the mechanistic basis of alpha-particle induction of lung cancer.

A principal justification for studying cancer cells in vitro, abstracted from the entire organism, is that a neoplasm is usually considered as arising from a single cell that has undergone a critical change. Evidence of that includes the fact that some malignancies can be propagated by a single cell, and many, but not all, tumors have been shown to be monoclonal in origin, in that every cell carries the same biochemical marker (for example, Pathak 1990). It is important to note that many steps are involved from the malignant transformation of a single cell to the development of an overt neoplasm, including tissue response and potential immunological factors (Nagarkatti and others 1996), and care must be taken in directly extrapolating exposure and dose-response relationships for cells exposed to low doses of high-LET particles to risk for the development of cancer.

Cellular and molecular research generally focuses on early changes induced by radon and attempts to understand the mechanisms involved in production and repair of these changes. Such mechanistic understanding is essential to evaluate the response of cells to environmental radon exposure in which only a small fraction of the cell population interacts with the alpha particles. However, the role and progression of these cellular and molecular changes in the development of disease also can be addressed with experimental-animal studies. It should be noted that although many of the studies discussed in this chapter used interactions of bronchial tissue with alpha particles as the experimental model, there is a much greater base of information on interactions of x rays with other target tissues. When appropriate, we draw inferences from these other experimental models, but such conclusions will inevitably be less certain than those derived from experiments with alpha particles and bronchial tissue, the target tissue for radon-induced damage.

As the cells of a cancer grow and divide, progressive stages, or steps, from preneoplasia to malignancy can be identified. Those steps have been described as initiation, promotion, and progression. The progressive nature of carcinogenesis has been known for many years; it was first described in phenomenologic terms for skin cancer in animals. With sputum cytology, it has been possible to use histologic changes in lung cells as a predictive measure of bronchogenic cancer (Saccomanno and others 1988). The progressive cellular changes suggest a multistage process during the development of radon-induced lung cancer. More-recent evidence of the multistep nature of cancer has come from studies of the clinical progression of colorectal cancer from polyp to metastatic cancer (see, for example, Fearon and others 1990). Those studies have demonstrated an association between the clinical progression of the cancer—from a benign state, through nonmalignant adenomas, to full-blown cancer—and the activation of oncogenes, the loss of antioncogenes, and other chromosomal changes.

Although the multistep nature of radiation carcinogenesis is almost certainly true, it is as yet only a qualitative observation. Our current state of knowledge precludes systematic quantitative understanding of all the various steps from early subcellular lesions to observed malignancy, and of the potential influence that these multiple steps can have on the shape of the dose-response relationship at low doses.

## RADIATION AND ONCOGENES

The identification of oncogenes and findings on their role in human cancer have made it possible to understand why agents as diverse as retroviruses, ionizing radiation, and chemicals can result in tumors that are indistinguishable from one another (Bishop 1983; Bishop and Varmus 1984). A retrovirus can insert a gene into a cell, and radiation and chemicals can produce a mutation in a gene that is already in the cell; all can activate oncogenes.

A central feature of oncogenes is that they act in a dominant fashion. The presence of a single copy of an activated oncogene in a cell is sufficient to produce a transformed phenotype, even in the presence of a normal copy of the gene (Lee and others 1987). Cells that are already immortal, such as NIH 3T3 mouse cells, can be transformed to a malignant state by transfection with a *ras* oncogene. Primary rat embryo fibroblasts, which are short-term cultured cells, are not transformed by the *ras* gene alone or by the *myc* gene alone, but can be transformed by transfection of the cells with both *myc* and *ras* (Land and others 1983). That is interpreted to mean that the *myc* gene confers immortality, whereas the *ras* gene produces the change reflected in morphology (Land and others 1983). Generally, at least 2 activated oncogenes in cooperation are needed to convert a primary cell to a tumorigenic line (Hunter 1991). Oncogene products that act in the nucleus cooperate most efficiently with products that act in the cytoplasm, as exemplified by the combination of *ras* and *myc*.

Over 100 oncogenes have been identified in human cancer; most belonging to the *ras* family. However, activated oncogenes are associated with 10-15% of human cancers and tend to be found more commonly in the leukemias and lymphomas and less commonly in solid tumors. Oncogenes have been shown to be activated by a range of genetic changes, for example by point mutations, as in *ras* (Bos 1990); deletions, as in *Nmo-1* (Petersen and others 1989); reciprocal translocations, as in *myc* (Dalla-Favera and others 1983); and gene amplification, as in *myc* (Brodeur and others 1984).

Ionizing radiation, including alpha radiation, is not particularly efficient at producing point mutations, but it does produce large interstitial deletions and reciprocal translocations with high efficiency (for example Evans 1991; Metting and others 1992; Searle and others 1976). Consequently, in assessment of the predominant initial radiation damage—the first of the many steps by which alpha particles can induce cancer—deletions or translocations seem to be the most likely candidates for the first changes.

Numerous experimental and epidemiologic studies have demonstrated that radiation can cause cancer (Martland 1931; Court-Brown and Doll 1958; Beebe and others 1962). That it does so via direct or indirect alterations to DNA is clear in in vitro studies, such as those of Borek and others (1987) in which DNA isolated from radiation-transformed C3H10T$^1$/$_2$ cells was shown to transform recipient cells after transfection.

The molecular mechanisms of radiation-induced transformation are unknown. Several studies have used indirect methods to attempt to identify oncogenes in radiation-transformed cells (Guerrero and others 1984; Shuin and others 1986; Hall and Hei 1990). One approach has been to search DNA isolated from radiation-transformed cells for mutations in known oncogenes. In that way, *K-ras* and *N-ras* were shown to be activated in some of the mouse lymphomas induced by gamma radiation (Newcomb and others 1988); it is not known, however, whether these are the initial radiation-induced changes. Another approach has been to determine whether any known oncogenes are overexpressed in transformed cells. This requires measuring mRNA in known oncogenes. Two studies used the method to examine gamma-irradiated C3H10T$^1$/$_2$ cells (Schwab and others 1983; Krolewski and Little 1989). Each used several overexpressed, cloned oncogenes as probes, but they could not identify an oncogene; both speculated on the possibility that gamma radiation could activate an as-yet-unidentified oncogene. A more-recent direct approach to the question has been to isolate the oncogenes present in the transformed cells. Such an approach was used in an attempt to isolate an oncogene from gamma-irradiated C3H10T$^1$/$_2$ cells (Hall and Freyer 1991). Many cloned oncogenes have been tested by hybridization and were negative so the gene has not yet been identified. Later experiments by Hei and colleagues (1994b) showed that a single small dose of alpha particles (30 cGy of absorbed dose), corresponding to an average of a few particles per cell nucleus, can cause human bronchoepithelial cells to become tumorigenic. A dominant gene is involved, inasmuch as the phenotype can be transmitted by transfection. Again, no known oncogene has been identified. The data support the speculation that one or more as-yet-unknown oncogenes can be involved in radiation-induced transformation.

## TUMOR-SUPPRESSOR GENES

Suppressor genes act recessively: both copies must be lost or inactivated for the cell to express the malignant phenotype. Stanbridge (1976) showed that if a hybrid was made by fusing a normal human fibroblast to a malignant HeLa cell, the normal cell suppressed the expression of malignancy by the HeLa cell. It was shown further that if during the repeated subculture of the hybrid cells, chromosome 11 was lost, the malignant phenotype was restored. It was inferred that chromosome 11 in the normal human fibroblast contains a gene capable of suppressing the malignant phenotype. In later experiments, Saxon and colleagues (1986) injected microcells containing a single human chromosome 11 into HeLa cells and found that it suppressed their malignant phenotype; if chromosome 11 was lost from the cell, the malignant phenotype was restored.

The importance of suppressor genes became evident from the work of Knudson (1971) with retinoblastoma. A familial form of retinoblastoma occurs at a high rate and a sporadic form at a very low rate. Knudson argued that in the familial form 1 mutant allele with lost function is inherited from the affected parent. A somatic event during embryogenesis inactivates the normal allele inherited from the unaffected parent. Almost all the children of such pairs of parents exhibit bilateral retinoblastoma. In sporadic retinoblastoma, 2 somatic mutations are necessary, the second in a descendant of a cell that suffered the first. Those double events are much less likely than a single event, so the incidence of the sporadic form of retinoblastoma is much lower. Knudson elaborated the "2-hit hypothesis" in the early 1970s (Knudson 1971). By the middle 1970s, the location of the relevant gene was identified on chromosome 13 (Cavanee and others 1985); in the 1980s, the Rb gene was cloned and sequenced (Lee and others 1987). The Rb gene is present in all cases of retinoblastoma and associated sarcomas; it is sometimes present in cases of other tumors, such as small-cell lung cancer, bladder cancer, and mammary cancer.

The action of radiation is a potential mechanism for deleting a suppressor gene. Alpha particles are particularly efficient at producing large deletions (for example, Metting and others 1992). Two radiation-induced breaks in the same arm of a chromosome can readily result in a deletion. Studies with defined restriction cuts in cellular and plasmid DNA have indicated that small deletions can also result from processing of single sites of DNA damage (Thacker 1994).

A suppressor gene acts recessively, so the deletion would have to occur in both chromosomes of a pair; this would be a very low-frequency event. In practice, the loss of the pair of suppressor genes often occurs by the process of somatic homozygosity (Cavanee 1989). One chromosome of a pair is lost, a deletion occurs in the other chromosome, and then the second chromosome and the deletion are replicated. Consequently, the cells in the tumor have 2 chromosomes that originated from the same parent. That has been shown to be a mechanism in retinoblastoma, small-cell lung cancer, and glioblastoma; the case of glioblastoma is particularly interesting, inasmuch as somatic homozygosity must occur in 2 different chromosomes for this high-grade tumor to be produced (Cavanee 1989).

The list of suppressor genes whose location and function are known is growing steadily. The 2 most common and most intensively studied are the Rb gene and the p53 gene, both of which are directly involved in cell-cycle checkpoint control (Kasten and others 1991; Smith and others 1994).

## GENOMIC INSTABILITY

The multistage nature of cancer is one of the most pervasive hypotheses in cancer research. The idea is over 60 years old and continues to derive support from research findings, such as recently from the work of Vogelstein and colleagues (Vogelstein 1990; Fearon and others 1990) with hereditary colon cancer. The progression from normal epithelium to metastatic cancer appears to involve a number of mutations in different oncogenes and tumor-suppressor genes and multiple chromosomal changes.

In the multistage formation of radiation-induced carcinogenesis, it is unclear as to how a single relatively small dose of radiation could result in mutations in so many different genes. The induction of multiple mutations seems highly unlikely, but data from the Japanese atomic-bomb survivors clearly show that a modest dose of radiation can induce many types of solid tumors, including those in the digestive tract. A more likely possibility is that radiation causes mutations in a gene responsible for the stability of the genome, which leads to a mutator phenotype. The multiple mutations and chromosomal changes follow as a cascade because of the induced instability as described below.

Both densely ionizing and sparsely ionizing radiation have been shown to induce chromosomal and mutational changes that appear in the progeny of exposed cells many generations after the initial exposure (Morgan and others 1996). The changes can occur in a high proportion of the surviving irradiated cells even after doses that give an average of only 1 alpha-particle traversal per cell. Examples of radiation-induced changes that are used as indicators of genomic instability are chromosomal aberrations, gene mutations, and even tumor induction in animal-model systems (Kennedy and Little 1984; Seymour and others 1986; Gorgojo and Little 1989; Chang and Little 1992; Kadhim and others 1992, 1994, 1995; Sabatier and others 1992; Martins and others 1993; Marder and Morgan 1993; Selvanayagam and others 1995). The high proportion of initially irradiated cells that transmit the instability phenotype and the variety of events observed suggest that this is not the result of a targeted effect of the initial radiation damage of specific genes, but rather a consequence of more-generalized damage to the cell; whether the initial damage is genetic or epigenetic is an unresolved question. Induced genomic instability is transmissible to progeny cells and can persist for multiple generations. Although this is an attractive hypothesis to account for carcinogenesis by low doses of high linear-energy-transfer radiation, typified by single-particle traversals, the case is far from proved.

## INDIVIDUAL AND GENETIC SUSCEPTIBILITY

There is much published evidence that many cancer-predisposing genes are present in the human genome (Sankaranarayanan and Chakraborty 1995). For some tumor types, changes in these genes are responsible for a large fraction of the total cancer frequency. For example, 40% of children with retinoblastoma carry a germ-line mutation in the RB1 gene (Vogel 1979; Cowell and Hogg 1992). The tumor-suppressor gene p53 has been associated either directly or indirectly with at least 50% of human cancers (Hollestein and others 1991), although a causal link is less clear.

There have been substantial breakthroughs in the molecular biology and mechanisms involved in the genetics of breast cancer and about 5-10% of breast cancers might be inherited (Newman and others 1988). The BRCA1 gene has been located on chromosome 17 (Hall and others 1990) and cloned (Miki and others 1994). About half the inherited breast cancer and more than half the ovarian cancers are thought to be associated with mutations in the BRCA1 gene. With linkage analysis, a second gene (BRCA2) involved in breast cancer has been identified (Wooster and others 1994). It has been suggested that 1 or both of those 2 genes might be responsible for up to 90% of all familial breast cancer cases (Sankaranarayanan and Charkraborty 1995). While the risk to the individual carrying the mutated gene is very high, such mutations account for only about 4% of the total breast cancer patients. A larger proportion, perhaps 9 to 18% of all breast cancer, is associated with carriers of the ATM gene (Swift and others 1991).

Another common cancer related to genetic breakthroughs is hereditary nonpolypotic colon cancer (Bodmer and others 1994). Genes associated with certain rare diseases such as ataxia telangiectasia (Shiloh 1995) and xeroderma pigmentosum (Kaur and Athwal 1989), have also been identified. Some suggestive evidence links lung cancer with several genes affecting carcinogen metabolism (such as CYP2D6, CYP1A1, and GSTM1), but the links are quite speculative (Caporaso and others 1995).

There is now considerable evidence that a substantial fraction of spontaneous cancers have a genetic basis (Cavenee and White 1995), and it has been estimated that the prevalence of cancer-predisposing disorders is about 16 per 1,000 live births (Sankaranarayanan and Chakraborty 1995). Some evidence has been presented (for example, Swift and others 1991; Lavin and others 1994) that ionizing radiation might interact with the genetic predisposition to increase the frequency of radiation-induced cancer. The current evidence for that hypothesis is still relatively weak (for example, Hall and others 1992), but if a radiation-sensitive subpopulation did account for most of the radiation-induced tumors of a specific type, this would profoundly influence risk estimates. Ataxia telangiectasia heterozygotes, who probably constitute more than 1% of the U.S. population (Swift and others 1986), are an example of such a relatively large subpopulation that could, at least in principle, be at increased risk for radiation-induced carcinogenesis (Swift and others 1991; Hall and others 1992; Lavin and others 1994).

The role of genetic susceptibility in the induction of cancer by environmental insults, including low-LET radiation and radon exposure, has been reviewed by Cox (1994a,b). There is evidence from transgenic mice that cancer predisposition increases the frequency and decreases the latency of cancer formation initiated by low-LET radiation (Kemp and others 1994). In addition, a study of patterns of inheritance in mice (Franko and others 1996) suggested a genetic component to radiation-induced pulmonary fibrosis. No such strong evidence has yet been found in human populations. For lung tumors, altered phenotypes and genotypes in several genes, such as the CYP family and GSTM1, have been associated with tobacco-related cancers (Anttila and others 1994; Kihara and others 1995), but the available data do not support a causal association between these markers and cancer risk (Alexandrie and others 1994; Raunio and others 1995; Caporaso and others 1995).

Even though there is ample evidence that many cancers have a strong genetic basis, the evidence that cells isolated from persons with cancer-predisposed genotypes are more sensitive to radiation than are normal cells seems to be mixed (Sanford and others 1989; Scott and others 1996). In addition, the current evidence that people with a cancer predisposition are at higher risk for radiation-induced cancers is limited. However, current knowledge of the functions of the cancer-predisposing genes, and of the consequences of their mutations constitutes sufficient grounds for assuming that among the genotypes of those predisposed to cancer there are some that also convey increased risk for radiation-induced cancers. There is also sufficient rationale for attempting to estimate quantitatively the effect of genotype-dependent differences in cancer predisposition on sensitivity to radiation-induced cancer (Sankaranarayanan and Chakraborty 1995). There is clear evidence of the existence of genes that are related to susceptibility to many forms of spontaneous cancer, and these genes could also be markers of an increase in susceptibility to radiation-induced cancer. This hypothesis remains to be proved. As genes such as the ATM gene for ataxia telangectasia are identified and sequenced, much attention will be focused on the possibility that some persons have a genetically based susceptibility to radiation-induced cancer (Sankaranarayanan and Chakraborty 1995), possibly including lung cancer induced by alpha particles. Present risk models do not include individual susceptibility.

Further insights into the role of genetic predisposition can be gained from comparison of the effect of radon in various animal species; there are marked species differences in the responsiveness of experimental animals to radon. Early studies in dogs (Cross and others 1986), mice (Morken 1973; Palmer and others 1973), and Syrian hamsters (Palmer and others 1973; Cross and others 1981) exposed to very high exposures of radon resulted in few lung tumors. The tumor incidence was 21% in dogs, zero in mice, and 1.3% in Syrian hamsters. In hamsters, there were no tumors at exposures below 108 Jhm$^{-3}$ [30,000 working-level months (WLM)]. Many of the exposures were high enough to result in marked life-shortening which can decrease tumor frequency and short-term pathologic changes. In contrast with dogs, mice, and Syrian hamsters, rats have a high incidence of respiratory-tract tumors after exposure to radon (Chameaud and others 1982; Cross and others 1984, 1986; Cross 1994a,b; Gray and others 1986).

The mechanistic bases of these interspecies differences are important to define, but the current evidence suggests that prima facie species-to-species extrapolations of absolute risk are unlikely to be useful since there are differences in response observed following the same insult delivered to different species. Thus, direct extrapolation of animal data to humans cannot be used to predict absolute risk. Data derived in humans can produce patterns of risk which might well be of use (Brenner and others 1995), in that the endpoint remains the same but only the radiation dose/dose rate/quality changes.

Research has been conducted to determine whether the resistance to radon in Syrian hamsters relative to that in rats was related to delivered dose or induced damage at the same level of exposure (Khan and others 1995). Rats and Syrian hamsters were exposed at the same time, which resulted in exposure to the same radon level and dose, and the frequency of micronuclei as an indicator of radiation dose was measured in deep-lung fibroblasts. It was determined that the exposure-response relationship for radon-induced micronuclei per $Jhm^{-3}$ (WLM) was higher in the Syrian hamster than in the rat. That suggests that the dose and damage to the lung cells were similar in the 2 species and that the amount of chromosomal damage initially induced might not be related directly to the differences in species sensitivity for the induction of lung cancer. Combining research on cellular and molecular changes with whole-animal exposures could provide some understanding of the basis of species and strain differences; these differences eventually might be related to individual changes in sensitivity for the induction of cancer.

## CELL-CYCLE EFFECTS

It is well established that ionizing radiation in general and alpha particles in particular produce a dose-dependent delay in progression through both the $G_2$ and the $G_1$ stages of the cell cycle (for example, Lucke-Huhle and others 1982; Kasten and others 1991). The $G_2$ delay has been postulated to give the cell time to repair damage before entering into mitosis (Maity and others 1994). The $G_1$ delay has been shown to depend on the function of the tumor-suppressor protein p53 (Kasten and others 1991) and to be controlled to some degree by Rb gene expression (White 1994). Tumor cells without p53 or with a mutated p53 have lost their ability to respond to cell-cycle arrest after exposure to gamma rays (White 1994). The molecular mechanisms associated with radiation-induced cell-cycle delay have been reviewed (Murnane 1995; Rowley 1996). Cell-cycle progression and delay constitute a multistep process that involves well-defined temporal and spatial changes in expression, phosphorylation, and complex interactions between the level and structure of proteins (Metting and Little 1995; Murnane 1995; Rowley 1996). The importance of DNA damage in producing cell-cycle delay response and the importance of the delay in repair of genetic and lethal damage have been demonstrated and reviewed for dividing mammalian cells (Murnane 1995).

The information available on the response of cells to high-LET radiation damage delivered in $G_0/G_1$ cells and in the role of cell-cycle delay in these cells as they move from $G_0$ into a cycling stage is far from complete. Consequently, the role of cell-cycle delay in altering response or affecting risk associated with indoor exposure to radon is not clear. However, most respiratory tract epithelial cells have rather long cell-turnover times of about 30 days (Adamson 1985), and spend only a small fraction of the total time in stages of the cell cycle that are most radiation sensitive. Inasmuch as the dose rate and number of traversals per cell are very low in the respiratory tract, the probability of alpha-particle traversal in a cycling cell is very low. In addition, the efficiency of cell killing by alpha particles might also decrease the relevance of cell-cycle delay to a risk assessment model. Those considerations make it likely, although not certain, that cell-cycle delay produced by environmental radon exposure plays a minor role in changing potential response or risk.

## APOPTOSIS

After exposure to ionizing radiation, mammalian cells die by one of 2 distinct processes. The classic form of death has been called "mitotic death"; cells die in attempting to divide as a consequence largely of complex chromosomal aberrations (Carrano and Heddel 1973). An alternative mode of death is by "apoptosis," or programmed cell death (Stewart 1994), which involves a characteristic progression of phenotypic changes, including induction of DNA fragmentation and the cell finally being phagocytosed by its neighbors. The relative importance of the 2 modes of cell death varies widely. For some cell types, apoptosis dominates; for others, apoptosis is seldom seen; in yet others, they are about equal. In most self-renewal tissues, apoptosis is a common mechanism to remove damaged or unwanted cells. Radiation-damaged cells are no exception. Failure of processes that lead to apoptotic death and removal of the damaged cells presents an alternative pathway to carcinogenesis for a radiation-damaged cell (Thompson 1995). While apoptosis is generally associated with doses significantly higher than doses usually attributed to radon progeny, apoptosis might be present at low doses.

## RADIATION-INDUCED PERTURBATIONS OF CELLULAR PROLIFERATION

It has been demonstrated that changes in regulation of cell proliferation play an important role in the development of cancer (Brooks and others 1982; Cohen and others 1992), and it has been suggested that changes in cellular proliferation can be used in risk assessment of exposures to carcinogens (Clayson and others 1989; Clifton and others 1991; Goldsworthy and others 1991). It has also been established that an increase in cell turnover in the upper and lower respiratory tract follows experimental inhalation of radon (Taya and others 1994) and that, in the nose and upper respiratory tract, this increase is related to the areas with the highest radiation dose (Atencio 1994).

The cell types and normal turnover rate in respiratory tract cells vary by the region of the respiratory tract and the cell type involved (Adamson 1985). Changes in cell kinetics in the respiratory tract have been demonstrated after internal deposition of radioactive materials (Sanders and others 1989), external radiation exposure (Adamson 1985), and inhalation of radon (Atencio 1994; Bisson and others 1994; Taya and others 1994).

Taya and colleagues (1994) demonstrated an increase in the labeling index (which reflects the proportion of cells synthesizing DNA) as a function of exposure at 0.42-3.465 Jhm$^{-3}$ (120-990 WLM) in rat alveolar, bronchiolar, bronchial, and tracheal epithelial cells over a range of times after exposure. The maximal increase in proliferation was at 14 days for all 4 regions of the respiratory tract. In studies of the nose and upper respiratory tract, Atencio (1994) demonstrated a similar time-dependent increase in cell proliferation after exposure to 0.595 Jhm$^{-3}$ (170 WLM) of radon progeny. The labeling index increased after the end of the exposure to a peak between 14 and 50 days and then returned nearly to background levels. The increase was observed only in the trachea, the nasal septum, and the middle section of the larynx. Several of these regions were calculated to have high deposition for vapors (Kimbell and others 1993) and small particles (James 1994). In rats, the bronchial region, which is calculated to be at greatest risk for cancer induction, also receives the highest dose and responds with the highest cell-proliferation response to inhaled radon. Overall, the findings suggest a relationship between initial dose and changes in cell proliferation.

However, in considering these results, it is important to recognize that overall exposure rates differ widely; in studies of rats, a few weeks to months, in miners, a few years to about half the lifetime, and in residential exposures, a lifetime.

Radon-induced tissue damage and cell killing increases cell turnover to replace damaged cells (Taya and others 1994). This radon-induced increase in cell proliferation can result in repair of tissue damage. Apoptosis can eliminate damaged cells directly and normal and enhanced cell proliferation can also eliminate damaged cells at mitosis (Carrano and Heddle 1973) potentially reducing the risk for cell transformation and cancer. On the other hand, changes in cell kinetics have the potential to increase clonal expansion of altered or mutated cells increasing risk. Cell proliferation is a required step during cancer induction without which cancer cannot form, thus, enhanced cell proliferation can be viewed as a mechanism of either tissue repair or promotion of the cancer process.

## CELLS AT RISK

To determine the dose, energy distribution, and cellular processes essential for radon-induced carcinogenesis, it is important to identify the respiratory tract cells at risk from radon exposure. In radon-inhalation studies, the cells of the respiratory tract receive the highest radiation dose and are presumably at the highest risk (NRC 1988).

The differences in tumor incidence at different locations in the respiratory tract might have a purely dosimetric explanation or might imply that some cells are more sensitive to transformation by radiation than others. For example, no tracheal tumors were reported in rats exposed to radon, even though it has been estimated that the tracheal epithelium receives a larger dose for a particular exposure than do fibroblasts in the deep lung (Brooks and others 1990b; Khan and others 1994). It also has been demonstrated in tracheal epithelial cells that radon inhalation can cause the induction of early stages of cell transformation by producing enhanced-growth variants—cells that continue to grow in selective media (Thomassen and others 1990); this suggests that tracheal epithelial cells have the potential to produce cancers. The lack of tumors in the trachea suggests differential cell- and tissue-specific responses to alpha particles.

Tumors in humans arise primarily in the segmental and subsegmental airways (Saccomanno and others 1996), in contrast with tumors in rats, which are found deeper in the lung. Identifying the cells at risk for the induction of cancer in humans is important from both a biologic and a dosimetric standpoint; but it is difficult, and our understanding is still uncertain (Masse and Cross 1989). Recent human studies (Saccomanno and others 1996) have examined the spatial distribution of lung cancer in ever-smokers that mined or did not mine uranium. A major observation was that the frequency of small-cell lung cancers in the central region of the lungs of uranium miners that smoked cigarettes was higher (30.8% of the total) than observed in the same region of non-miners (10.6%). This region receives the highest radiation dose to lung epithelial cells (appendix B).

Histogenesis of tumors in the tracheobronchial region suggests a common epithelial progenitor cell. It has long been assumed that the basal cell is most at risk (Ford and Terzaghi-Howe 1992a) because its role in repopulation and differentiation is seen as similar to that of the basal cells in other tissues, such as the skin. Ford and Terzaghi-Howe (1992a,b) showed that isolated basal cells from the rat trachea are the cells that are capable of growth in vitro and in vivo; that suggests that they are the precursor cells for the induction of tumors. Electron microscopy and immunohistochemistry have revealed that individual tumor cells coexpress features associated with several different tumor types (McDowell and Trump 1984). Uses of transfection techniques with oncogenes also has suggested that there is a common cell of origin for the induction of tumors in the respiratory tract (Pfeifer and others 1991; Amstad and others 1988).

Other lines of research have suggested that the major airway epithelial cell type at risk for radon-induced cancer is the secretory cell (Johnson and Hubbs 1990; Johnson 1995). It has been demonstrated that in rat trachea, secretory cells constitute a major progenitor cell compartment. Secretory cells proliferate in response to physical or chemical trauma and are involved in repair and maintenance of the tracheobronchial lining (Keenan and others 1983). Studies with denuded tracheal grafts showed that secretory cells that were isolated with a cell sorter could reestablish an epithelium composed of basal, secretory, and ciliated cells. When pure populations of basal cells were used, only basal and ciliated cells were found in the repopulated graft. Those findings suggest that secretory cells can differentiate to form all the cell types in the trachea, whereas basal cells have more-limited capacity to differentiate. The observations that both cell types can divide and differentiate, point to the potential role of the secretory cell in radon-induced cancer induction.

The cells involved in radiation-induced tumors in the pulmonary paren-chyma, as opposed to the airways, also are unidentified. Adenocarcinomas are thought to arise in the peripheral lung and display both mucous and serous cell differentiation. Bronchioloalveolar tumors, however, possess features of both Clara cells and alveolar type II cells. Clearly, cells can share common differen-tiation pathways during insult and the progression to cancer. These observations indicate that using tumor-cell structure as an indicator of the cells at risk in the peripheral lung might be misleading.

Robbins and Meyers (1995) have conducted extensive studies to define the cell populations in the human respiratory tract that are capable of cell division and thus might serve as the progenitor cells for respiratory tract cancer. They have determined that both basal and suprabasal cells are dividing in human tracheobronchial mucosa and that the suprabasal cells proliferate more frequently than the basal cells. From that observation, they developed a model that suggests two different stem-cell populations, which they call reserve stem cells (basal cells) and transient-amplifying stem cells. The transient-amplifying stem cells can, as demonstrated by Johnson and Hubbs (1990), give rise to all the different cell types, whereas the reserve stem cells renew the transient-amplifying stem-cell population and give rise to a narrower range of cell types.

## TARGET SIZE

A growing, although still small  body of scientific literature suggests that the number of cells that  respond to alpha-particle radiation is greater than the num-ber of cells traversed by alpha particles. Hickman and colleagues (1994) found that exposure of lung-epithelial cells to small doses of alpha particles from a $^{238}$Pu source caused an increase in expression of the p53 gene in many more cells than were calculated to have been traversed. They concluded that the hit cells had communicated with the cell population and caused the remainder of the cells to respond. Thus,  the target for interaction with alpha particles could be much larger than the cell nucleus.

It also has been demonstrated in both human and rodent cells that after exposure of cells to low doses of alpha particles, the number of cells with an increase in the frequency of sister-chromatid exchange (SCE) aberrations was larger than the number of cell nuclei traversed by alpha particles (Nagasawa and others 1990; Nagasawa and Little 1992; Deshpande and others 1996). Again, both intercellular communication and tissue-culture factors could be involved in these changes. It is important to note here, however, that SCE formation is unlikely to be relevant in carcinogenesis.

Brooks and his colleagues (1990a) have demonstrated that when cells were exposed simultaneously to very low doses of alpha particles (6 cGy) and x rays, the response to the x rays with respect to both cell killing and induction of micronuclei was markedly increased. That suggests that the alpha particles change the responsiveness of the whole cell population, even though only a small percentage of the total cell population was traversed by alpha particles. The studies were conducted with confluent lung epithelial cells, so the damaged cells could potentially transmit information to neighboring cells. Clastogenic factors have also been isolated from the blood of Chernobyl accident victims; these factors remained in the blood for years after the exposure (Emerit and others 1995).

The significance of cellular responses in "bystander" cells in the development of radiation-related disease and risk cannot be defined at this time. However, since cellular changes can be induced in more cells than are traversed by alpha particles, it might be necessary to redefine target size for alpha particles to include a much larger volume than the cell nucleus through which the radiation traverses.

Thus, the long-accepted assumption that causing a heritable effect, such as cancer, requires a track to pass through the nucleus of affected cells is no longer unequivocal, in that now there is limited evidence that an ionizing event in 1 cell can affect its neighbor(s). The data are intriguing, but there is no real evidence that cells that are adjacent to traversed cells are damaged in such a way as to be transformed to a malignant state, nor has a cellular or molecular mechanism been proposed for how this could occur. In any case, the so-called bystander effect is unlikely to affect the concept of linearity. For a low dose-rate alpha-particle exposure, the number of cells traversed, say per year, is a small fraction of the total and doubling the dose approximately doubles the number of cells traversed by a track (rather than doubling the number of tracks in each cell, as with x rays). Consequently, the linear relationship between dose and effect is still the most tenable, whether the carcinogenic event results from a direct effect on the cell concerned or from an indirect effect from damage to a neighboring cell.

## THE SPECIAL NATURE OF BIOLOGIC DAMAGE
## INDUCED BY ALPHA PARTICLES

Alpha particles, which produce a high density of ionizations along their path (high-LET), differ greatly from sparsely ionizing (low-LET) radiations, such as x rays in their microscopic, spatial, and temporal patterns of interactions with cells, subcellular structures, and molecules (see Figure 2-1). Insult from alpha particles is concentrated in a relatively small number of densely ionizing tracks that each traverse a cell in less than $10^{-12}$ s and deliver to it a large localized energy (about 10-50 cGy). In addition, in the case of alpha particles, relatively more of the individual damage is due to direct ionizations in the DNA rather than to hydroxyl-radical reaction, and it is more clustered on the scale of DNA and chromatin. For any radon exposures of practical relevance, a given cell in the lung is likely either to be unirradiated, to receive an instantaneous large energy deposition from a single alpha particle, or to receive a small number of such large energy depositions, each well separated in time by months or years.

For example, even for a miner in the highest-exposure cohort receiving 2.8 Jhm$^{-3}$ (800 WLM) over 5 years, a cell nucleus in the bronchial epithelium is estimated to receive on the average only about 1-5 alpha-particle traversals per year (Table 2-1). In contrast, a similar absorbed dose to the lung from gamma rays would imply that each cell nucleus had received a very large number (about $10^4$) of low-LET electron tracks, at the rate of about ten per day.

To calculate the dose to individual cells in the respiratory tract it is necessary to know the traversal number and the distance of the cells from the surface of the airways. The anatomy of the airway is discussed in appendix D on comparative dosimetry to illustrate the location of the cells of interest. To calculate the traversal number it is essential to define the location of the cell relative to the surface and the thickness of the mucus on the airway. Average values of mucus thickness at each of the anatomical locations were considered in the dose calculations. The hit numbers for different cellular locations and cell types are illustrated in Table 2-1. The variation of mucus thickness would be greater in the upper respiratory tract, nose and trachea, than in bronchus or bronchioles. Since the major site of cancers are in the smaller bronchus and bronchioles, the variation in mucus thickness was not considered as a major source of uncertainty in the dose calculations.

At the DNA level, also, there is a substantial difference between the spectrum of initial ionization damage from alpha particles and from gamma rays. Alpha particles produce more large clusters of multiple ionizations within the DNA and in adjacent molecules than do gamma rays (Goodhead and Nikjoo 1989). Considerations of track structure and radiation chemistry lead to the expectation that this will result in severe locally damaged sites (clustered damage) that the cell is less likely to be able to repair (Goodhead 1994; Ward 1994). Although the chemical nature of the individual single-strand breaks and base damages might not depend on the radiation type, their repairability could be influenced by the presence of the additional damage. The measured gross yields per Gy of initial DNA double-strand breaks (DSBs), irrespective of complexity, are about the same for alpha particles and gamma rays, according to recent measurements (Prise and others 1987; Charlton and others 1989; Brenner and Ward 1992; Jenner and others 1993; Prise 1994).

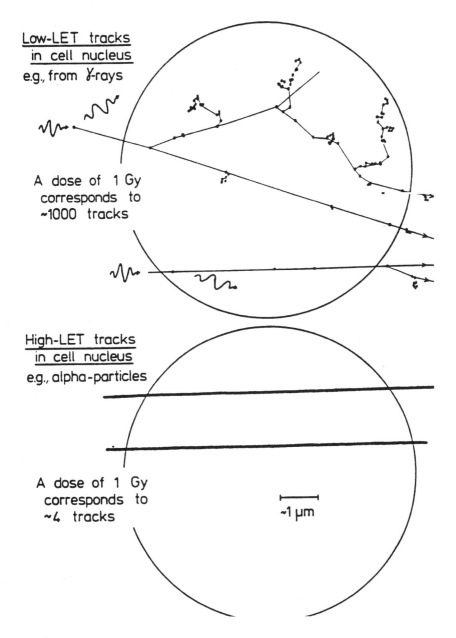

Low-LET tracks
in cell nucleus
e.g., from γ-rays

A dose of 1 Gy
corresponds to
~1000 tracks

High-LET tracks
in cell nucleus
e.g., alpha-particles

A dose of 1 Gy
corresponds to
~4 tracks

~1 μm

FIGURE 2-1    Illustration of the differences in ionization-track densities in a cell nucleus between low-LET and high-LET radiations.

TABLE 2-1  Summary of computed alpha-particle hits[a] for lung cells in different target locations

| Target location | 2.8 Jhm⁻³ (800 WLM) over 5 years in mine[b] | | 0.0267 Jhm⁻³ (7.6 WLM) over 1.1 years in mine[c] | | 50 Bqm⁻³ for 70 years in home[d] | | 150 Bqm⁻³ for 70 years in home[e] | |
|---|---|---|---|---|---|---|---|---|
| | Mean number hits | Mean hits per year | Mean number hits | Mean hits per year | Mean number hits | Mean hits per year | Mean number hits | Mean hits per year |
| Bronchial Basal cell nuclei | 6.7 | 1.3 | 0.063 | 0.058 | 0.15 | 0.0022 | 0.46 | 0.0066 |
| Bronchial Basal cell cytoplasm | 19 | 3.8 | 0.18 | 0.17 | 0.43 | 0.0062 | 1.3 | 0.19 |
| Bronchial Secretory cell nuclei | 26 | 5.3 | 0.25 | 0.23 | 0.60 | 0.0086 | 1.8 | 0.026 |
| Bronchial Secretory cell cytoplasm | 384 | 77 | 3.65 | 3.32 | 8.4 | 0.12 | 25.3 | 0.36 |
| Bronchiolar Secretory cell nuclei | 15 | 3.1 | 0.15 | 0.13 | 0.22 | 0.0031 | 0.65 | 0.0093 |
| Bronchiolar Secretory cell cytoplasm | 130 | 26 | 1.2 | 1.1 | 1.79 | 0.026 | 5.4 | 0.077 |

[a]These hit numbers represent the mean number of alpha-particle hits at a given cell (or nucleus) location in the lung during the specified time period. In general, many cells and nuclei will have occupied each location during the extended time period. The mean number of hits per cell cycle may be estimated as about 1/12 of the mean number of hits per year on the basis of a turnover time of about 30 days for the majority of respiratory tract epithelial cells (Adamson 1985). The hit numbers were evaluated at the request of the BEIR VI committee by A.C. James based on the lung model and reference value of ICRP 66 and the BEIR VI committee's reference work pattern and aerosol conditions for miners and reference aerosol conditions for homes (without cigarette smoke).

[b]Typical for highest exposure miner cohort (Colorado)—see appendix D.

[c]Typical for lowest exposure miner cohort (Radium Hill)—see appendix D.

[d]Approximate average concentration for homes in USA.

[e]Approximate USA action level concentration.

Reduced cellular repair of initial DNA DSBs caused by alpha particles and other high-LET radiations, relative to x rays has been observed (Ritter and others 1977; Coquerelle and others 1987; Blocher 1988; Fox and McNally 1990; Jenner and others 1993). Greater complexity and reduced repairability of alpha-particle induced breaks have been indicated also by studies with radical scavengers in mammalian cells (de Lara and others 1995) and in plasmid systems (Jones and others 1993; Hodgkins and others 1996). Association between separate DSBs produced by the same particle track can also occur because of the higher-order DNA-chromatin structure. It has been suggested that the high-LET particles can therefore yield an excess of DNA fragments of 0.1-2 kilobases (Holley and Chatterjee 1996) and also very large deletions, up to about 200 kb, have been observed (Löbrich and others 1996). Such associations might make repair more difficult and slower (Löbrich and others 1994).

Overall, even low-level exposure to alpha particles presents the "irradiated" proportion of the exposed cells with very large instantaneous burdens of damage, including severe clustered damage at the DNA level and multiple spatially correlated damage sites from the same alpha-particle track within the cell. Therefore, it is possible that biologic consequences of alpha-particle exposure differ from those of low-LET irradiation in both qualitative and quantitative respects (Goodhead 1988).

Low-dose alpha-particle induced radiation damage differs from damage caused by x rays in that the potential to transmit viable mutations or aberrations can be substantially reduced by cell death induced by the same single alpha particle. This single-track survival probability depends heavily on cell type and geometry, varying from essentially zero to over 80%. Such large variability makes it difficult to extrapolate relative mutagenic or carcinogenic effectiveness of alpha particles from effects of other kinds of radiation and from cell type to cell type.

It can be inferred from the average number of alpha-particle traversals per lethal event that the single-track survival probability for a given cell type depends on the shape of the nucleus at the time of exposure. The amount of energy deposited per nucleus and the extent of the biologic response depend on both cell and nuclear size and shape (Geard 1985; Raju and others 1993). For a variety of fibroblast cell lines, it was found that, for an alpha-particle of given energy, the incidence of a lethal event was roughly constant per unit track length through the cell nucleus at about 0.03-0.08 lethal events per micrometer (Goodhead and others 1980; Roberts and Goodhead 1987; Raju and others 1993). However, the probability is greater for radiosensitive lines and some other cell types, including hematopoietic cells (Lorimore and others 1993), and can in some cases lead to a negligible single-track survival probability (Griffiths and others 1994). Before the advent of microbeam irradiators (described below), it was not possible to measure directly the probability of cell survival after exposure to exactly 1 alpha particle.

Cellular and molecular studies have been conducted to define the nature of radon-induced damage. In many cases, other alpha-emitting radionuclide sources or particle accelerators have been used to expose cells to high-LET particles. A recent development has been the construction of alpha-particle microbeams capable of delivering exactly 1 alpha particle to each cell in a dish (Geard and others 1991; Braby 1992; Nelson and others 1996). Such particle research makes it possible to define the influence of LET and particle fluence on the induction of a wide variety of end points (Cox and others 1977b; Lloyd and others 1979, 1984; NCRP 1990; Goodhead and others 1991; Miller and others 1995). It has been well established that high-LET radiation, such as slow alpha particles, produces greater amounts of biologic damage per unit of dose or energy absorbed than low-LET radiation (Barendsen and others 1966; Brooks 1975; Goodhead and others 1980; Thacker and others 1982; Schwartz and others 1992; Evans 1993; Jostes and others 1993; Piao and Hei 1993; Brooks and others 1994; Jostes 1996; Simmons and others 1996). The derived values of relative biologic effectiveness (RBE) depend on the biologic effect under study, cell types, cell growth, and conditions of irradiation. A useful summary of the relationship between LET and RBE has been published by the National Council on Radiation Protection and Measurements (NCRP 1990). In comparison with normal cells, the RBE for cell inactivation by alpha particles is usually lower for cells that are very radiosensitive because of a genetic defect (Cox 1982).

A review of studies on chromosomal aberrations demonstrated a linear dose-response relationship for the induction of cytogenetic damage by high-LET radiation, including alpha particles, and a linear-quadratic function was needed to describe the induction of aberrations by acute exposure to low-LET radiation (Bender and others 1988; NCRP 1990). Dose fractionation or dose protraction had little effect on the induction of chromosomal aberrations following exposure to high-LET radiation (Brooks 1975; Bender and others 1989). At the chromosomal level, alpha particles produce initial DNA breaks that are more closely associated with each other than those produced by x rays. Those differences would be expected to result in increased yields of alpha-particle-induced aberrations of the exchange type, such as translocations, which are indeed observed (for example, Searle and others 1976). In addition, the spectrum of aberrations, including the occurrence of a high proportion of complex aberrations involving more than 2 chromosomes, can be substantially altered even after few alpha-particle traversals (Griffin and others 1994, 1995).

Several investigators (Bedford and Goodhead 1989; Cornforth and Goodwin 1991; Goodwin and Cornforth 1994; Loucas and Geard 1994) have used the premature-chromosome-condensation technique to investigate the initial yields of chromosomal damage after the passage of alpha particles and the kinetics of damage repair that produces "final" chromosomal damage. Showing trends similar to those in the data on DSB yields and repair, these experiments demonstrated that alpha particles produced no more than about twice the initial chromosomal breakage (about 5 min after exposure) as x rays. However, the yield of remaining damage measured after about 1 h was considerably larger after alpha-particle exposure.

A wide array of studies have been conducted to determine the frequency and nature of mutations induced by high- and low-LET radiation. In general, it has been found that alpha particles and other high-LET radiation produce greater frequencies of mutants per survivor per unit dose than does low-LET radiation and that RBE values tend to be greater for mutation induction than for cell inactivation (Thacker and others 1979, 1982; Chen and others 1984; Iliakis 1984; Metting and others 1992; Cox and Masson 1994b; Griffiths and others 1994; Bao and others 1995). The RBE for mutation by alpha particles, such as from radon progeny, is in the LET region of maximal effectiveness (Cox and Masson 1979).

In an in vivo study, a statistically significant correlation was reported between household radon levels and HPRT mutants in peripheral blood lymphocytes of residents (Bridges and others 1991). The observation that a mutant yield could be detected with significance relative to background implied a much greater effectiveness than had been found for alpha particles in high-dose in vitro experiments. However, the correlation was not confirmed in a larger follow-up study by the same research group (Cole and others 1996), and a negative correlation was reported in a separate smaller study (Albering and others 1994).

Conventional chromosome analyses for unstable aberrations in blood lymphocytes of persons living continuously in houses with very high radon concentrations have shown a significant increase in dicentrics and rings (Bauchinger and others 1994). Applying fluorescence in situ hybridization (FISH) techniques to these same samples showed slightly, but not significantly ($p<0.1$), raised overall frequencies of symmetric translocations, including significantly raised levels in males alone (Bauchinger and others 1996). The two sets of results are consistent and the authors suggest that the FISH measurements are less sensitive to discriminate radon exposure because control frequencies of translocations are much higher than those of dicentrics and radon doses to bone marrow, that should contribute most to stable aberrations, are lower than to mature lymphocytes. For more information see chapter 4.

The spectrum of mutants produced by ionizing radiation is wide and can depend on the gene, the host cell, the radiation quality, and the dose. Ionizing radiation in general and alpha particles in particular are most efficient at producing large genetic changes, deletions, and rearrangements, rather than point mutations. Several studies have suggested that densely ionizing radiations produce a higher proportion of gene mutations involving large deletions than do x rays (Chen and others 1990; Evans 1991; Hei and others 1994c; Schwartz and others 1994; Bao and others 1995; Kronenberg and others 1995; Zhu and others 1996). It has also been suggested that the frequency of large deletions increases as a function of dose for high-LET radiation (Hei and others 1993; Zhu and others 1996). Other investigators found no statistically significant difference between high- and low-LET radiation-induced mutagenesis in the fraction of mutations that lost the entire active gene (Thacker 1986; Kronenberg and Little 1989; Evans 1991, 1993; Jostes and others 1994), whereas in an episomal shuttle-vector model, deletions were observed to be more frequent after alpha-particle exposure than x-ray exposure but were of comparable size (Lutze and others 1992, 1994).

Those observed differences in the patterns of mutation induced by high-LET radiation in different cell lines seem to be partially resolved by the observation that in cell systems in which genes that were critical to cell survival were close to the marker gene, a rather low frequency of mutations was induced by deletions. In cell systems in which the marker gene was not near a critical gene and the marker gene was not involved in cell survival, a larger fraction of the total mutations were produced by large deletions after alpha-particle exposure than after low-LET radiation exposure (Evans 1991; Bao and others 1995).

A variety of additional differences have been reported between the spectra of deletions caused by high- and low-LET radiation. These include a higher proportion of smaller but more complex deletions and rearrangements caused by alpha particles (Bao and others 1995; Jin and others 1995; Thacker 1995; Chaudhry and others 1996) and differences in growth characteristics (Metting and others 1992; Chaudhry and others 1996; Amundson and others 1996). Hence, radiation produces a spectrum of mutations different from those which arise spontaneously or are caused by other agents, and the spectrum can depend on radiation quality.

Research at the sequence level is continuing in a number of laboratories to evaluate the molecular nature of mutations induced by alpha particles from radon (Jostes 1996). It should add to the understanding of the nature of the mutagenic lesions induced by high-LET radiation. One of the objectives of such research is to determine whether unique changes associated with high-LET damage can be used as a "signature" of alpha-particle-induced damage. Such a signature could help to identify environmental agents that are responsible for observed mutagenic damage (Schwartz and others 1994). Other studies also have proposed signatures for alpha-particle-induced biologic damage in the form of the induction of specific point mutations in the tumors of uranium miners (Taylor and others 1994; Vahakangas and others 1992); however, the results are not consistent, nor have they been confirmed by larger-scale animal studies (McDonald and others 1995; Kelly and others 1995).

## BIOLOGIC EFFECTS OF LOW EXPOSURE
## LEVELS TO ALPHA PARTICLES

The primary approach to radon risk estimation involves epidemiologic studies of underground miners whose mean exposure was typically much larger than average residential exposures. For example, the average radon-progeny exposure of the Colorado miner cohort was about 2.8 Jhm$^{-3}$ (800 WLM) over an average duration of 5 y. That implies that an average of about 7-26 alpha particles would have traversed each target-cell nucleus location in the segmental bronchial epithelium over this time (Table 2-1). Thus, very few target cells and their progeny in miners in this cohort are anticipated to have been traversed by only one particle. In the pooled miner data (Lubin and others 1994a), the average exposure was 0.567 Jhm⁻³ (162 WLM), which would correspond to about 1-5 traversals per cell nucleus.

It follows that past risk estimates based on all the miner data (Lubin and others 1994a) have been driven by data on miners in whom multiple traversals either dominated or were comparable in number with single traversals. For an average indoor exposure at a concentration of about 50 Bqm⁻³ (1.25 pCiL⁻¹) or an exposure of about 0.0007 Jhm⁻³/yr (0.2 WLM/yr), less than 1 in 400 basal cell nuclei (or less than 1 in 100 secretory cell nuclei) will be traversed per year by a single alpha particle, and less than 1 in $10^4$ cells will be traversed by more than one alpha particle (Table 2-1). Therefore, in extrapolating from miner exposures to environmental exposures, previous models, based primarily on highly exposed miners, translated effects of multiple traversals to the effects of single traversals. In chapter 3 we attempt to avoid that problem by focusing on the low-exposure miners.

However, the prima facie approach of performing alpha-particle experiments at both high and low doses to guide the extrapolation does not allow for a direct assessment of the effects of a single alpha particle. That is apparent if we consider an in vitro experiment designed to measure the effects of single alpha particles traversing, say, C3H10T$^1$/$_2$ cells growing as a monolayer with large projected nuclear areas. Assuming an LET of 150 keV/μm, and a low practical dose of about 0.1 Gy, on the average each cell nucleus will be traversed by a single alpha particle. However, because the number of traversals of a given cell is Poisson-distributed, about 37% of the cell nuclei will not be traversed by alpha particles, about 26% will be traversed by more than 1 particle, about 8% by more than 2, and about 2% by more than 3; this precludes a direct assessment of the effects of *exactly 1* alpha particle.

Two experimental approaches permit the effects of exactly 1 alpha particle to be investigated. One way is to unfold the Poisson distribution from experimental dose-response data mathematically (Brenner 1989), provided that the cell geometries and sensitivities are sufficiently uniform. A more-direct approach to the problem of single-particle traversals is to design experiments in which exactly one alpha particle is delivered to each of many cells. This microbeam-irradiator approach, previously mentioned, is being pursued in several laboratories around the world, including those at Columbia and Texas A & M universities and Gray Laboratory (Geard and others 1991; Braby 1992; Folkard and others 1995; Nelson and others 1996). The basic notion for this approach is that the locations of cells are determined and recorded by a computerized image-analysis system; then a highly collimated beam of low-intensity alpha particles is directed at each cell in turn, and a radiation detector is used to determine when one or any given number of alpha particles have passed through the cell. The dish is then moved so that the next cell is under the collimated beam. True single-particle irradiation should allow measurement of the effects of exactly 1 alpha-particle traversal relative to multiple traversals. These techniques should also allow evaluation of the effects of cytoplasmic traversals and traversals through nearby cells—the potential "bystander" effect discussed earlier.

## BIOLOGIC EFFECTS OF ALPHA PARTICLES
## AT LOW EXPOSURE RATES

In the last few years, it has become increasingly clear that densely ionizing radiation such as alpha particles can exhibit an inverse dose-rate effect for carcinogenesis (for example, Miller and others 1993); that is, for a given dose or cumulative exposure, as the dose rate is lowered, the probability of carcinogenesis increases. The phenomenon has come to be known as the inverse dose-rate effect because it is in marked contrast to the situation for sparsely ionizing radiation, which with protraction in delivery of a given dose, either by fractionation or by low dose rate, usually results in a decreased biologic effect.

The extent and consistency of published reports on the in vitro and in vivo inverse dose-rate effects, leave little doubt that such effects are real (Charles and others 1990; Brenner and Hall 1992). Of interest here is that the inverse dose-rate effect has been clearly demonstrated in miners exposed to radon-progeny alpha particles at different exposure rates. From comparisons of epidemiologic studies involving different average radon-progeny exposure rates, Darby and Doll (1990) inferred the existence of an inverse dose-rate effect. On the basis of epidemiologic studies, an inverse dose-rate effect was reported by Hornung and Meinhardt (1987) in Colorado uranium miners, by Ševc and colleagues (1988) and Tomášek and colleagues (1994a) in Czech uranium miners, and by Xuan and colleagues (1993) in Chinese tin miners. In a recent joint analysis of 11 cohorts of miners exposed to radon progeny, Lubin and colleagues (1995a) clearly demonstrated the existence of a significant inverse dose-rate effect.

Irrespective of the detailed mechanisms involved, and provided that they are confined to single independent cells, basic biophysical arguments imply that if a target cell or its progeny is hit by 1 or 0 alpha particles, it cannot show a dose-rate effect of any kind. Mechanistically (see, for example, Barendsen 1985; Goodhead 1988; Curtis 1989; Brenner 1994), a cell traversed only once by an alpha-particle cannot "know" or respond to any changes in dose rate. Thus, no inverse dose-rate effect would be expected at very low exposures, but such effects would be possible as the cumulative exposure increased to a point where multiple traversals of the targets become significant. The resulting overall effect therefore will be the result of an interplay between cumulative exposure and exposure rate (Brenner 1994). These considerations are summarized in Figure 2-2, which depicts a protraction effect that increases with increasing exposure and decreases with

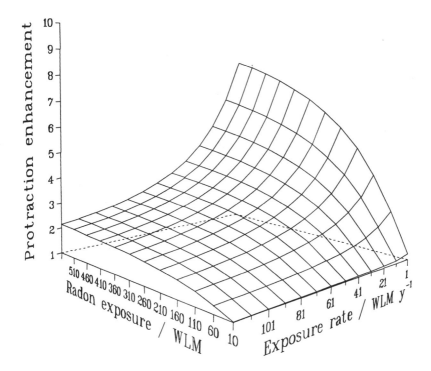

FIGURE 2-2    Schematic illustration of the relative effect of protraction, illustrating the interplay between exposure and exposure rate. Modified from NRC (1994).

increasing exposure rate. The particular quantitative values of the surface in the figure depend on the model and parameters, but its general features are likely to be largely independent of the model and have been shown to be consistent with the miner data (Lubin and others 1995a).

Possible mechanisms by which an inverse dose-rate effect could arise from exposures to single cells include

• High-dose-rate saturation of effect from multiple track traversals in a particularly sensitive phase of the cell cycle (Rossi and Kellerer 1986; Brenner and others 1993).
• Killing of initiated cells before they multiply.
• Enhancement of cellular repair (Burch and Chesters 1986).
• A correlation across the cell cycle between cell killing and oncogenic transformation induction (Elkind 1994; Brenner and others 1996).

Additional mechanisms in tissues for a low-dose-rate increase in effect could include

- Promotion of the transformation process or enhanced misrepair (Hill and others 1984).
- Enhancement of cellular proliferation (Moolgavkar 1993).
- Age-dependent host variations in sensitivity or second mutations in expanding initiated clones (Leenhouts and Chadwick 1994).

An average lifetime exposure from an indoor radon concentration of 200 $Bqm^{-3}$ (5.41 $pCiL^{-1}$), which is about 4 times the average indoor exposure, would result in on average about 1 alpha-particle traversal per bronchial epithelial cell nucleus location (0.6 for the location associated with the bronchial basal nuclei or 2-4 for the location of bronchial secretory cell nuclei; see Table 2-1). In most indoor-exposure situations, protraction would be expected to have little effect on risk unless there are large additional spatial and temporal factors, such as persisting long-range cell signaling or clonal expansion. In contrast, in the miner studies, even though exposure rates are higher, the higher exposures result in a statistically significant reduction in risk per unit of exposure. That conclusion, which is consistent with  the results from miner studies (see Figure 2-3 and Lubin and others 1995a), depends essentially on the notion that a dose-rate effect of any kind requires that autonomous target cells be exposed to multiple alpha-particle traversals.  It should also be noted that in tissues, cells may die and be replaced many times during a lifetime.  On the basis of these considerations, data on lower-exposure subset of the miner cohorts would be expected to yield  the most applicable estimate of residential risk.

Although the miner data show an inverse dose-rate effect, indoor-exposure data will probably show none, because of the low probability of multiple alpha-particle traversals in the low-exposure situation of a residence.  That is in accord with results on lung-cancer induction by radon in rats.  Specifically, results for high cumulative exposures over 3.5 $Jhm^{-3}$ (1,000 WLM) show a statistically significant inverse dose-rate effect:  for the same cumulative exposure, irradiation over longer periods resulted in significantly higher lung-cancer rates than irradiation over shorter periods (Cross 1992; Gilbert and others 1996).  As the exposure was decreased, no significant inverse dose-rate effect was observed (Gilbert and others 1996).  At exposures corresponding to less than 1 alpha-particle traversal per cell, 0.0875 $Jhm^{-3}$ (25 WLM), no increase in lung-cancer incidence was observed as the exposure rate was decreased (Morlier and others 1994).  All this is consistent with the pattern presented in Figure 2-3.

In contrast with the high-exposure studies,  some of the low-exposure studies in rats yielded evidence of a *decrease* in lung-cancer incidence with decreasing exposure rate (the "conventional" dose-rate effect).  Specifically, when 0.0875 $Jhm^{-3}$ (25 WLM) was protracted over 18 m, rather than over 4-6 m, a decrease in lung-tumor incidence was observed (Morlier and others 1994), although the statistical significance was marginal (p = 0.056).  Those studies suggest that biologic variables, such as fraction of the life span during the exposure and age at

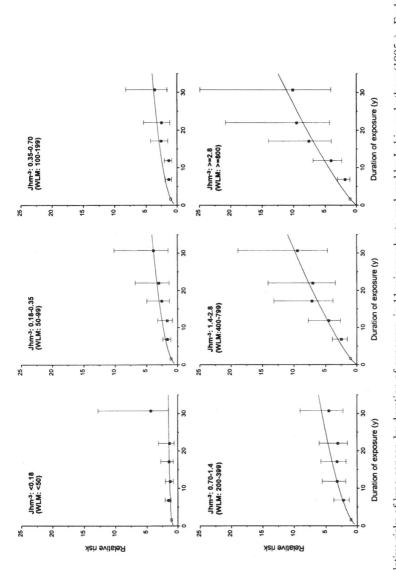

FIGURE 2-3   Relative risks of lung cancer, by duration of exposure, in 11 miner cohorts analyzed by Lubin and others (1995a). Each panel represents a different total exposure. For miners with the highest exposures of $> 1.4$ Jhm$^{-3}$ ($> 400$ WLM), there is a marked inverse dose-rate effect. The inverse dose-rate effect is less apparent for miner exposures between 0.18 and 1.4 Jhm$^{-3}$ (50 and 400 WLM) and it is essentially undetectable for exposures under 0.18 Jhm$^{-3}$ (50 WLM).

exposure, play an important role in the development of lung cancer, at least in experimental animals (Cross 1994a,b).

## INTERACTIONS BETWEEN LUNG CARCINOGENS

Radon is only one of the respiratory carcinogens to which humans are exposed. Tobacco-smoking is an extremely important risk factor for lung cancer in miners, as well as in the general population; and other lung carcinogens, such as arsenic, are also present in mines. A brief review of the in vitro and in vivo studies related to the issue of interactions between lung carcinogens follows.

Using an in vitro assay for oncogenic transforming C3H10T$^{1}/_{2}$ cells, Piao and Hei (1993) applied cigarette-smoke condensate (CSC) and observed a dose-dependent increase in the incidence of both cytotoxicity and oncogenic transformation. The frequency was significantly increased if the CSC was combined with a dose of either gamma rays or alpha particles. The transformation frequencies in cells treated with a combination of CSC and 0.5 Gy of alpha particles with energies selected to simulate radon-progeny alpha particles was consistent with the 2 agents acting in an additive manner, not a multiplicative manner.

The report of the BEIR IV committee (NRC 1988) reviewed the animal studies that included exposure to both radon progeny and cigarette smoke. The relevant studies included experiments involving rats conducted by the Compagnie Generale des Matieres Nucleaires (COGEMA) in France and experiments involving dogs conducted by Pacific Northwest Laboratories (PNL) in the United States. The report noted that the COGEMA experiments showed synergism (greater-than-additive effects) if the exposure to cigarette smoke followed the exposure to radon progeny but not if the smoke exposure preceded the radon-progeny exposure. In the PNL experiments, lung-tumor incidence was decreased if the animals were exposed to radon progeny and cigarette smoke on the same day, as opposed to sequentially.

Since the BEIR IV report, there have been several additional reports from COGEMA (Monchaux and others 1994) and PNL (Cross 1992). Cross and co-workers have reviewed the newer studies (Cross 1992, 1994a,b). The PNL group conducted initiation-promotion-initiation experiments with cigarette smoke and radon exposure (Cross 1992). Those experiments involved various sequences of exposure to smoking and radon progeny and splitting the dose of radon progeny. Only preliminary findings on lung tumors are available and the number of cancers has been very small. The findings of the COGEMA studies have been summarized recently (Monchaux and others 1994; Yao and others 1994). The extent to which lung-cancer incidence was increased by cigarette-smoke exposure after radon exposure was shown to depend on the duration of exposure to smoke. Decreasing duration was associated with decreasing lung-cancer incidence.

In spite of long-term research by 2 groups of investigators, the animal experiments on smoking and radon progeny have not yielded strong evidence on the combined effects of the 2 exposures. The findings are inconsistent and dependent on the sequence of exposures. In the residential setting, exposure to cigarette smoke and exposure to radon progeny occur essentially simultaneously throughout adulthood. Among the miners, smoking and radon exposure can take place simultaneously or radon exposure can begin before or after smoking has started (Thomas and others 1994). The unique pattern of sustained smoking by humans, which has not been replicated in the animal experiments, is an additional barrier to extending the findings of the animal studies to humans.

## THE DOSIMETRIC APPROACH TO RADON RISK ESTIMATION

The approach to domestic radon risk estimation taken in this report involves epidemiologic studies of people who have been exposed to radon. A different, and possibly complementary, approach is to estimate radon risks on the basis of people exposed primarily to sparsely ionizing radiation—largely the Japanese atomic-bomb survivor cohorts. This so-called dosimetric approach to radon risk assessment using ICRP quantities (ICRP 1991) has the following logic:

- Use physical models to estimate a bronchoepithelial dose per Jhm$^{-3}$ (WLM).
- Convert that lung dose, with the specified radiation weighting factor ($w_R$) (ICRP 1991), to an equivalent dose for radon-progeny alpha particles in the bronchial epithelium.
- Convert the equivalent dose to an effective dose, with the appropriate tissue weighting factor for lung.
- Use the best estimate for the lifetime-fatality probability coefficient per unit of effective dose to estimate the lifetime risk per Jhm$^{-3}$ (WLM).

A more direct dosimetric approach could be to apply an appropriate alpha-particle RBE factor specifically for lung, rather than the general radiologic protection quantity $w_R$ (or "quality factor," Q). Then one could estimate the lifetime risk per Jhm$^{-3}$ (WLM) from the lung-cancer-fatality probability coefficient per unit of absorbed dose of atomic-bomb survivor data extrapolated to low dose rate.

Assuming a quality factor of 20, Burchall and James (1994) used a dosimetric approach to estimate a risk from residential radon exposure and found the risk to be larger than estimated from the miner data (for example, Lubin and others 1994a) by a factor of 4-5. However, it is difficult to interpret the difference between the two approaches. In light of the uncertainties in many of the steps involved in arriving at both types of risk estimates, the difference is modest. One of the major uncertainties in the dosimetric approach is related to the current impossibility of estimating RBE directly in any realistic quantitative sense for relevant in vivo end points. Consequently, the rationale usually adopted is to estimate its values for a variety of in vitro end points that are considered to be relevant to cancer induction and to be adequately quantifiable and then to define from these data sets a single value that is judged to be applicable to human cancers overall. The use of in vitro oncogenic-transformation data as a basis for risk estimates for more complex end points, such as carcinogenesis in general in humans, has been discussed elsewhere (ICRU 1986). Essentially, the rationale, other than the pragmatic issue of quantifiability, is that the radiation weighting factor is used for predicting only *relative risks* (compared with risks associated with gamma rays or x rays) of one kind of radiation relative to another, rather than *absolute* risks. However, many data on in vitro effects or carcinogenesis in animals show that the RBEs for the same kind of radiation depend substantially on the biologic system and cancer type under study. The RBE for induction of lung cancer by radon-progeny alpha particles remains uncertain.

On the basis of in vitro data on the $C3H10T^{1}/_{2}$ oncogenic-transformation system of Brenner and colleagues (1995) and data on the induction of micronuclei in rat-lung fibroblasts and CHO cells by radon and gamma rays of Brooks and colleagues (1994), a quality factor of 10 seems appropriate for cells at depth in the bronchial epithelium. That is half the currently recommended radiation weighting factor (ICRP 1991). It would result in partial reconciliation of dosimetrically and epidemiologically based radon risk estimates, although it would probably be misleading to over-interpret the resulting level of agreement, because many assumptions are involved in both approaches.

## MECHANISTIC CONSIDERATIONS IN ASSESSING RISKS ASSOCIATED WITH RADON

The mechanistic considerations discussed above must be incorporated into the design of epidemiologic analyses to estimate radon risks. We summarize the main issues of relevance to the estimation of risks associated with radon progeny.

### Biologically-Based Risk Models

A biologically-based risk model is a formalism that potentially provides realistic quantification of all the relevant steps from energy deposition to the appearance of cancer. If such a model were available, epidemiologic data could be fitted to it, and the resulting parameter estimates could be used to quantify the different mechanistic steps in radiation carcinogenesis. Low-dose extrapolation could then be conducted with more confidence than for a situation in which data are fitted with a purely empirical formalism. However, epidemiologic data usually include only incidence and mortality. Biologic models need cell proliferation rates and other factors, and such information is not usually available. Such approaches to radiation risk estimation have been proposed and critically discussed by various authors (for example, Moolgavkar and others 1993; Crump 1994a,b; Little and others 1994; Moolgavkar 1994; Goddard and Krewski 1995; Little 1995).

Those approaches must be considered desirable, in the long term, as a framework for interpreting the radiation-epidemiologic data. Today, however, it is important to recognize the complexity of the processes involved in radiation carcinogenesis and the many gaps in our knowledge of the most-basic relevant processes. Although the use of biologically-based models provides valuable insights into the carcinogenic process, the models are not sufficiently well developed to be used for quantitative risk estimation; indeed, their use might lend more credibility to the resulting risk estimates than is warranted.

Although all the steps leading from the deposition of radiation energy to the development of cancer are not understood, some general trends have emerged from the considerations in this chapter, which can be used to guide specific assumptions of the epidemiologically based analysis of radon-induced lung cancer. These trends are discussed in the remaining part of this chapter.

### Extrapolation from High to Low Radon-Progeny Exposures

The challenge is to guide the extrapolation of risks from radon-progeny exposures at which effects can readily be observed and risks quantified down to lower exposures at which events might occur with probabilities too small to measure with sufficient precision in any human population. Low exposures and doses correspond to the traversal of cells by single alpha particles. As the dose is further decreased, the insult to cells that are traversed by an alpha particle remains the same, but the number of traversed cells decreases proportionately. There is good evidence that a single alpha particle can cause a substantial change in a cell. For example, a single traversal by an alpha particle with an LET of 120 keV/$\mu$m can result typically in about 10-20 double-strand breaks in a cell deduced from measured yields of about 30-40 dsb/Gy for low-LET radiation (Ward 1988; Stenerlow and others 1996), similar relative yields for alpha particles (Jenner and others 1993; Prise 1994), and a dose of about 0.2-0.5 Gy to a cell traversed by a single alpha particle, depending on cellular geometry. Even allowing for the substantial degree of repair that is known to take place, the passage of the particle most likely causes some irreparable damage or permanent change. Direct evidence of such clastogenic changes, based on single alpha-particle microbeam irradiation (Geard and others 1991; Braby 1992; Nelson and others 1996) has been reported. There is also convincing evidence that alpha particles are efficient at inducing genomic instability (Kadhim and others 1992, 1994, 1995; Sabatier and others 1994), so traversal by a single particle can potentially initiate a cascade of events that can lead to chromosomal aberrations or delayed mutations many generations later. The later effects can be in cells adjacent to those actually traversed.

Those observations, taken separately or together, provide a mechanistic basis for a linear relationship between alpha-particle dose and biologic effect at dose levels that correspond to 1 particle traversal per cell or less. At those exposures, varying the dose proportionately varies the number of cells traversed by alpha particles, but does not alter the level of damage sustained by cells that are traversed. That is the situation for exposure to alpha particles from radon progeny in a domestic environment where it is unlikely that any cell at risk in the bronchial epithelium is traversed by more than 1 alpha particle in a lifetime. As an example, an exposure of 0.0175-0.07 Jhm$^{-3}$ (5-20 WLM) would result in an extremely low probability that any cell would be traversed by more than 1 alpha particle. Within and below that exposure range, linearity is thus a reasonable assumption with the implication of no threshold in dose. At the minimum, this linearity refers to effects in single cells since varying the dose merely changes the number of cells traversed but does not alter the level of damage per cell. Of course, the development of a tumor involves many steps beyond oncogenic transformation in a single cell, but the weight of evidence for the clonal origin of most tumors (Wainscoat and Fey 1990) suggests that this argument also applies to tumor induction by radon. Indeed, over this range, linearity and a threshold seem to be mutually exclusive. It is important to note that we have considered only low-dose extrapolation of the effects of alpha particles, and these arguments do not necessarily apply to the effects of sparsely ionizing radiation, such as x or gamma rays.

At higher doses of densely ionizing radiation, various processes can, in principle, result in a nonlinear relation, as has often been observed (for example, Ullrich and others 1976; Ullrich 1983; Fry and others 1985; Chmelevsky and others 1984; NRC 1988; Furuse and others 1992). These high-dose effects include the following:

• Damage resulting from the interaction of chromosomal breaks from different alpha particles would be expected to result in a component of response that is quadratically related to dose.

• Decreased efficiency of repair of alpha-particle damage, which, in principle, could be reduced at very high doses through saturation.

• Cell killing and alterations in cell proliferation (from radiation and tobacco exposure) which are likely to influence the development of an initiated cell into a tumor.

• Production of diffusible clastogenic factors at high doses which might be able to contribute to generalized tissue response remote from the sites of individual alpha-particle decays.

• Inverse dose-rate effects which could change the slope of the dose-response relation at high exposure levels.

Although those factors could all potentially affect linearity, they require substantial doses. They might be important in the high-exposure miner data, in which multiple cell traversals occur with significant probabilities, but are unlikely to be important at low exposure levels that correspond to 1 particle or less per cell.

In summary, the weight of evidence from cellular and molecular studies strongly supports the concept of linearity with dose in cellular systems including cell lethality, mutation, or transformation with no threshold for low-dose alpha-particle irradiation but leaves open the possibility of a change in slope or a departure from linearity for cancer induction at higher doses. The overwhelming evidence for the monoclonal origin of most cancers suggests linearity without threshold would also apply to low-dose radon-induced carcinogenesis. This observation emphasizes the desirability of extrapolating to typical indoor exposures from the lowest exposure range that is practical in the miner data set.

## Effect of Changing Exposure Rate

Extrapolation from higher to lower radon exposures is also affected by the inverse dose-rate effect (Brenner 1994). In vivo and in vitro experiments have shown an inverse dose-rate effect for alpha-particle irradiation. Specifically, protracting a given total dose (experimentally, at least 0.10 Gy) of densely ionizing radiation, such as alpha particles, can increase oncogenic transformation in vitro or carcinogenesis in vivo. This dose-rate effect, whatever its underlying mechanism, operates at doses corresponding to multiple particle traversals per cell but is likely to disappear at low doses corresponding to an average of much less than 1 traversal per cell (Figure 2-2). A similar dependence of effect on exposure is clearly evident in the miner data (Figure 2-3).

Extrapolating radon risk from the full miner data to the low-exposure domestic situation involves extrapolating from a situation in which multiple traversals are common to one in which they are rare; consequently, such an extrapolation would be from a situation in which the inverse dose-rate effect might well be important to one in which it is likely to be unimportant. That presents a problem for the committee's risk assessment in that the mechanisms whereby inverse dose-rate effects operate are not yet established, although several mechanisms have been hypothesized. However, given that both experimental evidence and fundamental biophysical evidence suggest that the inverse dose-rate effect should be of little importance below about 0.35 Jhm$^{-3}$ (100 WLM), these considerations again underline the importance of assessing risks of radon in homes on the basis of miner data corresponding to as low an exposure as possible.

## Interaction of Radon Progeny with Other Agents

Experiments with a combination of alpha-particle and tobacco-smoke condensate exposure, using oncogenic transformation in vitro as a test assay, show that effects of the 2 agents are consistent with a purely additive interaction (Piao and Hei 1993). With such in vitro systems, large-scale experiments are possible that yield unequivocal results. However, such experiments might well be good models only for the initiation part of the carcinogenic process. Although the data are hard to interpret in animal experiments, alpha particles and tobacco smoke often appear to produce effects that are larger than additive; this observation can be understood in that tobacco smoke, as well as being a carcinogen, contains irritants that stimulate cell proliferation which is a known factor in oncogenesis. The experimental data are consistent with the supra-additive model which appears to be most useful in the human data (see chapter 3).

## Biologic Signatures of Alpha-Particle Cancers

Early attempts (Vahakangas and others 1992; Taylor and others 1994) to identify a molecular "signature" of prior alpha-particle damage through the identification of unusual point mutations have not yet proved useful (Rossi 1991; Hei and others 1994a; McDonald and others 1995; Bartsch and others 1995; Hollstein and others 1997). More mechanistic approaches based on larger-scale genomic alterations (for example, Brenner and Sachs 1994; Griffin and others 1995; Savage 1996) are more promising. The newer techniques currently require further experimental validation but offer hope for future molecular epidemiologic approaches to the radon problem.

## Individual Susceptibility

Animal experiments show significant but unexplained differences among species in susceptibility to lung cancer from radon progeny. For example, rats are susceptible to radon-induced lung cancer, but Syrian hamsters are extremely resistant. In addition, within a given species, different inbred strains show variations in susceptibility to ionizing radiation or chemical carcinogens. Cell lines or animals with specific repair deficiencies also show increased susceptibility to radiation-induced malignant or premalignant changes, although there have been few experiments specifically with alpha particles from radon progeny.

The sum of available evidence leads to the conclusions that for an outbred human population a broad spectrum of susceptibility to alpha-particle-induced carcinogenesis would be expected and that there could be a marked increase in susceptibility in people suffering from a genetic deficiency. Evidence of genes related to susceptibility for many forms of cancer is emerging, although there is as yet no convincing evidence of a gene that confers sensitivity to radiation-induced cancer on a heterozygotic population.

Much research is being focused on the possibility of subpopulations that might have a genetically based increased susceptibility to radiation-induced cancer. The existence of sub-populations that are highly sensitive to alpha-particle-induced lung cancer could substantially affect *individual* risk estimates but may have only minor impact on population risk estimates.

# 3

# Models and Risk Projections

## INTRODUCTION

This chapter presents the committee's risk models relating lung cancer to radon exposure and applies the models to exposures of the general population to estimate the burden of lung cancer due to exposure to indoor radon. We discuss both the committee's models describing lung-cancer risk in miners and the application of the models in projecting lung-cancer risks in the general population. We also describe prior risk models and the basis for our approach to developing new risk models. The committee decided to use primarily miner-based data for risk estimation and to use models in which risk is linearly related to dose at low doses. Those two decisions follow those of the BEIR IV committee. However, the rationale for our model is supported more strongly than was that of the BEIR IV committee, being grounded in the biologic considerations developed in chapter 2 and in the stronger body of observational evidence provided by the pooled data from the studies of underground miners, as well as a meta-analysis of the reported 8 case-control studies of residential radon exposure and lung cancer.

In this chapter, we provide risk projections that describe both the increment in lifetime risk of lung-cancer mortality for various exposure scenarios and the population burden of lung cancer attributable to exposure to indoor radon. This chapter also addresses uncertainties associated with the models and with risk projections based on the models. Appendix A describes the modelling and uncertainty analysis procedures in detail.

## RISK-ESTIMATION APPROACHES

This section briefly reviews alternative approaches to estimating lung-cancer risk associated at radon exposure levels typically found in homes and provides the rationale for the committee's selected approach. Figure 1-3 showed the alternative approaches considered and the related data sources.

### Dosimetric Approach

The dosimetric approach applies the well-characterized radiation data from human exposures to γ rays, in particular data from the atomic-bomb survivors, to derive estimates of the risk associated with exposure to radon (ICRP 1990). This approach has the following steps:

1. Use physical dosimetric models of the lung to estimate alpha-particle dose to lung-airway epithelium for indoor radon exposure.

2. Convert the alpha-particle dose to an equivalent low-linear-energy-transfer (low-LET) dose for low-LET radiations, using an appropriate weighting factor for radon-progeny alpha particles in the bronchial epithelium.

3. Convert the equivalent dose to an effective dose, using the appropriate tissue-weighting factor for lung (ICRP 1990). (It is possible to omit step 3 and use lung-specific γ-ray-based risk estimates in step 4).

4. Use risk coefficients per unit of effective dose, based primarily on atomic-bomb survivor data, to estimate the risk per unit of cumulative exposure to radon.

One strength of this dosimetric approach is its use of the wealth of data from the continuing epidemiologic study of the atomic-bomb survivors in Hiroshima and Nagasaki. Lung-cancer risk has been well characterized in that cohort in relation to dose. However, the approach is weakened by the need for scaling factors to convert from the acute, whole-body, primarily γ-ray exposure to the chronic, localized, alpha-particle exposure of the lung from indoor radon. In addition, the data from Hiroshima and Nagasaki are subject to uncertainty owing to limitations of the dosimetry and the need to extrapolate from an exposed population in Japan to other population groups with differing background cancer rates.

### Biologically Motivated Approach

Biologically motivated models are intended to provide realistic representation of the steps in radon carcinogenesis from energy deposition to the appearance of cancer. In this context, the parameters of the model have a direct biologic interpretation. One such model is the Moolgavkar-Venzon-Knudson 2-stage clonal expansion model, which incorporates both tissue growth and cell kinetics (Moolgavkar and Luebeck 1990). Such approaches to cancer risk estimation have been proposed and reviewed by various authors (for example, Little and others 1992, 1994; Moolgavkar and others 1993; Crump 1994a,b; Moolgavkar 1994; Goddard and Krewski 1995; Little 1995).

This committee did not pursue biologically motivated cancer-risk models for several reasons. First, the mechanisms of radon-induced carcinogenesis must be known with sufficient certainty before an appropriate biologically motivated model can be constructed. Despite the considerable amount of information summarized in chapter 2, the committee recognized that current knowledge of radiation cancer mechanisms remains incomplete and any postulated model would necessarily be an oversimplification of a complex process. Second, application of a fully biologically motivated model requires information on fundamental biologic events, such as mutation rates and cell kinetics, that is not readily available in the present application. Third, a comprehensive biologically motivated model involving many parameters, such as the 2-stage clonal-expansion model used by Moolgavkar and others (1993) to describe the Colorado miner data, cannot be fruitfully applied without comprehensive longitudinal data on personal exposures to both radon progeny and tobacco. When the various steps in radon-induced carcinogenesis are more fully understood, the biologically motivated approach might become the preferred approach. However, the committee considered an empirical approach to be preferable at present.

## Empirical Approach

Statistical methods for the analysis of epidemiologic data, particularly cohort data, have evolved rapidly since the 1970s. These statistical methods can be used to estimate lung-cancer risks directly from epidemiologic data, as done by the BEIR IV committee. To implement the now-common empirical approach, it is assumed that disease rates in narrow time intervals are constant, or at least can be accurately approximated by mean disease rates in the time intervals. Epidemiologic cohort data are summarized in a multidimensional table, in which each cell contains information on person-years at risk, number of events (lung-cancer deaths) occurring within the cell, and variables that identify the cell, such as age, cumulative exposure, and exposure rate. For each cell, the observed number of events is assumed to follow a Poisson distribution, with a mean equal to the underlying disease rate for the cell multiplied by the person-years at risk. Poisson events are assumed to be infrequent and have a distribution in which the variance equals the mean.

In the development of an empirical risk model to describe rates of radon-induced lung cancer in miners, several a priori assumptions are needed about either the shape of the exposure-response function or the factors that influence risk. In its most general implementation, empirical modeling is sufficiently flexible to offer some degree of biological plausibility with only minimal assumptions needed about the structure of the model. That generality, as well as the ability to model without assuming any underlying biologic mechanism of disease, leads to the characterization of the modeling approach as empirical or descriptive.  The empirical modeling approach also allows for evaluation of diverse factors that modify risk, such as attained age and exposure rate, through formal statistical testing.  Given the limitations of the available data and the resulting difficulty in discriminating among plausible alternative models, the empirical approach undoubtedly results in models that are relatively crude and at best yield rough approximations of actual patterns of risk.  While the committee relied on data on lung-cancer mortality in underground miners to construct its proposed risk models, a series of assumptions is needed to extend the miner-based model to the general population.  For example, the committee used models in which the exposure-risk relation is linear at low exposures, based on the mechanistic considerations discussed in chapter 2.  Other assumptions made in projecting population risks are described later in this chapter.

## RATIONALE FOR THE COMMITTEE'S CHOSEN METHOD FOR RADON RISK ESTIMATION

The committee critically assessed the principal approaches (see Figure 1-3) that could be used to estimate the risk associated with exposure to indoor radon, with respect both to sources of data for developing risk models and to techniques for modeling.  The combinations of data resources and risk estimation approaches of present interest are as follows:

1.  Biologically motivated analysis of miner data.
2.  Dosimetric approach using low-LET data (for example, atomic-bomb survivor data).
3.  Empirical analysis of miner data.
4.  Empirical analysis of data from residential case-control studies.

The strengths and limitations of the three different data sources are summarized in Table 3-1.

With regard to the first approach, the committee recognized that use of biologically motivated risk models is a highly desirable goal, but it felt that such models have not reached a stage at which they can be used for radon risk assessment.  Specifically, the complexity and multiplicity of the processes involved in radiation carcinogenesis were noted, as were the gaps in knowledge of the most-basic relevant processes.  The paradigms describing carcinogenesis in general and radiation carcinogenesis in particular are changing rapidly.  For example, the potential importance of delayed genomic instability (Chang and Little 1992; Kadhim and others 1992; Morgan and others 1996), not incorporated in currently formalized biologically motivated models, was not apparent until within the last few years.

The second approach, the dosimetric approach based on the atomic-bomb

TABLE 3-1   Relative strengths of data for alternative approaches for estimating the risks posed by indoor radon

| Criteria | Atomic-bomb survivor data | Residential data | Miner data |
|---|---|---|---|
| Exposure estimation | ••• | • | •• |
| Potential power of study | ••• | • | ••• |
| Dose range | • | ••• | • |
| Exposure time | • | ••• | •• |
| Women/children | ••• | ••• | • |
| Effects of smoking | • | •• | • |
| Scaling factors required | • | ••• | •• |

••• = adequate
•• = fair
• = problematic

survivors, has both strengths and weaknesses. Its strengths include the availability of estimates from a large cohort of men, women, and children exposed to a wide range of doses; the extensively characterized dose estimates for the survivors; and the 45-year period followup. For the present application to radon progeny, weaknesses include the very different types of radiation and exposure patterns to which the bomb survivors were exposed—acute whole-body doses of gamma rays and, to a lesser extent, neutrons. In particular, the radiation weighting factor needed to relate gamma-ray risks to alpha-particle risks is probably not known to better than within a factor of about 5 (Burchall and James 1994; Brooks and others 1994; Brenner and others 1995). The risk estimates from the study of atomic-bomb survivors are also subject to uncertainty (NRC 1990). The committee reasoned that the uncertainties in extrapolating risks from acute whole-body γ-ray exposure to prolonged, localized alpha-particle exposure were too great to justify use of this approach.

Over the last decade, considerable resources have been devoted to case-control studies of residential radon exposure. A number of studies have been completed, and some are still in progress. These studies are reviewed in appendix G. In principle, residential studies yield the most relevant risk estimates, because they relate directly to the population of interest. However, because of the very low risk associated with exposures at residential levels, risk estimates obtained from these studies, even estimates based on meta-analysis of several studies, are very imprecise. Furthermore, the residential studies offer little opportunity for evaluating with the modifying effects of such factors as smoking and time since exposure. For those reasons, the committee rejected a model based on the residential-radon studies (the fourth approach) as the primary source of risk estimation. However, the committee did compare risk estimates based on the residential data with the low-exposure risk estimates that it generated from the miner data.

Having thus considered the various alternative approaches, the committee chose to follow the general approach of the BEIR IV committee and of Lubin and others (1994a) to the analysis of pooled miner data and to base risk estimates on an empirically derived model (the third approach). This approach provided the committee with well-established databases and methods as a starting point.

Empirical or descriptive modeling of risk allows a unified approach for testing the validity of the form of the model and of the significance of model parameters. For example, the committee used a relative risk model rather than an absolute risk model to describe lung-cancer risk to radon exposure. It had been observed that a relative risk model, with time-varying covariates, provided a more parsimonious description of the miner data than an absolute excess-risk model (Lubin and others 1994a).

The flexibility of the modeling approach allowed the incorporation of specific biologically based patterns of risk. Two important choices in the committee's analysis are the incorporation of an inverse exposure-rate effect and the assumption of linearity of the exposure-response relationship at low cumulative exposure. For radon-induced lung cancer, those choices have a plausible biologic rationale, as well as some experimental justification (see chapter 2).

## PREVIOUS MODELS

A number of models have been previously developed for estimating lung-cancer risk posed by exposure to radon and its progeny. Models developed through the middle 1980s were described in the BEIR IV report (NRC 1988). These and other models are discussed in detail in appendix A to this report. With the exception of preliminary reports from 2 studies which later changed, these models have all assumed linearity of the exposure-response relationship. All models used risk estimates derived from the studies of miners.

The earliest risk models specified effects of exposure in terms of the absolute excess risk of lung cancer from radon-progeny exposure. The absolute (excess) risk model represents lung-cancer mortality as $r(x, z, w) = r_o(x) + g(z, w)$, where $r_o(x)$ is the background lung-cancer rate and $g(z, w)$ is an effect of exposure. (Here, $w$ denotes cumulative exposure, $x$ represents covariates that determine the background risk, and $z$ denotes covariates that modify the exposure-response relationship.) The model proposed by the BEIR III committee allowed the absolute excess risk to vary by categories of attained age with allowance for different minimal latent periods for each category. A descriptive model for the absolute excess lung-cancer risk, proposed by Harley and Pasternack (1981), served as the basis of risk estimates in Reports 77 and 78 of the National Council of Radiation Protection and Measurements (NCRP 1984a,b). That model assumed that exposure has no effect on risk before age 40 years and that, after a latent period, the absolute excess risk declines exponentially with time since exposure. The model was proposed specifically to address risk associated with radon-progeny exposure, and, although miner data were not used to define its form, published results of analyses of miner data were used to specify latent periods and parameter values thought to be reasonable and appropriate (NRC 1988).

A meta-analysis of miner-study results by Thomas and McNeil (1982), the BEIR IV committee analysis of pooled data from 4 miner cohort studies (NRC 1988), and the findings from a number of individual cohort studies of miners suggested that models of the relative risk (RR) were preferable to models of the absolute excess risk.

Recent descriptive models for lung-cancer risk associated with radon-progeny exposure have also modeled the relative risk rather than the absolute risk. Under the general relative risk model, the lung-cancer rate $r$ $(x,\ z,\ w)$ can be written as $r(x, z, w) = r_o (x)RR(z, w)$, where $r_o (x)$ is the lung-cancer rate among non-exposed, and $RR$ $(z,\ w)$ is the exposure-response function. Of particular interest is the linear relative risk model:

$$RR = 1 + \beta w, \qquad (1)$$

where $\beta w$ estimates the excess relative risk (ERR), $w$ is exposure, and $\beta$ estimates the increment in ERR for unit change in exposure $w$.

The simplest of the relative-risk models was proposed in Report 50 of the International Commission on Radiological Protection (ICRP 1987). The ICRP model for extrapolation to indoor exposures was a linear model for ERR in relation to cumulative exposure. The value of $\beta$ was derived by reducing a value thought to be representative of the miner studies to reflect differences in conditions between mines and homes. On the basis of findings in the atomic-bomb survivors and dosimetric considerations, $\beta$ was increased by a factor of 3 for exposures occurring before age 20 years. The assumption of a constant relative risk and the higher risks assigned to exposures in childhood can be questioned. Detailed analyses of miner data, however, have indicated that the exposure-response relationship is not constant but varies with other factors (NRC 1988; Thomas 1981). In addition, there is little evidence of enhanced effects of exposure at young ages in the miner data (Lubin and others 1994a).

The BEIR IV committee analyzed pooled data from 4 cohort studies of radon-exposed miners (NRC 1988). It found that the simple linear ERR model did not fit the data adequately and that the exposure-response parameter $\beta$ varied with time since exposure and attained age. Since its publication in 1988, the BEIR IV model has served as the primary basis for assessing risks for underground miners and the general population. Using the BEIR IV model as a starting point, Jacobi and others (1992) proposed a related "smoothed" model for the relative risk of lung cancer from radon-progeny exposure, which served as the basis of risk estimation in ICRP Report 65 (ICRP 1993). Expanding the analytic approach in the BEIR IV report, Lubin and others (1994a, 1995b) pooled data from 11 cohort studies of miners, including the 4 studies used in the BEIR IV analysis, and fitted similar types of models for the ERR. Lubin and colleagues

(1994a) again found that the exposure-response relation varied with time since exposure and attained age, but they also found variation with exposure rate. Lower exposure rates were associated with increased risk. The BEIR VI committee used the work of Lubin and colleagues (1994a) as a starting point for the analyses described in this chapter.

## BEIR VI RISK MODEL FOR LUNG CANCER IN MINERS

### Introduction

This section considers the sources of data, methods of combining data from diverse populations, and assumptions that underlie the lung-cancer risk model developed by the committee in its analysis of miner data. The committee used a relative-risk model that relates lung-cancer rate in miners to their occupational exposure to radon.

In the analysis, exposure refers to occupational exposure to radon progeny during employment in underground mines, and relative risks refer to the additional risks associated with occupational exposure to radon progeny beyond the background risk from lung cancer, which reflects other exposures, including indoor radon. Residential radon-progeny exposures of the miners are not considered in the analysis data and are implicitly assumed to be the same, on average, at all levels of occupational exposure. Any bias in the modeling due to ignoring nonmine exposures is likely to be small, because residential radon concentrations are generally much lower than mine concentrations.

The committee's model is based on a linear relationship between exposure and the relative risk of lung cancer. This linear relationship was based on an empirical evaluation of the 11 individual miner studies. In analyzing the miner data, Lubin and others (1994a) explored various models for describing the form of the relative risk in relation to radon exposure. Within the range of exposures in miners, linear models provided an adequate characterization of each cohort except the Colorado Plateau uranium miners. In the Colorado data, the authors found a relative-risk pattern that was concave at high cumulative exposures. Accordingly, in the analysis of pooled data, data from the Colorado study were limited to exposures below $11.2$ Jhm$^{-3}$ (3,200 WLM), below which relative risks were consistent with linearity.

### Sources of Data

Pooled data from 11 cohort studies of radon-exposed underground miners were used to develop the committee's risk models; these data were derived from all the major studies with estimates of exposure for individual miners (Table 3-2). Data were available from 7 studies in addition to those considered by the BEIR IV committee. These data are described in detail in appendixes D and E.

TABLE 3-2    Epidemiologic studies of underground miners used in the BEIR
VI analysis[a]

| Location | Type of mine | Number of miners | Period of follow-up | Data available on smoking |
|----------|--------------|------------------|---------------------|---------------------------|
| China | Tin | 17,143 | 1976-87 | Smoker: yes/no (missing on 24% of subjects, 25 (out of 907) nonsmoking lung-cancer cases) |
| Czechoslovakia | Uranium | 4,320 | 1948-90 | Not available |
| Colorado | Uranium | 3,347 | 1950-90 | Cigarette use: duration, rate, cessation (unavailable after 1969) |
| Ontario | Uranium | 21,346 | 1955-86 | Not available |
| Newfoundland | Florspar | 2,088 | 1950-84 | Type of product, duration, cessation (available for 48% of subjects, including 25 cases) |
| Sweden | Iron | 1,294 | 1951-91 | Type of product, amount, cessation (from 35% sample of active miners in 1972, supplemented by later surveys) |
| New Mexico | Uranium | 3,469 | 1943-85 | Cigarette use: duration, rate, cessation (available through time of last physical examination) |
| Beaverlodge | Uranium | 8,486 | 1950-80 | Not available |
| Port Radium | Uranium | 2,103 | 1950-80 | Not available |
| Radium Hill | Uranium | 2,516 | 1948-87 | Smoking status: ever, never, unknown (available for about half the subjects, 1 nonsmoking case) |
| France | Uranium | 1,785 | 1948-86 | Not available |

[a]Lubin and others 1994a.

Since the 1994 publication of the original pooled analysis by Lubin and
colleagues (1994a), data from 4 studies (Chinese tin miners and the Czechoslovakia,[1] Colorado and French uranium miners) have been updated or modified (Lubin
and others 1997). In assembling the original data for the China study, the original
investigators (Xuan and others 1993) assumed that all miners worked 285 d/yr
until the early 1980s, which corresponded to the end of the followup less the

---

[1]For historical reasons, the study is referred to as the Czechoslovakia or Czech cohort, although
the country is now 2 independent states, the Czech Republic and Slovakia. The mining area was
located in what is now the Czech Republic. About 25% of the miners were of Slovak origin and
most later returned to Slovakia.

5-yr lag period. Recent information has indicated, however, that miners worked 313 d/yr before 1953, 285 d/yr in 1953-1984, and 259 d/yr after 1984. Estimates of exposures have been updated accordingly. An extensive reevaluation of exposure histories and of follow-up and vital status has been carried out for the Czech cohort (Tomášek and others 1994a). There were 705 lung-cancer cases in the updated data, compared with 661 in the previous data set, and the cohort was enlarged from 4,284 to 4,320 miners. For the Colorado study, followup has been extended from December 31, 1987, through December 31, 1990 (Hornung and others 1995). In the updated data used by the committee, there were 336 lung-cancer deaths at exposures under 11.2 Jhm$^{-3}$ (under 3,200 WLM) in a total of 377 cases, compared with 294 lung-cancer deaths at exposures under 11.2 Jhm$^{-3}$ in a total of 329 total cases in the prior pooled analysis. For the French miner data, the investigators made small corrections in exposure estimates and in health outcomes other than lung cancer.

In addition to the data changes for those cohorts, there has been a reassessment of estimates of exposure of a nested case-control sample within the Beaverlodge cohort of uranium miners, including all lung-cancer cases and their matched control subjects (Howe and Stager 1996). For these Beaverlodge miners, exposure estimates were about 60% higher than the original values. Because of the computational difficulties of merging case-control data with cohort data, only the data from the Beaverlodge cohort study with the original exposure estimates were used in the committee's analysis.

## Analysis of Pooled Data from Different Studies

In the development of risk models, it is important to take account of the totality of evidence from all relevant studies. When data from many different sources are available, this is most effectively accomplished by analyzing combined or pooled data. The models developed by Lubin and others (1994a) were based on analyses of data from 11 miner cohorts. Other examples of analyses of pooled data are those by Cardis and others (1995) on cohorts of externally irradiated nuclear workers in the United States, the United Kingdom, and Canada and Lubin and others (1994b) on data from 3 case-control studies of indoor radon exposure and lung cancer.

Analyses of pooled data can provide more precise estimates of parameters than those based on individual studies—an advantage that is especially important for investigating  modifying factors, which requires comparing risks among subsets of the data. They can also test whether differences in findings among studies represent true inconsistency or simply result by chance. The application of similar methods to data from all studies and the presentation of results in a comparable format facilitate comparisons of results from different studies.

Analysis of pooled data from diverse sources must, however, be done with care because the data might not be fully comparable. In the present context, the cohorts differ with respect to the methods used to estimate radon-progeny exposure, the completeness of mortality follow-up, and the accuracy of disease diagnosis. The cohorts also differ with respect to demographic characteristics, other exposures encountered in the mines, and smoking patterns. Such differences can lead to heterogeneity in risk estimates. Heterogeneity can be partially addressed by adjusting for modifying factors, such as exposure rate, on which data are available. However, lack of adequate data on all covariates that affect risks and biases in the data can result in residual heterogeneity even with extensive adjustment for covariates. It is important to take account of heterogeneity in analyzing the data, particularly in expressing the uncertainty in the risk estimates obtained.

Statistical methods for analyzing data sets derived from different sources, taking into account heterogeneity among sources, are described in appendix A. Random-effects models (Davidian and Giltinan 1995) provide a natural statistical approach for combining data from different sources in the presence of heterogeneity. Specifically, heterogeneity is accommodated by allowing for random perturbations in parameter values from cohort to cohort, and this results in a random-effects distribution of parameter values across cohorts. The mean of the distribution constitutes an overall summary of the parameter value across cohorts and its variance describes the component of uncertainty due to unaccounted for differences between cohort studies. Two-stage statistical methods have also been used in analysis of pooled data from different studies. With the 2-stage approach, estimates of the model parameters specific for each cohort are derived, and an overall estimate is then obtained by an appropriately weighted linear combination of the cohort-specific estimates, taking into account variation within and between cohorts. The 2-stage approach was used in recent analyses by Lubin and others (1994a) and also by Burnett and others (1995) in combining data on air pollution and respiratory health in 16 Canadian cities.

Both the random-effects and 2-stage approaches were used to combine data from the 11 miner cohorts (see appendix A), but the results presented in this chapter are based on the 2-stage method. In simple modeling situations, the 2-stage and random-effects models were generally found to be in good agreement. In the more-complex modeling conducted by the committee, however, the random-effects approach proved to be computationally more burdensome, and convergence was not always obtained with the iterative numerical methods required in model fitting. Consequently, the committee relied primarily on the 2-stage method in conducting its combined analyses. The committee's 2-stage approach can be viewed as a simplification of, and an approximation to, the full random-effects approach.

The committee recognized that each of the 11 miner studies has certain unique characteristics that contribute to the observed differences in risk among cohorts. In the presence of such differences, the desirability of pooling data from heterogeneous populations can be questioned. Pooling makes maximal use of all relevant data in an objective manner and provides an overall summary measure of risk. Provided that cohort heterogeneity is acknowledged in the pooling process, the standard error of this overall risk estimate will be an appropriate measure of its statistical uncertainty. The committee considers that, if done carefully, analyses of pooled data can be informative. The committee also believes that, in the absence of clear reasons to exclude particular cohorts from the analysis, it is preferable to make use of all the available data for risk-assessment purposes.

## Model Based on Full Data Set

Selection of the committee's risk models was guided by the extensive analysis by Lubin and others (1994a) of the 11 miner cohorts. That analysis indicated that a linear model was sufficient to describe the miner data. Because of the present focus on residential exposures, which are generally at or below the low end of the range of exposures experienced by miners, a linear exposure-response model was considered appropriate for purposes of this report. This choice was supported by the committee's review of the biologic basis of radon carcinogenesis, set out in chapter 2.

Lubin and others (1994a) examined models that took into account factors that modify cancer risk, including time since exposure, attained age, duration of exposure, and intensity of exposure. Those models are described in detail in appendix A. Briefly, they found that models that took into account time since exposure, attained age, and either duration of exposure or concentration of radon progeny as an indicator of exposure rate provided equally good fits to the miner data. Models with fewer than 3 of the modifying factors did not provide comparable fits to the available data. For a given total exposure, duration of exposure is inversely related to the concentration at which exposure was received (the exposure rate), and either duration of exposure or average concentration provides an indication of average exposure rate.

Models with both categorical, that is, discrete, and continuous covariates were considered. The categorical models were preferred by the committee for purposes of risk projection. Both types of models led to comparable predictions of risk within the range of exposure experienced by the miners, but projections of risk to lower exposures based on categorical models depended less on observations at high doses than those from continuous parametric models.

On the basis of previous experience in modeling the miner data, the committee concluded that 2 categorical models (referred to as the exposure-age-duration and exposure-age-concentration models) provided the most-appropriate basis for risk assessment. The concentration model includes categories for exposure in three windows of time since exposure, for age, and for concentration; in the duration model, duration is the measure of exposure rate. Although the committee did not repeat the comprehensive model-selection exercise conducted by Lubin and others (1994a), it analyzed in detail the updated data from the miner studies with these and related models (see appendix A).

The committee's preferred models for predicting relative risk were of the form

$$RR = 1 + \beta(w_{5\text{-}14} + \theta_{15\text{-}24} \, w_{15\text{-}24} + \theta_{25+} \, w_{25+}) \, \phi_{age}\gamma_z, \tag{2}$$

where $\beta$ is the exposure-response parameter; total exposure, $w$, is partitioned into temporal exposure windows with $w_{5\text{-}14}$, $w_{15\text{-}24}$, and $w_{25+}$ defining the exposures incurred 5-14 yr, 15-24 yr, and 25 yr or more before the current age; and $\theta_{15\text{-}24}$ and $\theta_{25+}$ represent the relative contributions to risk from exposures 15-24 yr and 25+ yr or more before the attained age. The factor 1 in equation (2) represents the background *RR* for lung cancer without occupational exposure but with outdoor and indoor exposures. Note that $\theta_{5\text{-}14} = 1$ by definition for $w_{5\text{-}14}$. The parameters $\phi_{age}$ and $\gamma_z$ define effect-modification factors and represent, respectively, multiple categories of attained age ($\phi_{age}$) and of either exposure rate or exposure duration ($\gamma_z$). Details of the model fitting are given in appendix A.

Preliminary analyses indicated that the effects of time since exposure, attained age, and exposure duration or concentration were similar in most cohorts, so that these parameters were constrained to be the same in all 11 cohorts when the models were fitted to the pooled data. However, the parameter $\beta$ varied considerably across cohorts. Consequently, the overall estimate of $\beta$ was obtained by using the 2-stage method, so that the associated standard error reflects variation both within and between cohorts.

Estimates of the parameters in the committee's 2 models, based on the 2-stage approach, are given in Table 3-3. (Similar results were obtained with the random-effects approach discussed in appendix A.) Although the parameter estimates changed slightly with the updated miner data, the general pattern of effects was comparable with that observed in the original analysis by Lubin and others (1994a). For a given level of exposure, ERR declined with increased time since exposure ($\theta_{5\text{-}14} > \theta_{15\text{-}24} > \theta_{25+}$), and with attained age. Lung-cancer risk increased with either lengthening duration of exposure or decreasing exposure rate.

## Model Based on Exposure-Restricted Data

The mean exposure in the analysis of pooled data on the miners was 0.57 Jhm$^{-3}$ (162 WLM), about 10 times the exposure from lifetime occupancy in an average U.S. home. The mean duration of exposure for the miners was about 6 yr, about one-tenth the duration of residential exposures. Thus, mean exposure rates of miners were about 100 times those of residents of typical houses. Those differences in exposure profiles between the entire group of miners and the general population are a source of uncertainty in the model based on the full data set.

To reduce that uncertainty, the BEIR VI committee limited the miner data to exposures that approach those in typical residences; even with this restriction of the

TABLE 3-3    Parameter estimates from BEIR VI models based on original (Lubin and others 1994) and updated pooled (Lubin and others 1997) miner data

| | Exposure-age-duration model[a] | | | Exposure-age-concentration model[a] | | |
|---|---|---|---|---|---|---|
| | Original data | Updated data | | Original data | Updated data |
| $\beta^{b} \times 100$ | 0.39 | 0.55 | $\beta \times 100$ | 6.11 | 7.68 |
| Time-since-exposure windows | | | | | |
| $\theta_{5-14}$ | 1.00 | 1.00 | $\theta_{5-14}$ | 1.00 | 1.00 |
| $\theta_{15-24}$ | 0.76 | 0.72 | $\theta_{15-24}$ | 0.81 | 0.78 |
| $\theta_{25+}$ | 0.31 | 0.44 | $\theta_{25+}$ | 0.40 | 0.51 |
| Attained age | | | | | |
| $\phi_{<55}$ | 1.00 | 1.00 | $\phi_{<55}$ | 1.00 | 1.00 |
| $\phi_{55-64}$ | 0.57 | 0.52 | $\phi_{55-64}$ | 0.65 | 0.57 |
| $\phi_{65-74}$ | 0.34 | 0.28 | $\phi_{65-74}$ | 0.38 | 0.29 |
| $\phi_{75+}$ | 0.28 | 0.13 | $\phi_{75+}$ | 0.22 | 0.09 |
| Duration of exposure | | | Exposure rate (WL) | | |
| $\gamma_{<5}$ | 1.00 | 1.00 | $\gamma_{<0.5}$ | 1.00 | 1.00 |
| $\gamma_{5-14}$ | 3.17 | 2.78 | $\gamma_{0.5-1.0}$ | 0.51 | 0.49 |
| $\gamma_{15-24}$ | 5.27 | 4.42 | $\gamma_{1.0-3.0}$ | 0.32 | 0.37 |
| $\gamma_{25-34}$ | 9.08 | 6.62 | $\gamma_{3.0-5.0}$ | 0.27 | 0.32 |
| $\gamma_{35+}$ | 13.6 | 10.2 | $\gamma_{5.0-15.0}$ | 0.13 | 0.17 |
| | | | $\gamma_{15.0+}$ | 0.10 | 0.11 |

[a]Parameters estimated on the basis of the model $RR = 1 + \beta w^{*} \phi_{a} \gamma_{z}$ fit using the two-stage method where $w^{*} = w_{5-14} + \theta_{15-24} w_{15-24} + \theta_{25+} w_{25+}$. Here the subscript $a$ denotes categories of attained age and the subscript $z$ denotes categories of either exposure duration (in years) or radon concentration in WL.
[b]Units are $WLM^{-1}$.

data, the number of lung cancers exceeded the total number analyzed by the BEIR IV committee.  Models were developed on the basis of data on exposures under 0.175 $Jhm^{-3}$ (50 WLM) and under 0.350 $Jhm^{-3}$ (100 WLM) (Lubin and others 1997).  The remaining data include sufficient lung-cancer cases for analysis and cover the range of cumulative exposures for most of the general population.  For the exposures used, the inverse exposure-rate effect is not considered to be important (see chapter 2).  Table 3-4 describes the unrestricted and restricted data.  There were 274,161 person-years of observation among occupationally nonexposed workers, including 115 lung-cancer cases.[2]  For exposures under 0.350 $Jhm^{-3}$ (100 WLM), there were 564,772 person-years (64% of total exposed person-years) and

---

[2]When the unrestricted data are used, there were 266,547 person-years and 113 cases among "nonexposed" workers. The difference is due to the different categorizations used in the creation of the person-years tables.

TABLE 3-4    Numbers of lung-cancer cases and person-years and exposure information for cases in the pooled miner data and in data with restrictions of exposure

| | 0.175 Jhm$^{-3}$ (<50 WLM) | 0.35 Jhm$^{-3}$ (<100 WLM) | No restrictions |
|---|---|---|---|
| Cohort | | | |
| China | 77 | 116 | 980 |
| Czech | 15 | 77 | 705 |
| Colorado | 15 | 22 | 336 |
| Ontario | 180 | 231 | 291 |
| Newfoundland | 21 | 24 | 118 |
| Sweden | 17 | 36 | 79 |
| New Mexico | 8 | 11 | 69 |
| Beaverlodge | 42 | 49 | 65 |
| Port Radium | 20 | 25 | 57 |
| Radium Hill | 52 | 53 | 54 |
| France | 22 | 33 | 45 |
| | All data combined[a] | | |
| Lung-cancer deaths | | | |
| Nonexposed | 115 | 115 | 113 |
| Exposed | 353 | 562 | 2674 |
| Person-years | | | |
| Nonexposed | 274,161 | 274,161 | 271,457[b] |
| Exposed | 454,159 | 564,772 | 883,996 |
| Mean values for exposed lung-cancer cases | | | |
| WLM | 19.7 | 40 | 493.6 |
| WL | 0.9 | 1.2 | 4.1 |
| Years since last exposure | 17 | 17.4 | 13.8 |
| Duration of exposure, yr | 5.4 | 6.6 | 14.1 |
| Attained age, yr | 50 | 58.6 | 58.5 |

[a]Totals exclude 115 workers and 12 lung-cancer cases that were in both the Colorado and New Mexico studies. The cases had exposures in excess of 0.35 Jhm$^{-3}$ (100 WLM).

[b]The numbers of "nonexposed" person-years and lung-cancer cases differ because of the factors that define the person-years.

562 lung-cancer deaths (21% of total exposed cases); for exposures under 0.175 Jhm$^{-3}$ (50 WLM), there were 454,159 person-years (51% of total exposed person-years) and 353 lung-cancer deaths (13% of total exposed cases).

In addition to fitting of the full model (2), the constant linear *RR* model in equation (1) was fitted to the restricted data. For model (1), estimates of β with exposures restricted to less than 0.175 Jhm$^{-3}$ (< 50 WLM) and < 0.350 Jhm$^{-3}$ (< 100 WLM) were 3.343 /Jhm$^{-3}$ (95% CI, 0.571-7.143) and 2.286/ Jhm$^{-3}$ (95% CI, 0.857-4.000), respectively. For the unrestricted data, the estimate of β was 1.257/ Jhm$^{-3}$ (95% CI, 0.571-2.857). For exposures under 0.350 Jhm$^{-3}$, there was some suggestion of nonlinearity, but fitting a nonlinear model ($RR = 1 + \beta w^\gamma$) did not significantly improve the model fit (p = 0.30). In accord with the biophysical model for the inverse exposure-rate effect, the gradients of increasing effects with increasing duration of exposure and of decreasing effects with increasing exposure-rate suggest diminution of the inverse exposure rate effect at less than 0.350 Jhm$^{-3}$, compared with the results with the unrestricted data. Other patterns of modification of the exposure-response coefficient ($\beta$) in the restricted analyses were consistent with the patterns with the unrestricted data for all factors except attained age. The excess *RR* declined with time since exposure and exposure rate, and increased with exposure duration. Although the exposure-response relationship did not vary significantly with exposure rate and duration factors in the restricted data, it increased with attained age for exposures under 0.350 Jhm$^{-3}$, whereas it decreased with attained age in the unrestricted data. Lubin and others (1997) fit the exposure-duration and exposure-rate models to the unrestricted data. The models with parameters fixed were then applied to data restricted to < 0.175 Jhm$^{-3}$ (< 50 WLM). The deviances from these fits were similar to the deviances obtained using the simple linear excess *RR* model in WLM for restricted data. These results indicate that the exposure-duration model and the exposure-rate model obtained from the unrestricted data adequately described the low-exposure data.

## COHERENCE OF EVIDENCE FROM MINERS
## AND FROM THE GENERAL POPULATION

The committee also assessed the comparability of the miner-based models with the evidence from the case-control studies in the general population, an additional source of information on risks posed by indoor radon. A meta-analysis of 8 case-control studies, each having 200 or more lung-cancer cases and estimates of exposure, has been conducted (Lubin and Boice 1997; see appendix G). The committee also assessed the comparability of risk estimates from these data resources: the full miner data set, the exposure-restricted miner data set, and the residential case-control studies.

The miner-based models and the results from the meta-analysis are not strictly parallel; the miner-based models are time- and age-dependent, and the meta-analysis is based on a log-linear *RR* model for the time-weighted average exposure rate in an exposure window selected by each of the investigators. A log-linear model for *RR* gives estimates of risk comparable with those from the linear model (equation 1) with relative risks near unity. To compare estimates made with the different models, the *RR* was calculated for 30 yr of exposure at 148 Bqm$^{-3}$ (4 pCiL$^{-1}$). The results are as follows.

• Miner-based exposure-age-duration or exposure-age-concentration models, full data set: For a person 65-69 yr old, the estimated *RR* is 1.11 with the exposure-age-duration model and 1.26 with the exposure-age-concentration model.

• Miner-based constant-ERR model, data restricted set to under 0.175 Jhm$^{-3}$ (50 WLM): For a person in the same age category, the estimated *RR* is 1.17. This estimate is not age-dependent.

• Model based on meta-analysis of residential case-control studies: The estimated *RR* for a person living in a home for 30 years at 148 Bqm$^{-3}$ (4 pCiL$^{-1}$) is 1.14 (95% CI, 1.01-1.30). This estimate is not age-dependent.

The models derived from the full and restricted miner data provide similar *RR* estimates for the scenario of indoor exposure. That comparability supports the assumption of linearity of the exposure-response relationship across the range of exposures encompassed by the miner data. There is also coherence between the miner-based risk estimates and the estimate based on the meta-analysis of case-control studies, in which the exposures of participants span the range of typical indoor exposures (see Figure 3-1 and appendix G). The coherence further supports the extrapolation of the exposure-response relationship from the miner studies to the general population. However, the meta-analysis could not describe the exposure-response relationship at the lower end of residential exposure precisely. The comparison of the 3 models does not explicitly take into account any differential effects of smoking on radon risks in the 3 data sets—the full and restricted miner data and the case-control data. However, differences in proportions of ever-smokers in the 3 data sets were likely to have been small and were not likely to have introduced major bias in the comparison of radon risks.

Results from the indoor case-control studies do not provide direct information on lifetime risks posed by radon exposure. The excess risk of 14% at 148 Bqm$^{-3}$ corresponds to only 30 years of exposure in a house at a constant radon concentration and hence does not reflect the risk of lung cancer associated with lifetime exposure, where the estimated excess lifetime relative risk at 148 Bqm$^{-3}$ based on the miner models is 40-50% (Table 3-5). Estimated relative risks from indoor studies and from miner-based models reflect a 30-year exposure period at 148 Bqm$^{-3}$ and not lifetime exposures at this level. Thus, if exposures outside this 30-year period influence lung-cancer risk, as suggested by the miner data, then the 14% excess relative risk at 148 Bqm$^{-3}$ from indoor studies is a biased estimate of the lifetime relative risk at this concentration and therefore cannot be used to estimate attributable risks for a population.

## BEIR VI RISK ASSESSMENT FOR LUNG CANCER IN GENERAL POPULATION

### Introduction

To extrapolate the risk model from the BEIR VI analyses of miner risks to residential exposures, several assumptions must be made (Table 3-6). The assumed shape of the exposure-response relationship is critical. A linear relation-

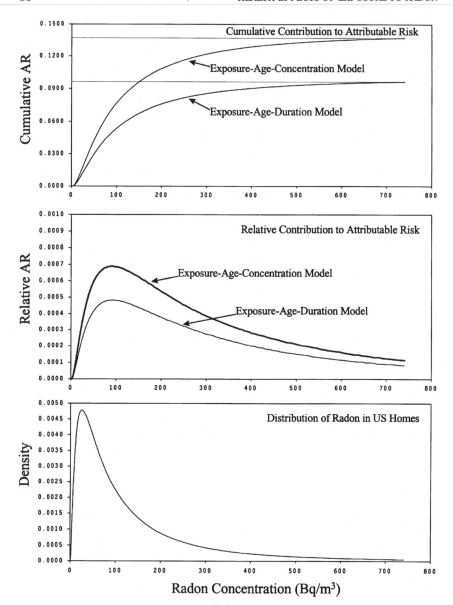

FIGURE 3-1   Contributions to the population attributable risk as a function of radon concentration in U.S. homes, based on the BEIR VI risk models.

TABLE 3-5 Estimated lifetime relative risk of lung-cancer risk associated with lifetime indoor exposure to radon

| Concentration | | Exposure | | | LRR — Exposure-age-concentration model | | | | Exposure-age-duration model | | | |
|---|---|---|---|---|---|---|---|---|---|---|---|---|
| | | | | | Male | | Female | | Male | | Female | |
| Bqm$^{-3}$ | pCiL$^{-1}$ | WL$^a$ | WLM/y$^b$ | Jhm$^{-3}$/y | Ever-smokers | Never-smokers | Ever-smokers | Never-smokers | Ever-smokers | Never-smokers | Ever-smokers | Never-smokers |
| **Multiplicative model:** | | | | | | | | | | | | |
| 25 | 0.7 | 0.003 | 0.10 | 0.00035 | 1.090 | 1.097 | 1.099 | 1.103 | 1.060 | 1.065 | 1.066 | 1.068 |
| 50 | 1.4 | 0.005 | 0.19 | 0.00067 | 1.179 | 1.194 | 1.197 | 1.206 | 1.120 | 1.130 | 1.131 | 1.137 |
| 100 | 2.7 | 0.011 | 0.39 | 0.00137 | 1.352 | 1.388 | 1.391 | 1.411 | 1.237 | 1.259 | 1.261 | 1.274 |
| 150 | 4.1 | 0.016 | 0.58 | 0.00203 | 1.521 | 1.582 | 1.582 | 1.616 | 1.352 | 1.389 | 1.390 | 1.410 |
| 200 | 5.4 | 0.022 | 0.78 | 0.00273 | 1.684 | 1.775 | 1.769 | 1.821 | 1.464 | 1.518 | 1.517 | 1.547 |
| 400 | 10.8 | 0.043 | 1.56 | 0.00546 | 2.290 | 2.542 | 2.490 | 2.637 | 1.892 | 2.033 | 2.012 | 2.091 |
| 800 | 21.6 | 0.086 | 3.12 | 0.01092 | 3.303 | 4.057 | 3.797 | 4.255 | 2.649 | 3.053 | 2.939 | 3.174 |
| **Submultiplicative model$^c$:** | | | | | | | | | | | | |
| 25 | 0.7 | 0.003 | 0.10 | 0.00035 | 1.081 | 1.194 | 1.089 | 1.206 | 1.054 | 1.130 | 1.059 | 1.137 |
| 50 | 1.4 | 0.005 | 0.19 | 0.00067 | 1.161 | 1.388 | 1.177 | 1.411 | 1.108 | 1.259 | 1.118 | 1.274 |
| 100 | 2.7 | 0.011 | 0.39 | 0.00137 | 1.318 | 1.775 | 1.352 | 1.821 | 1.214 | 1.518 | 1.235 | 1.547 |
| 150 | 4.1 | 0.016 | 0.58 | 0.00203 | 1.471 | 2.159 | 1.525 | 2.229 | 1.318 | 1.776 | 1.352 | 1.819 |
| 200 | 5.4 | 0.022 | 0.78 | 0.00273 | 1.619 | 2.542 | 1.694 | 2.637 | 1.420 | 2.033 | 1.466 | 2.091 |
| 400 | 10.8 | 0.043 | 1.56 | 0.00546 | 2.174 | 4.057 | 2.349 | 4.255 | 1.809 | 3.053 | 1.915 | 3.174 |
| 800 | 21.6 | 0.086 | 3.12 | 0.01092 | 3.120 | 7.008 | 3.549 | 7.440 | 2.507 | 5.058 | 2.760 | 5.317 |

$^a$Based on radon gas at 40% equilibrium with its decay products.
$^b$Based on 70% occupancy of the home.
$^c$ Based on the committee's preferred submultiplicative model for the joint effect of smoking and radon.

TABLE 3-6   Assumptions required for extrapolating lung-cancer risk
estimates from miners to general population

| Characteristic | Assumption |
|---|---|
| Shape of exposure-response function | Linear |
| Exposure rate | Risks at residential levels comparable with those in miners exposed at less than 0.298 $Jm^{-3}$ (0.5 WL) (exposure-rate model) or for durations longer than 35 yr (exposure-duration model) |
| Sex | Ratio of ERR to exposure is the same for males and females |
| Age at exposure | Ratio of ERR to exposure is the same for all ages at exposure |
| Tobacco smoking | Submultiplicative interaction of smoking and radon; on basis of analyses of ever- and never-smoking miners, the ratio of ERR to exposure for never-smokers is about twice that for ever-smokers |
| Dosimetry of radon progeny in the lung | No modification of risk required, because dosimetric K factor estimated to be 1 |
| Other differences between miners and those exposed in homes | Ratio of ERR to exposure not dependent on these differences |

ship was assumed in analysis of the miner data on the grounds that it is the simplest model that provides a satisfactory fit to the data in the range of exposures received by the miners. However, to estimate population risks at exposures outside the range of the miner data, a particular exposure-response relationship is assumed at exposures lower than those received by the miners. This assumption needs to be supported by underlying biologic mechanisms.

The committee chose to use a linear relationship between risk and low doses of radon progeny without a threshold. That choice is based primarily on the mechanistic considerations described in chapter 2. In brief, those considerations are related to the stochastic nature of the energy deposition by alpha particles; at low doses, a decrease in dose simply results in a decrease in the number of cells subjected to the same insult. That observation, combined with the evidence that a single alpha particle can cause substantial permanent damage to a cell and that most cancers are of monoclonal origin, provides the mechanistic basis of the use of a linear model at low doses. In addition, as discussed above, exposure-response relationships estimated from the observational data in miners with the lowest exposures, and from the case-control studies of indoor radon, are consistent with linearity (Figure 3-2).

Another critical issue in the extrapolation of risks to the general population is that exposure rates in homes are a thousand-fold to a hundred-fold less from those in most mines. At levels of exposure experienced by miners, biologic consider-

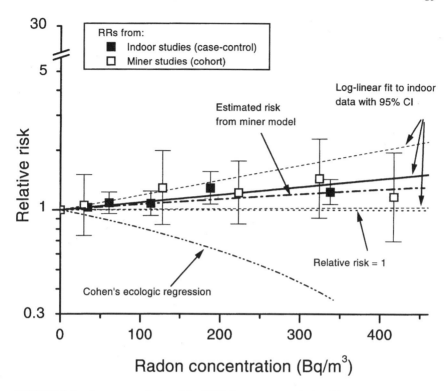

FIGURE 3-2  Summary relative risks (RR) from meta-analysis of indoor-radon studies and RRs from pooled analysis of underground-miner studies, restricted to exposures under 0.175 Jhm$^{-3}$ (50 WLM). Included are RR of 1, fitted exposure-response and its 95% confidence interval from indoor-radon studies, and estimated linear RR based on ecologic analysis by Cohen (1995).

ations reviewed in chapter 2 suggest that, for a particular total exposure, longer duration of exposure is associated with increased risks. However, for extrapolating from the lowest category of exposures in mines to typical residential exposures (Figure 3-1), any modification of risk by exposure rate is thought to be negligible because of the low probability of multiple alpha-particle traversals of epithelial cells. Exposure-rate effects are observed in the miner data (Lubin and others 1995a) and have been incorporated into the committee's models. The consistency of the risk estimates based on data at exposures lower than 0.175 Jhm$^{-3}$ (50 WLM) and on the meta-analysis offers some support for using the committee's models in adjusting for exposure-rate effects at residential doses.

In addition to differing in exposure levels and exposure rates, the general population differs from the miner cohorts in including females and persons exposed at all ages and in tobacco-smoking and other exposures. Assumptions regarding the possible modifying effects of those factors are required to extrapolate the miner-based risk estimates. These assumptions are discussed below.

As in the BEIR IV report, we applied the same risk model to baseline rates for males and females, assuming a multiplicative joint association for exposure and sex; that is, the ratio of ERR to exposure is assumed to be the same for males and females, given specific ages, exposure rates and times since exposure. The background lung-cancer risk for females is lower than that for males, so this assumption results in a smaller lifetime absolute excess risk for females attributable to radon exposure. There are no directly relevant data on modification by sex of the risk posed by radon exposure. Two somewhat contradictory analogies can be made. After occupational risk factors are adjusted for, the effects of cigarette-smoking, for given durations and intensities, are at least as great in females as in males (Doll and others 1980; Lubin and others 1984; Risch and others 1993); this indicates consistency with a multiplicative interaction (USDHHS 1990). In contrast, among the Japanese atomic-bomb survivors, the ERR per Gy for lung-cancer mortality was about 4 times greater in females than in males, although the absolute excess risk was only about 50% greater in females (Shimizu and others 1988). That pattern of sex differences suggests that the proportional translation of radiation effects from males to females could be incorrect. The relevance of the observation in the Japanese atomic-bomb survivors, who received whole-body acute exposure to gamma radiation and some neutron radiation, to people exposed to localized lung doses of alpha radiation throughout their lives is unclear.

The committee also assumed that the ERR per unit exposure does not vary with age at exposure. Analyses of the underground-miner data provided little evidence of variation of the ratio of ERR to exposure with age at the start of mining, but data on those exposed under the age of 20 yr are limited. The atomic-bomb survivor studies, which include many subjects exposed under the age of 20, also give little indication that age at exposure modifies lung-cancer risk, although estimates of parameters describing age effects for lung cancer are imprecise. In contrast, risks for several other radiation-induced cancers have shown strong age dependency in atomic-bomb survivors and other cohorts. There remains considerable uncertainty with regard to a difference in effects between radon exposure in childhood and in adulthood.

Analyses of the miner data indicate a combined effect of smoking and radon intermediate between additive and multiplicative (see appendix A). This type of combined effect indicates synergism between smoking and radon, but the degree of synergism could not be characterized with a high degree of precision. In estimating lifetime risk for ever-smokers and never-smokers, we used 2 approaches based on modeling the miner data. In the first approach, we assumed a multiplicative relationship (applying the risk model separately to lung-cancer rates in ever-smokers and in never-smokers), recognizing that the risks to ever-smokers might be overestimated and the risks to never-smokers underestimated. In the second approach, based on analysis of data from known ever-smoking and never-smoking miners, we observed a submultiplicative relationship for the joint association, wherein the ratio of ERR to exposure is greater in never-smokers than in ever-smokers. Acknowledging the limitations of the available data on smoking, we considered the submultiplicative relationship to be preferable for predicting population risk. This preference was based on consideration of model fit and on the higher ratio of ERR to exposure in never-smokers.

For the adjustment of age-specific lung-cancer mortality rates to reflect rates in ever-smokers and never-smokers, we used assumptions based on 1993 data (CDC 1995): relative risks for ever-smokers of 14 and 12 compared with never-smokers and percentages of ever-smokers in the population of 58% and 42% for males and females, respectively. By definition, ever-smokers include both current and former smokers. These assumptions imply that 95% of cases of lung cancer in men and 90% of cases in women are in ever-smokers; these percentages are consistent with recent data reviewed in appendix C. We assumed 18 yr as the age of starting to smoke regularly. Because of limitations of available data, risks were not estimated separately for former smokers.

Finally, extrapolation from mines to homes requires consideration of the factors that affect the relationship between exposure and dose to target cells in the lung, such as particle size, distribution, and bronchial structure (appendix B). We have incorporated the formalism proposed in BEIR IV (NRC 1988) and the later Panel on Dosimetric Assumptions (NRC 1991), in which the dose per unit exposure in homes is related to the dose per unit exposure in mines by a dimensionless factor $K$, described in detail in appendix B. Figure 1-2 gives the components of the extrapolation. The exposure-dose factors can be addressed by modeling the dose produced by a given exposure for relevant groups, including miners and the various groups of occupants of homes (men, women, children, and infants).

Because the estimated median $K$-values for males, females, and children are all close to 1, population risks have been projected with miner-based risk models under the assumption that the dosimetric $K$ is equal to 1. However, uncertainty analyses of population-risk projections were conducted to estimate the significance of deviations from 1.

## Measures of Risk

To assess the population lung-cancer risk posed by indoor radon, the model for the exposure-response relationship is applied to the observed distribution of residential radon exposures received by the U.S. population. This risk-characterization step yields the information needed by policy-makers and by stakeholders, primarily the general public in this instance. Previous estimates of the numbers of lung-cancer deaths per year from indoor radon are based on such calculations.

Several measures can be used for risk characterization: population-level indicators that describe the total risk to a population and individual-level indicators that describe the risks to individuals who have specific exposures and particular characteristics, such as, gender and smoking status. For describing the risk to the population, we have used the population attributable risk (AR), which describes the burden of lung-cancer deaths that, in theory, could be prevented if all exposures to radon were eliminated. AR estimates include cases in ever-smokers and never-smokers. The cases in ever-smokers include those resulting from the synergism between smoking and radon exposure; in principle these cases could have been prevented either by smoking prevention (the smokers remained never-smokers) or by reduction of indoor radon concentration to outdoor levels. Although the latter cannot be practically achieved, the committee provides AR estimates for several radon mitigation scenarios that might be feasible through programs to reduce exposures to indoor radon. At the individual level, the committee provides estimates of lifetime relative risk (LRR). Those estimates are based on a constant exposure rate over a lifetime for a cohort followed to extinction compared with a similar cohort exposed to no radon also followed to extinction. The risks have been corrected for competing causes of death by using standard lifetable methods as described in BEIR IV.

## Relative-Risk Estimates

### Lifetime Relative Risks

LRRs were computed with the committee's exposure-age-concentration and exposure-age-duration models (Table 3-5). The LRR describes the proportional increment in lung-cancer risk posed by radon exposure beyond the background level (exposures from outdoor air). In addition, LRR estimates were calculated with the BEIR IV model and a constant-relative-risk model based on exposure-restricted data for purposes of comparison. Estimates were computed for exposure scenarios that reflect indoor-radon exposure patterns of interest.

Table 3-5 shows the estimated LRRs for lifetime exposure at constant radon concentrations of 25, 50, 100, 150, 200, 400, and 800 Bqm$^{-3}$ (0.7, 1.4, 2.7, 4.1, 5.4, 10.8, and 21.6 pCiL$^{-1}$). Those concentrations were converted to annual exposures by assuming an equilibrium of 40% between radon and its progeny (appendix B) and assuming 70% of time spent at home. The LRRs are similar for the exposure-age-concentration model and the exposure-age-duration model. The LRRs estimated with the BEIR VI models and the BEIR IV model are also similar, even with the inclusion of exposure rate in the new model. As anticipated, the LRR increases with exposure. Women have a somewhat steeper increase in LRR with increasing exposure than men because of their lower background lung-cancer mortality. The higher LRRs in never-smokers than in ever-smokers also reflect differing background mortality rates.

## Population-Risk Estimates

### Estimation of Attributable Risk

Following Levin (1953) and Lubin and Boice (1989), the attributable risk (AR) of lung cancer due to ionizing radiation can be defined as the proportion of lung-cancer deaths attributable to exposure to radon progeny. The AR estimates indicate the proportion of lung-cancer deaths that theoretically may be reduced by reduction of indoor radon concentrations to outdoor levels. The AR of lung cancer from indoor exposure to radon can be estimated given the exposure-response relationship for radon and lung-cancer risk, the distribution of indoor radon concentrations, and mortality from lung cancer and from all causes. Mortality can be adjusted, as done by the USEPA (1992a), to reflect the mortality pattern of a hypothetical population not exposed to radon. However, the effect of such an adjustment would be small and, because of the uncertainty inherent in AR estimates, was not carried out.

ARs of lung cancer from indoor radon in the U.S. population were computed with the exposure-age-duration and exposure-age-concentration models with the parameter estimates from Table 3-3, the BEIR IV risk model, and the linear ERR model fitted to data at exposures under $0.175$ Jhm$^{-3}$ (50 WLM). AR calculations were described in the BEIR IV report (NRC 1988), and are given in appendix A. The BEIR VI risk assessment used the results of the National Residential Radon Survey, which were based on a statistical sample of U.S. residences (Marcinowski and others 1994). The ARs based on the committee's risk models and on the constant linear ERR model were similar (Table 3-7) and consistent with those reported previously (USEPA 1992b; Lubin and Boice 1989; Lubin and others 1995a; Puskin and Nelson 1989).

When the model was applied separately to males and females and to ever-smokers and never-smokers, the ARs were similar. The ARs for the various risk models used 1985-1989 mortality data (for both lung cancer and all causes of death) and the same distribution for domestic radon concentration as shown in Table 3-7. The computed ARs for the total population with the current models ranged from about 10 to 15%. The AR estimates based on the various models were similar both to each other and to the estimates based on the BEIR IV model.

In this analysis, we assumed a lognormal distribution for residential radon concentrations, with a median of 24.8 Bqm$^{-3}$ (0.67 pCiL$^{-1}$) and a geometric standard deviation of 3.11 (Marcinowski and others 1994). In Table 3-8 and in Figure 3-2, we present the ARs from each portion of the distribution of radon concentrations in U.S. homes. For males (results for females are similar), the overall AR based on the exposure-age-duration model is estimated to be 0.099, as compared to 0.141 with the exposure-age-concentration model, about 30% less. However, the percentage contributions to the overall distribution from different percentiles of the exposure are very close for the two models. The

TABLE 3-7   Comparison of attributable risks for indoor radon progeny exposure in the U.S. for ever-smokers and never-smokers, using the risk models with and without adjustment for a sub-multiplicative relationship between radon progeny exposure and smoking

| Model | Total population[a] | Multiplicative model[b] | | Sub-multiplicative model[c] | |
|---|---|---|---|---|---|
| | | Ever-smokers | Never-smokers | Ever-smokers | Never-smokers |
| **Males** | | | | | |
| Committee preferred models | | | | | |
| Exposure-age-concentration | 0.141 | 0.136 | 0.149 | 0.125 | 0.258 |
| Exposure-age-duration | 0.099 | 0.096 | 0.105 | 0.087 | 0.189 |
| Other models | | | | | |
| CRR[d] (< 0.175 Jhm$^{-3}$, | | | | | |
| < 50 WLM) | 0.109 | 0.105 | 0.117 | 0.096 | 0.209 |
| BEIR IV | 0.082 | 0.079 | 0.086 | 0.071 | 0.158 |
| **Females** | | | | | |
| Committee preferred models | | | | | |
| Exposure-age-concentration | 0.153 | 0.149 | 0.156 | 0.137 | 0.269 |
| Exposure-age-duration | 0.108 | 0.105 | 0.110 | 0.096 | 0.197 |
| Other models | | | | | |
| CRR[d] (< 0.175 Jhm$^{-3}$, | | | | | |
| < 50 WLM) | 0.114 | 0.111 | 0.117 | 0.101 | 0.209 |
| BEIR IV | 0.087 | 0.085 | 0.089 | 0.077 | 0.163 |

[a]No adjustment for smoking.
[b]Unadjusted risk model applied to each group, implying a joint multiplicative relationship for radon progeny exposure and smoking.
[c]Models adjusted by multiplying the baseline ERR/WLM by 0.9 for ever-smokers and by 2.0 for never-smokers.
[d]CRR = constant relative risk.

homes with higher radon concentrations contribute disproportionately to the AR; the 49.9% of homes with radon levels of 25 Bqm$^{-3}$ (0.68 pCiL$^{-1}$) or less account for only 12.8% (12.7% under the exposure-age-duration model) of the overall AR, whereas the 5.9% of homes with radon levels above 148 Bqm$^{-3}$ (4 pCiL$^{-1}$), the Environmental Protection Agency (EPA) action level, account for about 30% of the AR.

## Attributable Risk Accounting for Smoking Status

The committee's two risk models were initially developed without explicitly incorporating the available data on smoking status. For estimation of the number of lung cancers in the general population due to indoor exposure to radon progeny, the BEIR IV committee and other groups have recommended that the same

TABLE 3-8   Distribution of attributable risks for U.S. males from indoor residential radon exposure, based on BEIR VI models

| Exposure range, Bqm$^{-3}$ | Proportion of homes in Range, % | Contribution to AR | | | |
|---|---|---|---|---|---|
| | | Exposure-age-concentration model | | Exposure-age-duration model | |
| | | Actual | % | Actual | % |
| 0-25 | 49.9 | 0.018 | 12.8 | 0.013 | 12.8 |
| 26-50 | 23.4 | 0.026 | 18.5 | 0.018 | 18.4 |
| 51-75 | 10.4 | 0.020 | 14.2 | 0.014 | 14.2 |
| 76-100 | 5.4 | 0.015 | 10.5 | 0.010 | 10.5 |
| 101-150 | 5.2 | 0.020 | 13.9 | 0.014 | 13.9 |
| 151-200 | 2.4 | 0.013 | 9.2 | 0.009 | 9.2 |
| 201-300 | 1.8 | 0.014 | 9.6 | 0.010 | 9.7 |
| 301-400 | 0.7 | 0.007 | 5.2 | 0.005 | 5.3 |
| 401-600 | 0.4 | 0.006 | 4.5 | 0.005 | 4.6 |
| 601 + | 0.4 | 0.002 | 1.5 | 0.002 | 1.6 |
| Total | 100.0 | 0.141 | 100.0 | 0.099 | 100.0 |

risk model be applied to ever-smokers and to never-smokers, thereby assuming the joint relationship of radon-progeny exposure and smoking to be multiplicative (NRC 1988). However, the current analysis, like some previous analyses, indicates that, although the joint effect of radon-progeny exposure and smoking is consistent with a multiplicative model, the most likely relationship is intermediate between multiplicative and additive. This intermediate combined effect implies that estimates of the number of radon-associated lung-cancer deaths among ever-smokers and never-smokers predicted with a multiplicative assumption are too high for ever-smokers and too low for never-smokers.

Using the results of the analyses of the effect of radon-progeny exposure among never-smokers and ever-smokers, we can modify the exposure-age-concentration and exposure-age-duration models to account for smoking status. There were insufficient data to develop a risk model for never-smokers directly, but we can adjust the risk models on the basis of the relative difference in the exposure-response relationships for ever-smokers and never-smokers. When only data on miners for whom some smoking information was available were used, the overall ERR/0.0035 Jhm$^{-3}$ was estimated to be 1.02% among never-smokers (95% CI, 0.15-7.18%), and 0.48% among ever-smokers (95% CI, 0.18-1.27%). Among these same miners, the overall ERR/0.0035 Jhm$^{-3}$ ignoring smoking status was 0.53% (95% CI 0.20%,1.38%). An influence analysis, involving omission of data from each of the cohorts with smoking data one at a time, did not identify any one cohort as having a dominant effect on these estimates.

The estimated ratios of ERR per unit exposure based on the miner data on ever-smokers and on never-smokers are directly comparable only insofar as mean age, time since exposure, and exposure rate—factors known to modify the ratio—are similar for ever-smokers and never-smokers. Those modifiers were found to differ only slightly between the two groups: never-smokers were 1 yr older, their time since last exposure was 6 mo longer, and their exposure rate was 90% of that of ever-smokers. We therefore assumed that the estimated ratios approximate the relative effects of radon-progeny exposure in ever-smokers and never-smokers.

Among miners with smoking data, the proportional effect of exposure among ever-smokers relative to the overall effect without considering smoking status was 0.9 (0.48/0.53); among never-smokers, the relative effect was about 1.9 (1.02/0.53). To modify the risk models for smoking status, we adjusted the estimated baseline ERR/0.035 Jhm$^{-3}$ without altering the parameter estimates to take into account the various modifying factors. Specifically, in the exposure-age-concentration model, the overall estimate for $\beta$ of 0.0768 for all miners combined was reduced to 0.069 for ever-smokers and increased to 0.153 for never-smokers. In the exposure-age-duration model, the estimate of 0.0055 was reduced to 0.0050 for ever-smokers and increased to 0.011 for never-smokers. ARs for indoor radon-progeny exposure in the US for ever-smokers and never-smokers, based on risk models with and without this adjustment, are given in Table 3-9.

In 1995, about 157,400 lung-cancer deaths—95,400 in men and 62,000 in women—occurred in the United States (Boring and others 1995). Most lung-

TABLE 3-9    Estimated attributable risk (AR) for domestic exposure to radon using 1985-89 U.S. population mortality rates based on selected risk models

| Model | Population | Ever-smokers[a] | Never-smokers[a] |
|---|---|---|---|
| **Males** | | | |
| Committee's preferred models | | | |
|    Exposure-age-concentration | 0.141 | 0.125 | 0.258 |
|    Exposure-age-duration | 0.099 | 0.087 | 0.189 |
| Other models | | | |
|    CRR[b] (< 0.175 Jhm$^{-3}$, < 50 WLM) | 0.109 | 0.096 | 0.209 |
|    BEIR IV | 0.082 | 0.071 | 0.158 |
| **Females** | | | |
| Committee's preferred models | | | |
|    Exposure-age-concentration | 0.153 | 0.137 | 0.269 |
|    Exposure-age-duration | 0.108 | 0.096 | 0.197 |
| Other models | | | |
|    CRR[b] (< 0.175 Jhm$^{-3}$, < 50 WLM) | 0.114 | 0.101 | 0.209 |
|    BEIR IV | 0.087 | 0.077 | 0.163 |

[a]Based on the committee's preferred submultiplicative model for the joint effect of smoking and radon.

[b]CRR = constant relative risk.

cancer cases occur among ever-smokers: about 95% of cases in male ever-smokers and 90% in female ever-smokers (appendix C). This assumption places a higher proportion of cases in smokers than did the previous report of Lubin and colleagues (1994a), which assumed only 85% of cases to be in smokers. Table 3-10 shows the estimated lung-cancer deaths in the United States in 1995, including those attributable to indoor radon progeny exposure according to the preferred BEIR VI models.

## Effect of Radon Mitigation on Attributable Risk

The overall AR describes the anticipated consequences of virtual elimination of indoor radon exposures under the committee's risk models. A more-realistic assessment of the reduction of attributable risk due to radon exposure mitigation focuses on exposure-reduction scenarios that might actually be achieved. Lubin and Boice (1989) considered three such scenarios: all homes above 148 Bqm$^{-3}$ (4 pCiL$^{-1}$) are reduced to zero, that is, the outdoor level, all homes above 148 Bqm$^{-3}$ (4 pCiL$^{-1}$) are distributed at a lower concentration below 148 Bqm$^{-3}$

TABLE 3-10   Estimated number of lung cancer deaths in the U.S. in 1995 attributable to indoor residential radon progeny exposure

| Smoking status | Number of lung-cancer deaths | Lung-cancer deaths attributable to radon progeny exposure (No.) | | | |
| --- | --- | --- | --- | --- | --- |
| | | Exposure-age-concentration model | | Exposure-age-duration model | |
| **Males[a]** | | | | | |
| Total | 95,400 | 13,000[b] | 12,500[c] | 9,200[b] | 8,800[c] |
| Ever-smokers | 90,600 | 12,300 | 11,300 | 8,700 | 7,900 |
| Never-smokers | 4,800 | 700 | 1,200 | 500 | 900 |
| **Females[a]** | | | | | |
| Total | 62,000 | 9,300 | 9,300 | 6,600 | 6,600 |
| Ever-smokers | 55,800 | 8,300 | 7,600 | 5,900 | 5,400 |
| Never-smokers | 6,200 | 1,000 | 1,700 | 700 | 1,200 |
| **Males and Females** | | | | | |
| Total | 157,400 | 22,300 | 21,800 | 15,800 | 15,400 |
| Ever-smokers | 146,400 | 20,600 | 18,900 | 14,600 | 13,300 |
| Never-smokers | 11,000 | 1,700 | 2,900 | 1,200 | 2,100 |

[a]Assuming that 95% of all lung cancers among males occur among ever-smokers, and that 90% of all lung cancers among females occur among ever-smokers.  Percentages of ever-smokers in the population were 58% and 42% for males and females respectively.
[b]Estimates based on applying same risk model to ever-smokers and never-smokers, implying a joint multiplicative relationship for radon progeny exposure and smoking.
[c]Estimates based on applying a smoking adjustment to the risk models, multiplying the baseline ERR/WLM by 0.9 for ever-smokers and by 2.0 for never-smokers, the committee's preferred approach.

(using a truncated log-normal distribution), and all homes above 148 Bqm$^{-3}$ (4 pCiL$^{-1}$) are reduced to exactly 148 Bqm$^{-3}$ (4 pCiL$^{-1}$). These AR apply to a mixed population of male ever-smokers and never-smokers with exposures corresponding to the 1993 distribution of radon concentrations in U.S. homes. Because the distribution of radon concentration is approximately lognormal with the bulk of the population exposed to very low levels, the effective AR is substantially lower than the total AR. Effective ARs (EARs), which indicate the fraction of total lung-cancer deaths that would be eliminated by implementing one of these mitigation scenarios, are shown in Table 3-11. For example, under the exposure-age-concentration model, mitigating homes at or above 148 Bqm$^{-3}$ (4 pCiL$^{-1}$) would result in an estimated reduction in lung-cancer mortality of 4.2% if homes were mitigated to zero, 3.7% if homes were distributed across the range of levels below 148 Bqm$^{-3}$ (4 pCiL$^{-1}$) (on the basis of a truncated lognormal distribution), or 1.7% if homes were mitigated to exactly 148 Bqm$^{-3}$ (4 pCiL$^{-1}$). The second mitigation scenario is probably the most-appropriate characterization, and the first and third scenarios are limiting scenarios. Eliminating exposures at concentrations above 148 Bqm$^{-3}$ (4 pCiL$^{-1}$) results in an EAR of about 4%. Thus, 10-15% of all lung cancers are estimated as attributable to indoor radon, and eliminating exposures in excess of 148 Bqm$^{-3}$ (4 pCiL$^{-1}$) would reduce the lung-cancer burden from radon to 7-11% of all lung-cancer cases.

TABLE 3-11   Effective attributable risks (EAR$^a$) for lung cancer from residential radon exposure to radon using 1985-89 U.S. population mortality rates and the BEIR VI models

| | | | | Modified radon distributions$^b$ | | |
| Model | "Eliminate" all exposures | Cut-off in Bqm$^{-3}$ | Cut-off in pCiL$^{-1}$ | "0" | Truncated | Exact cut-off |
|---|---|---|---|---|---|---|
| Exposure-age-concentration | 0.141 | | | | | |
| | | 37 | 1 | 0.110 | 0.092 | 0.068 |
| | | 74 | 2 | 0.078 | 0.065 | 0.040 |
| | | 148 | 4 | 0.042 | 0.037 | 0.017 |
| Exposure-age-duration | 0.099 | | | | | |
| | | 37 | 1 | 0.077 | 0.065 | 0.049 |
| | | 74 | 2 | 0.055 | 0.047 | 0.028 |
| | | 148 | 4 | 0.031 | 0.027 | 0.012 |

$^a$EAR, the effective attributable risk, indicates the fraction of total lung-cancer deaths that would be eliminated by radon-exposure mitigation.
$^b$Population distribution of radon concentrations in homes based on Marcinowski and others (1994) new residential radon survey. All homes over cut-off are assumed mitigated: to zero exposure (denoted "0"); to levels under cut-off based on a truncated log-normal distribution (denoted "truncated"); or to exactly cut-off (denoted "Exact cut-off").

The rationale for and the biologic justification of the committee's choice of a linear, nonthreshold exposure-response relationship for the calculation of risk, particularly at low total exposures, is given in chapter 2. Results of the miner studies and of studies of indoor radon exposure are consistent with a linear relationship throughout the entire range of residential radon exposures, and there is no evidence of a threshold exposure below which exposure carries no risk. However, data on the lowest residential exposures are sparse, and a threshold exposure below which exposure does not increase risk is theoretically possible.

Table 3-12 shows the effects on the AR and the EAR of assuming a linear model, but with various threshold exposures below which there is no increased risk and from which threshold risk increases with the slope used in the non-threshold model. In this table, threshold exposures are specified in terms of concentrations: exposures at or below the indicated concentrations are assumed to not increase lung-cancer risk. With the assumed thresholds, the total estimated lung-cancer burden to the population is lower than with the committee's no-threshold model. For example, if the threshold were 74 $Bqm^{-3}$ (2 $pCiL^{-1}$), the total AR—the proportion of lung cancer eliminated on removal of all risk from residential radon exposure—is reduced from 0.141 to 0.083 according to the committee's exposure-age-concentration model, or from 0.099 to 0.058 according to the exposure-age-duration model. However, mitigating homes over 148 $Bqm^{-3}$ (4 $pCiL^{-1}$)—distributing high-radon homes to levels below this—increases the effective AR from 0.037 to 0.044 (exposure-age-concentration model) or from 0.026 to 0.031 (exposure-age-duration model). The increased effectiveness of mitigation is the result of lowering levels in homes with concentrations over

TABLE 3-12   Estimated attributable risk (AR) and effective attributable risk (EAR) for lung-cancer deaths in the U.S. from residential exposure to radon and it progeny and the consequences of assuming a threshold for exposure, below which there is no risk of lung cancer death[a]

| Threshold[b] | | Exposure-age-concentration model | | Exposure-age-duration model | |
|---|---|---|---|---|---|
| $Bqm^{-3}$ | $pCiL^{-1}$ | AR | EAR[c] | AR | EAR[c] |
| 0 | 0 | 0.141 | 0.037 | 0.099 | 0.026 |
| 37 | 1 | 0.113 | 0.040 | 0.079 | 0.028 |
| 74 | 2 | 0.083 | 0.044 | 0.058 | 0.031 |
| 148 | 4 | 0.048 | 0.048 | 0.033 | 0.033 |

[a]Estimates are based on the committee's exposure-age-concentration and exposure-age-duration models and assume constant lifetime exposure.

[b]Exposures at or below the indicated threshold level are assumed to carry no risk of lung cancer.

[c]Effective attributable risk estimates the proportion of radon-induced lung-cancer deaths which could be prevented if the distribution of radon concentration in houses was modified. In this example, EAR is based on reducing the radon concentration in all homes above 148 $Bq/m^3$ to a level between the outdoor level and 148 $Bq/m^3$ (4 $pCiL^{-1}$), based on a truncated log-normal distribution.

148 Bqm$^{-3}$ (4 pCiL$^{-1}$) to concentrations below 148 Bqm$^{-3}$ (4 pCiL$^{-1}$), which carry some risk assuming no threshold, to levels which carry no risk, assuming an exposure threshold. Thus, if there were a threshold exposure level, then the estimated total lung-cancer burden attributable to radon would decline, whereas the percentage reduction in the lung-cancer burden that could be achieved through any practical radon mitigation strategy would increase.

## SOURCES OF UNCERTAINTY

Quantitative estimates of human cancer risk are subject to a number of uncertainties, which need to be considered in risk management decision making. We discuss below sources of uncertainty in estimates of lung-cancer mortality associated with exposure to indoor radon in the United States. The initial discussion is in qualitative terms; the committee's quantitative treatment of uncertainty is described fully in appendix A and summarized later in this chapter.

Table 3-13 lists sources of uncertainty and categorizes them according to the steps in the risk-assessment process as discussed in appendix A. The material that follows discusses only uncertainties in the model relating lung-cancer risk to exposure (Table 3-13, part I). The model for addressing differences in radon-progeny dosimetry in mines and in homes (Table 3-13, part II) is discussed

TABLE 3-13   Sources of uncertainty in estimates of lifetime risk of lung cancer mortality resulting from exposure to radon in homes

I Sources of uncertainty arising from the lung-cancer risk model
  A  Uncertainties in parameter estimates derived from miner data
    1  Sampling variation in the underground miner data;
    2  Errors and limitations in the underground miner data;
      a) Errors in health effects data including vital status and information on cause of death;
      b) Errors in data on exposure to radon and radon progeny including estimated cumulative exposures, exposure rates and durations;
      c) Limitations in data on other exposures including data on smoking and on other exposures such as arsenic.
  B  Uncertainties in application of the model to the general
    1  Shape of the exposure/exposure rate response function for estimates at low exposures and exposure rates;
    2  Temporal expression of risks;
    3  Dependence of risks on sex;
    4  Dependence of risks on age at exposure;
    5  Dependence risks on smoking status.
II Sources of uncertainty arising from the exposure/dose model
III Sources of uncertainty arising from the exposure distribution model
    1  Estimate of the average radon concentration;
    2  Estimate of the average equilibrium fraction;
    3  Estimate of the average occupancy factor.
IV Sources of uncertainty in the demographic data used to calculate lifetime risk

briefly in appendix A and more thoroughly in appendix B. Uncertainties in the exposure distribution for the U.S. population (Table 3-13, part III) apply to AR estimates (Tables 3-7 through 3-11), but do not apply to estimates of LRR associated with specified exposure scenarios (Table 3-5). The exposure distribution used in calculating AR was taken from Marcinowski and others (1994) and is discussed in appendix G. Uncertainties in the demographic data (Table 3-13) are briefly discussed in appendix A. Puskin (1992) discusses sources of uncertainty in BEIR IV risk estimates; much of this discussion is relevant, at least in general terms, to the BEIR VI risk estimates.

## Uncertainties in Parameter Estimates Derived from Underground-Miner Data

### Uncertainty Due to Sampling Variation

Uncertainty resulting from sampling variation differs from uncertainty of most other sources in that it can be quantified using statistical methods. However, because estimates of lifetime risks are a complex function of the parameters of the risk model, Monte Carlo simulations are needed to obtain confidence intervals for such estimates. The committee's simulations, which include sampling variation as well as other sources of uncertainty, are described later in this section.

### Errors in the Underground Miner Data

Errors in the data from the 11 miner cohorts used to construct the committee's risk model contribute additional uncertainty in parameter estimates. Errors in determination of vital status and cause of death could also potentially result in bias. However, because the analyses of the miner cohorts were based on internal comparisons within each of the cohorts, the effect of any bias from errors in health outcome data is not likely to be large, unless the errors depended on the level of exposure, which seems unlikely.

Errors in estimates of miner exposures are more likely to have biased estimates of risk. These errors occurred because measurements of radon and radon progeny were limited for many of the mines, especially during the early periods of mine operation (see appendix F). These limitations could have led to both systematic and random errors in estimates of exposure of individual miners. With regard to systematic bias, historical accounts were not sufficient to determine the overall magnitude or even the overall direction of bias. It is possible that different types of systematic bias affecting the estimates might not negate each other. It is well known that random measurement error in the exposure estimates can lead to underestimation of risk (see appendix G). The complex nature of the error structure and the lack of adequate data on all sources of error make it difficult or impossible to quantify its impact on risk estimates.

For most of the miner cohorts, errors in exposure estimates were likely to have been largest in the earliest periods of operation, when exposure rates were highest and fewer measurements were made or, in some cases, simple estimates were made in the absence of measurements. For that reason, measurement errors might affect not only the estimates of the overall risk coefficient, but also estimates of parameters that describe the relationship of risk with time-related variables, such as exposure rate, time since exposure, and age at risk. However, analyses restricted to exposures below 0.175 Jhm$^{-3}$ and below 0.350 Jhm$^{-3}$ led to attributable risk estimates that were very similar to those obtained with the committee's recommended models based on the full miner data set. Miners exposed at those low levels were employed predominantly in later periods when exposure-assessment methods had improved substantially, and their exposure estimates were probably affected to a lesser degree by measurement error than those of the miners who worked in the earlier periods.

In addition to radon progeny, some underground miners were exposed to arsenic, silica, and diesel fumes (see appendix F). Because those exposures might be positively associated with radon exposures, there is a potential for confounding if no adjustment is made. The other exposures might also have enhanced the risk posed by radon for miners through synergism. Such potential synergism constitutes a further source of uncertainty in extending miner-based risk estimates to the general population, which is not exposed to those agents. Limitations in the available data on miners make it difficult to evaluate bias in risk estimates for radon progeny because these other agents are not adequately considered.

In spite of the differences among the cohorts, an influence analysis showed that the AR estimates did not reflect an undue influence by any particular cohort (Table 3-14). The AR estimate changed by a few percent at most as individual cohorts were removed from the analysis.

TABLE 3-14   Influence analysis for the population attributable risk, based on calculating the AR after omitting data from each of the 11 cohorts

| Cohort excluded | Population attributable risk (AR) | |
|---|---|---|
| | Exposure-age-concentration model | Exposure-age-duration model |
| None | 0.141 | 0.099 |
| China | 0.175 | 0.124 |
| Czechoslovakia | 0.137 | 0.101 |
| Colorado | 0.137 | 0.104 |
| Ontario | 0.151 | 0.099 |
| Newfoundland | 0.135 | 0.100 |
| Malmberget | 0.144 | 0.099 |
| New Mexico | 0.136 | 0.096 |
| Beaverlodge | 0.128 | 0.083 |
| Port Radium | 0.148 | 0.101 |
| Radium Hill | 0.129 | 0.081 |
| France | 0.150 | 0.106 |

Smoking is also of potential concern both as a confounder and as a modifier (appendix C). The high prevalence of smoking among underground miners is a potential source of uncertainty in extrapolating the BEIR VI models to the general population. Because the committee does not expect strong correlations between smoking and radon exposure, it is not greatly concerned about uncontrolled confounding of radon risk estimates by cigarette-smoking. Limitations of the available smoking data are of greater concern in evaluating the modifying effects of smoking on risk posed by radon-progeny exposure.

### Uncertainties in Specification of the Lung-Cancer Exposure-Response Model and Its Application to Residential Exposure of the General U.S. Population

Uncertainties in model specification and application are especially important for extrapolating risks from underground miners to persons exposed in homes.

#### Shape of the Exposure/Exposure-Rate Response Relations

Cumulative exposures and exposure rates were generally much higher in underground mines than in homes, consequently, it might be thought of that the most-critical aspect of the committee's model development process is the choice of method for extrapolating risks to residential exposures. However, the committee's recommended risk models included parameters that specifically estimated effects at average exposure rates less than 0.03 Jm$^{-3}$ (0.5 WL) and average exposure durations exceeding 35 yr, the rates and durations of principal interest for residential exposures. Furthermore, restricted analyses based on miners with lower cumulative exposures (less than 0.175 or 0.350 Jhm$^{-3}$) and the meta-analysis of data from case-control studies of indoor radon led to risk estimates that were very similar to those based on the committee's recommended models.

#### Temporal Expression of Risk

The committee's risk model provides for a decline in risk with time since exposure and with age at risk. Although those patterns were consistently identified in nearly all the underground-miner cohorts, estimates of the effects could have been biased by errors in exposure measurements and in classification of death as due to lung cancer, as well as changes in smoking habits over time. We note that lifetime risk estimates based on miner data restricted to concentrations below 0.175 or 0.350 Jhm$^{-3}$, and on the assumption that risks did not decline with time since exposure or with age at risk, were similar to estimates based on the committee's recommended models.

## Dependence of Risks on Sex

The cohorts used to develop the committee's model included only men. The committee has chosen to assume that risks to males and females are comparable on a multiplicative scale, which leads to risks for females that are about one-third those for males and about one-third of those which would have been obtained had risks been assumed to be comparable in males and females on an absolute scale. The committee had no basis for further quantifying the uncertainty associated with the choice of the multiplicative rather than the additive scale.

## Dependence of Risks on Age at Exposure

Using the limited data available, the committee did not find evidence that age at exposure modified lung-cancer risks in underground miners; the preferred models were therefore based on the assumption that the excess relative risk did not depend on age at exposure. Data from the miner cohorts on exposures in childhood are limited primarily to the Chinese tin miners; even in this cohort, data on miners exposed in early childhood are sparse. Clearly, the uncertainty in lung-cancer risks in adults resulting from exposure in childhood is much greater than in risks resulting from exposure in adulthood.

## Dependence of Risks on Smoking Status

Evaluation of the modifying effect of smoking on lung-cancer risk is limited by the extent of the available data on smoking. Because of the need to extrapolate from miners to the general population, with a lower proportion of ever-smokers, uncertainties in the characterization of the combined effect of smoking and radon exposure can affect estimates of overall and average population risks. Uncertainty in the combined effect has particular implications for the attribution of cases into ever-smoking and never-smoking groups. In the United States, about half of persons who have ever smoked have now stopped (USDHHS 1990). However, we did not calculate risks posed by radon exposure of former smokers, because we lacked data on changes in lung-cancer mortality in miners in relation to age and time since stopping smoking.

## UNCERTAINTY ANALYSIS

The previous section identified factors that can contribute to uncertainty in radon risk estimates. In addition to a qualitative discussion of sources of uncertainty, it is desirable to quantify the extent to which those sources contribute to uncertainty in lung-cancer risk estimates. Risk analysts recognize the importance of addressing uncertainty in risk assessment (NRC 1994b) and have developed methods for quantifying uncertainties (Bartlett and others 1996). Hoffman and

Hammonds (1994) have recently applied such methods in assessing the cancer risks associated with radionuclide exposure.

Methods for quantitative uncertainty analysis are discussed in detail in appendix A, including the general framework for the analysis of uncertainty developed by Rai and others (1996). That framework is applicable when the risk depends on a series of risk factors $X_1, \ldots, X_p$, each of which can be subject to uncertainty and might vary among individuals in the population at risk. Some risk factors, such as body weight, might be subject to little uncertainty but vary widely in the population of interest. Other risk factors, such as genetic susceptibility to particular types of cancer, might vary little in the population but be subject to appreciable uncertainty. If the uncertainty and the variability in each of the risk factors can be specified, the overall effect of uncertainty and variability in risk can be evaluated. It is also possible to identify which risk factors contribute most to overall uncertainty and variability.

Although not all potential sources of uncertainty were quantified, the committee conducted a number of limited analyses, which proved to be informative. A complete quantitative analysis of all sources of uncertainty in factors affecting radon lung-cancer risk is not feasible for two principal reasons. First, it is difficult to enumerate all factors that may influence the lung-cancer risk associated with environmental exposures to radon. Second, characterization of the extent of both interindividual variability and uncertainty of some of these factors may not be possible using existing information.

To address uncertainty in radon risk estimates, the committee focused on those factors included in the committee's two preferred models: the exposure-age-concentration model and the exposure-age-duration model. In addition to the four risk factors considered in each of these two models, the committee evaluated the impact of uncertainty in the K-factor on the uncertainty range of AR estimates.

The committee used the methods proposed by Rai and others (1996) to evaluate uncertainty in radon risk estimates. That approach is applicable whenever the risk can be expressed as a function $H(X_1, \ldots, X_p)$ of $p$ risk factors $X_1, \ldots, X_p$, and when both uncertainty and variability in each factor can be specified. The analysis is considerably simpler when the function $H$ is multiplicative, with $H = X_1 x \ldots x X_p$.

For purposes of the present report, the committee focused its attention on a quantitative uncertainty analysis of the population attributable risk, this being of most interest from the public health point of view. Because the AR is a measure of population rather than individual risk, the inter-individual variability in lung-cancer risk is effectively averaged out in this analysis. In other words, since the AR is calculated by jointly integrating over the distribution of radon exposures and the distribution of $K$ among individuals in the population, the AR is subject to uncertainty, but not variability.

The population attributable risk depends on the risk model used to describe the exposure-response relationship between radon and lung cancer, the distribution of radon concentration in U.S. homes, and the dosimetric K-factor. The methods adopted by the committee to evaluate uncertainty in the AR require specification of the prior uncertainty in each of the factors affecting risk. As detailed in appendix A, the uncertainty in the parameters in the BEIR VI risk models was described by log-normal distributions, with dispersion at least as great as the sampling error in the estimated parameter value. Variability among radon concentrations in U.S. homes was characterized by a log-normal distribution used to describe the results of the National Residential Radon Survey; uncertainty in individual radon measurements (measurement error) was not addressed in this analysis. Variability in the K-factor was also described by a log-normal distribution based on a sample of observations in U.S. homes; uncertainty in K-factors for specific homes was described by a log-uniform distribution. Although the committee exercised some judgment in specifying distributions, the result represents a best attempt to allow for some degree of uncertainty in a number of the critical factors affecting the AR.

Because the AR is not a simple multiplicative function, Monte Carlo methods were used to evaluate uncertainty under the committee's preferred risk models. As shown in Figure 3-3, this analysis leads to an uncertainty distribution reflecting the likelihood of different possible values for the AR, centered roughly at the best estimates of the AR given in Table 3-7. Because the uncertainties in the model parameters are largely statistical, the uncertainty distributions reflecting only uncertainty in the parameters of the committee's risk models (Figure 3-3, case I) can be used to obtain approximate confidence intervals for the AR. For males (Figure 3-3a), 95% of the mass of this distribution falls in the range 0.09-0.24 for the exposure-age-concentration model, and 0.07-0.16 for the exposure-age-duration model. For females (Figure 3-3b), the corresponding limits are similar: 0.10-0.26 for the exposure-age-concentration model and 0.08-0.18 for the exposure-age-duration model. In this analysis, a constant value of K = 1 was used.

A second uncertainty analysis was conducted in which variability in K was taken into account (Figure 3-3, case II). Allowing for variability in K does not increase the dispersion of the uncertainty distribution for the AR, but does shift the distribution to the right. For males, the 95% uncertainty intervals for the exposure-age-concentration and the exposure-age-duration models were 0.10-0.22 and 0.08-0.18, respectively. For females, the corresponding limits were 0.10-0.28 and 0.08-0.19, respectively.

The final analysis of uncertainty in the AR further acknowledged uncertainty in the observed radon concentrations in U.S. homes as well as uncertainty in K (Figure 3-3, case III). This increased the range of uncertainty in the AR under both risk models. For males, the 95% uncertainty intervals were 0.10-0.26 for the exposure-age-concentration model and 0.08-0.19 for the exposure-age-duration model. For females, the corresponding limits were 0.10-0.28 and 0.09-0.29, respectively.

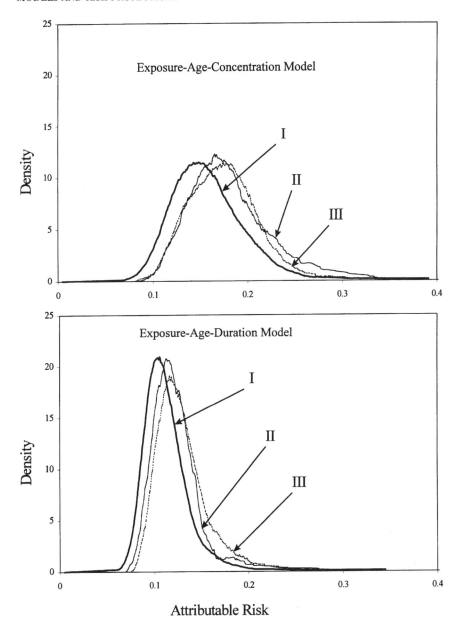

FIGURE 3-3a   Uncertainty distributions for the population attributable risk (AR) for males.  I: uncertainty in model parameters.  II: uncertainty in model parameters; variability in K; variability in radon levels.  III: uncertainty in model parameters; uncertainty/variability in K; variability in radon levels.

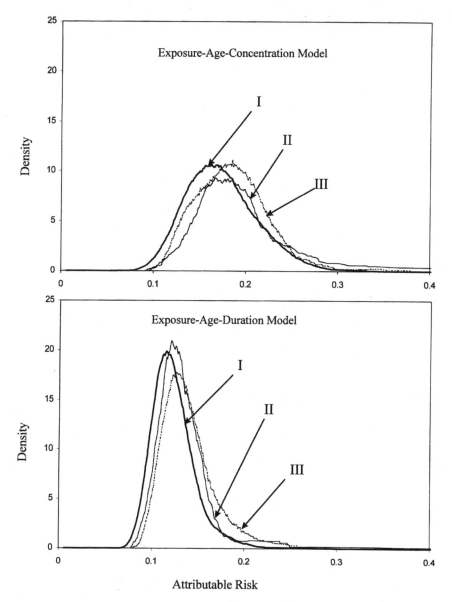

FIGURE 3-3b Uncertainty distributions for the population attributable risk (AR) for females. I: uncertainty in model parameters. II: uncertainty in model parameters; variability in K; variability in radon levels. III: uncertainty in model parameters; uncertainty/variability in K; variability in radon levels.

The uncertainty in the estimated values of the AR gives bounds to predictions of the number of lung-cancer cases attributable to residential radon exposure in the United States. For the exposure-age-duration model, the approximate 95% confidence limits on the AR imply a range of 11,400-26,200 lung-cancer cases for males and females combined. This range reflects statistical uncertainty in the central estimate of 15,400 cases given in Table 3-10. For the exposure-age-concentration model, the corresponding range is 14,800-38,600 cases, with a central estimate of 21,800 cases. The uncertainty ranges that take into account variability in K (case II in Figure 3-3) and uncertainty/variability in K (case III in Figure 3-3) are comparable in width but shifted slightly to the right because of the off-centering effect apparent in Figure 3-3 when variability in K is incorporated in the analysis. It is important to note that although these uncertainty limits encompass 95% of the mass of the uncertainty distributions in Figure 3-3, the uncertainty distributions place most of the mass nearer to the central values, indicating that values closer to the center of the distribution are most likely.

Because both the exposure-age-duration and exposure-age-concentration models fit the miner data equally well, the committee was unable to express a preference for either model. However, the committee noted that in as much as these 2 models are based on exposure levels in mines that generally exceed those in homes, model-based projections of the number of lung-cancer cases due to the presence of radon in U.S. homes are appropriate only if the models apply equally well at residential exposure levels.

To address that issue, the committee also calculated 95% uncertainty intervals for the projected number of lung-cancer cases attributable to residential radon exposure by using the constant-relative-risk (CRR) model restricted to exposures less than 0.175 Jhm$^{-3}$ (50 WLM) (Table 3-9). Because this simple CRR model involves only a single unknown parameter $\beta$ (estimated to be 0.0117/WLM), 95% confidence limits on $\beta$ (0.002-0.225/WLM) can be used to obtain corresponding confidence limits on the AR. This simple uncertainty analysis, which focuses on the subgroup of miners with exposure levels closest to those in U.S. homes, provided 95% confidence limits of 3,300-32,600 lung-cancer cases about the central estimate of 17,500 cases based on the estimates of the AR given in Table 3-9. Although these confidence limits are wider than those based on the committee's 2 preferred models because of the smaller sample, the CRR model is based on observations closest to residential exposure levels.

As discussed previously and in appendix A, other factors might contribute to uncertainty beyond those included in this analysis. Nonetheless, this limited analysis does indicate that the population AR of lung cancer due to radon in homes is subject to considerable uncertainty. The committee acknowledges that this analysis of uncertainty and variability depends on the specific assumptions made about uncertainty and variability in each of the factors affecting the AR. Because characterization of variability and especially uncertainty in the factors is difficult, these particular assumptions reflect to a large extent the committee's best judgement.

## COMPARISONS WITH BEIR IV

The BEIR VI committee's risk models are closely related to the model developed in the BEIR IV report. The BEIR IV committee combined data from 4 cohort studies of underground miners (Colorado Plateau, Ontario, Sweden, and Beaverlodge studies) and applied Poisson regression methods in model fitting. The starting point for the BEIR VI committee was the recent pooling of 11 studies by Lubin and others (1994a, 1995a), which included the same or updated data from the original 4 cohorts and data from seven additional cohorts. Closely comparable statistical methods were applied by both the BEIR IV and BEIR VI committees.

## BEIR IV AND BEIR VI RISK MODELS

The committee's models are a direct extension of the BEIR IV model, which included parameters for time since exposure and attained age, but not exposure rate or exposure duration, as in the BEIR VI models. The form of the BEIR IV model is obtained from the BEIR VI models by setting $\theta_{15-24} = \theta_{25+}$, $\phi_{65-74} = \phi_{75+}$ and $\gamma_z = 1$, that is,

$$RR(w^*) = 1 + \beta \, w \, \phi_{age}. \tag{3}$$

Parameter values for the BEIR IV model and the BEIR VI models show the same general declining patterns for increasing time since exposure and attained age. The decline in the ratio of ERR per unit exposure with attained age, however, is more pronounced in the current models. Note that the values for $\beta$ are not directly comparable, because the values reflect different baseline levels due to the inclusion of different modifying factors in the BEIR IV and BEIR VI models.

The overall ERR per unit of exposure in the absence of all modifying factors is not an adequate description of the relative risk from the miner studies and should not be used for formal comparisons. Nevertheless, the estimate of the overall ERR/Jhm$^{-3}$ for the 4 cohorts used in the BEIR IV report was 3.8 Jhm$^{-3}$ (0.0134/WLM), whereas the value was 1.4 Jhm$^{-3}$ (0.005/WLM) for the pooled analysis of the 11 miner cohorts (Lubin and others 1994a). Those values afford a somewhat crude comparison, suggesting that the combined risk for miners in the 11 studies was less than the BEIR IV estimate. Figures 3-4 through 3-6 show more-direct comparisons of estimated LRR for selected exposure patterns in miners. Figure 3-4 shows LRRs by exposure rate from 0-5.95 Jm$^{-3}$ (0-10 WL) for 5, 10, and 20 yr of exposure. BEIR IV estimates of LRRs for exposure rates of 0.60 Jm$^{-3}$ (1.0 WL) and greater were higher than estimates from current models. Similar patterns are seen in Figure 3-5, which shows LRRs by duration

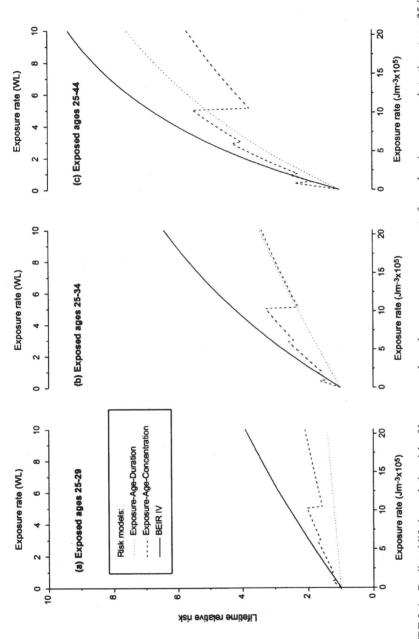

FIGURE 3-4 Predicted lifetime relative risk of lung cancer by radon progeny exposure rate for male miners exposed starting at age 25 for 5, 10, and 20 years.

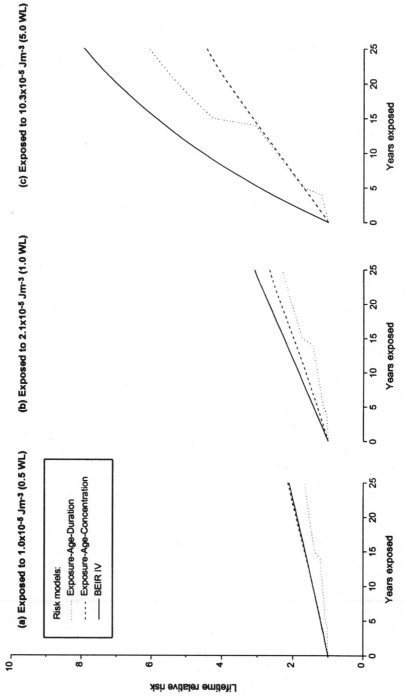

FIGURE 3-5  Predicted lifetime relative risk of lung cancer by duration of radon-progeny exposure for male miners exposed at various rates.

of exposure for exposures at constant rates of 0.30 Jm$^{-3}$ (0.5 WL), 0.60 Jm$^{-3}$ (1.0 WL), and 2.98 Jm$^{-3}$ (5.0 WL).

Figure 3-6 provides comparisons of the various projections of LRR for lifetime exposure to radon at concentrations found in homes. A K-factor of 1.0, an equilibrium ratio of 0.4, and 70% home occupancy are assumed. Slightly higher LRRs are estimated with the exposure-age-concentration model than with the exposure-age-duration and BEIR IV models; however, estimates of LRRs are generally similar for concentrations of 1000 Bqm$^{-3}$ (27.03 pCiL$^{-1}$) and below, levels that include the large majority of dwellings.

## SUMMARY AND CONCLUSIONS

Radon is one of the most extensively studied known human carcinogens. The series of cohort mortality studies of underground miners in countries throughout the

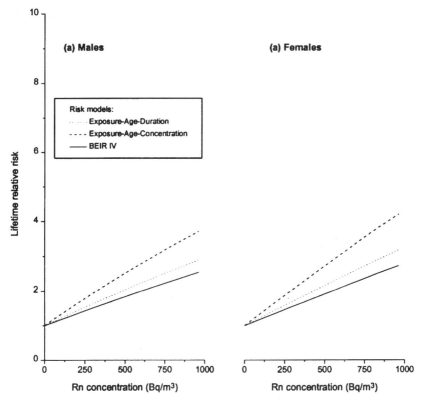

FIGURE 3-6 Predicted lifetime relative risk of lung cancer for males and females by "residential" radon concentration. Exposure occurs over a lifetime at a constant radon concentration.

world is highly informative with respect to the risk of lung cancer associated with exposure to radon. In each of those studies, miners have been shown to be at excess risk for lung cancer under past conditions of exposure. The quantitative estimates of exposures experienced by the miners, although subject to error, allow characterization of exposure-response relationships for radon and lung cancer. These data formed the basis for the development of the committee's risk models.

Case-control studies of residential radon exposure and lung cancer have also been conducted in various countries. Although also informative, the lower exposures of people in these studies and methodologic problems make it very difficult to identify the relationship between residential radon exposure and lung-cancer mortality in an individual study. However, the estimate of lung-cancer risk based on a recent meta-analysis of these 8 studies is in close agreement with the risk predicted on the basis of miner data.

The committee was fortunate to have available an update of the data on the 11 miner cohorts previously analyzed by Lubin and others (1994a). The most-recent data were used in developing the committee's risk models. The committee recognized that great care is needed in combining data from different cohorts of underground miners around the world. The levels of exposure to radon and other relevant covariates, such as arsenic and tobacco smoke, differed appreciably among groups of miners. The completeness and quality of the data available on relevant exposures also differed notably among the cohorts. Information on tobacco consumption was available for only 6 of the 11 cohorts; of these 6, only 3 had information on duration and intensity of exposure to tobacco smoke. Lifestyle and genetic factors that influence susceptibility to cancer might also account for heterogeneity among cohorts.

Despite those differences, the committee concluded that the best possible estimate of lung-cancer risk associated with radon exposure would be obtained by combining the available information from all 11 cohorts in a judicious manner. The committee used statistical methods for combining data that both allowed for heterogeneity among cohorts and provided an overall summary estimate of the lung-cancer risk. Confidence limits for the overall estimate of risk allow for such heterogeneity.

The committee's risk models described the ERR as a simple linear function of cumulative exposure to radon, allowing for differential effects of exposure during the periods 5-14 years, 15-24 years, and 25 years or more before lung-cancer death. The most weight was given to exposures occurring 5-14 years before death from lung cancer. The committee entertained 2 categorical risk models in which the ERR was modified either by attained age and duration of exposure or by attained age and exposure rate. The ERR decreased with both attained age and exposure rate and increased with duration of exposure. For cumulative exposures below 0.175 $Jhm^{-3}$ (50 WLM), a constant-relative-risk model without these modifying factors appeared to fit the data as well as the 2 models that allow for effect modification.

Lung-cancer risks associated with radon exposure were characterized in several ways. The LRR was used to describe the lifetime risk of lung cancer among people continually exposed to radon throughout the course of their lifetime relative to the risk among unexposed individuals.

The percentage of lung-cancer cases that can be attributed to residential exposure to radon is of particular interest for risk management. The committee used data from the National Residential Radon Survey in combination with its 2 categorical risk models to estimate the AR posed by residential radon exposures. The ARs were estimated to be in the range 10-15%. These estimates are somewhat higher than the estimate of about 8% based on the data and methods of BEIR IV. About 30% of the AR was associated with homes having concentrations above 148 Bqm$^{-3}$ (4 pCiL$^{-1}$).

Although the AR percentages were comparable for ever-smokers and never-smokers under the multiplicative model, the number of radon-related lung-cancer cases was much higher among ever-smokers than for never-smokers under the multiplicative model. Of the approximately 157,000 lung-cancer deaths occurring annually, radon was estimated to play a role in about 15,000 to 22,000 cases. Of these, 13,000 to 19,000 were in ever-smokers and 1,000 to 3,000 in never-smokers, depending on the choice of the model. These computed values represent the best estimates of the lung-cancer risk attributable to radon that can be made at this time.

The committee recognized that these estimates are subject to uncertainty, including kinds of uncertainty that are not captured by statistical confidence limits on risk estimates. Consequently, the committee attempted a quantitative analysis of the uncertainty associated with estimates of the population AR. This analysis was itself limited, inasmuch as characterization of such sources of uncertainty as exposure measurement error in the miner data is difficult. Using data whenever possible and expert judgment otherwise, the committee attempted to describe the sources of uncertainty in its 2 categorical risk models.

The best estimates of the population AR were in the range 10 to 14% on the basis of the committee's preferred risk models. The quantitative analysis conducted by the committee provided limits within which the AR was considered to lie with 95% certainty. For the exposure-age-concentration model, the uncertainty interval ranged from about 9 to 25%, with central estimates of about 14%. This reflects a substantial degree of uncertainty in the AR, although the uncertainty distributions indicated that values near the central estimates were much more likely than values near the upper and lower limits. For the exposure-age-duration model, the AR ranged from 7 to 17%, and centered at about 10%. The committee also computed uncertainty limits for the simple constant-relative-risk model fitted to the miner data below 0.175 Jhm$^{-3}$ (50 WLM), which is based on observations in miners closest to residential exposure levels. The latter analysis, which minimizes the degree of extrapolation outside the range of the miner data, led to uncertainty limits of 2-21%, with a central estimate of about 12%.

The committee also noted that its quantitative estimates of risks posed by residential radon exposure depend strongly on the assumption of a linear relationship, without a threshold, between low-dose exposure to radon and risk. As reviewed in chapter 2, that assumption is based on our current understanding of the mechanisms of radon-induced lung cancer, although it is recognized that this understanding is incomplete. The committee did not attempt to quantitatively address uncertainty due to the linear, no-threshold assumption, because specific mechanistically plausible alternative dose-effect relationships were not identified in the committee's review.

Despite the uncertainty, the committee concluded that the weight of evidence of the available data supports a finding that residential-radon exposure increases lung-cancer risk. The best estimate of risk that can be obtained at this time is based on the committee's analysis of the combined updated data on the 11 cohorts of underground miners. The committee noted that this estimate is consistent with that derived from a recent meta-analysis of summary relative risks from the 8 residential case-control studies conducted to date.

The committee questioned whether further residential studies are likely to clarify the uncertainty surrounding residential-radon lung-cancer risks. The case-control studies conducted to date have been somewhat inconsistent. While most are compatible with the hypothesis of an elevated cancer risk, their results could also be interpreted as compatible with the hypothesis of no increase in risk in light of inherent uncertainties in data. Further, the residential studies offer only very limited information on never-smokers. Clear evidence of an increased lung-cancer risk in miners, many of whom were exposed to levels of radon only about two-fold higher than associated with residence in some homes, played an important role in supporting the committee's conclusions about the likelihood of an increased lung-cancer risk due to residential radon exposures. The committee agreed with the recommendations of workshops conducted by the Department of Energy and the Commission of European Communities: further studies should not be initiated until studies now in progress are completed and the data are pooled from these studies and studies already completed are pooled.

The committee examined the effect of reductions in radon levels in U.S. homes on lung-cancer risk, assuming different scenarios of the efficiency of reduction. On the basis of the committee's categorical risk models, reducing radon concentration in all homes that are above 148 $Bqm^{-3}$ (4 $pCiL^{-1}$) to below 148 $Bqm^{-3}$ (4 $pCiL^{-1}$) is estimated to result in the avoidance of about 3 to 4% of lung cancers.

# 4

# Health Effects of Radon Progeny
# on Non-Lung-Cancer Outcomes

This report focuses on the lung-cancer risk associated with radon progeny. However, other health effects of exposure to radon progeny and to uranium mining in general have also been of concern. The health effects mentioned in the literature are nonmalignant respiratory diseases in underground miners, cancers other than lung cancer in miners and in the general population, and adverse reproductive outcomes of pregnancies in the wives of uranium miners and in communities adjacent to areas where uranium is mined and milled. The BEIR IV report addressed evidence available through 1987 on these potential health effects. Its appendix V, on nonmalignant respiratory and other diseases in miners, also covered effects on reproductive outcomes.

The evidence available to the BEIR IV committee on each of those issues was sparse. For nonmalignant respiratory diseases, animal studies provided some relevant findings, and a few epidemiologic studies had been reported. The committee noted that silicosis had been documented in silica-exposed populations of uranium miners; the evidence in other fibrotic lung disorders, particularly interstitial fibrosis, was limited. With regard to reproductive outcomes, the BEIR IV report found the data to be "sparse" and characterized the associations as "weak." The report provided little comment on cancers other than lung cancer.

Since the publication of BEIR IV, additional data on risks of cancers other than lung cancer have become available from an analysis of pooled data in underground miners (Darby and others 1995). Reports based on ecologic analyses have suggested that indoor radon is associated with diverse malignancies, including leukemias and some solid cancers (Henshaw and others 1990). Only limited information has been reported on nonmalignant respiratory disease and

*117*

reproductive outcomes. This chapter updates the BEIR IV report's coverage of outcomes other than lung cancer.

## DOSES TO ORGANS OTHER THAN LUNG

Interpretation of the epidemiologic findings on effects of radon other than in the lungs needs to be based on an understanding of the dosimetry of radon and its progeny. Dosimetric models used for this purpose have extended beyond deposition of progeny in the lung to distribution to and absorption in other organs. Although there has been little work on this aspect of radon dosimetry, the models suggest potentially important doses to skin and lymphocytes.

Table 4-1 shows estimated annual absorbed doses to various adult tissues from Rn-222 and its short-lived daughters, for a domestic concentration of 20 Bqm$^{-3}$ (0.54 pCiL$^{-1}$). The estimated dose for the basal cells in exposed skin depends heavily on assumed deposition velocity. The upper estimates are comparable with lung doses and might be responsible for a substantial number of skin cancers, most of which would be nonfatal. With assumptions of 125 $\mu$Gy y$^{-1}$ to exposed skin of the face and neck, 7.5 $\mu$Gy y$^{-1}$ to skin of all other regions, an RBE of 5 for alpha-

TABLE 4-1   Estimated annual absorbed doses to adult tissues from Rn-222 and its short-lived progeny for domestic radon concentration of 20 Bqm$^{-3}$ (0.54 pCiL$^{-1}$)

| Tissue | Annual dose ($\mu$Gy y$^{-1}$) | Reference |
|---|---|---|
| Lung | 500 | ICRP65 1993; Simmonds and others 1995 |
| Bronchial basal cell nuclei | 740 | James, personal communication |
| Bronchial secretory cell nuclei | 1,680 | James, personal communication |
| Skin (basal cells at 50 $\mu$m | 50-1,000 | Harley and Robbins 1992 |
| in exposed skin) | 85-850 | Eatough and Henshaw 1992 |
| Red marrow | 0.5-1.3 | Harley and Robbins 1992 |
| | 4-6 | Richardson and others 1991 |
| | 4.5 | Simmonds and others 1995 |
| | 2.2-2.6 | James 1992 |
| Bone surface | 0.4-1 | Harley and Robbins 1992 |
| | 1.2 | ICRP65 1993; Simmonds and others 1995 |
| | 4.4 | James 1992 |
| Breast | ~ 1.5 | Harley and Robbins 1992 |
| | 1.2 | ICRP65 1993; Simmonds and others 1995 |
| T-lymphocytes: | | |
| circulating | 1 | Harley and Robbins 1992 |
| in bronchial epithelium | 1,000 | Harley and Robbins 1992 |
| Blood | 1.1 | Harley and Robbins 1992 |
| Liver | 2.5 | James 1992 |
| Kidney | 14.4 | James 1992 |

particles, and ICRP59 risk factors for low-LET radiation, Eatough and Henshaw (1995) have estimated that about 1-10% of nonmelanoma skin cancers in the UK can be theoretically associated with domestic radon, including possible interaction with UV radiation. Calculated doses from thoron, available only for lung and bone marrow, tend to be lower than those from radon (Table 4-2).

Doses to red marrow from radon are substantially lower but nevertheless lead, by conventional risk estimation, to expectations of contributions to the natural incidence of leukemia. Richardson and others (1991) suggested that about 6-12% of myeloid leukemia in the U.K. could be from radon. Similarly, Simmonds and others (1995) have applied age-dependent radon doses and risks to a national study population of children in Seascale, U.K. From their evaluations, it can be deduced that some 34% of the expected natural childhood-leukemia incidence is attributable to natural radiation, made up of about 20% from natural low-LET radiation and about 14% from natural high-LET radiation (Simmonds and others 1995; COMARE 1996). About 3% of the 14% is due to radon and thoron. James (1992) has estimated from ICRP (1991) risk factors that 1.25% of the additional cancers caused by indoor radon and thoron are expected to be leukemia; the vast majority (98%) of the additional cancers are expected to be in the lung. However, no excess leukemia has been observed in the studies of the mining cohorts.

## NONMALIGNANT RESPIRATORY DISEASES

In addition to radon, potential causes of nonmalignant respiratory diseases in the underground miners include silica (well documented as causing silicosis) γ, blasting fumes, and, in some mines, diesel exhaust. Persons with silicosis are at

TABLE 4-2   Estimated doses to adult tissues from thoron and its short-lived decay products for typical domestic thoron concentration corresponding to $^{212}$Pb-progeny concentration of 0.3 Bqm$^{-3}$ (0.008 pCiL$^{-1}$)

| Tissue | Annual dose ($\mu$Gy y$^{-1}$) | Reference |
|---|---|---|
| Lung | 34 | Simmonds and others 1995 |
| Red marrow | 1.5 | Stather and others 1986 |
| | 0.5 | James 1992[a] |
| Bone surface | 22 | Simmonds and others 1995 |
| | 3.0 | James 1992[a] |
| Liver | 1.7 | James 1992[a] |
| Kidney | 10 | James 1992[a] |

[a]The doses presented here are one third of those tabulated in James (1992), so as to correspond to a typical PAEC for indoor thoron progeny of 21 nJm$^{-3}$ (1 mWL) (Nero and others 1990) and $^{212}$Pb concentration of 0.3 Bqm$^{-3}$ (0.008 pCiL$^{-1}$) (James, personal communication).

increased risk for tuberculosis and the development of silicotuberculosis, a fi-brotic disorder. On reviewing the epidemiologic evidence on nonmalignant respiratory diseases, the BEIR IV committee (NRC 1988) concluded that the effects of the various inhaled agents in the mines could not be readily separated, and it did not attribute nonmalignant respiratory diseases in underground miners to radon specifically. The principal evidence considered came from surveys of Colorado Plateau and New Mexico uranium miners.

A case series recently reported by Archer (1996) provides evidence that uranium miners may develop a fibrotic lung disease distinct from silicosis. About 400 uranium miners had been referred to Archer and colleagues for evaluation of lung disease and of exposure, either for special studies or for the purpose of seeking compensation, and 22 of these were selected as possible examples of radiation-related diffuse interstitial fibrosis. All had evidence of diffuse abnormalities in both lower lung fields. Lung-biopsy specimens from five of the 22 were examined. The findings were not those of silicosis. The 5 men had evidence of interstitial fibrosis and honeycombing, a pattern of destruction found in advanced lung disease. Birefringent crystals consistent with silica were present in trace quantities at most. Archer and others conclude that there is a radiation-caused diffuse interstitial fibrosis in uranium miners.

The New Mexico Miners' Outreach Program provides free screening to active and retired miners for mining-related disease (Mapel and others 1996). Cross-sectional data from 1,359 former uranium miners were analyzed by Mapel and colleagues, who examined predictors of lung function and of radiographic abnormality with regression methods. Duration of underground mining was negatively associated with lung function, but the association was statistically significant only for the American Indian miners. Underground uranium-mining experience was a significant predictor of having pneumoconiosis for Hispanic and American Indian former uranium miners. Quantitative estimates of radon-progeny exposure were not available, so the investigators could not assess further the contribution of radon-progeny exposure to the reduction of lung function associated with duration of underground mining.

## MALIGNANCIES OTHER THAN LUNG CANCER

The dramatic excess of lung cancer in underground miners exposed to radon progeny has focused emphasis in research, in risk assessment, and in risk management on this cancer. At the time of the BEIR IV report, there was limited information on cancers of sites other than the lung, and it came primarily from the cohort study of Colorado Plateau miners (Waxweiler and others 1976). Although statistical power related to this single cohort was limited, cancers were not in apparent excess for sites other than the lung (Waxweiler and others 1976). Nor had understanding of radon dosimetry led to concern about cancers at other sites.

Consequently, the findings of general-population-based ecologic studies indicating associations between population estimates of indoor radon exposures and rates for cancers other than lung cancer were not expected, and they were not readily explicable on a biologic basis. A 1989 letter to the editor by Lucie (1989) described an association between county-level leukemia incidence and average radon concentration. A 1990 report by Henshaw and colleagues (1990), indicating that radon might be associated with myeloid leukemia, cancer of the kidney, melanoma, and some childhood cancers brought public and scientific concern to the issue. Henshaw and collaborators conducted an ecologic analysis of cancer rates and estimated background radon exposure; positive and statistically significant associations were found. However, after those publications, methodologic biases affecting interpretation of the results were noted, and correspondence called for caution in interpreting the results. Furthermore, the highest radon levels in that report were quite low (near average indoor concentrations) and the study is subject to the flaws frequently associated with ecologic studies as described in appendix G. Additional ecologic analyses were reported, and case-control studies are now in progress. These findings provide a rationale for further consideration of malignancies other than lung cancer in the underground miner data, and Darby and others have evaluated these malignancies through combined analyses of data from 11 studies of underground miners.

## Studies of Underground Miners

Since the BEIR IV report, initial analyses of data on a number of cohorts of underground miners have been reported, thereby increasing the extent of information available on malignancies other than lung cancer. Reports of several studies have provided information on all malignancies (Morrison and others 1988; Darby and others 1995; Tomášek and others 1993; Tirmarche and others 1993). Tomášek and others (1993) reported on the Czech cohort of 4,320 miners. At an average of 25 years of followup, statistically significant excesses of deaths from cancers of the liver, gallbladder, and extrahepatic bile ducts were noted, but there were no excess deaths for other individual sites. There was not a significant excess of all cancers other than lung cancer (observed:expected ratio, or O/E, = 1.11; 95% confidence interval, CI 0.98-1.24). In the Newfoundland fluorspar miners, excess occurrence of deaths from cancers of the buccal cavity and pharynx and of the salivary glands was noted, but small numbers of cases limited dose-response analysis and interpretation of the findings (Morrison and others 1988). Tirmarche and others (1993) found significant excess occurrence of cancer of the larynx (O/E = 2.35; 95% CI, 1.37-3.76).

In the Cornish tin miners, there were nonsignificant excesses of deaths from cancer of the stomach (O/E, 1.41) and from leukemia (O/E, 1.73). These analyses of data from cohorts did not indicate any consistent patterns across the cohorts, and the possibility of false-positive findings when several associations are investigated in several cohorts makes the results difficult to interpret.

The most-informative analysis on cancers other than lung cancer was based on pooling of data from 11 studies of underground miners (Darby and others 1995). Darby and colleagues assembled data from 11 cohorts—10 of those included in the lung cancer pooled lung-cancer analysis and a replacement of the Radium Hill cohort with the Cornish tin miners because of the small number of deaths and incomplete follow-up in the Radium Hill group. For each cohort, the expected number of deaths was calculated on the basis of external comparison rates; the external comparisons excluded the China cohort because appropriate national or regional rates were not available. In addition, internal comparisons were made by evaluating the association for specific cancers with cumulative exposure to radon progeny. Both types of comparisons were carried out separately for time since first employment categories of less than 10 years and equal to or greater than 10 years. External comparisons were also made for the combined periods.

Overall, there was no excess of deaths from cancers other than lung cancer (O/E, 1.01; 95% CI, 0.95-1.07). Of the 28 individual cancer categories evaluated, there were significant excesses for stomach cancer (O/E, 1.33; 95% CI, 1.16-1.52) and for primary liver cancer (O/E, 1.73; 95% CI, 1.29-2.28) (Table 4-3). The ratios of observed to expected deaths were greater than unity for all leukemias combined and for each of the principal subtypes, although none of the excesses was statistically significant. Leukemia also showed a significant increase when analyses were restricted to less than 10 years since first employment. There were statistically significant decreases for some sites.

In comparisons by extent of exposure, none of the cancers showing excesses (stomach, liver, and leukemia) was positively associated with exposure, so increased rates for these cancers were judged unlikely to be related to exposure to radon and radon progeny. For the leukemia elevation in the first 10 years since start of employment, the possibility of exposure to gamma radiation in the mines was considered.

Both cancer of the pancreas and the category of all cancers other than lung showed statistically significant positive associations with cumulative exposure, although the latter association was found only for the first 10 years since start of employment. The finding for cancer of the pancreas was interpreted as likely to be spurious. The finding for all cancers other than lung was attributed to 2 deaths in workers with cumulative exposure exceeding 5.25 Jhm$^{-3}$ (1,500 WLM); these deaths were in the "other and unspecified cancers" category, and review of available records resulted in the conclusion that the true underlying cause of death was very likely lung cancer. With deaths in the "other and unspecified" category excluded, the category of all cancers excluding lung cancer was not found to be significantly associated with cumulative Jhm$^{-3}$ (WLM).

Overall, the study of Darby and others (1995) found no substantial evidence

TABLE 4-3    Numbers of deaths observed (O), ratio of observed to expected deaths (O/E), and 95% confidence interval (CI) for selected sites of cancer, analysis of pooled data on miners (all studies except China), by time since first employment

| Cancer site (ICD-9 code) | O | O/E[a] | 95% CI |
|---|---|---|---|
| Tongue and mouth (141, 143-145) | 11[b] | 0.52 | 0.26-0.93 |
| Salivary gland (142) | 4 | 1.41 | 0.39-3.62 |
| Pharynx (146-149) | 9[c] | 0.35 | 1.16-0.66 |
| Esophagus (150) | 45 | 1.05 | 0.77-1.41 |
| Stomach (151) | 217[c] | 1.33 | 1.16-1.52 |
| Colon (152-153) | 95[b] | 0.77 | 0.63-0.95 |
| Rectum (154) | 60 | 0.86 | 0.66-1.11 |
| Liver, primary (155.5, 155.1) | 50[c] | 1.73 | 1.29-2.28 |
| Liver, unspecified (155.2) | 3 | 0.43 | 0.09-1.26 |
| Gallbladder (156) | 19 | 1.23 | 0.74-1.92 |
| Pancreas (157) | 91 | 1.05 | 0.85-1.29 |
| Nose (160) | 3 | 0.69 | 1.14-2.02 |
| Larynx (160) | 38 | 1.21 | 0.86-1.67 |
| Bone (170) | 10 | 1.04 | 0.50-1.91 |
| Connective tissue (171) | 5 | 0.82 | 0.27-1.91 |
| Malignant melanoma (172) | 18 | 0.92 | 0.54-1.45 |
| Other skin (173) | 9 | 1.60 | 0.73-3.03 |
| Prostate (185) | 83 | 0.88 | 0.70-1.09 |
| Testis (186) | 6 | 0.72 | 0.26-1.57 |
| Bladder (188, 189.3-189.9) | 39 | 0.85 | 0.61-1.16 |
| Kidney (189.0-189.2) | 44 | 0.91 | 0.66-1.22 |
| Brain and central nervous system (191, 192) | 52 | 0.95 | 0.71-1.25 |
| Thyroid gland (193) | 2 | 0.47 | 0.06-1.71 |
| Non-Hodgkin's lymphoma (200, 202) | 36 | 0.80 | 0.56-1.10 |
| Hodgkin's disease (201) | 17 | 0.93 | 0.54-1.48 |
| Multiple myeloma (203) | 26 | 1.30 | 0.85-1.90 |
| Leukemia (204-208) | 69 | 1.16 | 0.90-1.47 |
| Leukemia excluding chronic lymphatic (204-208 except 204.1)[d] | 36 | 1.11 | 0.78-1.54 |
| Myeloid leukemia (205-206)[d] | | | |
| Acute myeloid leukemia (205.0, 205.2, 206.0, 206.2)[d] | 27 | 1.41 | 0.93-2.05 |
| | 12 | 1.16 | 0.60-2.02 |
| Other and unspecified | 118 | 1.12 | 0.93-1.35 |
| All cancers other than lung (140-161, 163-208) | 1,179 | 1.01 | 0.95-1.07 |

[a]Expected deaths calculated from national or local mortality rates.
[b]$0.05 \geq P > 0.01$
[c]$P \leq 0.001$ (2-sided tests).
[d]For each study, only the period for which the 8th or 9th ICD revisions was in use nationally is included.
Source: Darby and others (1995).

of increased risks of cancers other than lung cancer in the 11 miner cohorts. The authors concluded that the study provided strong evidence that high levels of exposure to radon do not cause a "material risk" of mortality from cancers other than lung cancer and that protection standards for radon should continue to be based on consideration of lung-cancer risk alone.

### Studies of the General Population

The hypothesis that radon might cause cancers other than lung cancer in the general population originated in ecologic studies reported after the BEIR IV report. Those analyses were conducted with units of analysis ranging from counties to countries (Table 4-4). Lucie (1989) published one of the initial reports, showing a positive correlation between acute myelogenous leukemia incidence for counties in the U.K. and county average radon concentration. Henshaw and colleagues (1990) followed with their report which provided estimates of radiation dose to the red marrow that indicated a significant dose to the marrow at typical indoor radon concentrations. For example, at an exposure of about 185 $Bqm^{-3}$ (5 $pCiL^{-1}$), the estimated dose to marrow from radon was estimated to be similar to that from low-LET radiation. Henshaw and colleagues described ecologic correlations between estimated mean exposures of residents of 15 countries and incidence of leukemias, childhood cancers, and selected additional cancers in these countries. Further analyses were reported for provinces of Canada. Henshaw and others did not find correlations for lung cancer or for skin cancer.

Bridges and colleagues (1991) later reported a correlation between frequency of a mutation of the hypoxanthine guanine phosphoribosyl transferase gene (hprt) and indoor radon exposure, supporting the plausibility of the ecologic associations found by Lucie and Henshaw and colleagues. Bridges and colleagues selected 20 persons, mostly never-smokers, from homes that had been monitored for radon; they were selected to provide a range of exposure from below about 111 to 740 $Bqm^{-3}$ (3 to 20 $pCiL^{-1}$). The logarithm of the mutation frequency was significantly associated with radon concentration in the homes, which was measured twice—for 1 month and for 3 months. This result suggested a mutation frequency that was much greater than would be expected from estimated radon dose to blood and previous in vitro hprt mutation data in numerous cell types. However, in a larger followup study by the same group (Cole and others 1996), no significant association was found. The later study was of 65 persons from 41 houses in the same small rural town, with measured radon of 19-484 $Bqm^{-3}$ (0.51-13.08 $pCiL^{-1}$). BCL-2 t (14;18) translocations, chromosomal events associated with leukemia and lymphoma, also showed no association with radon exposure.

Bauchinger and others (1994) performed conventional chromosomal analyses in blood lymphocytes of 25 persons living continuously in houses with indoor radon concentrations exceeding the average in German houses of 50 $Bqm^{-3}$ (1.35

TABLE 4-4  Ecologic studies of cancer other than lung cancer

| Reference | Location | Exposure measure | Outcomes | Findings |
|---|---|---|---|---|
| Lucie 1989 | United Kingdom | County average radon concentration | AML[a] incidences for 23 countries | Positive correlation |
| Henshaw and others 1990 a,b | 15 countries and regions in countries | National survey data | Cancer and leukemia incidences | Positive correlation with myeloid leukemia, melanoma, kidney cancer, prostatic cancer, and some childhood cancers |
| Alexander and others 1990 | United Kingdom | County average radon concentration | Leukemia and lymphoma incidences | Positive correlation with multiple outcome groups |
| Butland and others 1990 | Various countries | County average radon concentration | Incidence, of leukemia and selected cancers | Positive, significant correlation for all childhood cancers |
| Muirhead and others 1994 | United Kingdom | County and district average concentrations from surveys | Incidence, of leukemia and NHL[b] | Variable pattern of correlation comparing county and district-level analyses |

[a]Acute myelogenous leukemia.
[b]Non-Hodgkin's lymphoma.

pCiL$^{-1}$) by a factor of 5-60. The mean frequencies of dicentrics and rings per cell ($1.5 \pm 0.4 \times 10^{-3}$) were significantly higher than control levels ($0.54 \pm 0.11 \times 10^{-3}$). Grouping the persons by estimated cumulative exposure showed a tendency for $\lambda$ an exposure-effect relationship. Estimated radon exposures ranged from 700 to 6300 Bqm$^{-3}$ (18.92 to 170.3 pCiL$^{-1}$). Subsequently, painting of 3 chromosomes by fluorescence in situ hybridization (FISH) techniques was carried out on the same lymphocyte samples (Bauchinger and others 1996). The mean frequency of symmetrical translocations was slightly (1-5 fold), but not significantly ($p < 0.1$), raised in the radon group compared to the controls. Only for males separately did the raised level reach statistical significance. For dicentric chromosomes, scoring of FISH-painted chromosomes gave results that were consistent (3-fold increase) with those previously obtained by the conventional methods. The apparent lower sensitivity of the FISH-translocation measurements to discriminate radon exposure was ascribed to the much higher control frequencies for translocations compared to dicentrics and to the lower doses received by the hemopoietic compartments such as bone marrow that should contribute most to stable symmetrical aberrations, as compared to mature blood lymphocytes that are the direct target cells for observed dicentric aberrations.

The report of Henshaw and colleagues (1990) attracted substantial attention and criticism, as shown in correspondence to the *Lancet* (Mole 1990; Bowie 1990; Prentice and Copplestone 1990; Baverstock 1990; Butland and others 1990) and other journals (Peto 1990; Wolff 1991). Mole (1990) raised questions concerning the dosimetric model, and Butland and others (1990) noted that even Henshaw's high estimate of dose to the red marrow from radon is only about 1% of that to the lung. Butland and colleagues repeated the analysis of Henshaw and colleagues but used the data from cancer registries that they regarded as most satisfactory. They found a significant positive correlation for all childhood cancers combined. Results for other sites and for the leukemias were positive but not statistically significant. However, the numbers of countries were small and power was limited. None of the analyses showed a significant association with lung cancer. Other critics raised concerns about confounding and biologic plausibility. Wolff and Stern (1991) presented analyses that suggested that the observed correlations might have resulted from confounding by socioeconomic factors.

Muirhead and others (1992) conducted an ecologic analysis of childhood leukemia and non-Hodgkin's lymphoma rates based on small areas (districts) of the U.K. and found no significant associations with radon exposure, even though analyses by aggregated areas (counties) showed a significant positive correlation with radon levels. The correlation between districts within counties was negative; that between counties was positive. The Henshaw and others (1990) analysis of U.K. data was similar to that based on the aggregated areas.

For the United States, Cohen (1993) examined ecologic correlations between estimated average radon concentrations in 1,600 counties and cancer mortality, in unspecified years, in males and females. Positive and statistically significant associations were found for a number of sites, including, in men, cancer of the lip, salivary gland, nasopharynx, nose, nasal cavity, middle ear, breast, eye, thyroid, thymus and endocrine glands, and multiple myeloma and, in women, cancer of the salivary gland, nasopharynx, large intestine, liver, gallbladder, bile ducts, larynx, bone, connective tissue, kidney and uterus, eye, thymus, and endocrine glands, and lymphosarcoma, reticular sarcoma, Hodgkin's disease, multiple myeloma, and leukemia. The consequences of adjusting for smoking at the state level were also examined.

Mifune and colleagues (1992) described cancer mortality for 1952-1988 in inhabitants of the Misasa spa area in Japan, comparing standardized mortality ratios based on national data with those in a control area. In the spa area, there are 90 hot-spring sources, and average radon concentration was reported to be 26 $mBqL^{-1}$ in outdoor air and 35 $mBqL^{-1}$ in indoor air. Routes of exposure included use of the hot springs for bathing and the medical treatment of patients. Overall, there was no excess of all cancers, and the risk of lung-cancer death in the Misasa area was only 55% of that in the control area.

## REPRODUCTIVE OUTCOMES

The committee identified only a single new investigation relevant to concern about reproductive outcomes. Shields and others (1992) conducted a case-control study of congenital abnormalities, stillbirths, developmental disorders, and deaths from causes other than injuries. The case series included 266 cases and an equal number of controls with a normal birth. Exposure variables included the occupations of the parents and grandparents; the nearness of the subject's residences to uranium mines, mine dumps, and mill tailings; and living in a home constructed with uranium-mine rock. There was no evident effect of the fathers' being employed in a uranium mine or mill. There was increased risk for adverse pregnancy outcome if the mother lived near tailings or mine dumps. Interpretation of the findings is limited by lack of statistical power, particularly within the categories of specific adverse outcomes.

## CONCLUSIONS

Although it has been nearly 10 years since the BEIR IV committee reviewed the evidence on health effects of radon-progeny exposure other than lung cancer, the database is still limited. Nevertheless, the committee found several conclusions to be warranted. In regard to cancers other than lung cancer, the committee interpreted the pooled analysis reported by Darby and colleagues (1995) as not indicating excess risk for cancers other than cancer of the lung in radon-exposed miners. Although 95% confidence limits are wide for some sites, the data provide evidence that radon and its progeny are not a major cause of nonlung cancers and leukemias in the general population, as suggested by some ecologic studies.

The committee concluded that the findings in the miners could be reasonably extended to the general population; there is no basis for considering that effects would be observed in the range of typical exposures of the general population that would not be observed in the underground miners exposed at generally much higher levels. The studies in the general population are ecologic and subject to many potential biases. The committee agrees with the conclusion of Darby and others that there is no need to consider cancers other than lung cancer in setting protection standards and guidelines for radon. The dose calculations suggest that radon and progeny could contribute to some proportion of skin-cancer cases.

Only 2 new studies had been reported on nonmalignant respiratory diseases. The report from New Mexico again documented that uranium mining adversely affects lung function (Mapel and others 1996). Archer and colleagues (1996) described an intriguing series of cases that support the possibility that exposure to radon progeny cause fibrosis of the pulmonary interstitium, often referred to as pulmonary or interstitial fibrosis. However, this clinical case series is insufficient to establish the link to radon progeny specifically, and there is a need for more research on the persistent question of the existence of radon-related pulmonary fibrosis.

The new case-control study of reproductive outcomes in Shiprock, New Mexico, was limited by sample size and the possibility of measurement error because of the reliance on self-reported exposure measures. The committee was unable to reach any conclusion with regard to adverse effects of radon exposure on reproductive outcome.

# Appendix A

# Risk Modeling and Uncertainty Analysis

## INTRODUCTION

Epidemiologic studies of underground miners exposed to radon have convincingly established that radon-decay products are carcinogenic and that exposure to these products at levels previously found in mines increases lung-cancer risk (NCRP 1984a,b; NRC 1988; Samet 1989; Lubin and others 1994a). Lubin and others (1994a) conducted a pooled analysis of data from 11 major studies of underground miners (identified as the studies from Colorado, Czechoslovakia, China, Ontario, Newfoundland, Sweden, New Mexico, Beaverlodge, Port Radium, Radium Hill, and France). That analysis included over 2,700 lung-cancer cases among 68,000 miners representing nearly 1.2 million person-years of observation. Lubin and others (1994) found that the relationship between the relative risk (RR) of lung cancer and cumulative exposure to radon progeny was generally consistent with linearity within each cohort. However, estimates of the excess relative risk (ERR) due to exposure to radon progeny varied substantially among the cohorts. For example, the ERR following exposure to 100 working level months (WLM) varied from 0.16 in China to 5.06 in Radium Hill. The precision of these estimates was highly variable from cohort to cohort.

The combined effect of radon exposure and tobacco smoke on lung-cancer risk has been discussed in the literature (Lundin and others 1971; Whittemore and McMillan 1983; Moolgavkar and others 1993; Lubin 1994) (See also appendix C.) Although the Colorado Plateau uranium miners study has revealed a synergistic effect between exposure to radon and cigarette smoking, this interaction is not well characterized (NRC 1988; Lubin 1994). Data from two large studies, the

China tin miners study and the Colorado Plateau uranium miners study, indicated that the lung-cancer risk associated with the combined exposure is greater than the sum of the risks associated with each factor individually, evidence of a synergistic effect between radon and tobacco smoke in the induction of lung cancer. Data on tobacco use, available for 6 of 11 cohorts, are summarized in Table A-1. In the Colorado, Newfoundland and New Mexico studies, detailed data on tobacco use including duration, intensity, and cessation are available, whereas studies in China and Radium Hill identify individuals only as ever-smokers or never-smokers.

Since publication of the report by Lubin and others (1994a), studies of the Chinese tin miners, and of the Czech, Colorado, and French uranium miners have been updated or modified (Lubin and others 1997). Modifications of these four data sets are described in Table A-2. In addition, there has also been a reassessment of exposure for a nested case-control series within the Beaverlodge cohort of uranium miners, including all lung-cancer cases and matched control subjects (Howe and Stager 1996). For the Beaverlodge miners, exposure estimates were about 60% higher than the original values. Because of the computational and conceptual difficulties of merging case-control data with cohort data, only the data from the Beaverlodge cohort study with the original exposure estimates were used in the BEIR VI analysis.

In the first part of this appendix, we consider models for describing the relationship between exposure to radon and lung-cancer risk. We begin with a review of risk models developed by other investigators. In order to lay the foundation for the committee's risk model, we then discuss methods for combining data from different sources, including random-effects and two-stage methods. Those methods are then used in a combined reanalysis of the updated data from the 11 miner cohorts considered previously by Lubin and others (1994a).

By using a random-effects model, the overall effect of radon on lung-cancer risk can be described by fixed regression coefficients, and variation across cohorts characterized by random regression coefficients (Wang and others 1995). Two-stage regression analysis represents an alternative to random-effects methods, which benefits from an element of numerical simplicity. Although both methods are considered, the emphasis in the report is on the computationally simpler two-stage method.

In the second part of the appendix we focus on uncertainties in predictions of risk. There are many sources of uncertainty in health-risk assessments. Epidemiologic data on exposed human populations can be subject to considerable uncertainty. Retrospective exposure profiles are difficult to construct, particularly with chronic diseases such as cancer for which exposure data many years prior to disease ascertainment are needed. For example, radon measurements in homes taken today may not reflect past exposures because people change residences, make building renovations, or change their lifestyle such as sleeping with the bedroom window open or closed, and because of inherent variability in radon

TABLE A-1  Characteristics of 11 underground miner studies[a]

| Location | Type of mine | Number of miners | Period of follow-up | Data available on smoking |
|---|---|---|---|---|
| China | Tin | 17,143 | 1976-87 | Smoker: yes/no [missing on 24% of subjects, 25 (out of 907) non-smoking lung cancer cases.] |
| Czechoslovakia | Uranium | 4,284 | 1952-90 | Not available |
| Colorado, USA | Uranium | 3,347 | 1950-90 | Cigarette use: duration, rate, cessation [unavailable after 1969, 25 (out of 294) non-smoking cases.] |
| Ontario, Canada | Uranium | 21,346 | 1955-86 | Not available |
| Newfoundland, Canada | Fluorspar | 2,088 | 1950-84 | Type of product, and duration, cessation (available for 48% of subjects, including 25 cases.) |
| Sweden | Iron | 1,294 | 1951-91 | Type of product, and amount, cessation (from 35% sample of active miners in 1972, supplemented by subsequent surveys) |
| New Mexico, USA | Uranium | 3,469 | 1943-85 | Cigarette use: duration, rate, cessation (available through time of last physical examination) |
| Beaverlodge, Canada | Uranium | 8,486 | 1950-80 | Not available |
| Port Radium, Canada | Uranium | 2,103 | 1950-80 | Not available |
| Radium Hill, Australia | Uranium | 2,516 | 1948-87 | Smoke status: ever, never, unknown (available for about half subjects, 1 non-smoking case.) |
| France | Uranium | 1,785 | 1948-86 | Not available |

[a]Lubin and others (1994a).

TABLE A-2  Summary of new information on 5 miner cohorts

| Cohort | Updated information | Related reference |
|---|---|---|
| Chinese tin miners | New information indicated miners worked 313 days/yr before 1953, 285 days/yr from 1953-84, and 259 days/yr from 1985. | Unpublished information |
| Czech uranium miners | Exposure histories re-evaluated and follow-up improved. There were 705 lung-cancer cases, compared to 661 in the previous analysis.  Cohort was enlarged from 4,284 to 4,320 miners, including all miners who entered 1948-59. | Tomášek and others 1994 |
| Colorado uranium miners | Follow-up extended from December 31, 1987 to December 31, 1990. In updated data, there were 336 lung-cancer deaths < 3,200 WLM used in the pooled analysis (and 377 total cases), compared to 294 lung-cancer deaths < 3,200 WLM (and 329 total cases). | Hornung and others 1995 |
| Beaverlodge uranium miners[a] | Re-calculation of WLM exposures, but limited to a nested case-control sample. | Howe and Stager 1996 |
| French uranium miners | Corrections of some (non-lung cancer) outcomes and exposure data.  Changes were not extensive. | Unpublished information |

[a]The most recent update of the Beaverlodge data was not included in the BEIR VI analysis.

measurements. Exposures in prospective studies may also be uncertain. In addition to errors in exposure ascertainment, errors in disease diagnosis are also possible. In epidemiologic studies using computerized record linkage to link exposure data from one database with health status in another database, even vital status can be in error (Bartlett and others 1993).

In addition to identifying sources of uncertainty, the committee attempted a quantitative analysis of uncertainty in radon risk estimates. This analysis is conducted within the general framework developed by Rai and others (1996) for quantitative uncertainty analysis in health risk assessment. Since not all sources of uncertainty and variability could be fully characterized, the committee acknowledges that this analysis is necessarily incomplete. Nonetheless, the committee felt that this analysis is informative, and provides a basis for further research in this area.

## PREVIOUS RISK MODELS

A number of different exposure-response models may be used to describe the relationship between lung-cancer risk and exposure to radon (Krewski and others 1992). Analyses of miner cohorts have been largely based on empirical models that describe risk as a linear or linear-quadratic function of exposure. Those analyses have formed the basis for estimates of risks of exposure to radon prepared by the National Research Council (NRC), the National Commission on Radiological Protection and Measurements (NCRP), and the International Commission on Radiological Protection (ICRP). Previous estimates of risk prepared by those organizations are reviewed here.

### Empirical Models

Numerous studies of lung cancer in radon-exposed underground miners have been published, although until recently the number of distinct populations had been small and the total follow-up time and numbers of lung-cancer cases limited (NRC 1988). Those analyses are described in detail in appendix D. The relatively small numbers of lung-cancer cases in individual studies have hindered the evaluation of temporal patterns and other determinants of risk. Early efforts at risk modeling were limited by the lack of data and, as a result, investigators relied on summaries of the miner studies. A complete review of the earlier risk estimates was provided in the BEIR IV Report (NRC 1988). In the current report, we review the most relevant efforts.

### National Research Council (1980)

The NRC BEIR III committee based their modeling efforts on results of miner studies (NRC 1980). The model assumed a linear relationship between exposure and the absolute excess risk of lung cancer. The absolute or excess risk (ER) model represents lung-cancer mortality as $r(x,z,w) = r_o(x) + g(z,w)$, where $r_o(x)$ is the background lung-cancer rate, and $g(z,w)$ is the effect of exposure. Here, $w$ denotes cumulative exposure, $x$ is a vector of covariates which affect the background lung-cancer rate, and $z$ is a vector of covariates that may modify the exposure-response relationship. The excess risk varied by categories of attained age, <35, 35-49, 50-65, >65 yrs, with 0, 10, 20, 30 excess cases per $10^6$ person-years per unit of exposure in WLM. In addition, the model specified a minimum latent period of 15-20 yr for those exposed at ages 15-34 or 10 yr for those exposed above age 34. The derivation of this model and the method of combining the available miner data were not described. The model did not directly account for the effects of smoking.

## National Council on Radiation Protection and Measurements (1984)

The National Council on Radiation Protection and Measurements (NCRP) committee's Report 77 (NCRP 1984a) and its Report 78 (NCRP 1984b) adopted the excess-risk model of Harley and Pasternak (1981). That model was based on the following assumptions.

• The latent interval is 5 yr for persons first exposed at ages 35 yr and older, and $(40-u)$ yr for persons exposed under age 35 yr, where $u$ is the age at first exposure.
• Following a latent interval, disease rate declines exponentially with time since exposure.
• Lung cancer is rare before age 40 yrs.
• The median age at lung-cancer occurrence for miners is age 60 yr for never-smokers and
• 50 yr for ever-smokers.
• The minimum time from initial cell transformation to clinical detection is 5 yr.

For an annual exposure at age $u$, the excess lung-cancer risk at age $t > u$ (and $t > 40$ years of age) is taken to be

$$A(t,u) = Re^{-m(t-u)}S(t) / S(u), \qquad (1)$$

where $R$ is the excess-risk coefficient per WLM, $S(u)$ is the probability of survival to a specified age, and $m$ is the rate of removal of transformed stem cells due to repair or cell death. The NCRP committee fixed $m = ln(2)/20\,yr^{-1}$, corresponding to a 20-yr half-life for exposure effects. The exponential term reduces the exposure effect with time after exposure, while the survival probability adjusts for competing causes of mortality. Lifetime risk for exposure at age $u$ is obtained by integrating over age ($t$) from age 40 yr to some specified life span. For chronic exposures in years $u_1, \ldots, u_n$, lifetime risks are obtained by summing risks for each of the annual exposures.

Parameters for the NCRP model were values assumed to be "reasonable" from published data, but not based on a direct evaluation of the miner data. Although the choices of $R$ and $m$ are critical in applying the model, the NCRP provides little guidance on their selection. The 20-yr half-life was selected as being "representative for extrapolation." $S$ was taken from 1978 World Health Organization tables for the U.S. population. NCRP 78 had the first model to incorporate a reduction of risk with time-since-exposure.

In the NCRP model, the joint effects of radon-progeny exposure and smoking were considered additive. That is, radon exposure had the same effect on the excess risk, regardless of smoking status. The increased absolute excess risk with radon-progeny exposure was added to the background lung-cancer risk in ever-smokers or in never-smokers.

### International Commission on Radiological Protection (1987)

Report 50 of the International Commission on Radiological Protection presented a risk model for indoor radon exposure (ICRP 1987). The ICRP model was based on a simple constant excess relative risk model of the form

$$RR(w) = 1 + \beta w, \tag{2}$$

where $w$ is total cumulative exposure allowing for a lag interval of 10 years. No accommodation for variation in the exposure-response parameter $\beta$ with other factors was included. The value $\beta$ was taken as 0.7% per WLM, a representative value for the excess relative risk of the available miner studies after adjustment for exposure-dose differences between mines and homes. Based on the study of atomic-bomb survivors and on dosimetric considerations, the model assumed a greater effect for radon-progeny exposure at young ages; $\beta$ was set at 2.1% per WLM for exposures occurring under the age of 20 years.

In the ICRP model, the joint effects of radon-progeny exposure and smoking were assumed multiplicative. Thus, for smoking status-specific risk estimation, the radon relative-risk model was applied separately to ever-smokers and to never-smokers. Similarly, the same model was applied to the background rates for males and for females.

### Thomas and Others (1985)

In recent years, the number of data sets on radon-exposed miners has increased and follow-up has lengthened for the cohorts developed initially. Thus, direct synthesis of multiple miner studies has become increasingly important and informative for defining the form of the risk model and estimating the values of its parameters. The first joint analysis of results of epidemiologic studies using modern statistical methodology was carried out by Thomas and others (1985). (See also Thomas and McNeill 1982, an earlier and more detailed report on which the more recent work was based.) They carried out a meta-analysis of 5 miner studies. They fit relative risk and excess (or attributable) risk models, including various "cell killing" and "non-linear" models to summary data from cohort studies from Czechoslovakia (now the Czech Republic), Colorado, Ontario, Newfoundland, and Sweden. Original data from these cohorts with more extensive follow-up are included in the analysis conducted by Lubin and others (1994a) and by this BEIR VI committee. Thomas and others found no significant deviation from a linear exposure-response relationship, although inferences on curvilinearity were somewhat dependent on the choice of referent population. Fitting a linear excess relative-risk (ERR) model, Thomas and others (1985) estimated the ERR to increase by 2.28% per WLM, implying a doubling of risk at 44 WLM. The ERR per WLM was found to vary with attained age. The risk with combined exposure to smoking and to radon progeny, while consistent with a multiplicative model, was most consistent with a relationship intermediate between additive and multiplicative. The analyses were necessarily limited by the extent of the data and not having access to original data. However, many of these results presaged subsequent work.

### National Research Council (1988)

A comprehensive assessment of risk from underground exposure to radon progeny was carried out by the National Research Council's Biological Effects of Ionizing Radiation IV committee (BEIR IV). That committee conducted a pooled analysis of data from four cohort studies of underground miners including the studies of Colorado, Ontario, Beaverlodge, and Sweden (NRC 1988). The Beaverlodge and Swedish data sets were the same as in the current analysis, while less extensive data were available from the other 2 studies. The BEIR IV committee used regression methods similar to those used in this report. The committee found that excess risk did not increase in a simple fashion with exposure, either in direct proportion to background (a constant ERR model) or at a constant level above background (that is, a constant attributable or excess risk [ER] model). Rather, the risk varied with two time-dependent factors: time since exposure and attained age. The analysis showed that lung-cancer risk increased linearly with cumulative exposure to radon progeny, and that the exposure-response trend declined with attained age and with time since exposure. The committee evaluated other potentially important covariates, such as age at first exposure and exposure duration, but risk patterns were not consistent across studies. Although the constant ERR model was not compatible with the data, the ERR per WLM was estimated to be 1.34% (corresponding to a doubling of risk at 75 WLM) under this model.

Analyses of the combined effects of radon-progeny exposure and smoking in the Colorado and New Mexico (a case-control subset of data included in the current analysis) miners were also presented in the BEIR IV Report. Results indicated that while a multiplicative-risk relationship between the two factors could not be excluded, an intermediate relationship between additive and multiplicative was most consistent with the data.

### International Commission on Radiological Protection (1993)

In a 1993 report on risks from radon-progeny exposure in homes and at work (ICRP 1993), ICRP did not provide its own risk model, but used the so-called "GSF model" developed by Jacobi and others (1992). That model was related to a "smoothed" version of the BEIR IV model (NRC 1988). Compared to the BEIR IV model, the GSF model provided a monotonic variation of the excess relative risk with age. The GSF model had the same general structure as the BEIR IV model, with the effect of exposure adjusted by time since the exposure occurred, but with the exposure-response relationship determined by age at exposure, as opposed to attained age. For attained age $a$, exposure $w_e$, occurring at age $a_e$, and time since exposure $f$, the ERR at age $a$ was defined as:

$$ERR(a) = s(a_e)w_e\phi(f). \tag{3}$$

Exposures within 4 years ($f = a - a_e \leq 4$) were assumed to have no impact on the *RR* of lung cancer. The function $s$ was not explicitly defined in the ICRP Report, but was described as a decreasing function of age at exposure, taking values of 0.036 per WLM for age at exposure of 20 years and 0.017 per WLM for age at exposure of 60 years. The effect of time since exposure on risk was modeled through the function $\phi$ defined as

$$
\begin{aligned}
\phi(f) &= 0 & f \leq 4 \text{ years,} \\
&= 0.25\ (f-4) & \text{for } 4 < f < 8 \text{ years,} \\
&= 1 & \text{for } 8 \leq f \leq 12 \text{ years,} \\
&= [(^1\!/_2)(f-12)/10]+1 & \text{for } f > 12 \text{ years.}
\end{aligned}
\tag{4}
$$

According to the ICRP report, risk projections based on the GSF model were similar to those of the BEIR IV model (ICRP 1993).

The ICRP approach was notable in two respects. First, in contrast to the earlier ICRP risk model, which was based on a constant relative risk in cumulative exposure (ICRP 1987), no modification to the exposure-response relationship was included for exposures received at ages 20 years and under. The previous ICRP model postulated a 3-fold greater exposure-response for young ages. Second, the ICRP model assumed an equal (absolute) excess risk in males and females. Thus, the model assumed that the radon-related excess risk in males should be directly added to the background lung-cancer rate in females. This additive feature of the model results in a markedly greater relative risk in females than in males.

## National Cancer Institute (1994)

The analysis and risk model published by Lubin and others (1994a, 1995b), which served as the starting point for the current report, utilized methods similar to those of the BEIR IV committee (NRC 1988). This approach assumes that the time to death from lung cancer is distributed in a fashion so that follow-up time to a key event was piece-wise exponential, that is, death rates are constant within fixed time intervals and exposure categories. Note that death times can be censored due to loss to follow-up or study termination. This assumption was considered appropriate for two reasons: 1) variability in lung-cancer mortality rates within each time interval and exposure category was small relative to the variability between intervals; and 2) disease rates within time intervals and exposure categories were well-characterized by the average rate. The method allowed for use of external referent rates (although they were not used in Lubin and others 1994a, 1995b or in the current analysis) and for the modeling of excess disease rates. A full discussion of these models, including regression of the standardized mortality ratio, is given in Breslow and others (1983) and Breslow and Day (1987).

Relative-risk regression procedures were applied to data summarized in a multi-way table, consisting of events, person-years, and summary variables for each cell of the cross-tabulation. Analyses were conducted using the EPICURE package of computer programs (Preston and others 1991). Data were cross-classified by various factors, depending on the cohort and on the variables being analyzed. For a typical cohort, data were cross-classified by attained age (< 40, 40-44, . . ., 65-69, ≥ 75 years), calendar period (< 1950, 1950-54, . . ., 1980-84, ≥ 1985), estimated exposure (0, 1-49, 50-99, 100-199, 200-399, 400-799, ≥ 800 WLM), duration of radon-progeny exposure (<5, 5-9, 10-14, ≥15 years), age at first radon progeny exposure (< 10, 10-19, 20-29, ≥ 30 years) and other mining experience (no, yes). For each cell of the table, the number of observed lung-cancer deaths, the number of person-years, and the mean (weighted by person-years) for the cross-classification variables, such as cumulative exposure, exposure duration, attained age, and age at first exposure, were computed. For pooling purposes, data were further cross-classified by cohort. A 5-year lag period was assumed.

Whenever possible, similar categories for variables that specified dimensions of the person-year tables were established across the cohorts; however, because of intrinsic differences among the cohorts, this was not always possible. For example, some exposures in the Newfoundland cohort were as high as 21 Jhm$^{-3}$ (6,000 WLM), while all exposures in the Radium Hill cohort were under 0.35 Jhm$^{-3}$ (100 WLM).

The risk model was developed with the following approach. Suppose the lung-cancer mortality rate is given by $r(x,z,w)$, and depends on cumulative exposure, $w$, a vector of covariates, $x$, which described the background lung-cancer rate, and a vector of covariates, $z$, which may modify the exposure-response relationship. The relative risk, $r$, was expressed as a product of the background disease rate among nonexposed, denoted $r_0(x)$, and an exposure-response function, $RR(z,w)$. The background rate $r_0$ depends on $x$ while the exposure-response function RR depends on $z$, which may include one or more components of $x$, as well as $w$. This general relative risk model can be written as

$$r(x,z,w) = r_0(x)RR(z,w). \tag{5}$$

The background lung-cancer rate was modeled as $r_0(x) = exp(ax)$, with $x$ being the vector of controlling variables and a the corresponding parameter vector. Components of $x$ typically included indicator variables for age group, calendar period, and cohort, as well as variables describing other mine exposures. Main effects and all higher order interactions were included.

Specific models were fit for RR. A linear RR model in $w$ was fitted, namely,

$$RR = 1 + \beta w. \tag{6}$$

Here, $\beta$ is a parameter which describes increase in ERR per unit increase in $w$ (ERR/exposure). More generally, a model for the assessment of a broad range of exposure-response relationships was defined as

$$RR = (1 + \beta w^k)e^{\theta w}. \tag{7}$$

Again, $\beta$ reflects the overall ERR/exposure, the parameter, $\theta$ measures the exponential deviation from linearity (sometimes referred to in the radiation effects literature as a "cell killing" parameter), and $k$ is a parameter to describe departure from linearity. This general model includes the linear ERR model ($k = 1$, $\theta = 0$), the linear-exponential model ($k = 1$) and the "non-linear" model considered by Thomas and others (1995) ($\theta \neq 0$). Tests for improvement in model fit in relation to $\theta$ and $k$ were carried out using likelihood ratio procedures.

An important goal of the analysis was to examine variations of the exposure-response trend with other variables, that is, to test whether $\beta$ varied within categories of other factors, such as attained age, age at first exposure, duration and rate of exposure, and time since last exposure. In epidemiologic terms, they evaluated the components of the covariates vector $z$ as an effect modifier. Suppose a particular covariate $z$ had $J$ categories with values $z_1, \ldots, z_J$. Variation in the exposure-response relationship within levels of $z$ was assessed by fitting model (6) and comparing its deviance with model (8) below which included $J$ exposure-response parameters, namely,

$$RR = 1 + \beta_j w, \tag{8}$$

where $\beta_j$ was the ERR/exposure within category $z_j$. Under the null hypothesis of no effect modification, the difference in the model deviances was approximately

$\chi^2$ with $J$-1 degrees of freedom.  A significant $p$-value indicated that the effect of exposure on lung-cancer mortality was not homogeneous across levels of $z$.

As in the BEIR IV Report (NRC 1988), cumulative exposure to radon progeny was divided into time-since-exposure windows, although the analysis in the National Cancer Institute report (Lubin and others 1994a) included additional categories.  For each year of age, cumulative exposure (minus the 5-year lag interval) $w^*$, was expressed as a weighted combination of three exposures:

$$w^* = \theta_{5-14}w_{5-14} + \theta_{15-24}w_{15-24} + \theta_{25+}w_{25+}, \tag{9}$$

where $w_{5-14}$ was the cumulative exposure received 5-14 years prior to the specific age, $w_{15-24}$ was the cumulative exposure received 15-24 years prior, and $w_{25+}$ was exposure 25 or more years ago.  Model (6) was extended as

$$RR = 1 + \beta w^* \tag{10}$$

For identifiability, $\theta_{5-14} = 1$, and the quantity $w^*$ is interpretable as the "effective exposure" to radon progeny, with $\theta_{15-24}$ and $\theta_{25+}$ defining the relative contributions to the total exposure from the corresponding time periods.

## Biologic-Based Models

Moolgavkar and others (1993) have applied a biologic-based exposure-response model to the Colorado uranium miners' data.  Specifically, the 2-stage clonal expansion model discussed by Moolgavkar and Luebeck (1990) was used in a first attempt to describe cancer mortality rates among the Colorado miners in terms of a mechanistic stochastic model of carcinogenesis.

Biologic-based models of carcinogenesis are useful for several reasons (Goddard and Krewski 1995).  First, they provide a convenient framework within which to describe the process of carcinogenesis.  Carcinogenesis is formulated as a multistage process in which the kinetics of cell division and cell death play important roles.  The currently available biologic-based models incorporate these features of carcinogenesis.  Second, the parameters of a biologic-based model have biologic meaning.  With the two-stage clonal expansion model, separate parameters are used to describe the first and second stage mutation rates as well as the birth and death rates of initiated cells.  Third, a validated biologic-based model is likely to enjoy greater acceptance when used for the quantitative estimation and prediction of cancer risks.  And fourth, description of the temporal aspects of cancer risk is facilitated by the use of biologic-based models of carcinogenesis.  In fact, these models provide a flexible family of hazard functions for analyses of time-to-tumor data that allow for incorporation of age and time dependent covariates in a natural way without increasing the number of parameters to be estimated.  Empirical models offer less flexibility in this regard, using crude temporal indictors of risk such as age at first exposure, duration of exposure, and cumulative lifetime exposure. Biologic-based models integrate important biologic aspects of carcinogenesis with rigorous statistical methods for data analyses (Moolgavkar and Luebeck 1990).

Although multi-stage biologic models are available, the 2-stage model represents a useful starting point. The 2-stage clonal expansion model is based on the following assumptions. First, there is a pool of stem cells within the tissue of interest susceptible to malignant transformation. Second, malignant tumors are clonal in origin, arising from a single transformed progenitor cell. Third, malignant transformation is the result of 2 specific rate-limiting mutations. Once a malignant cell is generated, it will give rise to a histologically detectable cancerous lesion following a certain lag time. This model has been found to satisfactorily describe a variety of toxicologic and epidemiologic data (Krewski and others 1992).

In chemical carcinogenesis, the occurrence of the first mutation is identified with initiation, whereas the second mutation is associated with malignant conversion of an initiated cell or completion. Carcinogenic chemicals and radionuclides may act by increasing the first or second stage mutation rates, or the rate of expansion of the initiated cell population. Clonal expansion of the initiated cell population is referred to as promotion.

### Application to Colorado Uranium Miners' Data

Moolgavkar and others (1993) have re-analyzed data on lung-cancer mortality from the Colorado Plateau miners' cohort and British doctors' cohort within the framework of the two-stage clonal expansion model. The Colorado uranium miners' data used in this analysis were described by Hornung and Meinhard (1987), and included the information on the age at which exposure to radon progeny and cigarette smoke began, the ages at which these exposures stopped, the cumulative exposure to radon progeny, the number of cigarettes smoked per day, the age at last observation or death, and information on whether or not the individual had died of lung cancer by that time. A lag period of 3.5 years between malignant transformation of lung tissue and death from lung cancer was assumed. Consequently, exposures occurring within 3.5 years of disease diagnosis were not included in the analysis.

Lung-cancer mortality among a large cohort of British doctors was used to obtain information on the baseline lung-cancer risks in men not exposed occupationally to radon or to tobacco. (Since there were only eight miners who were not exposed to radon and who did not smoke, the background parameters cannot be precisely estimated from the Colorado data alone.) This cohort is of particular significance because data on tobacco consumption were obtained prospectively over the course of follow-up. Following Doll and Peto (1978), analysis was restricted to lung-cancer mortality among men aged 40-79 years who either were never-smokers or regularly smoked no more than 40 cigarettes per day. Former smokers were not considered.

In fitting the two-stage model to the data, it was assumed that $X(t)$ is the number of susceptible cells at age $t$ and that $\mu(d_r, d_s)$ is the rate of the first mutation as a function of the exposure rate of radon progeny, $d_r$, measured in WLM/month (WLM/m), and the exposure rate of cigarette smoke, $d_s$, measured in cigarettes per day. Specifically, intermediate or initiated cells are generated from normal ones as a nonhomogeneous Poisson process with intensity $\mu X$. The intermediate cells divide with rate $\alpha$, die or differentiate with rate $\beta$, and divide into one intermediate and one malignant cell with rate $v$. All of these transition rates may be influenced by exposure to radon or cigarette smoke. The manner in which rates of mutation and cell proliferation depend on exposure to these two agents is discussed below.

The Colorado miners' data and the British doctors' data were first analyzed separately. An analysis in which certain model parameters were assumed to be the same in both data sets was then conducted. It was found that assuming equal effects of tobacco on the mutation rates in the two data sets produced a likelihood that was virtually identical to the likelihood resulting from the separate analyses. Thus the data were consistent with the hypothesis that the spontaneous- and tobacco-related mutation rates are identical in the two cohorts.

The following exposure-response functions were used for each of the parameters:

$$\mu(d_r, d_s) = a_0 + a_s d_s + a_r d_r, \tag{11}$$

$$v(d_r, d_s) = b_0 + b_s d_s + b_r d_r \tag{12}$$

where $v$ is the second-mutation rate and

$$(\alpha - \beta)(d_r, d_s) = c_0 + c_{s1}(1 - exp(-c_{s2} d_s) + c_{r1}(1 - exp(-c_{r2} d_r) \tag{13}$$

in the Colorado miners' data, and

$$(\alpha - \beta)(d_r, d_s) = e_0 + e_{s1}(1 - exp(-e_{s2} d_s) + e_{r1}(1 - exp(-e_{r2} d_r) \tag{14}$$

in the British doctors' data.

In both data sets $\beta/\alpha$ = constant, independent of the level of exposure to radon or tobacco. In this model, referred to as model A, 15 parameters were estimated from the data. Estimates of the spontaneous mutation rates $a_0$ and $b_0$ were almost equal, and the second mutation rate $v$ appears to be unaffected by either radon progeny or cigarette smoke. Since the likelihood is little changed by setting $a_0 = b_0$ and $b_s = b_r = 0$, a reduced form of model A with only 12 parameters was therefore considered.

A second model B in which all model parameters common to the Colorado miners and British doctors' cohorts were assumed to be equal was also fit. This model eliminates the three parameters in Model A having to do with cell proliferation in the British doctors' data by taking $c_0 = e_0$, $c_{s1} = e_{s1}$, and $c_{s2} = e_{s2}$, leaving only 9 parameters to be estimated. Comparisons of the numbers of observed and expected lung-cancer deaths within specified exposure categories suggested that both models fit the data reasonably well.

The qualitative conclusions based on the separate and joint analyses (models A and B) were similar. Both radon progeny and cigarette-smoking appear to affect the first mutation rate and the kinetics of intermediate cell division. The second mutation rate was found to be independent of radon progeny and cigarette-smoke exposures. With both models, the age-specific relative risks associated with joint exposure were supra-additive but sub-multiplicative, confirming previous findings by Whittemore and McMillan (1983) based on an empirical analysis of the Colorado miners' data. However, since addition of interaction terms to the exposure-response functions for any of the model parameters was not significant, there was no suggestion of any interaction between radon progeny and cigarette smoke at the cellular level. The analysis also confirmed the inverse exposure-rate effect in the Colorado miners' cohort.

The National Research Council (1988) examined risk assessment methods for the analysis of complex mixtures, including simple binary mixtures of the type of interest here. The NRC concluded that interactive effects between two carcinogens are likely to be negligible when the level of exposure to both agents is low. Further theoretical support for this finding was provided by Kodell and others (1991) within the context of the two-stage clonal expansion model of carcinogenesis. Application of the two-stage model to the Colorado uranium miners data provided empirical confirmation of these latter theoretical results.

The two-stage model has also been applied to data on the incidence of lung tumors in experiments conducted at Battelle Pacific Northwest Laboratories (Moolgavkar and Luebeck 1993; Luebeck and others 1996). Those experiments involved 3,750 rats subjected to cumulative radon-progeny exposures ranging from 0.07 to 35 $Jhm^{-3}$ (20 to 10,000 WLM) at different exposure rates. The analysis of the data evaluated the dependence of the mutation and cell proliferation rates on the radon-progeny exposure rate, as well as exposure to uranium ore dust, which was a component (at a constant concentration) in all exposures. The two-stage model was found to provide an adequate fit to these data.

The analyses yielded results that were similar to those obtained from the application of the two-stage model to the Colorado miners' data discussed previously. Specifically, exposure to radon progeny was found to affect the first-stage mutation rate, but not the second-stage mutation rate. The estimated rate of the first mutation was consistent with rates measured experimentally in vitro. The authors found evidence of an inverse exposure-rate effect, which they attributed primarily to promotion of intermediate lesions.

## METHODS FOR COMBINING DATA FROM SEVERAL COHORTS

In this section, we describe the models and methods used in the committee's combined re-analysis of the 11 miner cohorts. In developing the BEIR VI model, modified Poisson regression methods were used to analyze data from the 11 miner studies. Following Lubin and others (1994a), the death rates are assumed to be constant within fixed time intervals and exposure categories. Data entering into regression analyses are in the form of a multi-way person-years table consisting of lung-cancer deaths, person-years, covariates of interest, and potential confounders. Under the Poisson regression model, the observed number of cases is assumed to follow a Poisson distribution for which the variance is equal to the mean (Breslow and Day 1987). Specifically, the expected number of deaths is $N_{jk}r_{jk}(x,z,w)$. Here, $N_{jk}$ denotes the number of person-years at risk in the $j$th category $(j = 1, \ldots, J_k)$ in the $k$th cohort $(k = 1, \ldots, K)$ and $r_{jk}(x,z,w)$ denotes the corresponding mortality rate depending on the cumulative exposure $w$ to radon progeny, a vector of covariates $z$, which can modify exposure-response relationship, and a vector of potential confounders $x$.

Although Poisson regression is widely used in the analysis of cohort mortality data, the existence of extra-Poisson variation cannot be ruled out. Although not affecting point estimates of the model parameters, such overdispersion, should it exist, can lead to overstatement of precision.

Under a general relative-risk model, the mortality rate can be expressed as the product

$$r_{jk}(x,z,w) = r_{0jk}(x)RR_{jk}(z,w), \qquad (15)$$

where $r_{0jk}(x)$ and $RR_{jk}(z,w)$ denote the background mortality rate and relative risk for the $j$th state in the $k$th cohort, respectively. Covariates considered include attained age (years), rate of radon progeny exposure (WL), and duration of radon-progeny exposure (years). Age and other mining experience involving exposure to arsenic and gold are considered to be potential confounders. Six of the eleven cohorts include information on tobacco-smoking by the study subjects.

## Relative Risk Model

The relative risk (RR) of lung cancer associated with tobacco-smoking and exposure to radon progeny can be described quite generally in terms of a mixture of multiplicative and additive relationships (Breslow and Clayton 1993). Following Lubin and others (1994a), we consider a multiplicative model to describe the joint effects of radon and tobacco. The relative risk of lung cancer associated with radon-progeny exposure in the $j$th category of the $k$th cohort is modeled as

$$RR_{jk} = 1 + \beta_k w_{jk} \phi_a \gamma_z \qquad (16)$$

where $\beta_k$ is the excess relative risk of lung cancer associated with exposure to radon progeny for the $k$th cohort, $w_{jk}$ denotes the cumulative radon exposure within the $j$th stratum of the $k$th cohort, $\phi_a$ denotes the modifying effect of attained age, and $\gamma_z$ denotes the effect of either exposure duration or exposure concentration. The joint effects of radon and cigarette-smoking on lung-cancer risk are as described:

$$RR_{jk}(\alpha_k) = (1 + \beta_k w_{jk} \phi_a \gamma_z) \theta_k \qquad (17)$$

where $\theta_k = 1$ for never-smokers and $\theta_k > 1$ for ever-smokers. Although we do not fit this multiplicative model directly to the miner data (due to limited information on tobacco consumption patterns among the miners), it forms the basis for an indirect adjustment for tobacco due to Lubin and Steindorf (1995) described later.

## Random-Effects Model

Heterogeneity across cohorts can be described by a random-effects model (Rutter and Elashoff 1994), in which the overall effects and variation among individual cohorts are characterized by fixed and random regression coefficients respectively. Specifically, to describe heterogeneity across cohorts, the parameter $\beta_k$ is decomposed into two parts:

$$\beta_k = \beta + b_{\beta,k}, \qquad (18)$$

where $\beta$ is the fixed effect for all cohorts and $b_{\beta,k}$ is the random effect specific to the $k$th cohort. More generally, the parameters $\alpha_k = (\beta_k, \phi_{c,k}, \gamma_{z,k})$ in model (16) can be written as

$$\alpha_k = \alpha + a_k, \qquad (19)$$

where $\alpha = (\beta, \phi_a \gamma_z)$ denotes the vector of fixed effects, and a vector of random-effects $a_k = (a_{\beta,k}, a_{\theta,k}, \alpha_{\gamma,k})$, specifies the deviation from the overall effect associated with the $k$th cohort. This generalization allows for inter-cohort variability in the parameters $\phi_a$ and $\gamma_z$, although this was not necessary for the miner data used in the present analysis.

Different statistical methods can be used to fit nonlinear random-effects regression models. Because exact methods for model fitting can be computationally intensive, Burnett and others (1995) proposed the use of a locally linear first order approximation to simplify the calculation. Wang and others (1995) provide a detailed discussion of how this approximation random effects model may be fit to data from the 11 miner cohorts using generalized estimating equations (GEEs). GEEs are robust in the sense that only the first two moments of the distribution of random effects need be specified (Zeger and others 1988) rather than the complete distribution needed to apply likelihood-based methods.

Even with these simplifications, fitting nonlinear random-effects models proved to be computationally difficult with the large number of categories in the multi-way person-years table. Consequently, two-stage regression methods were also used to fit the nonlinear models of interest here.

## Two-Stage Regression Analysis

The two-stage regression method represents a simplification of and an approximation to the random-effects models. With the two-stage method, the model of interest is first fit separately within each cohort. An overall estimate of $\beta$ is then obtained as a simple linear combination of the cohort-specific estimates.

Details of the two-stage regression method have been given by Laird and Mosteller (1990) and Whitehead and Whitehead (1991). Without loss of generality, we will use the linear model

$$RR_k(z, w, \alpha_k) = 1 + \beta_k w \tag{20}$$

to illustrate how the two-stage analysis is conducted.

**Stage 1.** In the first stage, model (20) is fitted to each cohort. Let $\hat{\beta}_k$ be the estimate of model parameter $\beta_k$ and $s_k$ be the estimated variance of $\hat{\beta}_k$.

**Stage 2.** Define

$$\overline{\beta} = \frac{\sum_k s_k^{-1} \hat{\beta}_k}{\sum_k s_k^{-1}} \tag{21}$$

$$\hat{\tau} = \frac{\sum_k s_k^{-1} (\hat{\beta}_k - \overline{\beta}_k)^2 - (K-1)}{\sum_k s_k^{-1} - \frac{\sum_k s_k^{-2}}{\sum_k s_k^{-1}}} \tag{22}$$

and

$$\omega_k = \frac{(\hat{\tau} + s_k)^{-1}}{\sum_k (\hat{\tau} + s_k)^{-1}}.$$

The pooled estimate $\hat{\beta}$ of the overall effect is then given by

$$\hat{\beta} = \sum_k \omega_k \hat{\beta}_k. \tag{24}$$

The variance in the estimate of the overall effect is estimated by

$$Var(\hat{\beta}) = \left( \sum_k (\hat{\tau}_k + s_k)^{-1} \right)^{-1}. \tag{25}$$

Heterogeneity among cohorts is reflected in positive values of $\tau$, which increase the estimated variance of the overall effect $\hat{\beta}$.

A statistical test for homogeneity of the $\hat{\beta}_k$ among cohorts is given by

$$\chi^2_{homog} = \sum_k s_k^{-1} (\hat{\beta}_k - \overline{\beta})^2 \tag{26}$$

which has a chi-square distribution with $K - 1$ degree of freedom.

Provided $\hat{\tau} > 0$, the shrinkage estimator of the cohort-specific effect $\beta_k^*$ is

$$\hat{\beta}_k^* = \frac{s_k\hat{\beta} + \hat{\tau}\hat{\beta}_k}{s_k + \hat{\tau}}, \tag{27}$$

with the deviation from the overall estimate given by

$$\hat{\delta}_k = \hat{\beta} - \hat{\beta}_k^* \tag{28}$$

An estimate of the variance of this deviation is given by

$$Var(\hat{\delta}_k) = \frac{\hat{\tau}s_k}{\hat{\tau} + s_k}. \tag{29}$$

Heterogeneity between different studies is taken into account in estimating overall risks with both the random-effects and two-stage analyses. This is particularly important when there are significant differences among the cohorts, that is, when the fixed-effects model fits the data poorly (Greenland 1994). Fitting nonlinear regression models is considerably easier with the two-stage method than with the random-effects method, particularly with large datasets and many parameters. The two stage regression method is applied by using well-established computer software to analyze each cohort separately; combining information across cohorts is then done by means of a simple linear combination of the parameter of interest. With random-effects methods, on the other hand, the data from all cohorts are analyzed simultaneously. Because of the computational complexity in fitting the random effects model, a first-order approximation simplifies the calculations. Even with this approximation, however, convergence was difficult to obtain in some situations. Consequently, the committee focused primarily on two-stage regression methods for model fitting.

## Combined Analysis of Miner Cohorts

The updated data on the 11 miner cohorts were summarized in the form of a multi-way table prior to analysis. The categorizations were essentially the same as those used by Lubin and others (1994a). All models considered by the committee were fitted by Poisson regression using EPICURE (Preston and others 1991). The most recent release of EPICURE allows for extra-Poisson variation, but was not available to the committee during the course of its analysis. Cohort effects were included in the single model by stratifying the background disease risk by cohort, as well as age group, and other occupational exposures and ethnicity. Cohort-specific estimates of the ERR/WLM ($\beta_j$) were obtained from the single model fit to all the data. Since preliminary analyses indicated that the effect of time since exposure, attained age and exposure duration or radon concentration were similar in most cohorts, these parameters were considered to be the same in all cohorts. However, the parameter $\beta$ did vary considerably across the cohorts. Consequently, the overall estimate of $\beta$ was obtained using the two-stage method with associated standard errors reflecting variation within and between cohorts. We also used the random-effects method to obtain the overall estimate of $\beta$. With the random-effects method, we used the background parameters obtained previously from the EPICURE (Preston and others 1991) fit of the corresponding model to all the data. The methods used to fit the random-effects model are identical to those of Wang and others (1996), with the exception that the parameter $\beta$ was obtained from the transformation $\beta = exp\{\beta^*\}$ after first estimating $\beta^*$. This same transformation was used with the two-stage method in order that the sampling distribution of the estimator $\hat{\beta}$ be closer to normal. The standard error of $\hat{\beta}$ shown here for the two-stage method assumes that the $\hat{\beta}_k$ are independent; taking the covariance forms into account leads to a slight reduction in the standard error of $\hat{\beta}$. These covariance forms are included in the more comprehensive uncertainty analysis discussed later in this appendix.

The results of fitting the simple linear model

$$RR = 1 + \beta w \qquad (30)$$

to the miner data are shown in Table A-3. In addition to results for both the random-effects and two-stage methods, the results obtained by analyzing each cohort separately are also shown. The ERR per unit exposure is given by the

TABLE A-3    Estimated ERR/WLM(%)[a] based on two-stage and cohort-specific analyses

| Cohort | Two-stage analysis | Cohort-specific analysis |
|---|---|---|
| Combined | 0.76 (1.86)[b] 0.59[c] (1.32)[b] | |
| China | 0.17 | 0.17 |
| Czechoslovakia | 0.67 | 0.67 |
| Colorado | 0.44 | 0.42 |
| Ontario | 0.82 | 0.89 |
| Newfoundland | 0.82 | 0.82 |
| Sweden | 1.04 | 1.25 |
| New Mexico | 1.58 | 2.84 |
| Beaverlodge | 2.33 | 2.95 |
| Port Radium | 0.24 | 0.19 |
| Radium Hill | 2.75 | 4.76 |
| France | 0.51 | 0.09 |

[a]ERR/WLM is the parameter $\beta$ in the model $RR = 1 + \beta w$, where $w$ denotes cumulative radon progeny exposure.

[b]Multiplicative standard error; $\exp \sqrt{\text{var}(\log \hat{\beta})}$

[c]Based on random-effects model.

parameter β. All methods provide estimates of this parameter separately for each cohort; the random-effects and two-stage methods also provide overall estimates of β. Note that the estimates of β for individual cohorts based on the two stage-method differ from the cohort-specific estimates. In general, the adjusted estimates of β obtained using the two-stage method tend to "shrink" towards the overall estimate of β. The overall estimate of the ERR/WLM based on the random-effects analysis is 0.59% with a multiplicative standard error of 1.32. The two-stage analysis leads to an estimate of 0.76% with a multiplicative standard error of 1.86. (The difference in standard errors for the two-stage and random-effects methods is due to the different approaches used to estimate the standard error, and the small number of cohorts involved in the analysis.) The cohort-specific estimates of the excess relative risk per unit exposure are shown graphically in Figure A-1.

To evaluate overall patterns of risk, the committee investigated a number of other models of the form (16) with covariates including attained age, rate of exposure to radon progeny, time since exposure, and duration of exposure. This limited model selection process led the committee to the same two risk models favored by Lubin and others (1994a). These two models represent the committee's preferred risk models, and will be referred to as the exposure-age-duration model and exposure-age-concentration model, respectively.

Estimates of the parameters in the committee's two preferred models are shown in Table A-4. These estimates were obtained by fitting each of these two models to the data for all eleven cohorts simultaneously, constraining the covariates to be the same in all cohorts, but allowing β to vary among cohorts. Again, an overall estimate of β was obtained using the two-stage method. An overall estimate of β was also obtained using the random-effects method. However, since convergence could not be obtained when attempting to estimate the covariate values using the random-effects method, the covariate values were fixed at their previously estimated values, leaving only the ERR per unit exposure to be determined in this simplified random-effects analysis.

The pattern of modifying effects of the covariates on the exposure-response relationship was similar to that observed in the original analysis of the 11 miner cohorts by Lubin and others 1994. Specifically, the exposure-response relationship decreased with time since exposure, attained age, and exposure rate, but increased with exposure duration.

In order to determine whether the overall estimate of the ERR per unit exposure was unduly affected by the data from any one cohort, the committee conducted an influence analysis in which the parameter β was estimated after omitting, in turn, data from each individual cohort (Table A-5, Figure A-2). This influence analysis showed that the omission of any one cohort did not have a strong impact on the overall estimate of β. Some caution is required in the interpretation of these results, since differences in age, exposure rate, exposure

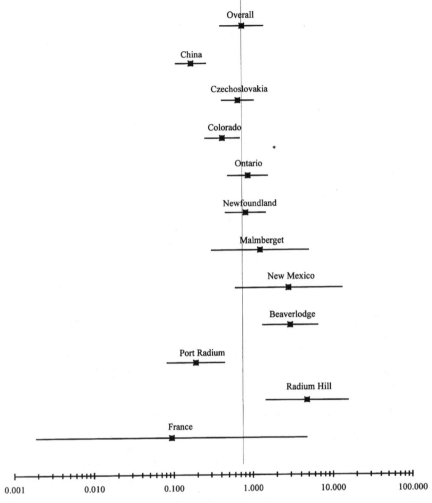

FIGURE A-1    Estimate of the ERR per unit exposure ERR/WLM (%) obtained by fitting a constant excess relative risk model to each cohort separately (overall estimate based on the 2-stage method) and 95% confidence intervals.

duration, time since exposure, and smoking habits among cohorts could explain some of the inter-cohort variation in the $\beta$'s observed in this analysis.

## Smoking and Radon Progeny Exposure

Data on tobacco-smoking in the various cohorts were limited: either lacking entirely; not readily comparable among cohorts because of the method or the amount of information collected; or incomplete for workers within particular cohorts. Six of the eleven cohorts included some data on smoking (China, Colorado, Newfoundland, Sweden, New Mexico and Radium Hill) although among these six cohorts, a substantial number of workers lacked smoking information. Moreover, surveys of smoking practices were not usually undertaken on a regular prospective basis, necessitating assumptions about the stability of smoking practices over time. The impact of incomplete information on tobacco is not clear, including the misclassification of ex-smokers as current smokers.

Smoking rates in Western countries have generally declined, although it is uncertain if smoking rates among miners have declined and, if so, to what extent. Among Chinese workers, smoking rates have been relatively stable, although tobacco use practices (cigarettes and bamboo water pipes) have changed, especially among younger workers. Company policies regarding smoking, such as prohibiting smoking while underground, are also evolving. In recent years, smoking, in some cohorts, may have been under-reported by workers concerned about health compensation issues.

TABLE A-4    Estimates of the parameters in the committee's two preferred risk models

| Exposure-age-duration model[a] | | Exposure-age-concentration model[a] | |
|---|---|---|---|
| $\beta \times 100$ | $0.55^{b}$ $(2.03)^{d}$ | $\beta \times 100$ | $7.68^{b}$ $(1.94)^{d}$ |
| | $0.45^{c}$ $(1.31)^{d}$ | | $7.44^{c}$ $(2.63)^{d}$ |
| Time since exposure windows | | | |
| $\theta_{5\text{-}14}$ | 1.00 | $\theta_{5\text{-}14}$ | 1.00 |
| $\theta_{15\text{-}24}$ | 0.72 | $\theta_{15\text{-}24}$ | 0.78 |
| $\theta_{25+}$ | 0.44 | $\theta_{25+}$ | 0.51 |
| Attained age | | | |
| $\phi_{<55}$ | 1.00 | $\phi_{<55}$ | 1.00 |
| $\phi_{55\text{-}64}$ | 0.52 | $\phi_{55\text{-}64}$ | 0.57 |
| $\phi_{65\text{-}74}$ | 0.28 | $\phi_{64\text{-}74}$ | 0.29 |
| $\phi_{75+}$ | 0.13 | $\phi_{75+}$ | 0.09 |
| Duration of exposure | | Exposure rate (WL) | |
| $\gamma_{<5}$ | 1.00 | $\gamma_{<0.5}$ | 1.00 |
| $\gamma_{5\text{-}14}$ | 2.78 | $\gamma_{0.5\text{-}1.0}$ | 0.49 |
| $\gamma_{15\text{-}24}$ | 4.42 | $\gamma_{1.0\text{-}3.0}$ | 0.37 |
| $\gamma_{25\text{-}34}$ | 6.62 | $\gamma_{3.0\text{-}5.0}$ | 0.32 |
| $\gamma_{35+}$ | 10.20 | $\gamma_{5.0\text{-}15.0}$ | 0.17 |
| | | $\gamma_{15+}$ | 0.11 |

[a]Parameters estimated based on the fitted model: $RR = 1 + \beta\, w^{*}\phi_{a}\, \gamma_{z}$ where $w^{*} = w_{5\text{-}14} + \theta_{2}\, w_{15\text{-}24} + \theta_{3}\, w_{25+}$. Here, $\phi_{a}$ denotes attained age $a$ in years $\gamma_{z}$ denotes either exposure duration in years or radon progeny concentration categories in WL.
[b]Two-stage method.
[c]Random-effects method.
[d]Multiplicative standard error: $\exp \sqrt{\mathrm{var}(\log \beta)}$

TABLE A-5    Estimates of β (ERR/WLM) based on the data from all cohorts except one

| Omitted Cohort | Exposure-age-duration model | | | Exposure-age-concentration model | | |
|---|---|---|---|---|---|---|
| | β[a] | 95%C.I. | | β[a] | 95%C.I. | |
| None | 0.553 | 0.271 | 1.125 | 7.681 | 3.969 | 14.864 |
| China | 0.714 | 0.398 | 1.280 | 9.884 | 6.322 | 15.455 |
| Czechoslovakia | 0.565 | 0.249 | 1.278 | 7.374 | 3.553 | 15.304 |
| Colorado | 0.586 | 0.263 | 1.305 | 7.359 | 3.553 | 15.243 |
| Ontario | 0.551 | 0.246 | 1.235 | 8.248 | 3.883 | 17.519 |
| Newfoundland | 0.560 | 0.251 | 1.252 | 7.240 | 3.520 | 14.890 |
| Malmberget | 0.555 | 0.259 | 1.188 | 7.773 | 3.821 | 15.810 |
| New Mexico | 0.535 | 0.248 | 1.153 | 7.280 | 3.663 | 14.468 |
| Beaverlodge | 0.456 | 0.234 | 0.888 | 6.785 | 3.451 | 13.339 |
| Port Radium | 0.566 | 0.257 | 1.246 | 8.057 | 3.878 | 16.741 |
| Radium Hill | 0.440 | 0.228 | 0.850 | 6.855 | 3.462 | 13.571 |
| France | 0.598 | 0.289 | 1.237 | 8.169 | 4.207 | 15.862 |

[a]Combined estimate based on two-stage procedure.

## Adjustments for Smoking Status

The data on smoking are generally too sparse to model the joint effects of smoking and radon exposure. However, using the results of analyses of the effect of radon-progeny exposure among never-smokers and among ever-smokers, it is possible to adjust the committee's preferred models to account for smoking status. In this regard, the committee adopted the approach introduced by Lubin and Steindorf (1995), based on the relative difference in the exposure-response relationship for ever-smokers and never-smokers. Restricting data to miners for whom some smoking information was available (China, Colorado, Newfoundland, New Mexico, and Radium Hill), the overall ERR/WLM was estimated to be 1.02% (95% CI: 0.15-7.18%) among never-smokers and 0.48% (95% CI: 0.18-1.27%) among ever-smokers. Among these same miners, the overall ERR/WLM, ignoring smoking status, was 0.53% (95% CI: 0.20-1.38%). The effects of sequentially omitting a single cohort from this analysis are shown in Table A-6 and Figure A-3.

Estimates of the ERR/WLM for ever-smokers and never-smokers are comparable only to the extent that the mean age, time since exposure, and exposure rate (factors known to modify the ERR/WLM) are similar for ever-smokers and never-smokers. Analysis revealed that these modifiers differed only slightly between the two groups: never-smokers were one year older, six months further from time of last exposure, and exposed at a rate 0.9 WL less than ever-smokers. We therefore assumed that the observed ERR/WLM estimates approximate the relative effects of radon-progeny exposure in ever-smokers and never-smokers.

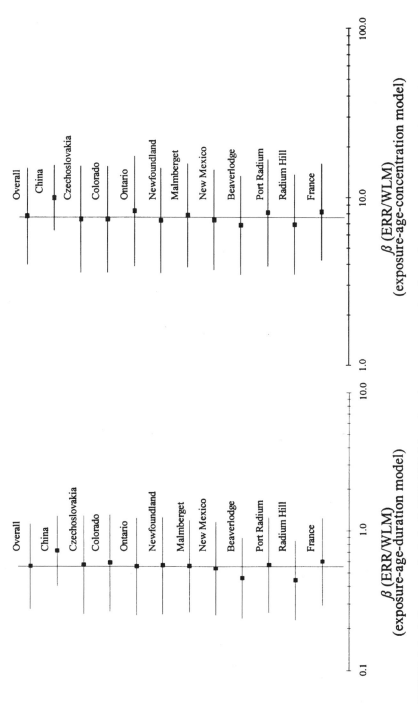

β (ERR/WLM)
(exposure-age-duration model)

β (ERR/WLM)
(exposure-age-concentration model)

FIGURE A-2   Influence analysis for β (ERR/WLM) based on the exposure-age-duration and exposure-age-concentration models.

TABLE A-6   Influence analysis of smoking correction factors[a]

| Omitted cohort[b] | β ever-smokers/β overall | β never-smokers/β overall |
|---|---|---|
| None | 0.916 | 1.937 |
| China | 0.921 | 1.121 |
| Colorado | 0.929 | 1.220 |
| Newfoundland | 0.864 | 2.448 |
| Malmberget | 0.952 | 2.584 |
| New Mexico | 0.897 | 2.651 |

[a]Based on cohorts for which smoking data were available (see Table A-1).
[b]The data for Radium Hill were too sparse to obtain a meaningful estimate.

Based on this analysis, the effect of exposure among ever-smokers, relative to the overall effect, ignoring smoking status, was 0.9 (0.48/0.53), whereas among never-smokers the relative effect, was approximately 2-fold (1.02/0.53). The influence analysis summarized in Table A-6 indicates that no one cohort has an inordinate effect on these two ratios. These crude correction factors were applied to the committee's preferred models to obtain estimates of risk for ever-smokers and never-smokers separately. Specifically, we adjusted the estimate of the baseline ERR/WLM β in the exposure-age-duration and exposure-age-concentration models, while leaving the parameter estimates for the modifying factors unchanged. In the exposure-age-concentration model, the estimate of 0.0768 was reduced to 0.069 for ever-smokers and increased to 0.153 for never-smokers, while in the exposure-age-duration model the estimate of 0.0055 was reduced to 0.0050 for ever-smokers and increased to 0.011 for never-smokers. The present adjustment differs somewhat from that obtained by Lubin and Steindorf (1995) due to use of updated information, particularly for the China and Colorado cohorts.

The estimates of β based on the data for all 11 cohorts (Table A-3) differed slightly from the estimates based on the six cohorts with data on tobacco consumption (Table A-7). However, within the six cohorts for which smoking information is available, the ERR/WLM was nearly twice as large in never-smokers as compared with ever-smokers. This analysis indicates that the effects of ever-smoking and radon progeny exposure are not incorporated multiplicatively, but as a sub-multiplicative mixture.

## QUALITATIVE UNCERTAINTY ANALYSIS

In the remainder of this appendix, we focus on sources of uncertainty and variability in radon risk estimates. Clearly, lack of accurate information on a number of variables that affect radon risk, including critical variables such as radon exposure and tobacco consumption, confers uncertainty on committee projections of risk. Individual risks probably also vary within the population, de-

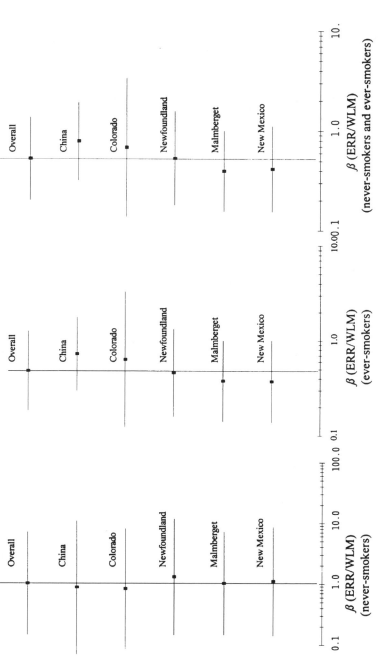

FIGURE A-3 Influence analysis for β (ERR(WLM)) in cohorts with smoking information based on the constant relative risk model $RR = 1 + β(ERR/WLM) \omega$.

TABLE A-7   Influence analysis of β for six cohorts with information on smoking[a]

| Omitted Cohort | β[b] (%) | 95% Confidence Interval (%)[c] | |
|---|---|---|---|
| *Never-smokers* | | | |
| None | 1.021 | 0.145 | 7.180 |
| China | 0.880 | 0.069 | 10.889 |
| Colorado | 0.831 | 0.085 | 8.113 |
| Newfoundland | 1.298 | 0.144 | 11.701 |
| Malmberget | 1.019 | 0.145 | 7.164 |
| New Mexico | 1.097 | 0.138 | 8.685 |
| *Ever-smokers* | | | |
| None | 0.483 | 0.183 | 1.272 |
| China | 0.724 | 0.296 | 1.769 |
| Colorado | 0.633 | 0.122 | 3.286 |
| Newfoundland | 0.458 | 0.157 | 1.335 |
| Malmberget | 0.375 | 0.141 | 1.001 |
| New Mexico | 0.371 | 0.137 | 1.007 |
| *Never-smokers and ever-smokers* | | | |
| None | 0.527 | 0.202 | 1.375 |
| China | 0.785 | 0.320 | 1.927 |
| Colorado | 0.681 | 0.137 | 3.386 |
| Newfoundland | 0.530 | 0.178 | 1.575 |
| Malmberget | 0.394 | 0.154 | 1.006 |
| New Mexico | 0.414 | 0.153 | 1.122 |

[a]The data for Radium Hill was too sparse to obtain useful estimate of β for never-smokers.
[b]Based on the constant relative risk model $RR = 1+\beta w$ fit using the 2-stage method.
[c]Confidence limits based on multiplicative standard error.

pending on radon-exposure patterns, tobacco consumption, and other (possibly unknown) factors.

It is important to distinguish clearly between uncertainty and variability. Uncertainty represents the degree of ignorance about the precise value of a particular parameter, such as the body weight of a given individual (Morgan and others 1990; NRC 1994b; Hattis and Burmaster 1994). On the other hand, variability represents inherent variation in the value of a particular parameter within the population of interest. In addition to being uncertain, body weight also varies among individuals. A parameter such as body weight, which can be determined with a high degree of accuracy and precision, may be subject to little uncertainty, but can be highly variable. Other parameters may be subject to little variability but substantial uncertainty. Yet others may be both highly uncertain and highly variable.

The committee attempted to identify the main sources of uncertainty in assessing the lung-cancer risk from radon exposure. The committee also conducted a limited quantitative analysis of the uncertainty in estimates of both relative risk and attributable risk. The committee did not find it feasible to evaluate the combined effects of all sources of uncertainty affecting radon-risk estimates, but a discussion follows of sources that were included and those that were not. This discussion focuses on sources of uncertainties in estimates of risk resulting from radon exposure in somewhat general terms, and gives an indication of the likely magnitude of the uncertainty for many of the sources.

## Measures of Risk

In this report, radon risks have been characterized in two ways. First estimates of relative risk (RR) were obtained using the committee's preferred risk models (chapter 3, Table 3-6). These estimates reflect the lifetime relative risk (LRR) of lung-cancer mortality at specified (constant) levels of exposure, for ever-smokers and never-smokers. Such estimates are subject to a number of uncertainties, which vary among individuals depending on the level of exposure to radon, smoking status, and other factors.

In order to gauge the public health impact of residential radon exposure, the population attributable risk (AR) is also considered (chapter 3, Table 3-7). It is important to note that such population-based measures of risk are subject to uncertainty, but not variability. Factors, such as the level of exposure to radon, or the dosimetric K-factor which vary among individuals within the population, are effectively integrated out when calculating the AR.

The type of risk estimate and purposes for which it is to be used affect the uncertainties in the risk. Estimates that are intended to reflect current U.S. exposure conditions are subject to uncertainty in estimating these conditions, whereas estimates intended to apply to hypothetical exposure scenarios are not.

The population to which the risk estimate is to be applied is also relevant; for example, because the underground miner studies included only males, estimates for females are less certain than those for males. In addition, because all factors that modify risks may not have been identified and included in risk models, estimates of individual risks are more uncertain than estimates of total (or average) population risks.

## Categorization of Uncertainties

Sources of uncertainty may be categorized in different ways. One approach is to consider uncertainties arising at each step in the risk assessment process. This report focuses largely on the development of an exposure-response model that expresses the dependence of lifetime lung-cancer risks on radon exposure and demographic variables. This model was based on analyses of data from 11 underground miner cohorts. Risk estimates for persons exposed in residences also require a model for estimating differences in radon dosimetry in mines and in homes, referred to here as the "exposure/dose conversion model". In addition, risk estimates intended to reflect current U.S. exposure conditions require estimates of these conditions. Demographic information, including baseline lung-cancer risks, is also needed. Thus, uncertainties are first categorized as indicated by the Roman numerals in chapter 3, Table 3-13. The material here primarily addresses uncertainties arising from the exposure-response model relating lung-cancer risk to radon exposure. The exposure/dose conversion model is discussed in appendix B, while estimation of the exposure distribution is discussed in appendix G. Uncertainties in the exposure/dose model and in the demographic data are briefly discussed later in this section.

Uncertainties in the model relating lung-cancer risk to exposure are further categorized by distinguishing uncertainties in the parameter estimates from uncertainties in specification of the model, and in its application to residential exposure of the U.S. population. This categorization is similar to that used by Morgan and others (1990), by the NRC committee on Risk Assessment of Hazardous Air Pollutants (NRC 1994b), and by the BEIR V committee (1990). As Morgan and others note, the distinction between parameter and model uncertainties is somewhat blurred, since the selected model can be viewed as a special case of a much richer model with many of the risk factors omitted. For example, the linear model used in this report is a special case of more general non-linear models, and was chosen because more general models did not provide significantly better fits to the underground miner data (Lubin and others 1994a) and because of the radiobiological considerations discussed in chapter 2. Morgan and others point out that every model is necessarily an oversimplification of reality, and, in this sense, all models are approximations. The distinction between parameter and model uncertainties is nevertheless a useful one, since the latter are especially important for extrapolating from exposure in mines to exposure in U.S. homes.

Because the risk model was developed from analyses of underground miner data, it provides a reasonably good description of the average risk for these miners. The U.S. population includes groups (for example, females) and conditions (for example, very low exposures) that are not represented in the underground studies, and, even for groups that are represented, the relative contributions of these groups differ (for example, the underground miner cohorts include a larger proportion of ever-smokers than the general U.S. population). A fully adequate model needs to take account of all modifying factors that differ between miners and the general population. To the extent that such factors are omitted from models, bias in risk estimates may result.

Rai and others (1996) distinguish between uncertainty and variability in risk assessment. Uncertainty represents ignorance about the values of model parameters, while variability represents the inherent variation in the values of parameters among individuals in the population of interest. Whereas population-based measures of risk such as the AR are not subject to variability, individual measures of risk such as the LRR do vary among individuals in the population. Models that take account of factors that modify risk and allow risk estimates for subgroups of the population can reduce variability within these subgroups. For example, there may be less variability in separate estimates of risk for ever-smokers and never-smokers than in a combined estimate; however, there might still be considerable variability among ever-smokers if the model did not account for the amount of smoking or distinguish between current and former smokers. For many factors, there is little basis for estimating variability. For example, both the shape of the exposure-response function and the pattern of risks over time probably vary from person to person, but data for quantifying this variation are scarce. Past efforts to quantify uncertainties in estimates of risk from radon or from low-LET radiation have addressed only uncertainty, and have not attempted to quantify variability (NRC 1988, 1990; Puskin 1992).

## Uncertainties in Parameter Estimates in the Lung Cancer Exposure-Response Model Derived from Underground Miner Data

### Sampling Variation in Underground Miner Data

Uncertainty resulting from sampling variation differs from most other uncertainty sources in that it can be quantified using rigorous statistical approaches. Sampling uncertainty in the fitted coefficients of models based on analyses of data from the 11 miner cohorts can be described in terms of the variance-covariance matrix of these coefficients, which takes account of both random error and heterogeneity among cohorts. The assumption of multivariate normality is widely used and generally valid in large samples. However, because the population attributable risk is a complex function of the estimated parameters, and because the distributions of the statistics involved may not be adequately approximated by multivariate normal distributions, the committee conducted Monte Carlo simulations to obtain uncertainty distributions and confidence intervals for attributable risks as described later in this appendix.

### Errors in the Underground Miner Data

In the three sections that follow, errors in the data from the 11 miner cohorts used to determine the committee's risk model are discussed. Ecological and case-control studies of people exposed to residential radon are also subject to biases from their data, but these are not discussed here because these data were not used directly in developing the risk model. These uncertainties are however discussed in appendix G.

### Errors in the Underground Miner Data on Health Effects

Vital status and information on cause of death were determined from a variety of information sources in the 11 miner cohorts, including local and national cancer and vital statistics registries, company records, electoral rolls, driving license records, telephone directories, and death certificates. In most cohorts, follow-up methods were thorough, but undoubtedly some misclassification of vital status occurred because of identifier inaccuracies and subject mobility. Failure to determine that a subject had died would lead to incorrectly including person-years after death had occurred, and might also lead to omitting some deaths due to lung cancer.

Errors in assignment of cause of death may also have occurred. Lung-cancer deaths that were incorrectly diagnosed as other diseases may have been missed, some deaths counted as lung cancers may have in fact been due to other causes, and secondary and primary lung cancers may sometimes have been confused. Several of the studies relied on death certificate diagnoses, and these were not always verified. Percy and others (1981) assessed the validity of death certificates for cancer in the United States in the 1970s by comparing the cause of death listed on death certificates with hospital records. For lung cancer, death certificates were both sensitive (95% of lung cancers were detected) and specific (93.9% of death certificates coded with lung cancer as the cause of death were confirmed by hospital records). These results would not necessarily apply to data from other countries, but most of the countries contributing data have high standards of medical care and should have high confirmation rates. It is worth noting that China and Czechoslovakia, the two cohorts with the largest numbers of lung-cancer deaths, used special methods to ascertain cases, and did not rely on death certificate information.

Because analyses of miner cohorts were based on internal comparisons within each of the cohorts, bias from errors in health endpoint data would be largest if these errors depended on the level of exposure; dependencies of this type do not seem likely. Because non-lung-cancer deaths that were incorrectly classified as lung cancers would not be related to exposure, such deaths could lead to slight underestimation of risk even if the misclassification were non-differential. Non-differential errors that resulted in missing some lung-cancer deaths would reduce power, but should not lead to bias. Relative to other sources of error, bias from errors in health outcome data is not thought to be large, and no attempt has been made to quantify it.

### Errors in the Underground Miner Data on Exposure to Radon and Radon Progeny

The exposure estimates used in the committee's analyses of the 11 miner cohorts are subject to many sources of error, as discussed in detail in appendix F. These errors occur because measurements of radon and radon progeny were limited for many of the mines, especially during the early period of mine operations. When measurements were not available, it was necessary to estimate exposures on the basis of data from earlier or later years, from measurements in proximal mines, or based on mine conditions. Even when measurements were available, they may not always have been entirely representative of the typical conditions to which workers were exposed, especially if the measurements were made for regulatory purposes. In some cases, especially in earlier years, measurements were made of radon rather than radon progeny, and these could not be used to estimate radon-progeny exposure accurately unless information on equilibrium was available.

In addition, the number of hours spent in the mine and in various locations were not fully documented. These limitations could have led to both systematic and random errors in estimates of radon-progeny exposure for individual miners. As indicated in appendix F, there are many factors that could lead to bias in exposure estimates. Data were not sufficient to determine the overall magnitude or even the overall direction of bias, and there is no assurance that various systematic biases "cancel" each other. Thus, these potential biases increase the uncertainty in risk estimates. Random exposure-measurement error can be expected to lead to underestimation of risk, and can also distort exposure-response relationships. Statistical methods are available for adjusting for such errors, but require that the nature and magnitude of the errors be specified (see appendix E and Thomas and others 1993). These methods are often difficult to implement, especially when the error structure is complex. The result of a limited simulation study in which the impact of exposure assessment error is illustrated in given in appendix G. This study also illustrates how these effects can be mitigated using adjustments for measurement error, provided the measurement error model is known.

Errors may also vary by calendar-year period and by the specific method used to estimate each worker's exposure. Some sources of error are independent for estimates for different workers while others may be correlated. Even if the details of these errors were fully understood, accounting for them would not be a simple task.

It is likely that some unknown portion of the variation in risk estimates among the 11 cohorts results from variation in the direction and magnitude of systematic biases and random errors in exposure measurements. A portion of the uncertainty resulting from exposure-measurement error may thus be included in the heterogeneity component of the variance for the risk coefficient based on the combined analyses. However, it is unlikely that this takes account of all uncertainty from this source, and no account has been taken of the tendency for random error to bias risk estimates downward. For most of the cohorts, exposure measurement errors are likely to be greatest in the earliest periods of operation when exposures were largest and fewer measurements were made. For this reason, measurement errors not only affect the estimates of the overall risk coefficient, but may also bias estimates of parameters that describe the relationship of risk with other variables such as exposure rate, time since exposure, and age at risk. For example, the underestimation of risk due to random exposure errors may be more severe at high exposure rates than at low exposure rates, leading to exaggeration of the effects of factors that modify the dependence of risk on exposure rate.

At this time, it is difficult to quantify bias and uncertainty resulting from exposure-measurement error. For the Czechoslovakian cohort, improvements in both exposure measurements and follow-up data increased the ERR/WLM from 0.37% to 0.61% with most of the increase attributed to the exposure-measurement changes (Tomášek and others 1994a). This does not reflect the full effect of measurement error in the earlier data since the revised data were still subject to error. Because exposure-measurement methods were not the same for all cohorts, the impact of exposure-measurement errors undoubtedly varies considerably among the eleven cohorts. Some efforts have been made to evaluate the impact of exposure measurement errors in individual miner cohorts, as described in appendix E.

In chapter 3, analyses with a focus on miner data below 0.175 Jhm$^{-3}$ (50 WLM), are described. These restricted analyses did not involve estimation of an exposure-rate parameter. Furthermore, miners exposed at these low levels were predominantly miners who were employed in more recent periods when exposure-assessment methods had improved substantially over those employed earlier. More than a third of the lung-cancer deaths in the exposure restricted analyses came from the Ontario cohort, where exposure assessment methods were among the best of the 11 cohorts. Lifetime risk estimates based on models developed from these analyses were very similar to those obtained using the committee's recommended models.

## Limitations in the Underground Miner Data on Other Exposures

In addition to the radon-progeny exposure, underground miners were also exposed to arsenic, silica, and diesel fumes. Because these exposures may be positively correlated with radon-progeny exposures, there is potential for confounding if no adjustment for such exposures is made. Even in the absence of such a correlation, these other exposures may have enhanced the risk due to radon progeny, leading to larger risks in miners than would result from exposure to radon progeny alone. Quantitative data on arsenic exposure were available only for China and Ontario, although Colorado, New Mexico, and France had data on whether subjects had previously worked in underground mines other than uranium mines. Adjustment for arsenic exposure reduced the estimated ERR/WLM for China from 0.61% to 0.16%, but had little effect on risk estimates for the other cohorts. Limitations in the available data make it very difficult to evaluate the potential bias in the overall risk estimate resulting from inadequate adjustment for arsenic exposure. Data on individual miner exposures to silica or to diesel fumes were not available. The possible effects of all three exposures are discussed in appendix F.

Smoking is another exposure that is of concern both as a confounder and as a risk modifier. The high prevalence of smoking in underground miners represents a source of uncertainty in extending the risk estimates from mines to the general population. Although tobacco-smoking is a very strong risk factor for lung cancer, data on smoking are rarely sufficiently detailed to allow adequate adjustments for smoking in epidemiological studies of lung cancer. Of the 11 cohorts used to develop the risk model, only 6 of the cohorts had data on smoking. These data were not always quantitative, and detailed data on changes in smoking status over time were not available for any of the cohorts except for some limited data on changes in smoking rates and dates of occurrence for miners in the Colorado cohort. Because there is no compelling reason to expect a correlation between smoking and radon progeny exposure, inadequacies in smoking data are of greater concern in evaluating the modifying effects of smoking on radon risks than in evaluating smoking as a confounder. This issue is discussed further below and in appendix C.

## Uncertainties in the Specification of the Lung Cancer Exposure-Response Model and in Its Application to the General U.S. Population

As noted above, these uncertainties are most important for extrapolating risks from underground mines to persons exposed in homes.

### Shape of the Exposure/Exposure-Rate Response Function

Cumulative exposures and exposure rates were generally much higher in underground mines than those encountered in homes. Perhaps the most fundamental aspect of the committee's model is the choice of method for extrapolating risks from occupational to residential exposure levels. Estimates of the ERR per unit exposure obtained using the committee's preferred models are based on data on miners with average exposure rates below 0.5 WL. In addition, both analyses restricted to miners with low cumulative exposures $< 0.175$ Jhm$^{-3}$ or $0.35$ Jhm$^{-3}$ ($< 50$ or $<100$ WLM) and a meta-analysis of studies of persons exposed in residences gave risk estimates that were similar to those obtained with the committee's recommended models. Nevertheless, the possibility of non-linearity, even in this low exposure range, cannot be entirely excluded. This issue is discussed further in chapters 2 and 3.

### Temporal Expression of Risks

The committee's risk models provide for a decline in risk with time since exposure and with age at risk. Although these patterns were consistently identified in nearly all of the underground miner cohorts, it is possible that the estimates of these effects could have been biased by time-dependent errors in both exposure measurements and health-endpoint data, and also by possible changes in smoking habits over time. To estimate these effects, it was necessary to include the high exposure miner data; unfortunately, data were inadequate to effectively investigate whether such effects might vary by level of exposure. The statistical uncertainty in estimating the parameters quantifying age at risk effects was included in the Monte Carlo simulations that were conducted, but the uncertainty in estimating the parameters quantifying the time since exposure parameters was not included.

## Dependence of Risks on Sex

The cohorts used to develop the committee's risk models included only male miners. Case-control studies of residential radon exposure include both males and females, but data are not yet sufficient to investigate the possible modifying effects of sex. Lung-cancer death rates in the United States are about three or four times higher for males than for females, which could affect risks resulting from radon exposure. It is not known with certainty whether risks for the two sexes follow the pattern of baseline risks, and are thus comparable on a multiplicative scale, or if they are independent of baseline risks, and thus are more comparable on an absolute scale. The committee has chosen to assume that risks are comparable on a multiplicative scale. However, the calculations were also made under the alternative assumption that risks for males and females were comparable on an absolute scale. This latter assumption increased the estimated excess lifetime risk for females by a factor of about 2.5.

A large portion of the difference in lung-cancer rates for males and females is undoubtedly due to differences in smoking habits. A multiplicative interaction between radon and smoking would thus support the use of a multiplicative treatment of sex, whereas an additive interaction would not. Analyses of the underground miner data indicated an interaction intermediate between multiplicative and additive, and this might suggest such an intermediate approach for addressing the modifying effects of sex. Also, lung-cancer risks in male and female A-bomb survivors were more comparable on a absolute scale than on a multiplicative one. See chapter 3 and Puskin (1992) for further discussion.

## Dependence of Risks on Age at Exposure

Lubin and others (1994a) did not find evidence that age at exposure modified lung-cancer risks in underground miners. Consequently, the committee's preferred models are based on the assumption that the excess relative risk did not depend on age at exposure. Studies of cancer risk in A-bomb survivors and other populations (UNSCEAR 1994) have also failed to provide evidence of modification of lung-cancer risk by age at exposure, although a decline in the excess relative risk with increasing age at exposure has been observed for other cancer categories including all solid tumors, digestive cancers, and cancers of the breast and thyroid. In most cases, the strongest evidence for a dependence of risk on age at exposure was based on a comparison of risks in those exposed as children (under age 20 years) and those exposed later in life.

Data from the miner cohorts on exposures in childhood are limited primarily to the China cohort. Of 813 lung-cancer deaths occurring in miners initially exposed under age 20 years, 735 occurred in this cohort, and all 54 of the lung cancers occurring in miners exposed under age 10 yrs were from the China cohort. In fact, a large percentage of miners in the China cohort were first exposed at very young ages, with only 25% (245) of the lung cancers occurring in those with first exposure at age 20 yrs and older. Analyses by Lubin and others (1994a) show statistically significant variations in ERR per unit exposure with age at first exposure in the China cohort, although the pattern was not consistent. Clearly the uncertainty in the lung-cancer risks in adults resulting from exposure in childhood is much greater than for risks resulting from exposure in adulthood Not only is there the possibility that the ERR per unit exposure for childhood exposures might differ from that predicted by the committee's models, it is also possible that the pattern of decline in risks differs from that observed for adult exposures. It is noted, for example, that the committee's preferred models are based on the assumption that risks persist for a lifetime, but there are no data to validate this assumption for exposure in childhood.

**Dependence of Risks on Smoking Status**

Limitations in the available data on smoking make it difficult to evaluate the modifying effect of smoking on radon risks. Because of the need to extrapolate from miners with a high proportion of ever-smokers to the general population with a lower proportion of ever-smokers, uncertainties in adjustments for smoking may even affect estimates of average population risks. Such uncertainties have a particularly strong effect on estimates that specifically address risks in ever-smokers and never-smokers. Because of limitations in available data, it was not possible to develop a model that took account of the amount smoked, degree of inhaling, and changes in smoking habits over time.

**Uncertainties in the Model for Estimating Differences in Radon-Progeny Dosimetry in the Mines and in Homes**

Several factors that affect the lung dosimetry of radon progeny differ between mines and homes. These differences must be accounted for in using risk estimates based on underground miners to estimate risks for persons exposed in homes. The parameter summarizing these differences is often referred to as the K-factor. This factor can be expected to vary among individuals, and its average value may also be subject to uncertainty. Uncertainty in the K-factor is discussed in more detail in appendix B (dosimetry). Variability in the K-factor was included in the committee's Monte Carlo simulations.

Uncertainty in the K-factor values arises from several sources. These include the measurement error in the size-dependent concentrations of airborne radioactivity collected in the sampling devices, the error in deconvoluting a size distribution for the measured activity fractions, and the uncertainty in the relative fractions of time assigned to the various locations in the mine. The error in determining the collected activity concentration depends on the statistics of the counts used to estimate the amounts of the 3 decay products, and the variation in the pump flow. Thus, there is variation in the in-home measurements since the $^{222}$Rn concentrations ranged from about 30 Bqm$^{-3}$ up to 800 Bqm$^{-3}$, depending on the home being studied. For typical airborne activity concentrations, the uncertainties in the individual concentrations are of the order of 5 to 10% and for PAEC, the errors are 3 to 5%.

These errors then propagate in a non-linear manner since they are input values to the algorithms used to estimate the activity-weighted size distributions. In general, this inversion process is ill-posed since it is an undetermined problem for which a unique solution is not possible. It is known from simulation exercises (Ramamurthi and others 1990) that these algorithms can find acceptable solutions although not necessarily the "true" solution. Thus, it is not possible to definitively determine the overall uncertainties in the size distributions. Similarly, it is also extremely difficult to precisely determine the uncertainty in the times for various mining activities. Although an exact uncertainty cannot be assigned to each K-factor value, it is estimated that the values in the central portion of the distribution should not have errors in excess of 25%. Thus, the variability in K is larger than its uncertainty.

## Uncertainties Relating to Background Exposures

The risk models developed by the committee based on its analysis of data from the 11 miner cohorts are based on occupational radon exposures. However, miners were also exposed to radon in their homes and outdoors. Although these additional exposures contribute to their lung-cancer risk, the residential and ambient exposures experienced by miners are much lower than their occupational exposures. Consequently, the impact of these non-occupational exposures on the coefficients in the committee's risk model is expected to be negligible.

Ambient exposures also need to be considered when evaluating residential radon risks, since people are exposed to radon in outdoor air as well as in their homes. Again, since ambient exposures are not widely available, it is not possible to adjust for the effects of outdoor exposures on residential radon risks. However, since ambient exposures are much lower than typical indoor exposures, any adjustment for outdoor exposures when evaluating residential radon risks is likely to be small.

## Uncertainties in the Demographic Data Used to Calculate Lifetime Risks

An additional source of uncertainty in lifetime risks involves the application of the committee's risk model to obtain estimates of risks for the U.S. population. These calculations require assumptions about the age distribution of the population, life expectancy, and baseline age and sex-specific lung-cancer mortality rates. As described in chapter 3, current U.S. life table and vital statistics data were used for this purpose. It was assumed that these data were appropriate both now and in the future. Rather than evaluate uncertainty from this source, it seems preferable simply to state that the lifetime risk estimates presented in this report are appropriate only for a population with these demographic characteristics. If changes occur in the future, or risk projection for other populations are desired, these estimates will need to be recalculated to reflect these modifications.

## QUANTITATIVE UNCERTAINTY ANALYSIS

Currently, there is a trend in risk assessment towards a more complete characterization of risk using quantitative techniques for uncertainty analysis (Bartlett and others 1996; Morgan and others 1990; NRC 1994b). The results of these analyses can be summarized in the form of a distribution of possible risks within the exposed population, taking into account as many sources of uncertainty and variability as possible. This distribution gives an indication of maximal and minimal risks that might be experienced by different individuals, and the relative likelihood of intermediate risks between these two extremes.

Finley and Paustenbach (1994) discuss the benefits and disadvantages of probabilistic exposure assessment compared to using point estimates. Point estimates are simple to interpret, but provide no indication of level of confidence. Probability distributions provide risk analysts with a more complete picture of the possible range of exposure, but are more complex to determine and to use in decision making. Edelmann and Burmaster (1996) show that distributions with the same 95th percentiles could have dramatically different shapes, and that risk-management decisions based on these distributions may be different. Recently, considerable effort has been extended in estimating distributions of risk for use in health-risk assessment (Finley and others 1994; Ruffle and others 1994).

Uncertainty exists in all stages of the risk-assessment process (Small 1994; Dakins and others 1994). An integrated environmental health-risk assessment model includes source characterization, fate and transport of the substance, exposure media, biological modeling, and estimation of risk using dose-response modeling. Each component of the integrated risk assessment is subject to uncertainty. Often what is known about input parameters may be admissible ranges, shape of the distribution, or the type of data. To estimate uncertainty in the output, uncertainty distributions are associated with the input parameters. Distributions characterizing the uncertainty associated with each adjustment factor are developed using scientific knowledge to the extent possible  These distributions can then be sampled using Monte Carlo methods to provide estimates of the output distributions.  Monte Carlo methods have been used for estimating the impact of uncertainty in adjustment factors on estimation of human population thresholds for non-carcinogens (Baird and others 1996).  Though convenient and easy to use with modern computing technology, Monte Carlo methods should be carefully monitored to ensure the integrity of the results (Burmaster and Anderson 1994).

Distinguishing between uncertainty and variability is necessary but sometimes difficult (Hoffman and Hammonds 1994; Bogen 1995).  One approach for describing uncertainty and variability in lognormal random variables is to plot a range of probability distributions in two dimensions, representing uncertainty and variability, thereby permitting the impact of uncertainty to be visualized (Burmaster and Korsan 1996). Uncertainty can sometimes be reduced by collecting more information, but variability cannot.  Information on metabolic activation, detoxification, and DNA repair was recently considered in evaluating inter-individual variability (Hattis and Barlow 1996; Hattis and Silver 1994).  It was shown that empirical studies of biological parameters are useful in establishing uncertainty factors for heterogeneity in individual risk.  It was also noted that the inter-individual variability tends to be overstated due to the presence of measurement error; adjustments can be made using empirical Bayes shrinkage estimators (Goddard and others 1994) and other statistical techniques.

## Quantitative Analysis of Uncertainty and Variability

Rai and others (1996) have developed a general framework for the analysis of uncertainty and variability in risk.  In the most general case, the risk $R$ is defined as function

$$R = H(X_1, X_2, \ldots, X_p) \tag{31}$$

of $p$ risk factors $X_1, \ldots, X_p$. Each risk factor $X_1$ may vary within the population of interest according to some distribution with probability density function $f_i(X_i \mid \partial_i)$, conditional upon the parameter $\partial_i$. Uncertainty in $X_i$ is characterized by a distribution $g_i(\partial_i \mid \partial_i^0)$ for $\partial_i$, where $\partial_i^0$ is the true value of the parameter. If $\partial_i$ is a vector valued, $g_i$ is a multivariate distribution. Here, it is assumed that the forms of the distributions $f$ and $g$ are known.  The case in which the form of $f$ or $g$ is unknown introduces another level of complexity which remains to be addressed.

If $\partial_i$ is a known constant and the distribution $f_i$ is not concentrated at a single point, $X_i$ exhibits variability only.  On the other hand, $X_i$ is subject to both uncertainty and variability if both $\partial_i$ and $f_i$ are stochastic.  When $f_i$ is concentrated at a single point $\partial_i$, and $\partial_i$ is stochastic, $X_i$ is subject to uncertainty but not variability.  Consequently, the variables $X_1, \ldots, X_p$ can be partitioned into three groups: variables subject to uncertainty only, variables subject to variability only, and variables subject to both uncertainty and variability.

Rai and Krewski (1998) consider the special case of a multiplicative risk model

$$R = X_1 X_2 \ldots X_p. \tag{32}$$

The multiplicative model is applicable in many situations encountered in practice, and affords a number of simplifications in the analysis. In particular, the multiplicative risk model in (32) is simplified by applying the logarithmic transformation. $X_i = log\, X_i\ (i = 1, \ldots p)$. After transformation, the multiplicative model can be re-expressed as an additive model

$$R^* = \sum_{i=1}^{p} X_i^*, \tag{33}$$

where $R^* = log\, R$. Assume that each $X_i$ has mean $\mu_i$ and variance $Var(X_i)$, where

$$\mu_i = E\{E(X_i|\partial_i)\} \tag{34}$$

and

$$Var(X_i) = E\{Var(X_i|\partial_i)\} + Var\{E(X_i|\partial_i)\}. \tag{35}$$

The expected risk on logarithmic scale in $E(R^*) = \sum_{i=1}^{p} E(X_i^*)$, with variance $Var(R^*) = Var\left\{\sum_{i=1}^{p} X_i^*\right\}$.

## UNCERTAINTY IN POPULATION ATTRIBUTABLE RISK

In the present application, uncertainty in estimates of the population attributable risk of lung cancer due to residential radon exposure is of primary interest. The general approach to uncertainty analysis proposed by Rai and others (1996) is used for this purpose. As discussed previously, the attributable risk (AR) is not subject to variability, since it is a measure of population rather than individual risk.

Following Levin (1953) and Lubin and Boice (1989), the attributable risk of lung cancer due to radon exposure is defined as the proportion of lung-cancer deaths attributable to radon progeny. For continuous risk factors, the AR can be written as

$$AR = \frac{\int_{0}^{\infty} R(w) f_w(w) dw - R(0)}{\int_{0}^{\infty} R(w) f_w(w) dw} = \frac{\int_{0}^{\infty} \{RR(w) - 1\} f_w(w) dw}{\int_{0}^{\infty} RR(w) f_w(w) dw}. \tag{36}$$

Here, $f_w$ is the marginal probability density function of the exposure distribution, reflecting variability in residential radon concentrations. $R(w)$ is the lifetime risk of lung cancer for a lifetime exposure to radon progeny at a yearly level $w$ in the presence of competing risks, and

$$RR(w) = \frac{R(w)}{R(0)} \tag{37}$$

is the lifetime relative risk. Note that lifetime excess relative risk, $ERR(w) = RR(w)–1$, appears in the integrand (36).

The K-factor also influences the lifetime risk of lung cancer. To accommodate the effect of the K-factor, equation (36) can be modified as

$$AR = \frac{\int\limits_0^\infty \int\limits_0^\infty [R(w,k) - R(0,k)] f_{w,k}(w,k) dw\, dk}{\int\limits_0^\infty \int\limits_0^\infty R(w,k) f_{w,k}(w,k) dw\, dk}. \tag{38}$$

Here the joint marginal distribution of $w$ and $K$, $f_{w,k}(w,k)$, is the product of the distributions of $w$ and $K$, since we assume that these two variables are statistically independent.

Let $h_i$ and $h_i^*$ be the lung cancer and overall mortality rates for age group $i$, respectively, in a referent population. Furthermore, let $e_i$ be the excess relative risk due to exposure $w$ to radon progeny for age group $i$. Here, we consider two types of models for $e_i$:

$$e_i = \beta\phi(i)w^*(i)\gamma_{wl}(w_i)K \tag{39}$$

and

$$e_i = \beta\phi(i)w^*(i)\gamma_{dur}(i)K, \tag{40}$$

where

$$w^*(i) = w_{5-14}(i) + \theta_{15-24}w_{15-24}(i) + \theta_{15-24}w_{25+}(i). \tag{41}$$

The factors $w_{5-14}, w_{15-24}$ and $w_{25+}$ represent cumulative radon-progeny exposures received 5-14, 15-24 and more than 25 years prior to disease diagnosis, respectively. Note both (39) and (40) are multiplicative models of the type (32).

The remaining risk factors in (39) and (40) are redefined as

$$\phi(i) = \begin{cases} \phi_1 \ \textit{for } i < 55 = 1 \\ \phi_2 \ \textit{for } 55 \leq i \leq 64 \\ \phi_3 \ \textit{for } 65 \leq i \leq 74 \\ \phi_4 \ \textit{for } i \geq 75, \end{cases} \tag{42}$$

$$\gamma_{wl}(w) = \begin{cases} \gamma_{w1} & for\ w < 0.5 = 1 \\ \gamma_{w2} & for\ 0.5 \le w \le 0.9 \\ \gamma_{w3} & for\ 1.0 \le w \le 2.9 \\ \gamma_{w4} & for\ 3.0 \le w \le 4.9 \\ \gamma_{w5} & for\ 5.0 \le w \le 14.9 \\ \gamma_{w6} & for\ w > 15, \end{cases} \tag{43}$$

and

$$\gamma_{dur}(i) = \begin{cases} \gamma_{d1} & for\ (i-5) < 5 = 1 \\ \gamma_{d2} & for\ 5 \le (i-5) \le 14 \\ \gamma_{d3} & for\ 15 \le (i-5) \le 24 \\ \gamma_{d4} & for\ 25 \le (i-5) \le 34 \\ \gamma_{d5} & for\ (i-5) > 35. \end{cases} \tag{44}$$

Models (39) and (40) express the age-specific excess relative risk using a risk factor $\gamma_{wl}$ for exposure concentration and an alternative risk factor $\gamma_{dur}$ for exposure duration, respectively. These correspond to the committee's exposure-age-concentration and exposure-age-duration models.

The lifetime risk of lung cancer is given by the sum of the risk of lung cancer death each year:

$$R(w,k) = \sum_{i=1}^{110} \frac{h_i(1+e_i)}{h_i^* + h_i e_i} S(1,i,e_i)[1 - q_i \exp(-h_i e_i)] \exp\left(\sum_{k=1}^{i-1} h_k e_k\right), \tag{45}$$

Here,

$$S(1,i,e_i) = \prod_{k=1}^{i-1} q_k(e_k) \tag{46}$$

and

$$q_i(e) = \exp[-(h_i^* + h_i e)] \tag{47}$$

is the probability of surviving year $i$ for an individual with exposure $w$ given that the individual survived up to year $i$-1.

The model for the AR in (38) depends on the uncertainty and variability distributions of $w$ and $K$. We assume that $w$ has lognormal distribution with geometric mean 24.8 Bqm$^{-3}$ and geometric standard deviation 3.11, and has uncertainty only in the geometric standard deviation. We also assume that $K$ has a lognormal distribution with geometric mean 1.0 and geometric standard deviation 1.5, and has no uncertainty. Since there is no closed form expression for the integrand in model (38), we approximate the integral by summing over the ranges of values for $w$ and $K$.

This approximate version of the attributable risk model depends on the risk factors in either (39) or (40). The uncertainty in these factors is characterized by lognormal or log-uniform distributions as summarized in Table A-8. These uncertainty distributions were based on the committee's judgment as to the likely range of values for each of these factors. By postulating an uncertainty distribution for each of the factors in the model, the committee acknowledges that all factors are subject to some degree of uncertainty. Although statistical uncertainty in the estimates of the parameters in the committee's performed models is included, distributions also reflect other sources of uncertainty as discussed previously in this appendix.

Uncertainty distributions of the AR were obtained with Monte Carlo sampling (Rai and others 1996). Although computationally intensive, the analysis is straightforward. First, a set of model parameter values was obtained by sampling from a multivariate normal distribution with mean equal to the estimated parameter values and the values in the covariance matrix given in Table A-9 and based on the standard statistical software package S-Plus. (The Monte Carlo simulation was done on a log scale, with each risk factor $X_i$ in (39) and (40) replaced by $exp[Ef(X^*_i)]$.) Second, the attributable risk was calculated as described previously in this appendix. Repeating this procedure 10,000 times in case I and 1,000

TABLE A-8a    Uncertainty and variability distributions for risk factors in the exposure-age-concentration model for excess relative risk

| Risk factor | Variability | Uncertainty |
|---|---|---|
| Model parameters | | |
| $\alpha = (\beta, \phi_a, \gamma_z)$ | Constant | $\alpha \sim N(\mu, \Sigma)^a$ |
| Exposure to radon w | $LN^b$ ($gm^c = 24.8$, $gsd^d = 3.11$) | |
| K-factor | $LN$ ($gm = 1$, $gsd = 1.5$) | $gm = 1.00$, $gsd \sim LU^e$ (1.2, 2.2) |

TABLE A-8b    Uncertainty and variability distributions for risk factors in the exposure-age-duration model for excess relative risk

| Risk Factor | Variability | Uncertainty |
|---|---|---|
| Model parameters | | |
| $\alpha = (\beta, \phi_a, \gamma_z)$ | Constant | $\alpha \sim N(\mu, \Sigma)^f$ |
| Exposure to radon w | $LN$ ($gm = 24.8$, $gsd = 3.11$) | |
| K-factor | $LN$ ($gm = 1$, $gsd = 1.5$) | $gm = 1.00$; $gsd \sim LU(1.2, 2.2)$ |

[a]Multivariate normal distribution with $\mu$ and $\Sigma$ specified in Table A-9a.
[b]*LN*: Log-Normal.
[c]*gm*: geometric mean.
[d]*gsd*:geometric standard deviation = defined as $exp_\sigma$, where $\sigma$ denotes the standard deviation of $log_eX$.
[e]*LU(a,b)*: Log-uniform distribution, with, with $log_eX$ uniformly distributed between $log_ea$ and $log_eb$.
[f]Multivariate normal distribution with $\mu$ and $\Sigma$ specified in Table A-9b.

TABLE A-9a  Parameters for uncertainty distributions for risk factors in exposure-age-concentration model[a]

### I. Estimated Values of Parameters[b]

| Parameter | $\beta$ | $\theta_{15\text{-}24}$ | $\theta_{25+}$ | $\phi_{55\text{-}64}$ | $\phi_{65\text{-}74}$ | $\phi_{75+}$ | $\gamma_{0.5\text{-}1.0}$ | $\gamma_{1.0\text{-}3.0}$ | $\gamma_{3.0\text{-}5.0}$ | $\gamma_{5.0\text{-}15}$ | $\gamma_{15.0+}$ |
|---|---|---|---|---|---|---|---|---|---|---|---|
| Value | -2.57 | 0.77 | 0.51 | -0.56 | -1.23 | -2.38 | -0.72 | -0.98 | -1.13 | -1.80 | -2.21 |

### II. Covariance Matrix[c]

| | $\beta$ | $\theta_{15\text{-}24}$ | $\theta_{25+}$ | $\phi_{55\text{-}64}$ | $\phi_{65\text{-}74}$ | $\phi_{75+}$ | $\gamma_{0.5\text{-}1.0}$ | $\gamma_{1.0\text{-}3.0}$ | $\gamma_{3.0\text{-}5.0}$ | $\gamma_{5.0\text{-}15.}$ | $\gamma_{15.0+}$ |
|---|---|---|---|---|---|---|---|---|---|---|---|
| $\beta$ | 9.47 | | | | | | | | | | |
| $\theta_{15\text{-}24}$ | -0.36 | 0.77 | | | | | | | | | |
| $\theta_{25+}$ | -0.04 | 0.24 | 0.42 | | | | | | | | |
| $\phi_{55\text{-}64}$ | -2.87 | -0.10 | -0.15 | 5.71 | | | | | | | |
| $\phi_{65\text{-}74}$ | -3.18 | -0.17 | -0.33 | 2.85 | 10.87 | | | | | | |
| $\phi_{75+}$ | -3.44 | -0.19 | -0.54 | 2.90 | 3.20 | 87.65 | | | | | |
| $\gamma_{0.5\text{-}1.0}$ | -5.57 | -0.10 | -0.02 | 0.14 | 0.42 | 0.83 | 8.24 | | | | |
| $\gamma_{1.0\text{-}3.0}$ | -6.36 | -0.12 | -0.11 | 0.15 | 0.53 | 0.97 | 5.88 | 6.93 | | | |
| $\gamma_{3.0\text{-}5.0}$ | -6.58 | -0.16 | -0.10 | 0.18 | 0.59 | 1.08 | 5.83 | 6.69 | 7.30 | | |
| $\gamma_{5.0\text{-}15.0}$ | -6.90 | -0.05 | -0.09 | 0.26 | 0.61 | 0.81 | 5.69 | 6.51 | 6.67 | 7.84 | |
| $\gamma_{15.0+}$ | -7.04 | -0.02 | -0.08 | 0.27 | 0.54 | 0.50 | 5.63 | 6.44 | 6.64 | 7.33 | 8.59 |

[a]Inter-individual variability in both the level of exposure to radon and the K-factor is also characterized by log-normal distributions. The parameters of the two distributions were determined from national data on the distribution of radon in U.S. homes and from data on a sample of homes used to estimate the K-factor.

[b]Except for $\theta_{15\text{-}24}$ and $\theta_{25+}$ values are on log scale.

[c]Except for $\theta_{15\text{-}24}$ and $\theta_{25+}$ values are on log scale. All values were multiplied by 100.

TABLE A-9b  Parameters for uncertainty distributions for risk factors in exposure-age-duration model

I. Estimated Values of Parameters[a]

| Parameter | $\beta$ | $\theta_{15\text{-}24}$ | $\theta_{25+}$ | $\phi_{55\text{-}64}$ | $\phi_{65\text{-}74}$ | $\phi_{75+}$ | $\gamma_{5\text{-}14}$ | $\gamma_{15\text{-}24}$ | $\gamma_{25\text{-}34}$ | $\gamma_{35+}$ |
|---|---|---|---|---|---|---|---|---|---|---|
| Value | -5.20 | 0.72 | 0.44 | -0.65 | -1.29 | -2.07 | 1.02 | 1.49 | 1.89 | 2.32 |

II. Covariance Matrix[b]

| | $\beta$ | $\theta_{15\text{-}24}$ | $\theta_{25+}$ | $\phi_{55\text{-}64}$ | $\phi_{65\text{-}74}$ | $\phi_{75+}$ | $\gamma_{5\text{-}14}$ | $\gamma_{15\text{-}24}$ | $\gamma_{25\text{-}34}$ | $\gamma_{35+}$ |
|---|---|---|---|---|---|---|---|---|---|---|
| $\beta$ | 7.98 | | | | | | | | | |
| $\theta_{15\text{-}24}$ | -0.30 | 0.98 | | | | | | | | |
| $\theta_{25+}$ | -0.01 | 0.25 | 0.44 | | | | | | | |
| $\phi_{55\text{-}64}$ | -2.07 | -0.11 | -0.21 | 4.32 | | | | | | |
| $\phi_{65\text{-}74}$ | -2.16 | -0.20 | -0.39 | 2.10 | 9.60 | | | | | |
| $\phi_{75+}$ | -2.43 | -0.24 | -0.59 | 2.15 | 2.43 | 95.37 | | | | |
| $\gamma_{5\text{-}14}$ | -5.06 | -0.21 | -0.14 | 0.31 | 0.37 | 0.54 | 4.60 | | | |
| $\gamma_{15\text{-}24}$ | -5.56 | -0.39 | -0.23 | 0.31 | 0.54 | 0.93 | 4.67 | 5.94 | | |
| $\gamma_{25\text{-}34}$ | -5.66 | -0.40 | -0.15 | 0.20 | 0.45 | 0.83 | 4.73 | 5.60 | 6.75 | |
| $\gamma_{35+}$ | -5.65 | -0.37 | -0.12 | -0.15 | 0.18 | 0.65 | 4.76 | 5.61 | 6.03 | 7.26 |

[a]Except for $\theta_{15\text{-}24}$ and $\theta_{25+}$ values are on log scale.
[b]Except for $\theta_{15\text{-}24}$ and $\theta_{25+}$ values are on log scale.  All values were multiplied by 100.

times in cases II and III (which require considerably more computational effort) produced uncertainty distributions for the AR.

The median of the uncertainty distribution for the AR is shown in Table A-10 along with 95% uncertainty limits covering the central mass of distribution. These limits range from 9.1-28.2% for the exposure-age-concentration model and from 6.8-21.0% for the exposure-age-duration model.

TABLE A-10a    Impact of uncertainty and variability on uncertainty intervals for population attributable risk for males

| Source | Exposure-age-concentration model | | Exposure-age-duration model | |
|---|---|---|---|---|
| | *AR* | 95%U.I. | *AR* | 95%U.I. |
| Uncertainty when K = 1 | 0.148 | (0.091, 0.238) | 0.103 | (0.068, 0.158) |
| Uncertainty incorporating variability in K | 0.150 | (0.097, 0.224) | 0.106 | (0.077, 0.178) |
| Uncertainty incorporating variability and uncertainty in K | 0.159 | (0.095, 0.259) | 0.111 | (0.081, 0.194) |

TABLE A-10b    Impact of uncertainty and variability on uncertainty intervals for population attributable risk for females

| Source | Exposure-age-concentration model | | Exposure-age-duration model | |
|---|---|---|---|---|
| | *AR*[a] | 95%U.I. | *AR*[a] | 95%U.I. |
| Uncertainty when K = 1 | 0.160 | (0.099, 0.256) | 0.111 | (0.079, 0.179) |
| Uncertainty incorporating variability in K[b] | 0.169 | (0.104, 0.278) | 0.119 | (0.084, 0.192) |
| Uncertainty incorporating variability and uncertainty in K[c] | 0.173 | (0.104, 0.282) | 0.125 | (0.088, 0.210) |

[a]Median of uncertainty distribution
[b]K ~ LN(gm = 1 , gsd = 1.5)
[c]K ~ LN(gm = 1,  gsd ~ LU(1.2, 2.2))

# Appendix B

# Comparative Dosimetry

## INTRODUCTION

In order to extrapolate the lung-cancer risk derived from the epidemiological analysis of the underground miner data to the general population, a number of steps are needed. The approach to dealing with smoking, gender, and age at exposure were discussed in chapter 3. In addition, it is necessary to account for the differences in exposure conditions between mines and homes, in breathing rates for different activities in mining and typical home behavior, and in respiratory physiology between men, women, children, and infants. These factors determine the relationships between exposure and dose and the possibility of differing relationships between exposure and dose for miners and for the general population needs to be considered in extrapolating a risk model from miners to the general population. The comparative dosimetry approach used by the committee follows that described by the Panel on Dosimetric Assumptions Affecting the Application of Radon Risk Estimates (NRC 1991) and the BEIR IV Committee (NRC 1988).

In chapters 2 and 3, the committee reviewed the biological and epidemiological data suggesting that lung-cancer risk varies directly with exposure to radon and its decay products It is now assumed that lung-cancer risks varies directly with the dose of alpha energy delivered to the appropriate cellular targets, and that the dose can be estimated from the exposure using a dosimetric model. A dosimetric model is employed to estimate the dose received by particular classes within the general population such as adult males (not miners), adult women, children, and infants as well as to adult male miners. The dosimetric model takes into account the exposure conditions in homes and in mines as well as the relevant physiological characteristics of the population groups. The ratio of the dose of alpha energy per unit exposure for a particular population group (men, women, children, infant) as given by the radon concentration to the dose per unit radon concentration to the miners is given by K:

$$K = [Dose_{home}/Exposure_{home}]/[Dose_{mine}/Exposure_{mine}] \qquad (1)$$

The K-factor includes diverse environmental and physiological factors and the use of this double ratio greatly simplifies the risk assessment for indoor radon. This chapter addresses dosimetry of radon progeny in the lung and presents the calculated K-factor values by reviewing the information available on exposure conditions in homes and mines, presenting the dosimetric-model used and then presenting the resulting distributions of K-factor values that were calculated.

## EXPOSURE

### Introduction

The formation and decay sequence for $^{222}$Rn was shown in Figure 1-1. Because $^{222}$Rn has an almost 4-day half-life, it has time to penetrate through the soil and building materials into the indoor environment where it decays into its progeny. There is some recent evidence that in spite of its short half-life, 55 seconds, $^{220}$Rn can also penetrate into structures in significant amounts. However, the data are limited and the extent of the thoron problem is quite uncertain as discussed in a subsequent section of this chapter.

The short-lived decay products, $^{218}$Po (Radium-A), $^{214}$Pb (Radium-B), $^{214}$Bi (Radium-C), and $^{214}$Po (Radium-C′), represent a rapid sequence of decays that result in two $\alpha$-decays, two $\beta$-decays and several $\gamma$-emissions following the decay. To illustrate the behavior of the activity of the radioactive products of the radon decay, the activity of each of the short-lived isotopes is plotted as a function of time for initially pure $^{222}$Rn in Figure B-1. Because $^{222}$Rn has a longer half-life than either of the four short-lived products, the progeny reach the same activity (number of decays per unit time) as the radon. The mixture then decays with the 3.8 day half-life of the radon. Each $^{222}$Rn decay results in four progeny decays so that the total activity is then the sum of these individual decay-product concentrations. The activity is the product of the decay constant (ln 2/half-life) times the number of radioactive atoms. Thus, the short-lived $^{218}$Po can have an activity equal to the $^{222}$Rn because a large decay constant times a small number of atoms becomes equal to a small decay constant times a large number of atoms.

If the products formed by the decay of the radon were to remain in the air, then there would also be equal activity concentrations of $^{218}$Po, $^{214}$Pb, and other progeny. The resulting mixture is said to be in secular equilibrium. As used in the monitoring of uranium mines, an equilibrium mixture of these decay products

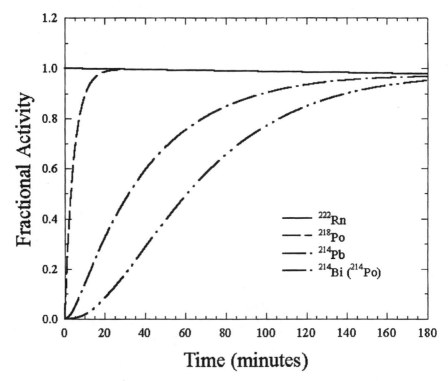

FIGURE B-1  Normalized in-growth of decay-product activities of an atmosphere initially containing only $^{222}$Rn.

at 3700 Bqm$^{-3}$ (100 pCiL$^{-1}$) is called a working level (WL).  Thus, a 370 Bqm$^{-3}$ (10 pCiL$^{-1}$) equilibrium mixture represents 0.1 working level.

The cumulative exposure to such activity can be expressed as the amount of activity in WL multiplied by time of exposure.  In occupational exposure assessments for miners, this cumulative exposure has historically been given in Working Level Months, WLM, where it is assumed there are 170 hours in a working month.  The WLM is calculated as

$$\text{Cumulative Exposure in WLM} = \sum_{i=1}^{n} (\text{WL})_i \bullet \left(\frac{t_i}{170}\right) \qquad (2)$$

where $(\text{WL})_i$ is the average concentration of radon decay products during exposure interval expressed in WL and $t_i$ is the number of hours of the exposure.  The cumulative exposure when spending all of the time in a house at a given decay product concentration is more than four times that for occupational exposure (8766 compared to 2000 hours worked on an annual basis).

More appropriately, the activity of the radon decay products is described by a quantity called the Potential Alpha Energy Concentration (PAEC). The total airborne potential alpha energy concentration, PAEC, is calculated as follows:

$$C_p(Jm^{-3}) = (5.79C_1 + 28.6C_2 + 21.0C_3) \times 10^{-10} \qquad (3)$$

where $C_p$ is the PAEC in $Jm^{-3}$, and $C_1$, $C_2$, $C_3$ are the activity concentrations of $^{218}Po$, $^{214}Pb$, and $^{214}Bi$ in $Bqm^{-3}$, respectively. This quantity incorporates the deposition of energy into the air. Exposure can be then expressed as the PAEC multiplied by the length of exposure in hours. This exposure is reported in the scientific units of Joule-hours per cubic meter ($Jhm^{-3}$). Although much of the prior epidemiological studies on the radon risk as observed in underground miner populations have used exposure measured in WLM, these values can be easily converted into proper SI units by multiplying the original estimates of exposure in WLM by $3.5 \times 10^{-3}$ $Jhm^{-3}$ $WLM^{-1}$.

In either homes or mines, decay products are lost from the air by attachment to environmental surfaces. In homes these surfaces include walls, floors, furniture, and the people in the room. The decay products also attach to airborne particles. The processes that control the airborne concentrations of the decay products are shown in Figure B-2. The attachment to the airborne particles maintains the radioactivity in the air since submicron particles typically remain suspended in the air over long times. The ratio of decay products to radon, termed the equilibrium factor, F, ranges from 0.2 to 0.8 with typical values of 0.35 to 0.4 (Hopke and others 1995).

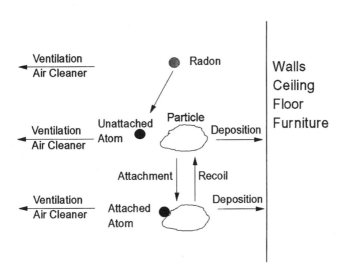

FIGURE B-2   Behavior of radon decay products in indoor air.

Another concept relating to the decay products is that of the "unattached" fraction. Although it is now known that the decay-product atoms are ultrafine particles (0.5 to 2 nm in diameter), an operationally defined quantity called the "unattached" fraction has been used to describe this activity. However, there is no general agreement on a procedure for estimating its value. The "unattached" fraction should not be considered as single atoms of progeny; we now know that the unattached fraction actually comprises small clusters of molecules. These unattached decay products have much higher mobilities in the air than the attached activity and can more effectively deposit in the respiratory tract. Thus, the "unattached" fraction has been given emphasis in estimating the health effects of radon decay products. Typically most of the "unattached" activity is $^{218}$Po and the value of unattached fraction, $f_p$, is usually in the range of 0.01 to 0.10 in indoor air, although it may be higher if the concentrations of particles in the air are very low. One does not actually measure unattached particle numbers but rather unattached decay-product activity in $Bqm^{-3}$. Thus, the PAEC can be obtained directly from the unattached activity.

## AIRBORNE PARTICLE PROPERTIES

The particles in indoor air are quite different in size from those that are typically encountered in the outdoor atmosphere. There are also many types of particle sources in a home such as open flames of a gas stove or candles, cooking, aerosol-spray products, and tobacco-smoking. In general, particles larger than 1 μm cannot penetrate into a house. A typical distribution of room air particles is shown in Figure B-3. The points are the average of 10 separate measurements. Particle sizes in typical room air are around 50 to 100 nanometers (nm). The size distributions for particles generated by gas stoves and by cigarette-smoking are shown in Figures B-4 and B-5. These distributions show that there are much higher particle concentrations when particle sources are operating. The high temperatures in the gas stove flame produces small particles with a peak in the distribution at about 20 nanometers while the lower temperatures in cigarette burning generates particles in the 100 to 300 nanometer (0.1 to 0.3 μm) range. The particle concentrations decrease over time because of surface deposition and ventilation. To obtain estimates of the activity-weighted size distributions, the number distributions are multiplied by the probability of attachment for the unattached radon decay product (Porstendörfer and others 1979).

### Exposure in Normally Occupied Homes

Over the past decade, improvements in measurement technology have allowed the direct measurement of the activity-weighted size distribution of the radon decay products in normally occupied homes. Methods have been developed by which the entire radioactive aerosol size distribution can be

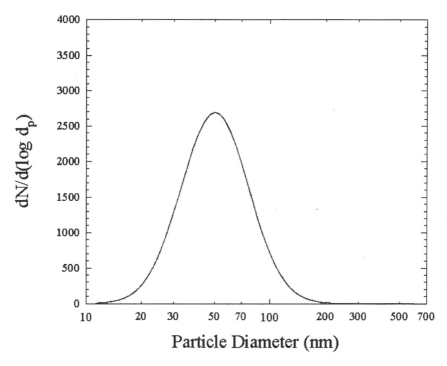

FIGURE B-3   Particle size distribution for room air.  The curve is a fit to averages (points not shown).  Each average derived from 10 separate measurements.  Figure taken from Li and Hopke (1993).

deduced from the measured activity collected on or penetrating through a series of screens.

The first activity-size measurements in indoor and ambient air were made by Sinclair and others (1977) using a specially designed high-volume-flow diffusion battery.  They observed bimodal distributions with activity mode diameters of 7.5 and 150 nm in indoor atmospheres and 30 and 500 nm outdoors in New York City.  Similar results were reported by George and Breslin (1980).  In Göttingen, West Germany, Becker and others (1984) only observed the larger mode using a modified impactor method with a minimum detectable size of 10 nm.   More extensive measurements in New York City by the group at the Environmental Measurements Laboratory (EML) have been reported by Knutson and others (1984) using several different types of diffusion batteries as well as cascade impactors.  They again observed modes around 10 nm and 130 nm in the PAEC-weighted size distribution measured with a low-volume screen diffusion battery.  Four samples taken with a medium-volume (25 L min$^{-1}$) screen diffusion battery showed a major mode at 80 to 110 nm and a minor mode containing 8 to 9% of

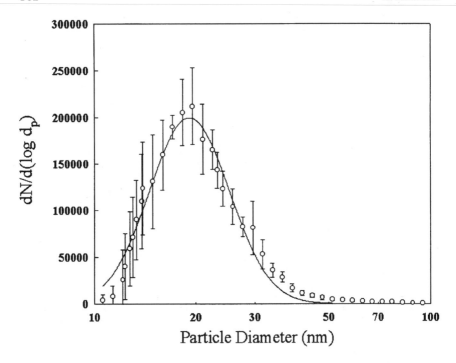

FIGURE B-4    Particle size distribution from a gas stove burner. Each point represents the average of 10 separate measurements. Figure taken from Li and Hopke (1993).

the PAEC with a diameter < 5 nm. Finally, the same group made measurements at Socorro, NM (George and others 1984). They report that the major mode was only slightly different from that found in New York. However, the minor mode was always < 5 nm, distinctly smaller than in the New York distributions.

One of the problems with the extension of screen diffusion batteries to smaller particle sizes is the substantial collection efficiency of the high mesh-number screens typically used in diffusion batteries designed to cover the range of particle size from 5 to 500 nm. At normally used flow rates, a single 635-mesh screen has greater than 90% efficiency for collecting 1 nm particles, the size of "unattached" [218]Po having a diffusion coefficient of the order of 0.05 cm$^2$s$^{-1}$. A theory of screen penetration (Cheng and Yeh 1980; Cheng and others 1980; Yeh and others 1982) had been developed in the late 1970s and this theory was validated to 4 nm by Scheibel and Porstendörfer (1984). The limitations of high mesh-number screens were recognized and it then became possible to examine new types of diffusion battery designs that could be extended to smaller particle diameters.

Reineking and others (1985), Reineking and Porstendörfer (1986) and Reineking and others (1988) use the high-volume flow diffusion batteries de-

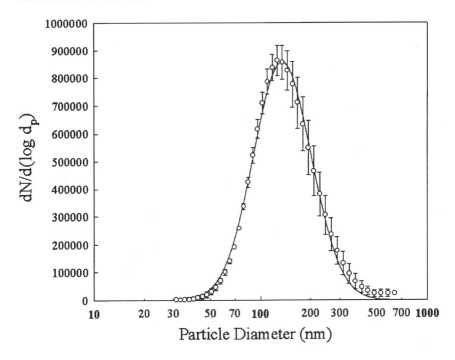

FIGURE B-5    Particle size distribution for sidestream cigarette smoke. Each point represents the average of 10 separate measurements. Figure taken from Li and Hopke (1993).

scribed in Reineking and Porstendörfer (1986) to obtain activity-size distributions. They obtained their size distributions by fitting log-normal distributions using an algorithm that searches the solution space for an acceptable fit (SIMPLEX). Size distributions of indoor air in rooms without and with the addition of particles from running an electric motor are presented in Figures B-6 and B-7.

Holub and Knutson (1987) report the development of low-flow diffusion batteries with low mesh number screens and extension of the EML batteries to smaller sizes. Tu and Knutson (1988a,b) have used the 25 L min$^{-1}$ screen diffusion batteries to measure the $^{218}$Po-weighted size distributions in the presence of several specific aerosol sources. The results of these measurements are presented in Figures B-8 and B-9. The presence of a mode around 10 nm is again observed in curve 1 in Figure B-8. Only in curve 1 (no active aerosol sources) in Figure B-9 is a mode at 1 nm observed. In all of the other cases, the activity is attached to the aerosol present in the house. The attachment was confirmed by independently measuring the aerosol-size distributions using an electrical aerosol analyzer (Liu and Pui 1975) and the attachment coefficients recommended by Porstendörfer and others (1979). There was a high level of agreement between the measured and calculated activity-weighted size distributions.

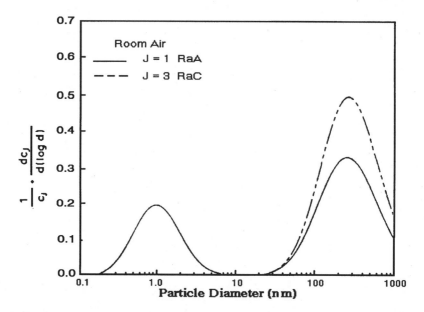

FIGURE B-6    Activity-weighted size distribution of the indoor aerosol in a closed room without active aerosol sources.  Figure taken from Reineking and Portstendörfer (1986).

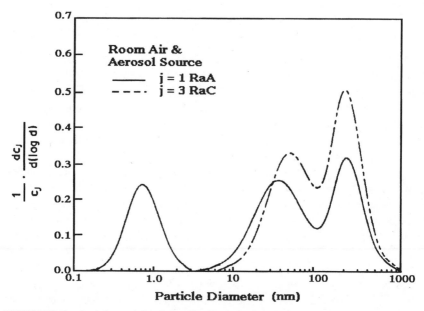

FIGURE B-7    Activity-weighted size distribution of the indoor aerosol in a closed room with an operating electric motor.  Figure taken from Reineking and Portstendörfer (1986).

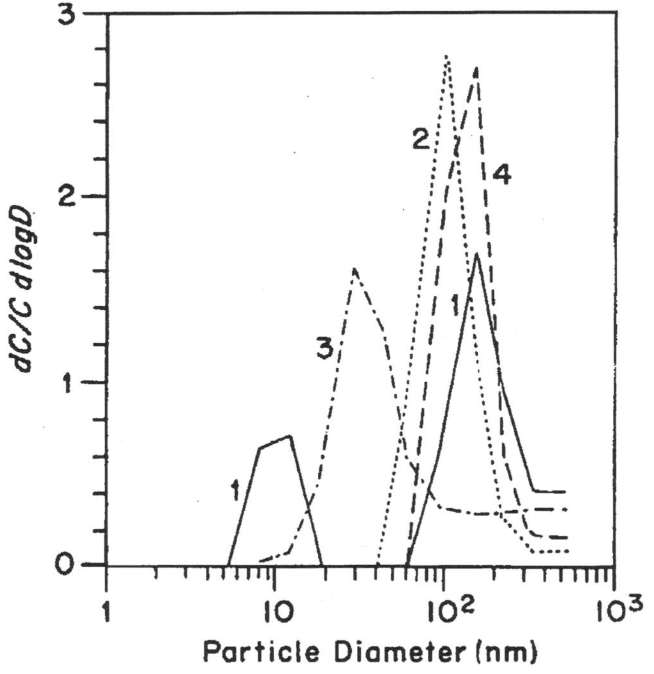

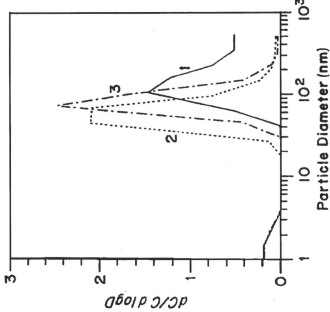

FIGURE B-9 ²¹⁸Po-weighted size distributions measured in house II after a kerosene heater was operated for: 1.0 min; 2.80 min; 3.20 min. Figure taken from Tu and Knutson (1988a).

FIGURE B-8 ²¹⁸Po-weighted size distributions measured in house I. 1. Cooking (5 min.); 2. Frying food; 3. Cooking soup; 4. Cigarette smoldering. Figure taken from Tu and Knutson (1988a).

Several other groups including the National Radiation Protection Board (NRPB) of the United Kingdom and the Australian Radiation Laboratory (ARL) have also developed these graded-screen diffusion batteries for activity-weighted size distribution measurements. Intercomparison experiments between the measurement methods have been performed (Hopke and others 1992). Generally good agreement was obtained.

Several automated systems to make use of this methodology have been developed. Strong (1988) used 6 sampling heads containing 0, 1, 3, 7 18, and 45 stainless steel, 400-mesh wire screens. He has measured the size distributions in several rooms in two houses at two times of the year. The size distributions observed in the kitchen are presented in Figure B-10.

These results are summarized in Table B-1. It should be noted that in the "kitchen" curve in Figure B-10, a trimodal distribution is observed; a true "unattached" fraction at 1 nm, a nuclei mode at 10 nm, and an accumulation mode at 100 to 130 nm. In the table, the "unattached" fractions presented are the integrated values from the size distributions. A problem of interpretation then arises regarding the "unattached" fraction since Strong integrates the distribution up to > 10 nm to obtain the fraction that he attributes as being "unattached." For the distribution in a kitchen during cooking, the activity median diameter for the "unattached" fraction is given as 11 nm. This designation is a clear departure from the original purpose for defining an "unattached" fraction. The advent of these more sophisticated and sensitive measurement systems has necessitated either a more precise definition of the meaning of the "unattached" fraction or cessation of its use (Hopke 1992).

Subsequent to these original measurements, Strong (1989) modified his system by changing the screens to one 200-mesh screen and 1, 4, 14, and 45 400-

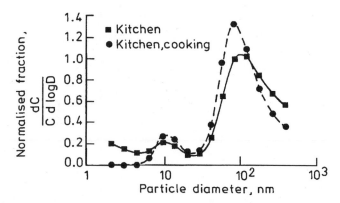

FIGURE B-10 PAEC size distributions measured in a rural house kitchen by Strong (1988).

TABLE B-1  Summary of activity size measurements made by Strong (1988) in row houses in the United Kingdom

| Site | Ambient Aerosol Median Diameter (nm) | GSD | $N(cm^{-3})$ | Attached AMD (nm) | GSD | $f_p$ (%) | Unattached AMD (nm) |
|---|---|---|---|---|---|---|---|
| Rural (summer) | | | | | | | |
| Bedroom | 42 | 2.0 | 5000 | 130 | 2.4 | 17 | 2.0 |
| Living Room | 30 | 2.0 | 5100 | 150 | 2.1 | 17 | 2.0 |
| Kitchen | 33 | 1.7 | 11000 | 130 | 2.0 | 18 | 6.0 |
| Kitchen (cooking) | 30 | 1.7 | 470000 | 110 | 1.9 | 11 | 11.0 |
| | | | | | | | |
| Rural (winter) | | | | | | | |
| Living Room | 32 | 1.7 | 4700 | 130 | 2.1 | 20 | 2.0 |
| | | | | | | | |
| Urban (winter) | | | | | | | |
| Living Room | 30 | 2.1 | 15000 | 110 | 2.1 | 20 | 3.5 |
| | | | | | | | |
| Mean of all sites | 33 | 2.0 | 8200 | 130 | 2.1 | 18 | 3.1 |

mesh screens as well as the open channel. This modification provides a stage such that there is higher resolution for the smallest sized particles extending the range of the system to 0.5 nm. The effective resolution cut-off of the original battery was about 2 nm. With the new battery, tri-modal distributions are clearly observed (Figure B-11). Although these measurements were made in the living room, the kitchen is adjacent and cooking with a gas stove was being performed at the time of these measurements. These results show the need to measure the size distribution and determine the actual exposure of individuals to airborne radon-progeny activity in order to estimate doses.

A similar system has been developed at the Australian Radiation Laboratory by Solomon (1989). It is designed for measurements in the size range of 2 to 600 nm. The measurements have been extended to a smaller size range (0.5 to 100 nm) using a manual, serial single-screen array sampling at 1 to 6 L min$^{-1}$ and to lower concentrations in the same size range using larger screens (9.5 cm diameter) and a 100 L min$^{-1}$ flow rate. Dr. Solomon has examined both the Twomey (1975) and Expectation Maximization (Maher and Laird 1985) algorithms for deconvoluting the size distributions from the screen-penetration data. For both algorithms, he has developed a Monte Carlo method for determining the stability of the inferred size distributions. New input values for the concentrations found on each stage are chosen from a normal distributions using the measured radon decay-product activity as the mean value and the measured uncertainty as the standard deviation of the distribution. This process can be repeated a number of times to provide a measure of the precision and robustness of the estimated size distributions.

A semi-continuous automated system has been developed by Ramamurthi and Hopke (1991). This system has now been used extensively to measure the exposure in a series of 6 homes in the northeastern United States and southeastern Ontario, Canada (Hopke and others 1995). The locations and characteristics of these houses are presented in Table B-2. The results for several of these houses

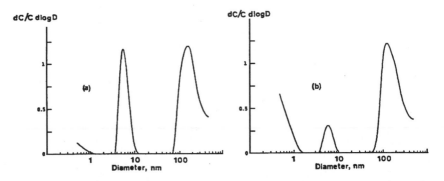

FIGURE B-11   Tri-model activity size distributions measured by Strong (1989) under conditions of a) F = 0.36, CN = 10,000 cm$^{-3}$; b) F = 0.26, CN = 5000 cm$^{-3}$.

TABLE B-2 Summary of the sampling campaigns and locations that provided activity-weighted size distribution in normally occupied homes

| Location | House ID | Sampling Period | No. of sample | Smoker | Heating System | Stove |
|---|---|---|---|---|---|---|
| Northfield, CT | | March-April, 1990 | 30 | No | Hot water baseboard, Oil-fired | Gas pilot light |
| Princeton, NJ | 31 | November 1990 | 61 | Yes | Forced Air, Gas-Fired | Gas pilot light |
| Princeton, NJ | 31 | February 1991 | 52 | Yes | Forced Air, Gas-Fired | Gas pilot light |
| Princeton, NJ | 41 | April 1991 | 30 | Yes | Forced Air, Gas-Fired | Gas without pilot light |
| Arnprior, ON | | May to July 1991 | 208 | No | Electric baseboard | Electric |
| Parishville, NY | | February to March 1992 | 59 | No | Forced Air, Oil-Fired | |
| Princeton, NJ | 51 | March 1992 | 46 | No | Forced Air, Gas-fired | |
| Princeton, NJ | 51 | July 1992 | 21 | No | Forced Air, Gas-Fired | |
| Princeton, NJ | 51 | March 1993 | 58 | No | Forced Air, Gas-Fired | |

have been gathered in order to provide baseline data for a series of studies of the behavior of air cleaners; Northford, CT (Li and Hopke 1991), Arnprior, Ontario (Hopke and others 1993), and Parishville, NY (Hopke and others 1994). Preliminary results, including 2 Princeton, NJ houses and one week of the Arnprior data, have been reported by Wasiolek and others (1992). Three other houses have been studied, but the measurements were made in the basement and may not represent the living areas of these homes. In two other homes, measurements were made while they were unoccupied and particles were produced through various activities such as cooking, vacuuming, and smoking. However, these results are not representative of the activity-weighted size distributions produced in normally occupied homes. In all cases radon concentrations were also measured using standard continuous radon monitors.

The measurements in the houses were combined in order to estimate the distribution of exposures and resulting doses. There are differing numbers of measurements in these various houses, but in order to combine the data, it was assumed that all of the measurements come from a single distribution. The results for these measurements are presented as cumulative distribution functions. The distribution of the equilibrium factor, F, based on the complete set of results (565 samples) is presented in Figure B-12. Summary statistics for this distribution are presented in Table B-3. The geometric mean value of F is 0.374 and the geometric standard deviation is 1.57, while the arithmetic mean is 0.408 with an arithmetic standard deviation of 0.159. Thus, the value of F in this sample of North American homes is lower than the previous estimate of 0.50 used by the Environmental Protection Agency in its risk estimates (USEPA 1992b) and is similar to the values measured in German homes by Reineking and Porstendörfer (1990).

Activity-weighted size distributions have been measured in 6 homes. In several cases, there was a person who lived in the home that smoked. The distributions of airborne radioactivity for each class of homes have been presented. In all of these cases, the activity-weighted size distributions are available so that doses can be calculated for each type of individual under consideration.

For a given radon concentration, and assuming other parameters to be constant, variation of F translates into a proportional variation in lung dose-rate. Wasiolek and others (1992) had earlier reported that the presence of a smoker in a home substantially affected the exposure and estimated dose. Thus, the size distributions were segregated into a non-smoking group (422 samples) and a smoking group (143 samples). The corresponding distributions for equilibrium factor for the non-smoking and smoking houses are given in Figures B-13 and B-14, respectively. The summary statistics for these distributions are also presented in Table B-3. These results will provide the basis for the activity-weighted size distributions that will be considered typical of exposure in normally occupied homes and used by the committee. For these measurements, the median dose is essentially the same in the homes with and without a smoker. This result is

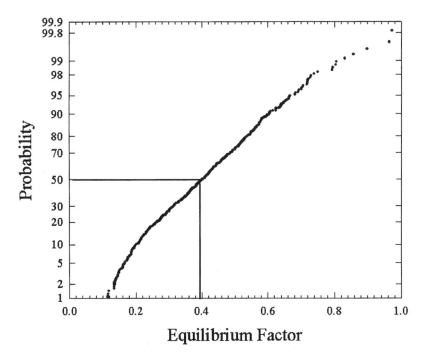

## Equilibrium Factor

FIGURE B-12  Cumulative frequency distribution for equilibrium factor based on the complete set of 565 measurements in 6 homes (Hopke and others 1995).

caused by the compensating factors of particle size and concentration. With a smoker present, there is a higher particle concentration leading to higher F values. However, the activity is found in particle sizes around 200 to 300 nm that are not deposited in the respiratory tract at a high rate and thus produce low doses per unit exposure (as will be presented later in this chapter).

In the case of the separated distributions, the influence of the smoker on the exposure conditions can be observed in the values in this table. The equilibrium factor in the homes with smokers is higher than in the non-smoker homes, where the arithmetic mean value in the smoker homes is 0.481 and geometric mean is 0.469 while the non-smoker values are 0.383 and 0.346.

### EXPOSURE IN MINES

#### Introduction

In the 1991 NRC report of the panel on Dosimetric Assumptions, an effort was made to estimate the aerosol characteristics that prevailed for the typical conditions that existed in mines. There are several major complicating factors in

TABLE B-3    Summary statistics for the measured or calculated distributions

|  | Arithmetic Mean | Arithmetic Standard Deviation | Geometric Mean | Geometric Standard Deviation |
|---|---|---|---|---|
| All Houses |  |  |  |  |
| Equilibrium Factor | 0.408 | 0.159 | 0.374 | 1.57 |
| Dose Rate per Unit Exposure |  |  |  |  |
| (nGy h$^{-1}$/ Bqm$^{-3}$) | 6.97 | 2.55 | 6.50 | 1.47 |
| Houses With Smokers Present |  |  |  |  |
| Equilibrium Factor | 0.481 | 0.108 | 0.469 | 1.24 |
| Dose Rate per Unit Exposure |  |  |  |  |
| (nGy h$^{-1}$/ Bqm$^{-3}$) | 6.77 | 1.45 | 6.62 | 1.24 |
| Houses Without Smokers Present |  |  |  |  |
| Equilibrium Factor | 0.383 | 0.166 | 0.346 | 1.61 |
| Dose Rate per Unit Exposure |  |  |  |  |
| (nGy h$^{-1}$/ Bqm$^{-3}$) | 7.04 | 2.83 | 6.346 | 1.54 |

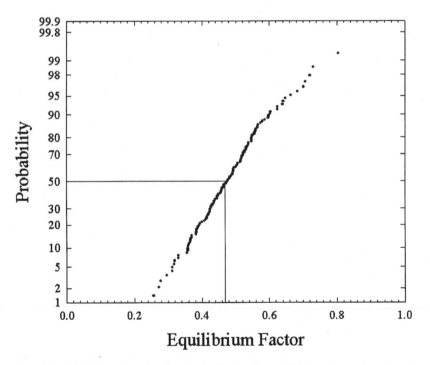

FIGURE B-13    Cumulative frequency distribution for equilibrium factor based on the set of measurements for houses with a smoker (Hopke and others 1995).

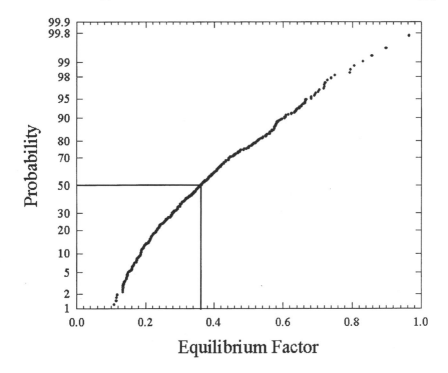

FIGURE B-14   Cumulative frequency distribution for equilibrium factor based on the set of measurements for houses without a smoker (Hopke and others 1995).

making the assessment of the activity-weighted size distributions. For example in the United States prior to 1961, there were no regulations of any kind with respect to the concentration of airborne activity and undoubtedly there were very high particle concentrations, particularly in the small mines if normal operations such as blasting and slushing were conducted. The history of ventilation in U.S. mines is outlined in Table B-4 (Also see workshop report in appendix E annex 2).

## Early Mining Operations

There are only 2 reports of any measurements in the time period prior to 1962 (Simpson and others 1954; HASL 1960). In both of these reports, efforts were made to measure activity-size distributions using a Casella cascade impactor (May 1945). Simpson and others examined the airborne activity in the Beaverlodge Mine. They found that essentially all of the activity was associated with particles less than 1 μm. When the dust concentrations were "high," there was more activity associated with larger particles. However, dust measurements were made with a konimeter. Particles in 5 cm$^3$ of air which impacted on a glass slide

TABLE B-4    History of Work Practices in U.S. Mines[a]

| Year | Description |
|------|-------------|
| 1947-1950 | No ventilation except to clear blasting smoke in old radium and vanadium mines. |
| 1950-1953 | As new mines were opened, there was more natural ventilation from vent holes and adjacent mines. |
| 1955 | Seven state conference on uranium mining health hazards. |
| 1957 | Union Carbide started extensive air sampling. |
| 1958 | Union Carbide adopted a 10 WL shut down. |
| 1960 | Governors' conference on radon hazards in uranium mines. |
| 1961 | Colorado adopted a 10 WL shut down. |
| 1962 | Union Carbide went to a 5 WL shut down. |
| 1963 | Colorado adopted a 3 WL limit. |
| 1967 | U.S. Department of Labor recommended a limit of 12 WLM per year. |
| 1969 | U.S. Department of Labor reduced limit to 3.4 WLM per year. |

[a]Courtesy of W. Chenoweth.

which had been heated to 1000°F before and after acid treatment, were counted at a magnification of 150× against a dark field. Counts were made of particles less than 5 μm diameter with a lower level of visibility being 0.25 μm. The results were reported as particles per cubic centimeter (ppcc). It is clearly very difficult to relate these values to total airborne particulate mass and thereby define "high" dust concentrations.

Measurements were made by HASL (1960) in 3 mines in western Colorado. Again 81 to 97% of the activity was attached to particles less than 1 μm in aerodynamic diameter. There was no separation of particles by size below 1 μm and it is difficult to use these data to estimate the dose per unit radon concentration. There were no dust concentrations given in this report.

### Later Measurements in Mines

There were a series of measurements made in 4 mines of the Grants Mineral Belt in New Mexico by George and others (1975, 1977). At the time, it was not possible to fully reconstruct the activity-weighted size distributions. However, subsequently Knutson and George (1992) have reanalyzed the data to obtain the full information from the original data. Typical distributions are shown in the next section. These results were chosen by the committee as the best representation of the aerosol conditions in typical working mines. Measurements were made in a variety of locations in the mines. Table B-5 presents a summary of the results of these measurements.

These data were grouped into 4 categories of work activities: slusher, stopes, drifts and haulageways, and machine shops as the representative regions within the mine. The hourly doses per unit radon concentration were calculated for each

TABLE B-5  Summary of the 1971 grants, NM mine measurements

| Mine | Date | Description | CNC[a] | $^{222}$Rn (kBqm$^{-3}$) | $^{218}$Po | $^{214}$Pb | $^{214}$Bi | F |
|---|---|---|---|---|---|---|---|---|
| A | 06/22/71 | Slushers Position 1 | 120,000 | 13.3 | 0.45 | 0.18 | 0.13 | 0.19 |
| A | 06/22/71 | Main Drift | 66,000 | 12.2 | 0.66 | 0.24 | 0.10 | 0.23 |
| A | 06/23/71 | Machine Shop | 140,000 | 4.1 | 0.57 | 0.23 | 0.12 | 0.22 |
| A | 06/23/71 | Slushers Position 2 | 110,000 | 10.0 | 0.34 | 0.13 | 0.05 | 0.12 |
| A | 06/23/71 | Main Drift | 93,000 | 6.3 | 0.63 | 0.18 | 0.06 | 0.18 |
| B | 06/24/71 | Stope 1 | 200,000 | 26.6 | 0.40 | 0.15 | 0.07 | 0.14 |
| B | 06/24/71 | Secondary Drift 2 | 210,000 | 44.4 | 0.27 | 0.05 | 0.01 | 0.06 |
| C | 06/25/71 | Machine Shop | 17,000 | 3.7 | 0.33 | 0.22 | 0.16 | 0.20 |
| C | 06/25/71 | Slushing Position 1 | 7,000 | 4.8 | 0.38 | 0.12 | 0.12 | 0.15 |
| C | 06/25/71 | Main Drift 1 | 260,000 | 6.3 | 0.54 | 0.18 | 0.06 | 0.18 |
| C | 06/28/71 | Main Drift 2 | 400,000 | 14.1 | 0.49 | 0.21 | 0.18 | 0.23 |
| C | 16/28/71 | Stope 2 | 320,000 | 17.8 | 0.66 | 0.25 | 0.11 | 0.23 |
| C | 06/28/71 | Stope 3 | 100,000 | 20.7 | 0.41 | 0.14 | 0.06 | 0.14 |
| D | 06/29/71 | Stope 1 | 40,000 | 13.0 | 0.48 | 0.25 | 0.22 | 0.27 |
| D | 06/29/71 | Drift 1 | 12,000 | 17.4 | 0.32 | 0.18 | 0.17 | 0.19 |
| D | 06/29/71 | Drift 2 | 5,000 | 18.5 | 0.52 | 0.19 | 0.13 | 0.20 |
| D | 06/29/71 | Drift 3 | 15,000 | 19.6 | 0.20 | 0.14 | 0.14 | 0.15 |
| D | 06/30/71 | Stope 2 | 80,000 | 19.6 | 0.37 | 0.14 | 0.09 | 0.15 |
| D | 06/30/71 | Stope 3 | 70,000 | 16.7 | 0.67 | 0.27 | 0.17 | 0.27 |
| D | 06/30/71 | Stope 4 | 40,000 | 19.2 | 0.43 | 0.18 | 0.14 | 0.19 |
| D | 07/01/71 | Main Haulage 2 | 26,000 | 9.3 | 0.25 | 0.07 | 0.04 | 0.08 |
| D | 07/01/71 | Exause Draft | 70,000 | 22.2 | 0.50 | 0.17 | 0.10 | 0.18 |
| D | 07/01/71 | Near Shaft | 10,000 | 5.6 | 0.21 | 0.05 | 0.01 | 0.05 |
| B | 06/24/71 | Secondary Drift 1 | 25,000 | 5.9 | 0.40 | 0.06 | 0.02 | 0.06 |
| C | 06/28/71 | Stope 1 | 3,000 | 7.0 | 0.33 | 0.15 | 0.14 | 0.17 |
| C | 06/28/71 | Slusher Position 2 | 1,000,000 | 7.8 | 0.41 | 0.14 | 0.06 | 0.14 |

[a]CNC = condensation nuclei counted.

group based on their average activity-weighted size distributions. The dose per unit radon for the miner was then estimated using a distribution of time as 40% in the slushing position, 40% in stopes, 10% in drifts and haulageways, and 10% in machine shops. It is assumed that the exposure in drifts and haulageways was received in getting to and from the work locations while the machine shop time would also reflect lunchrooms and other places away from active mining areas.

### Reference Size Distributions

In order to estimate the dose in homes and in mines, activity-weighted size distributions are needed. The size-dependent dose conversion coefficients can then be multiplied by the activity at each particle size and summed to yield the total dose. The reference size distributions in homes were taken from those measured by Hopke and others (1995) which were described earlier in this chapter. The activity-weighted distributions at the median dose for the houses with and without a smoker present were chosen as the representative distributions for homes. Figures B-15 and B-16 show these distributions.

The mine size distributions were derived from the constructed size distributions of Knutson and George (1992) who provided their data to the committee for this purpose. The size distributions for the various positions in mines are shown in Figure B-17.

### DOSE

### Dosimetry

Dosimetry is defined as the measurement of the amount, type, rate, and distribution of radiation emitted from a source of ionizing radiation and the calculation of both spatial and temporal patterns of energy absorption in any material of interest as a result of ionizing radiation. For a general review, see NRC 1991. Instruments and techniques exist to measure fields of penetrating radiation such as x rays or gamma rays that are external to the body, and provide means for directly quantifying the amount of energy deposited per unit mass of material (air, tissue, water). These dose measurements can then be related to a person present in the radiation field and the radiation dose that he or she would receive. In the case of internally-deposited radionuclides, however, direct measurement of the energy absorbed from ionizing radiation emitted by the decaying radionuclide is rarely, if ever, possible. Therefore, one must rely on dosimetric models to obtain estimates of the spatial and temporal patterns of energy deposition in tissues and organs of the body. In the simplest case, when the radionuclide is uniformly distributed throughout the volume of a tissue of homogeneous composition and when the size of the tissue is large compared with the range of the particulate emissions of the radionuclide, then the dose rate within the tissue is

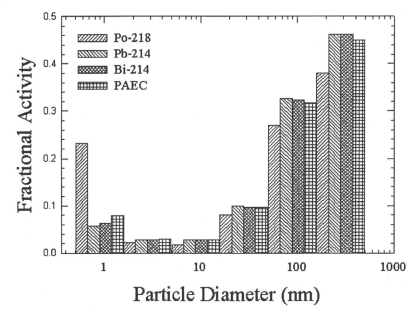

FIGURE B-15    Representative activity-weighted distributions for homes with a smoker present.

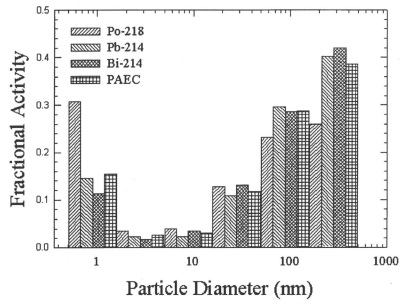

FIGURE B-16    Representative activity-weighted distributions for homes without a smoker present.

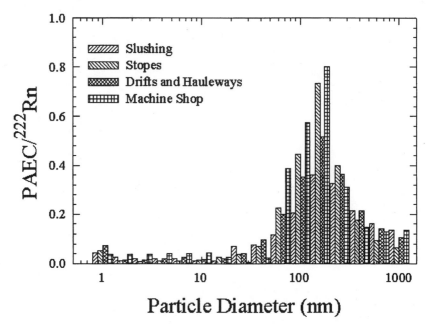

FIGURE B-17   Activity-weighted size distributions for the 4 types of locations in New Mexico mines.

also uniform and calculation of absorbed dose can proceed without complication. However, nonuniformities in the spatial and temporal distributions of the radionuclide, coupled with heterogeneous tissue composition, make the calculation of absorbed radiation dose complex and uncertain, as in the case of inhaled radon and radon progeny in the respiratory tract.

## Biodosimetry

To adequately link radiation exposure to risk it is essential to understand the relationship between the radiation exposure and the effective radiation dose to respiratory tract cells. A number of mathematical models that relate exposure to dose have been developed and they will be discussed in the next section of this document. Many of these models, as shown in this chapter, depend on knowledge of a range of physical factors such as attached fraction, particle size, respiratory rate, and airway morphology. This section briefly reviews the use of biodosimetry as a means to provide an additional check on physical models to evaluate the relationships between exposure in $Jhm^{-3}$ and dose in mGy in respiratory tract cells.

The biodosimetric approach has two major uses. First, to evaluate dose delivered to individuals in radiation accidents when no physical dosimetry is present. This area has been and continues to be an important part of radiation protection. The major cell type used for these studies has been the blood lymphocyte because its response represents an average dose to the individual, blood lymphocytes are easy to obtain, these cells divide very slowly or not at all in vivo and can be stimulated to divide in vitro which makes them a useful integrator of radiation dose with time, and finally the dose-response relationship for these cells has been carefully developed for a number of different radiation types. A useful review of the state of the art for biodosimetry using blood lymphocytes has been published (Bender and others 1988).

The second use of biodosimetry is to validate models of dosimetry for cells at risk in organs or systems where it is difficult or impossible to have direct measures of dose to target cells or tissues (Brooks and others 1997). This area is currently being developed for the respiratory tract. The approach has been to develop an in vitro/in vivo experimental animal respiratory tract model cellular system to understand the relationships that exist between cytogenetic damage in different populations of respiratory tract cells and radiation dose. The damage in these cell types has been measured as cell-killing (Thomassen and others 1992), chromosome aberrations (Brooks and others 1990b), or as the frequency of micronuclei induced in binucleated cells (Khan and others 1994, 1995; Brooks and others 1997; Bao and others 1997). These studies have measured damage using five different cell types from the respiratory tract; nasal, tracheal, and deep-lung epithelial cells (Brooks and others 1992, 1997; Thomassen and others 1992), deep-lung macrophages (Johnson and Newton 1994; Bisson and others 1994), and deep-lung fibroblasts (Khan and others 1994, 1995).

To characterize the relationships which exist between exposure, dose and cytogenetic damage it was necessary to understand how both physical and biological variables impact the response of model systems to radiation exposure. The influence of physical parameters including; total dose, dose rate, radiation type, LET, and aerosol characteristics have been characterized in deep-lung fibroblasts on dose-response relationships. Biological factors can also influence the response per unit of exposure or dose. The biological factors including cell turnover, species, strain, sex, cell cycle, and cell type have to be considered in these studies. Both of these sets of factors make biological dosimetry difficult to apply without adequate background information on the system to be used.

Research on physical factors demonstrated that in lung fibroblasts over the range of doses studied the radiation exposure increased the frequency of micronuclei as a linear function of radiation dose for both high- and low-LET radiation (Brooks and others 1994; Bao and others 1997). To determine the influence of dose and estimate an RBE for radon-induced micronuclei, CHO cells or rat deep-lung fibroblasts were given acute in vitro exposure to $^{60}$Co or radon. The in vitro dose was calculated according to the methods of Hui and others (1993). The slopes of the dose-response relationships were compared and resulted in RBEs of from 10-12 following acute radiation exposure (Brooks and others 1994).

To determine how dose-rate influenced dose-response relationships, rats were exposed to graded doses of [60]Co gamma rays for 4 or 67 hours and the frequency of micronuclei in deep-lung fibroblasts compared to that induced by acute exposure to [60]Co (Brooks and others 1995). When these values were compared to those following acute in vivo exposures a dose-rate effectiveness factor of 6.1 was derived. Exposure of the animals to [60]Co or inhalation of radon protracted over the same period of time (67 hrs) resulted in a linear increase in micronuclei as a function of dose. A comparison of the dose-response relationships for these exposure conditions resulted in an RBE of 65 for protracted exposures to radon vs low-LET gamma radiation (Brooks and others 1995).

In addition to physical factors, there are many biological factors that can influence observed dose and exposure response relationships. To extrapolate information from experimental animals to man it is essential to understand the dose-response relationships for in vivo and in vitro radiation exposure. When rat deep-lung fibroblasts were exposed either before or after isolation from the animal and the frequency of micronuclei determined, there were no differences observed in the slopes of the dose-response relationships (Khan and others 1994). Studies on the influence of cell proliferation on radon-induced micronuclei demonstrated that the sensitivity of proliferating and non-proliferating cells was not significantly different (Khan and others 1994). Studies on the induction of micronuclei in rat lung macrophages in vivo demonstrated that the maximum frequency of micronuclei was observed at 21 days after the radon inhalation. This is the time required in vivo for an optimum number of cells to undergo a single cell division and express the micronuclei induced by the radiation (Johnson and Newton 1994). The same length of time was used between exposure and scoring micronuclei for macrophages grown in vitro following exposure to alpha particles from [239]Pu.

Understanding how between-species differences for the induction of early cytogenetic damage relates to species differences for cancer induction is essential for risk extrapolation. Research has been conducted in a range of different species and strains to determine if the sensitivity for early cytogenetic damage reflects cancer risk. Studies were conducted to evaluate the clastogenic potency of radon in Wistar rats, Syrian hamsters, and Chinese hamsters (Khan and others 1995) and two strains of mice (A/J and C57BL6J) (Groch and others 1997). These animals were selected because of their sensitivity to radiation and chemically induced lung cancer. Rats are very sensitive to the induction of lung cancer from radon while Syrian hamsters, and mice have been reported to be very resistant to radon-induced cancer (Cross 1994a,b). The two strains of mice are very different in their response to lung cancer induced by ethyl carbamate. The A/J strain is very sensitive to chemically-induced lung cancer and the C57Bl6J is resistant. Our studies demonstrated that the order of sensitivity between species for radon-induced micronuclei was mice>Chinese hamsters>Syrian hamster> rats. There was no significant difference between the two strains of mice in the frequency of micronuclei induced by the radon inhalation or by ethyl carbamate injection. It was also determined that there was no influence of sex or strain of rats (Wistar vs Fischer 344) on [60]Co-induced micronuclei in respiratory epithelial or fibroblast cells (Bao and others 1997) even though these two strains of rats show a marked difference in the frequency of [239]Pu-induced lung cancer (Sanders and Lundgren 1995). These data suggest that there is little relationship between radiation sensitivity for the induction of lung cancer and micronuclei induction in respiratory-tract cells. There was on the other hand linear exposure- and dose-response relationships for the induction of micronuclei in all the animal species studied. This would support the concept that micronuclei are a very good indicator of dose but not a good indicator of cancer risk and extrapolation from dose or damage to risk should be done with caution.

It may be more important to understand the influence of cell type on cancer induction and genetic change than it is to understand how strain, sex, or species influences cancer risk. The basic assumption in using a relative risk model rather than the absolute risk model is that the induced cancer risk is related to the spontaneous cancer risk. The spontaneous cancer risk is very different in different organs and cells of the body. Studies related the initial radon-induced cellular damage to the cells at risk for radon-induced lung cancer. In these studies the frequency of micronuclei was measured in the deep-lung fibroblasts, deep-lung epithelial cells, and in nasal and tracheal epithelial cells following exposure to either [60]Co or radon inhalation. The [60]Co studies were designed to maintain a constant dose to each cell type and determine if there was a difference in radiation sensitivity between cells (Bao and others 1997). The radon study would have both dose and response varied for each of the cell types being evaluated. It was determined that following exposure to [60]Co the deep-lung fibroblasts and lung epithelial cells were from 2-3 times as sensitive for the induction of micronuclei as were the nasal or tracheal epithelial cells. The nose and trachea are much more resistant to radon-induced cancer than the deep-lung epithelial cells. When rats were exposed to radon and the frequency of micronuclei evaluated in all four cell types it was observed that the frequency of micronuclei in deep-lung epithelial cells (0.64 micronuclei/binucleated cell/WLM) was about 2-3 times as high as when fibroblasts were scored from the same animal (0.24 micronuclei/binucleated cell/WLM). This study demonstrated that even though these two tissues had similar sensitivity for the induction of micronuclei by low-LET radiation, they had different amounts of damage done by inhalation of radon. These observations suggest that the effective cellular dose (mGy) may be very different per WLM of exposure in the same animal.

By combining the data from the inhalation of radon in vivo and the dose-response relationship for micronuclei induced by radon exposure in vitro, it was possible to use micronuclei as a biodosimeter to convert exposure to dose in mGy (Khan and others 1994). Table B-6 shows the estimates of dose/exposure for a range of different tissues and cell types following inhalation of radon. Most of

TABLE B-6    Summary of biodosimetry

| End Point | Exposure (WLM) | Dose (mGy) | Exposure/Dose (mGy/WLM) |
|---|---|---|---|
| Cell Survival (T.E.) | Radon | $^{238}Pu$ | 2.1 |
| Chr. Abs (T.E.) | Radon | $^{238}Pu$ | 2.0-3.1 |
| Micro (L.M.) | Radon | $^{238}Pu$ | 9.8 |
| Micro (L.M.) | Radon | Gamma Rays | 0.75-2.5 |
| Micro (D.L.E.) | Radon | Radon | 0.47-1.1 |
| Micro (D.L.E.) | Radon | Calculation | 0.96-1.34 |
| Micro (N.E.) | Radon | Calculation | 0.18-0.33 |
| Micro (T.E.) | Radon | Calculation | 1.34-0.67 |

T.E. = Tracheal epithelial cells (Thomassen and others 1992, Brooks and others 1992, 1997)
L.M. = Deep-lung macrophages (Johnson and others 1994, Bisson and others 1994)
D.L.E. = Deep-lung epithelial cells (Brooks and others 1997)
N.E. = Nasal epithelial cells (Brooks and others 1997)

the tissues had between 0.2-3.0 mGy/WLM. The major exception was for the deep-lung macrophages (Johnson and Newton 1994) where values of about 10 mGy/WLM were reported.

These studies indicate that careful characterization of model cell systems is necessary before they can be used in biological dosimetry. Following such characterization, it has been demonstrated that biological dosimetry may be useful in determining dose when physical dosimetry is not available and for evaluating the distribution of dose and damage in the respiratory tract following inhalation of radon or other environmental pollutants. Additional studies are needed to better determine how dose distribution relates to exposure parameters. This research points out some of the problems associated with using short-term tests to predict risk and demonstrates that many of these tests are better measures of dose and initial damage than risk. However, at this time it is clear that biodosimetry cannot be used to estimate the exposure or dose to either miners or people living in their homes. Thus, a dosimetric model will be needed to relate exposure to dose.

## Dosimetric Models

Various models have been used to evaluate doses to the lungs from inhaled radon progeny (for example, Jacobi and Eisfeld 1980; Harley and Pasternack 1982; Hofmann 1982; NEA 1983; James 1988). These models divided the activity into two size categories: attached and unattached. In these models, a higher dose per unit exposure was assigned to the unattached fraction. However, the exact range of sizes that are assigned to this fraction in measurements by different investigators has varied widely (Ramamurthi and Hopke 1989). This imprecision in definition has necessitated more detailed assessments in which the dose conversion coefficients are calculated as continuous functions of particle size. The

1991 NRC report of the Panel on Dosimetric Assumption (NRC 1991) provides a detailed introduction to dosimetric models as well as a comprehensive evaluation of the conversion factors between exposure to PAEC and absorbed doses. The Panel's dosimetry model incorporated knowledge of model parameters of lung ventilation, aerosol deposition behavior, and anatomical descriptions of target cells as of 1990. This treatment was then further refined by the International Commission for Radiological Protection (ICRP 1994) based on subsequently developed information. Both of these dosimetry models consider the nuclei of secretory cells as well as those of basal cells as potentially sensitive targets in the respiratory tract. The ICRP's new respiratory tract deposition model also incorporates more accurate values for the filtration efficiency of the nasal passages (Swift and others 1992) than did the NRC model (NRC 1991). As described by James and others (1991) and James (1992), the new values for nasal filtration efficiency resulted in decreased doses to the bronchial tree from very small particles (below 3 nm) by a factor of around two in comparison with previous estimates. To use the newly adopted ICRP lung model to evaluate the equivalent lung dose as a function of the radon progeny activity-size distribution, the software code LUDEP (Jarvis and others 1993), which calculates regional lung deposition of activity per unit exposure, was extended to calculate tissue doses from the short-lived radon progeny as described by Birchall and James (1994). Subsequent to the committee's modeling and calculations, an additional model has been proposed by Harley and others (1996).

## Respiratory Tract Deposition of Particles

Deposition is the process that determines what fraction of the inspired particles is caught in the respiratory tract and, thus, fails to exit with expired air. It is likely that all particles that touch a wet surface are deposited; thus, the site of contact is the site of initial deposition. Distinct physical mechanisms operate on inspired particles to move them toward respiratory tract surfaces. Major mechanisms are inertial forces, gravitational sedimentation, Brownian diffusion, interception, and electrostatic forces. The extent to which each mechanism contributes to the deposition of a specific particle depends on the particle's physical characteristics, the subject's breathing pattern, and the geometry of the respiratory tract. Radon progeny are unusual in that they tend to be smaller than most aerosols of concern to health, and thus, Brownian diffusion dominates.

Detailed treatments of particle deposition have been given by Brain and Valberg (1979), Lippmann and others (1980), Heyder and others (1982, 1986), Raabe (1982), Stuart (1984), and Agnew (1984). Comprehensive treatises on aerosol behavior are also available (Fuchs 1964; Davies 1967; Mercer 1981; Hinds 1982; Reist 1984).

The behavior of a particle in the respiratory system is largely determined by its size and density. Particles of varying shape and density may be compared by their aerodynamic equivalent diameter ($D_{ae}$). Aerodynamic diameter is the diameter of the unit density (1 g/cm$^3$) sphere that has the same gravitational settling velocity as that of the particle in question. Aerodynamic diameter is proportional to the product of the geometric diameter and the square root of density.

## Diffusional

Radon progeny undergo Brownian diffusion—a random motion caused by their collisions with gas molecules; this motion can lead to contact and deposition on respiratory surfaces. Diffusion is significant for particles with diameters of less than 1 μm; only then does their size approach the mean free path of gas molecules. Thus, this is probably the dominant mechanism for radon progeny deposition since most of the alpha activity resides on such small particles. Unlike inertial or gravitational displacement, diffusion is independent of particle density; however, it is affected by particle shape (Heyder and Scheuch 1983). For these particles, size is best expressed in terms of a thermodynamic equivalent diameter, $D_i$, the diameter of a sphere that has the same diffusional displacement as that of the particle. The probability that a particle would be deposited by diffusion increases with an increase in the quotient $(t/D_t)^{1/2}$, where $t$ is the residence time and $D_t$ is the particle diffusion coefficient. Deposition is only weakly dependent on the inspiratory flow rate. Diffusion, like sedimentation, is most important in the peripheral airways and alveoli, where dimensions are smaller. However, as particle size becomes very small, diffusion may become an important mechanism even in the upper airways.

## Gravitational

Gravity accelerates falling bodies downward, and terminal settling velocity is reached when viscous resistive forces of the air are equal and opposite in direction to gravitational forces. Respirable particles reach this constant terminal sedimentation velocity in less that 0.2 ms. Then, particles can be removed if their settling causes them to strike airway walls or alveolar surfaces. The probability that a particle will deposit by gravitational settling is proportional to the product of the square of the aerodynamic diameter ($D_{ae}^2$) and the residence time. Thus, breathholding enhances deposition by sedimentation. Sedimentation is most important for particles larger than about 0.2 μm and within the peripheral airways and alveoli, where airflow rates are slow and residence times are long (Heyder and others 1986).

## Inertia

Inertia is the tendency of a moving particle to resist changes in direction and speed. It is related to momentum—the product of the particle's mass and velocity. High linear velocities and abrupt changes in the direction of airflow occur in the nose and oropharynx and at central airway bifurcations. Inertia causes a particle entering bends at these sites to continue in its original direction instead of following the curvature of the airflow. If the particle has sufficient mass and velocity, it will cross airflow streamlines and impact on the airway wall. The probability that a particle will deposit by inertial impaction, therefore, increases with increasing product of $D_{ae}^2$ and respired flow rate. Generally, inertial impaction is an important deposition mechanism for particles with aerodynamic diameters larger than 2 µm. Thus, it is probably unimportant for radon progeny in indoor air. It can occur during both inspiration and expiration in the extrathoracic airways (oropharynx, nasopharynx, and larynx) and central airways.

As particle size decreases, inertia and sedimentation become less important, but diffusion becomes more important. For example, a 2-µm-unit-density spherical particle is displaced by diffusion (Brownian displacement) by only about 9 µm in 1 s. It is important to know that this displacement varies with the square root of time. It settles via gravity by about 125 µm in the same period. However, as the particle size drops to 0.2 µm, the diffusional displacement in 1 s increases to 37 µm whereas gravitational displacement drops to only 2.1 µm. At 0.02 µm, gravitational displacement is only 0.013 µm/s while diffusional displacement in 1 s has soared to 290 µm.

## Anatomy of Respiratory System

The anatomy of the respiratory tract affects deposition through the diameters of the airways, the frequency and angles of branching, and the average distances to the alveolar walls. Furthermore, along with the inspiratory flow rate, airway anatomy specifies the local linear velocity of the airstream and the character of the flow. A significant change in the effective anatomy of the respiratory tract occurs when there is a switch between nose and mouth breathing. There are inter- and intraspecies differences in lung morphometry, even within the same individual, the dimensions of the respiratory tract vary with changing lung volume, with aging, and with pathological processes.

The human respiratory tract is pictured in Figure B-18. In order for the radon decay products to cause lung cancer, they must enter the body through the nose or mouth, pass down the trachea and then reach the bronchial region. Here the airways split and become smaller. The smaller branches are called the bronchioles. They conduct the air to the alveolar region where the oxygen and carbon dioxide exchange takes place. The surface area of the lung increases greatly as the air moves down the airways to the alveolar region.

Particle deposition in the various portions of the respiratory tract varies with particle size, breathing rates, relative amounts of oral and nasal breathing, and the dimensions of the various regions of the respiratory system. The parameters listed in Table B-7 for the various populations to be modeled are then the input.

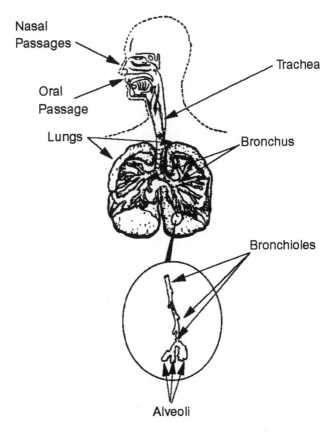

FIGURE B-18  Schematic diagram of the human respiratory tract. Adapted from *Handbook of Air Pollution*, USPHS 999-AP-44 (1968).

TABLE B-7  Parameters used in the ICRP (1994) lung-deposition model

| | |
|---|---|
| Functional Residual Capacity | Tracheal Diameter |
| Extrathorasic Dead Space | First Bronchiolar Diameter |
| Bronchial Dead Space | Ventilation Rate ($m^3$ $h^{-1}$) |
| Bronchiolar Dead Space | Frequency |
| Tidal Volume | Volumetric flow rate |
| Height | Nasal/oral fraction |

Recently, the ICRP model has been incorporated into a computer program that includes the respiratory tract deposition model. The model is based on average dimensions of the parts of respiratory tract based on a number of measurements. Recent studies have measured the deposition of particles using physical models of the nose and mouth as well as the initial portions of the bronchial tree. Using magnetic imaging technology, images of the noses and mouths of various people can be obtained and from these images, models can be built. Measurements have been made on other input parameters that are needed for the mathematical lung model. The results of the calculation using the ICRP model for a typical male breathing through his nose with a breathing rate with a typical mix of sleeping, resting, and light exercise ($0.779$ m$^3$h$^{-1}$) are presented in Figure B-19.

The smallest-sized activity deposits efficiently by diffusion and is largely removed in the nose and mouth. Thus, the unattached fraction (particles around 1 nm), long considered as delivering a proportionally greater dose per unit exposure to lung cells, has limited penetration to the bronchi. However, if these smaller particles increase in size slightly by chemical or physical processes in the

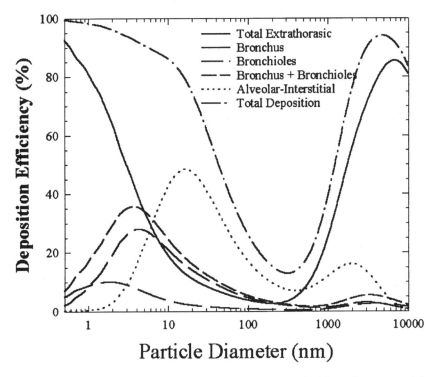

FIGURE B-19   Percent of particles of a specific size deposited in various parts of the human respiratory tract.

indoor air, they reach sizes around 5 nm. A 5 nm particle can effectively penetrate the nasal or oral passages, but still deposit in the bronchial region. Thus, the amount of radiation that reaches the lung tissue greatly increases in the 3 to 10 nm size range.

As the size continues to increase toward 100 nm, the effectiveness of the particle deposition decreases because the particle can move less effectively by diffusion. In the size range above 10 nm, most of the deposition occurs in the deep lung (alveolar region). In this region of the lung, there is a very large area so that the activity per unit surface area of the tissue is very small. For particle sizes above 500 nm, deposition starts to increase again because impaction can now help to deposit particles in the respiratory tract. For larger particles (> 2.5 μm) the deposition is mainly in the nose and mouth and very few of those particles reach the critical target cells of the lung where they could cause cancer. Thus, the size of the particle containing the radioactivity in both the home and the mines is very important in determining its effectiveness in delivering dose to the tissue.

The dose from radon progeny depends not only on the amount of aerosol deposited but also on the amount retained in the lungs over time. Retention is the amount of material present in the lungs at any time and equals deposition minus clearance. An equilibrium concentration is reached during continuous exposure to radon progeny when the rate of deposition equals the rate of clearance. The location where particles deposit in the lungs determines which mechanisms are used to clear them and how fast they are cleared. This influences the amount retained over time. Examples of the implications of particle characteristics on integrated retention were given by Brain and Valberg (1974) using a model developed by the Task Force on Lung Dynamics. They showed that the total amount as well as the distribution of retained dose among nose and pharynx, trachea and bronchi, and the pulmonary and lymphatic compartments were dramatically altered by particle size and solubility.

Particles that deposit on the ciliated airways are cleared primarily by the mucociliary escalator. Those particles that penetrate to and deposit in the peripheral, nonciliated lung can be cleared by many mechanisms, including translocation by alveolar macrophages, particle dissolution, or movement of free particles or particle-containing cells into the interstitium and/or the lymphatics. These pathways are relevant for materials that have a long biological half-life, such as silica, asbestos, or plutonium, but have little importance for radon progeny which are short-lived because of their short radioactive half-lives. The majority of the energy produced by alpha-particle emission is dissipated during the first hour following deposition. Thus, the movement of $^{222}$Rn progeny very soon after deposition is relevant to dosimetry. Reviews of clearance processes have been prepared by Kilburn (1977), Pavia (1984), and Schlesinger (1985).

The amount of particles retained within a specific lung region over time is a key determinant of dose. Mucous moves up the bronchial and bronchiolar regions and carries deposited material out of the lungs and into the gastrointestinal tract. Because of the short half-lives of the radon decay products, and the integration of their resulting dose over regions of the respiratory tract (bronchus, bronchioles, etc.), the radon decay products generally decay and cause damage before they can be cleared as discussed in detail by ICRP 66 (1994). However, for the other important isotope of radon, $^{220}$Rn, the second decay product, $^{212}$Pb, has a half-life of about 11 hours so that much of the radioactive $^{212}$Pb can be cleared before it decays. Thus, the thoron decay products have been estimated to produce a lower dose for the same exposure as the $^{222}$Rn progeny. Consequently, they have received little consideration in previous risk estimations.

The assumption that the tracheobronchial region is completely cleared by a rapid process has been challenged by a number of investigators beginning with Davies (1980). The main evidence for slow clearance in humans comes from a series of experiments by Stahlhofen and coworkers (Stahlhofen and others 1980, 1986a,b, 1987a,b, 1990, 1994; Stahlhofen 1989; Scheuch 1991; Scheuch and others 1993). The results of these studies are discussed in detail in ICRP 66 (1994) which concludes that a slow-clearance phase should be included from a complete lung dose model. Thus, calculations based on the ICRP 66 model include a fraction of slow clearance of about 80%, based on the work of Stahlhofen and coworkers. It should be noted that the presence of a slow-clearance fraction may be controversial. A recent NCRP lung model report (NCRP 125) was not available for use in this BEIR VI study. The presence of a slow-cleared fraction would increase the effectiveness of the $^{220}$Rn decay products in delivering dose to the cells at risk in the bronchial epithelium. However, there are insufficient exposure data for either mines or homes to permit estimation of the potential contribution of $^{220}$Rn decay products to the risk.

### Radon Progeny Dosimetry

The radioactive decay of inhaled short-lived radon progeny in the respiratory tract results in the deposition of alpha-energy in the cells of epithelial tissue. The dosimetric quantity measuring the radiation energy absorbed in tissue is the absorbed dose, D, expressed in grays (Gy).

Deposition of radon progeny in the respiratory tract causes alpha irradiation of several epithelial tissues; in the nasal passages, the bronchi, the bronchioles, and the alveolated part of the lungs (NRC 1991). Each of these tissues is potentially at risk for carcinogenesis (ICRP 1994). However, because the majority of lung cancers have been observed only in the bronchial and bronchiolar regions of the lung (Saccomanno and others 1996), only the dose to the bronchial and bronchiolar regions will be considered in the calculation of the dose per unit exposure.

### Dose Calculation

The conversion coefficients between airborne radon-progeny concentration and the equivalent dose-rate to the lungs as a whole, calculated as functions of monodisperse particle size, that are given by ICRP's lung dosimetry model are presented in Figures B-20(a) through B-20(c), for $^{218}$Po, $^{214}$Pb, and $^{214}$Bi, respectively. A breathing rate of 0.779 m$^3$ h$^{-1}$ is assumed to represent the average breathing rate of a reference subject (adult male) over the 7000 h yr$^{-1}$ assumed to be spent indoors at home (ICRP 1994a). This average breathing rate is estimated as 55% of time sleeping (0.45 m$^3$ h$^{-1}$), 15% sitting (0.54 m$^3$ h$^{-1}$), and 30% light exercise (1.5 m$^3$ h$^{-1}$). These dose conversion coefficients will be the basis for comparing the effects of exposure received from radon progeny arising from different exposure scenarios.

## Comparative Dosimetry

The objective of the comparative dosimetry is to calculate the differences in dose that are deposited in the respiratory tract between males working underground and males under more normal conditions as given above, women, children at the age of 10, and infants at the age of 1. The principal differences in the deposited doses for the same exposure conditions will be due to differences in the lung deposition. According to ICRP (1994), the depth and distribution of cells at risk are identical for all ages and both genders. Thus, the lung-deposition module of LUDEP was used to calculate particle deposition as a function of particle size for a series of different input parameters characterizing the different subpopulations of interest. The values for these parameters have been taken from ICRP (1994) and are presented in Tables B-8, B-9, and B-10. The results of the deposition modeling are the total particle deposition as a function of size summed over bronchial and bronchiolar regions of the lung for infants (1 year), children (10 years), men, and women in domestic environments. These values will be ratioed to the deposition in males at occupational breathing rates. The miner breathing rate value was chosen to be 1.25 m$^3$ h$^{-1}$, a value in agreement with the average value of 1.28 ± 0.26 m$^3$ h$^{-1}$ derived by Ruzer and others (1995) from measurements of underground metal miners in Tadjikistan. These values are substantially lower than the 1.8 m$^3$ h$^{-1}$ that had been used in the 1991 Dosimetry Panel Report (NRC 1991). ICRP (1994) suggests 1.69 m$^3$ h$^{-1}$ for heavy work. However, unpublished data from the South African Chamber of Mines Research Organization reported by ICRP (1994) estimated a mean value of around 1.3 m$^3$ h$^{-1}$. From the ratio of deposition and the dose-conversion coefficients calculated using the dose model described above, dose-conversion coefficients could be calculated for each type of individual (man, woman, child, and infant).

## K-FACTOR CALCULATION

To calculate the K factor, the typical dose rates to a member of the general population for each of the 2 exposure conditions (smoking/non-smoking environ-

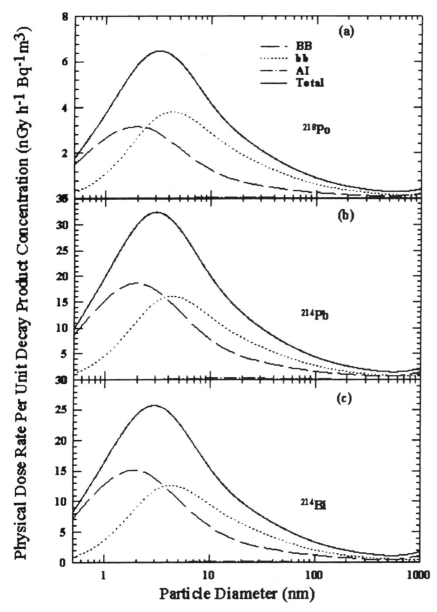

FIGURE B-20 Physical Lung Dose Conversion Factor between airborne concentration of the short-lived radon progeny (in $Bqm^{-3}$ and dose rate (in $nGy\ h^{-1}$), as a function of carrier particle diameter (in nm). The overall equivalent dose rate to the lungs as a whole is shown, together with the contributions from dose received by the bronchial region (BB), the bronchiolar region (bb), and the alveolar-interstitial region (A-I); (a) for $^{218}Po$, (b) $^{214}Pb$, and (c) $^{214}Bi$.

TABLE B-8  Respiratory reference values for a general Caucasian population at various levels of activity (taken from ICRP 1994)

| Population Subgroup | Sleeping | | | Sitting | | | Light Activity | | | Heavy Activity | | |
|---|---|---|---|---|---|---|---|---|---|---|---|---|
| | $V_T$[a] (dm) | B[b] (m³ h⁻¹) | $f_R$[c] (min⁻¹) | $V_T$ (dm) | B (m³ h⁻¹) | $f_R$ (min⁻¹) | $V_T$ (dm) | B (m³ h⁻¹) | $f_R$ (min⁻¹) | $V_T$ (dm) | B (m³ h⁻¹) | $f_R$ (min⁻¹) |
| Infant, 1 yr | 0.074 | 0.15 | 34 | 0.102 | 0.22 | 36 | 0.127 | 0.35 | 46 | N/A | N/A | N/A |
| Child, 10 yr | 0.304 | 0.31 | 17 | 0.333 | 0.38 | 19 | 0.583 | 1.12 | 32 | 0.841[d] 1.84 | 0.667[e] 44 | 2.22 36 |
| Female adult | 0.444 | 0.32 | 12 | 0.464 | 0.39 | 14 | 0.992 | 1.25 | 21 | 1.364 | 2.70 | 33 |
| Male adult | 0.625 | 0.45 | 12 | 0.750 | 0.54 | 12 | 1.250 | 1.50 | 20 | 1.923 | 3.00 | 26 |

[a] Tidal Volume.
[b] Ventilation Rate.
[c] Respiration Frequency.
[d] Male.
[e] Female.

TABLE B-9   Relative distribution of time in various activities for the time spent indoors at home (taken from ICRP 1994 and LUDEP default settings from Jarvis and others 1993)

| Population Subgroup | Sleeping | Sitting | Light Activity | Heavy Activity |
|---|---|---|---|---|
| Infant, 1 yr | 58.33% | 12.50% | 25.00% | — |
| Child, 10 yr | 41.67% | 12.28% | 30.56% | — |
| Female adult | 55.00% | 15.00% | 30.00% | — |
| Male adult | 55.00% | 15.00% | 30.00% | — |
| Miner adult | 0.00% | 31.30% | 68.80% | — |

TABLE B-10   Morphometric, physiologic, and lung volumes for children and adults (taken from ICRP 1994)

|  | Infant, 1 yr | Child, 10 yr | Adult Male | Adult Female |
|---|---|---|---|---|
| Height (cm) | 75 | 138 | 176 | 163 |
| Weight (kg) | 10 | 33 | 73 | 60 |
| TLC$^a$ (mL) |  |  |  |  |
|   Mean | 548 | 2869 | 6982 | 4968 |
|   Standard Deviation | 61.5 | — | 700 | 600 |
| FRC$^b$ (mL) |  |  |  |  |
|   Mean | 244 | 1484 | 3301 | 2681 |
|   Standard Deviation | 26 | 311 | 600 | 500 |
| $V_C{}^c$ (mL) |  |  |  |  |
|   Mean | 377 | 2326 | 5018 | 3551 |
|   Standard Deviation | 47 | — | 560 | 420 |
| $V_D{}^d$ (mL) |  |  |  |  |
|   Mean | 20 | 78 | 146 | 124 |
|   Standard Deviation | — | 14.9 | 22.5 | 21.0 |
| Anatomical Dead Space (mL) |  |  |  |  |
|   $V_{BB}{}^e$ | 6.81 | 25.10 | 50.00 | 40.08 |
|   $V_{bb}{}^f$ | 4.70 | 26.45 | 47.06 | 40.19 |
|   $V_{ET}{}^g$ | 8.73 | 26.49 | 48.75 | 44.18 |
| Smallest Bronchiole (cm) | 0.1068 | 0.1429 | 0.1651 | 0.1587 |
| Largest Bronchus (cm) | 0.7502 | 1.3114 | 1.6500 | 1.5342 |

$^a$Total lung capacity.
$^b$Functional residual capacity: volume of air that remains in the lungs after exhalation.
$^c$Vital capacity: maximum volume of air breathed in during inspiration.
$^d$Total anatomical dead space: the volume in which no gas exchange occurs.
$^e$Bronchial dead space volume.
$^f$Bronchiolar dead space volume.
$^g$Extrathoracic dead space volume.

ments) were calculated from activity concentrations associated with the activity-weighted size distributions for the home environment and the dose-conversion coefficient. This physical dose-rate per unit radon concentration is then divided by the estimated average dose to a miner based on the exposure results presented above. The miner dose-rate per unit radon concentration is 7.0 nGy h$^{-1}$/ (Bqm$^{-3}$). The dose-rates per unit radon are calculated for men, women, children, and infants based on the activity-weighted size distributions measured. Each value of the dose per unit radon concentration in homes is divided by the dose per unit radon concentration in mines to yield a series of K values. The distributions for the K values for houses with a smoker are presented in Figures B-21 to B-24. The distributions for the K values for homes without a smoker are shown in Figures B-25 to B-28.

The median values for the K factor for the various classes of individual in the 2 exposure scenarios (smoker and non-smoker) are given in Table B-11. It can be seen that all of these values are near to 1 so that the dose per unit exposure per unit $^{222}$Rn concentration are essentially the same in mines and homes. The measurements in homes (Hopke and others 1995) showed that the median dose per unit radon was relatively insensitive to the aerosol conditions in the homes.

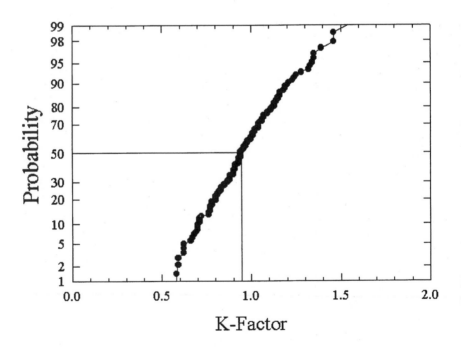

FIGURE B-21   Cumulative frequency distribution for the K factor for adult males based on the set of measurements in smoker homes.

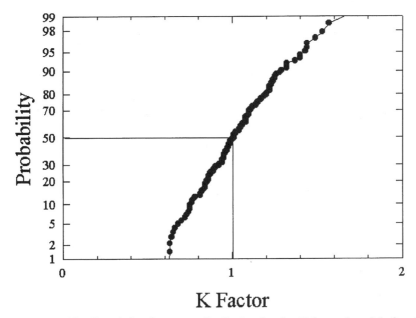

FIGURE B-22   Cumulative frequency distribution for the K factor for adult females based on the set of measurements in smoker homes.

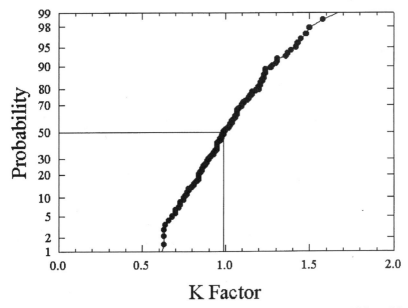

FIGURE B-23   Cumulative frequency distribution for the K factor for children (10 yrs of age) based on the set of measurements in smoker homes.

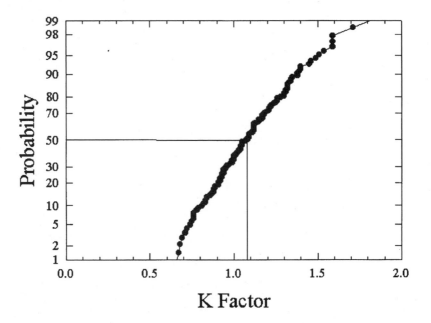

FIGURE B-24   Cumulative frequency distribution for the K factor for 1 year old infants based on the set of measurements in smoker homes.

In the case of the homes with a smoker, there is increased radon-progeny activity per unit radon gas concentration as shown by the increased equilibrium factor, but because the activity tends to be carried by larger particles, it is less effective in delivering dose to the respiratory tract. In the case of the non-smoker homes, the number of the larger particles is smaller and the amount of airborne progeny activity per unit radon gas concentration is also less, but because it tends to be carried by smaller particles, the average dose rate per unit activity concentration is higher.

The previous NRC analyses of lung dosimetry in mines and homes (NRC 1991) found K factor values around 0.75 for males and females and somewhat higher for children and infants. In that earlier NRC report, a sensitivity analysis was performed to ascertain the effects of changes in input parameters to the dosimetric model on the resulting K values. The dosimetric model used here is very similar to that used in that report and a new sensitivity analysis was not undertaken.

In that analysis, it was found that changes in the breathing rate had the largest effects on the dosimetry. Although breathing rate has some effect on the amount and the location of activity deposition in the lungs, the largest effect of changing the breathing rate is the resulting change in the total volume of air inhaled and thus the total amount of PAEC inhaled. The median K factor in the new calculation has increased from previously calculated values to close to 1. This change

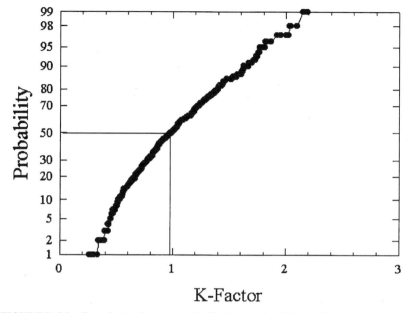

FIGURE B-25 Cumulative frequency distribution for the K factor for adult males based on the set of measurements in non-smoker homes.

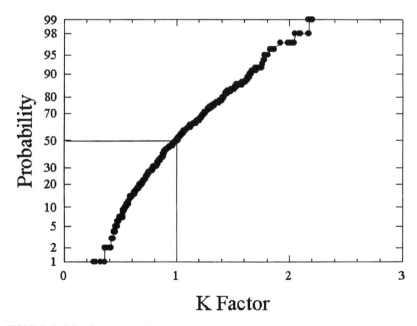

FIGURE B-26 Cumulative frequency distribution for the K factor for adult females based on the set of measurements in non-smoker homes.

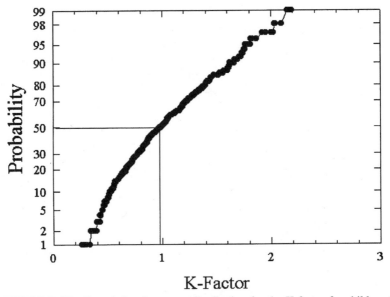

FIGURE B-27    Cumulative frequency distribution for the K factor for children (10 yrs of age) based on the set of measurements in non-smoker homes.

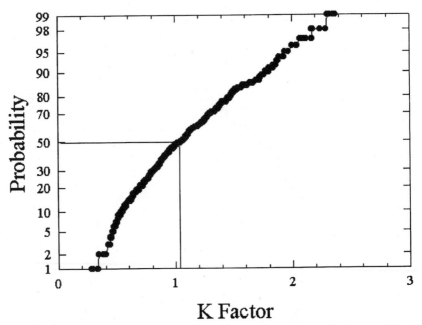

FIGURE B-28    Cumulative frequency distribution for the K factor for 1 year old infants based on the set of measurements in non-smoker homes.

TABLE B-11    Values of the K Factor

| Subject Type | Home with a Smoker | Home without a Smoker |
| --- | --- | --- |
| Male | 0.94 | 0.95 |
| Female | 1.01 | 1.00 |
| Child | 0.99 | 0.98 |
| Infant | 1.08 | 1.05 |

primarily reflects a reduction of the miner breathing rate from the 1.8 $m^3$ $h^{-1}$ to 1.25 $m^3$ $h^{-1}$ in this present study. A comparison of the inputs to the dosimetric modeling that was used to derive the K factor is presented in Table B-12. As discussed above, the major difference between the 2 results was the change in breathing rates based on Ruzer and others (1995) and the South African Miner Data quoted by ICRP (ICRP 1994). In this analysis, the committee was able to utilize the more extensive set of the measurement in homes by Hopke and others (1995) and the more complete reconstruction of the activity-weighted size distributions in mines provided by Knutson and George (1992) and George (1993). For the earlier study, the growth of particles when respirated into the high humidity of the lungs was estimated to be a factor of 2. In the intervening time, a number of measurements on room air and a variety of materials (Li and Hopke 1993, 1994; Dua and others 1995; Dua and Hopke 1996) have suggested that this growth factor is quite reasonable for typical room air and since much of aerosol in the ventilation air in the mine is ambient aerosol, particularly in the size range below 1 μm where the attached activity is typically observed, this factor of 2 is also quite reasonable for mines as well.

## THORON

There are 2 other naturally occurring isotopes of radon, $^{220}$Rn and $^{219}$Rn, commonly referred to as thoron and acton. $^{220}$Rn is a decay product of $^{232}$Th and $^{219}$Rn is a decay product of $^{235}$U. $^{219}$Rn has a half-life of 3.6 s and thus, it is not transported far after its formation to produce measurable exposure even to miners. However, even with a half-life of 55.6 s, $^{220}$Rn can be found present in the atmosphere (Schery 1992). Thoron undergoes α-decay through short-lived $^{216}$Po (0.15 s) which also undergoes α-decays to $^{212}$Pb with a half-life of 10.64 h. The decay chain, including half-lives and the energies of the emitted radiation, is outlined in Table B-13.

### Homes

There have only been limited measurements of $^{220}$Rn and its decay products in indoor air. In the measurement of $^{220}$Rn there can be substantial variation

TABLE B-12  Comparison of input parameters[a] to comparative dose modeling

| Parameter | 1991 Dose Panel | BEIR VI |
|---|---|---|
| Respiratory Physiology | ICRP66 Defaults[b] | ICRP66 Defaults |
| Breathing Rates (Miners) | 1.80 m³ h⁻¹ | 1.25 m³ h⁻¹ |
| Extrathoracic Deposition of Ultrafine Activity | ~50% for "unattached" (Breslin and others 1969) | ~85% for 1 nm (Swift and others 1992) |
| Breathing Rates (Domestic) | ICRP66 Defaults[b] | ICRP66 Defaults |
| Exposure Conditions in Mines | Limited size distributions (Cooper and others 1973) | 30 distributions recalculated from 1970 data (Knutson and George 1992) |
| Exposure Conditions in Homes | Measurements with created exposure scenarios (Li and Hopke 1991) | 565 measurements in normally occupied homes (Hopke and others 1995) |
| Hygroscopic Growth | Estimated factor of 2 | Factor of 2 based on series of published measurements (Li and Hopke 1993, 1994; Dua and others 1995; Dua and Hopke 1996) |

[a]Parameters are listed in tables B-8 through B-10 earlier in this appendix.
[b]Although ICRP officially adopted the parameter values after the 1991 NRC report was issued, the parameter values were the same.

TABLE B-13   Properties of the $^{220}$Rn decay chain

| | Type of Radiation | Half-Life | Radiation Energies (MeV) |
|---|---|---|---|
| Radon-220 (Th) | α | 55.6s | 6.2883 |
| Polonium-216 (ThA) | α | 0.15s | 6.7785 |
| Lead-212 (ThB) | β | 10.64h | 0.331, 0.569 |
| | γ | | 0.23863, 0.30009 |
| Bismuth-212 (ThC) | β (64%) | 60.6m | 0.67, 0.93, 1.55, 2.27 |
| | γ | | 0.7272 |
| | α (36%) | | 6.051, 6.090 |
| Polonium (ThD) | α | 0.298 ms | 8.7844 |
| Thallium-208 (ThC') | β | 3.053 min | 1.796, 1.28, 1.52 |
| | γ | | 2.6146, 0.5831, 0.5107 |
| Lead-208 | — | Stable | |

across the volume of a room because of the short half-life. Consequently, it is difficult to define representative values of $^{220}$Rn. Thus, most of the available measurements have been made of the $^{220}$Rn progeny rather than $^{220}$Rn. The available data for the ratio of PAEC arising from $^{220}$Rn decay products to that from $^{222}$Rn are summarized in Table B-14. Rannou (1987) has examined the variation of the PAEC($^{220}$Rn) as the PAEC($^{222}$Rn) changes and found that

$$PAEC(^{220}Rn) = PAEC\ (^{222}Rn)^{0.4} \qquad (4)$$

so that the exposure from the $^{222}$Rn progeny increases more rapidly than the $^{220}$Rn decay products. Thus, for dwellings with high $^{222}$Rn concentrations, it appears

TABLE B-14   Approximate ratio of PAEC for $^{220}$Rn progeny to $^{222}$Rn progeny at various locations (Schery 1990)

| Location | $\dfrac{PAEC\ (^{220}Rn)}{PAEC\ (^{222}Rn)}$ |
|---|---|
| Italy (Latium), 50 dwellings | 1.3 |
| Canada (Elliot Lake), 90 dwellings | 0.3 |
| Hungary, 22 dwellings | 0.5 |
| Norway, 22 dwellings | 0.5 |
| Germany (western), 150 dwellings | 0.8 |
| Germany (southwestern), 95 dwellings | 0.5 |
| France (Finistere), 219 dwellings | 0.3 |
| United States (20 states), 68 measurements | 0.6 |
| United States (Colorado), 12 indoor locations (Martz and others 1990) | 0.3 |
| Hong Kong, 10 indoor sites | 0.8 |

that the $^{220}$Rn progeny will not be an important additional source of exposure and dose.

In the measurements of the activity-weighted size distributions described above, the system used to make the activity measurements included $\alpha$-spectroscopy so that $^{220}$Rn progeny could have been observed. Such activities were not observed and thus, for these homes, only $^{222}$Rn decay products provided exposure and dose.

Steinhäusler (1996) has recently reviewed the information available regarding thoron exposure and dose. He concludes that the extent of information on $^{220}$Rn and its decay products is analogous to the state of affairs for $^{222}$Rn before the large-scale national radon surveys of the past 15 years. Few health studies have been conducted on the possible health effects of inhaling thoron decay products. Among residents in high-background areas of Brazil, China, and India, statistically significant increases have been observed for chromosome aberrations. The study in China found no increase in lung cancer for thoron exposure at a mean concentration of 168 Bqm$^{-3}$ (Wei and others 1993).

## Mines

Exposure measurements for $^{222}$Rn based on gross-counting methods can be affected by the presence of $^{220}$Rn decay products. For most of the mines which employed the workers used in the epidemiological studies, measurements are not available to determine the presence or absence of $^{220}$Rn progeny. For the mines in the Colorado Plateau and the Grants Mineral Belt of New Mexico, negligible concentrations of $^{220}$Rn decay products have been found. There are substantial concentrations of $^{220}$Rn decay products in the mines of the Elliot Lake, Ontario (DSMA Atcon 1985), where the PAEC arising from $^{222}$Rn and $^{220}$Rn were approximately equal. Corkill and Dory (1984) made a retrospective study of exposure in the fluorspar mines of Newfoundland. However, they do not mention $^{220}$Rn or its decay products. A similar lack of information exists for the other mining environments. Bigu (1985) has examined theoretical models for estimating the PAEC levels arising from the presence of both radon isotopes. The relative amounts of the radon isotopes and their progeny can be estimated based on the weight ratio of $^{238}$U to $^{232}$Th. For the Ontario uranium mines, this ratio is assumed to be 1. Results were then calculated based on a number of models that had been previously developed to predict the airborne activity concentrations. Bigu found that a substantial number of the measurements fall outside of the theoretical bounds and there are substantial differences among the results calculated by the various available models. Thus, more sophisticated models will need to be developed to permit adequate estimation of the exposure to $^{220}$Rn progeny. Thus, it is not possible to assess the uncertainty from a concentration of $^{220}$Rn progeny to lung dose in the miner-based risk estimates.

Steinhäusler (1996) also summarized the information available on occupationally exposed individuals (iron and rare-earth miners, workers in Th-processing plants and the monazite industry totaling 1557 persons). Increases in respiratory diseases, pancreatic cancer, and chromosome aberrations were found to be statistically significant. Among niobium miners in a Th-rich area, an increase in lung-cancer rates (observed/expected lung cancer cases = 11.3) was found (Solli and others 1985). Among approximately 53,000 European and Japanese patients treated with thorium oxide injections, an increase in lifetime excess cancer risks for liver and bone cancer as well as leukemia has been observed. However, there were no excess lung cancers. To reduce the uncertainties concerning the effects of $^{220}$Rn decay products, we will need to better characterize dosimetry of thoron progeny and obtain more data on exposures.

## CONCLUSIONS

In this chapter we have reviewed the basis for the dosimetric extrapolation of risk from miners to the general population, taking into account differences in the aerosol characteristics, breathing rates, and the respiratory physiology of the various segments of the population (women, children, infants, as well as men). The constants have changed from the values of about 0.70 to 0.75 reported by the NAS Dosimetry Panel (NRC 1991) to values very close to 1. The differences in input parameters between the calculations are summarized in Table B-13. The primary cause of the shift in K value is attributable to the reduction in the miner breathing rate based on measurements of actual miners in Tajikistan and South Africa. The risks for infants are slightly higher than for miners, but for the other groups, the risks per unit exposure are essentially identical in homes and mines. This value around 1 is reasonable given the compensatory factors of particle concentration raising the airborne concentrations, but in sizes that are only weakly deposited in the lung. Thus, the doses per unit radon concentration are essentially the same although in each case, specific unusual aerosol conditions could substantially increase the dose per unit exposure.

# Appendix C

# Tobacco-Smoking and Its Interaction with Radon

## INTRODUCTION

### Smoking as a Cause of Lung Cancer

During the 20th century, tobacco-smoking has become the dominant cause of lung cancer throughout the world (USDHHS 1989; Wu-Williams and Samet 1994; Peto and others 1992). This rapidly fatal malignancy, once an infrequent cause of death, is now one of the leading causes of cancer death in most countries of the developed world. Extensive epidemiologic data along with complementary toxicologic information have established the causal link of tobacco-smoking with lung cancer (IARC 1986; U.S. Department of Health Education and Welfare 1964). The epidemiologic evidence indicates that the risk of lung cancer is determined strongly by the extent of exposure to mainstream smoke, increasing with the number of cigarettes smoked and with the duration of smoking. Together, number of cigarettes smoked and the duration of smoking can be used in a multistage model to predict the general pattern of lung-cancer risk associated with smoking (Wu-Williams and Samet 1994; Doll and Peto 1978). For former smokers, risk declines with the duration of smoking cessation (USDHHS 1990). The pattern of inhalation of the tobacco smoke and the types of cigarettes smoked have a lesser influence on the risk of smoking (Wu-Williams and Samet 1994).

Passive smoking, the inhalation of the mixture of sidestream smoke and exhaled mainstream smoke, has been found to cause lung cancer among persons who have never smoked (USDHHS 1988; USEPA 1992c). This conclusion has been based on epidemiologic studies, primarily of the risk of lung cancer among nonsmoking women married to current smokers, and on supporting biological evidence including the extensive data base on active smoking.

## Combined Effect of Smoking and Radon as a Continued Source of Controversy in Risk Assessment

Because of this predominant role of cigarette-smoking as a cause of lung cancer, an understanding of the joint effect of smoking and radon exposure is needed for assessing the risks of radon exposure for the general population, which includes never-smokers, current smokers, and former smokers. Incomplete understanding of the combined effect of these two carcinogens remains a key uncertainty in assessing the risk of indoor radon, and the consequences of synergism between radon and smoking in making quantitative risk estimates have not been universally appreciated. The underground miners who were participants in the epidemiologic studies that are the basis for currently used risk estimates were primarily smokers, and epidemiologic data from the miners' studies have not provided a precise characterization of the lung-cancer risk arising from radon exposure in never-smokers. However, active cigarette smokers are now a minority in the adult population of the United States (USDHHS 1989) and risk estimates are needed for both smokers and never smokers.

This chapter addresses the joint effect of smoking and radon. It begins by describing trends of lung-cancer occurrence during the century and the implications of these trends for evaluating the role of indoor radon as a carcinogen. The chapter subsequently considers the numbers of cases of lung cancer in smokers and never-smokers as risk projections have been made separately for these two groups. It then summarizes the evidence on the combined effect of smoking and radon, drawing on in vitro systems, animal exposures, and epidemiologic studies of miners and the general population. The BEIR IV Report (NRC 1988) also reviewed the evidence on the joint effect of smoking and radon in an appendix chapter. We begin with the conceptual basis for considering the combined effect of smoking and radon.

## Concepts of Combined Effects

The 1985 Report of the U.S. Surgeon General (USDHHS 1985) set out a broad conceptual framework for considering the joint effect of cigarette-smoking with an occupational agent. The levels of potential interaction between the two agents are broad, ranging from molecular to behavioral (Table C-1). Some of the points of interaction would impact exposure, others—the exposure-dose relation, and others—the dose-response relation of radon progeny with lung-cancer risk. The epidemiologic data do not provide evidence relevant to assessing each of these potential points of intersection of radon-progeny exposure with cigarette-

TABLE C-1   Levels of interaction between smoking and radon

**Exposure**
- Work assignments of smokers and nonsmokers are different (for miners)
- Absenteeism rates differ for smokers and nonsmokers (for miners)

**Exposure-Dose Relationships**
- Differing patterns of physical activity and ventilation for smokers and nonsmokers
- Exposures of smokers and nonsmokers differ in activity-size distributions
- Differing patterns of lung deposition and clearance in smokers and nonsmokers
- Differing morphometry of target cells in smokers and nonsmokers

**Carcinogenesis**
- Alpha particles and tobacco smoke carcinogens act at the same or different steps in a multistage carcinogenic process

smoking. At most, the information on smoking in the studies provides some indication of elements of the smoking history, such as number of cigarettes smoked and age of starting to smoke; at a minimum, there is information on whether the participants had ever been regular cigarette smokers. Analyses of the epidemiologic data to characterize the joint effect of smoking and radon-progeny exposure simplify the multiple mechanisms by which the two agents could interact to a mathematical representation or model that typically includes a term for smoking and a term for radon-progeny exposure and a term for their joint action (NRC 1988); alternatively, data have been separately analyzed for ever-smokers and never-smokers (Lubin and others 1994a).

The terminology and methods used to characterize the combined effects of two or more agents have been poorly standardized with substantial blurring of concepts derived from toxicology, biostatistics, and epidemiology (Mauderly 1993; Greenland 1993). The terms "antagonism" and "synergism" refer to combined effects less than or greater than the effect predicted by the sum of the individual effects, respectively. In assessing the presence of synergism or antagonism, a model is assumed to predict the combined effect from the individual effects; lacking sufficient biologic understanding to be certain of the most appropriate model, the choice is often driven by convention or convenience.

## Epidemiologic Concepts

The effect of a risk factor for a disease may be considered as acting on an absolute scale or on a relative scale. In the absolute risk model, the risk ($r(x)$) of disease associated with some factor ($x$) can be expressed in a simple linear relationship as:

$$r(x) = r_0 + \beta x \qquad (1)$$

while in a relative risk relationship, risk is given by:

$$r(x) = r_0 (1 + \beta x) = r_0 + r_0\beta x \tag{2}$$

where $r_0$ is the background disease rate in the absence of exposure and $\beta$ describes the increment in risk per unit increment in exposure to x. Under a relative risk characterization of disease rate, the impact of an exposure on disease risk, $r_0\beta x$, depends on the background rate. In the specific case of lung cancer and radon, the background rate would incorporate the risk associated with smoking, unless smoking-specific risks were considered. In the absolute risk model, the effect of exposure on disease risk, $\beta x$, does not depend on the level of $r_0$. The selection of the risk model, absolute or relative, thus has substantial implications for interpreting the combined effects of two agents and additionally for extending risks observed in one population to another population which may not have comparable $r_0$ because of differing patterns of risk factors other than the exposure of interest.

The models can be readily extended to address the effects of multiple causes of disease. In the example of two exposures, $x_1$ and $x_2$ (for example, radon and smoking), disease risk ($r(x_1, x_2)$) under a relative risk model is given by
Additive model:

$$r(x_1, x_2) = r_0 + r_0\beta_1 x_1 + r_0\beta_2 x_2 \tag{3}$$

Multiplicative model:

$$\begin{aligned} r(x_1, x_2) &= r_0 (1 + \beta_1 x_1)(1 + \beta_2 x_2) = \\ &r_0 + r_0\beta_1 x_1 + r_0\beta_2 x_2 + r_0\beta_1 x_1\beta_2 x_2 = \\ &r_0 + r_0\beta_1 x_1 + r_0\beta_2 x_2(1 + \beta_1 x_1) \end{aligned} \tag{4}$$

Comparison of these two models makes clear the differing dependence of the effect of $x_2$ on $r_0$ and $x_1$. In assessing the role of $x_2$ on disease risk, a multiplicative model implies that the effect of $x_2$ on disease risk depends not only on $r_0$ but on the effect of $x_1$. In contrast, under the additive model, the effect of $x_2$ depends on $r_0$ but not on the effect of $x_1$.

In considering the combined effect of multiple causes of a disease, epidemiologists use the term "effect modification" to refer to interdependence of the effects of the agents (Last 1983). Synergism and antagonism describe the net consequence of the multiple risk factors in relation to the effect predicted by their independent action. Epidemiologists describe the effect of exposures in causing disease as either a difference measure on an absolute scale or a ratio measure on a relative scale (see Table C-2, for example). Table C-2 provides a hypothetical example for assessing combined effects of radon and smoking. The assessed lung-cancer rate (200 per 100,000) is compared to expectations under additive and multiplicative combined effects. The preference has been primarily for ratio measures (for example, the relative risk which compares risk in the exposed to risk in a referent group, typically the unexposed).

TABLE C-2   Example of epidemiological data on smoking and radon
exposure in a hypothetical cohort study of underground miners

Average Annual Age-Adjusted Incidence Rate For Lung Cancer (per 100,000)

|  |  | Radon-exposed | |
| --- | --- | --- | --- |
|  |  | Yes | No |
| Ever- | Yes | 200 | 100 |
| Smoked | No | 30 | 10 |

**On Absolute Scale:**
200 > 100 + 30 − 10

**On Multiplicative Scale:**
200 < 30 × 100 − 10

Effect modification is considered to be present when the combined effect of
two or more variables is larger or smaller than the anticipated effect predicted by
the independent effects, based on the measure used (Greenland 1993). Current
analytic approaches compare the combined effect to predictions based on either
additivity or multiplicativity of the individual effects. A factor may be an effect
modifier under additivity and not an effect modifier under multiplicativity. Epidemiologists have recognized that the appropriate scale for assessing the combined effect depends on the intent of the analysis (Rothman and others 1980).
For public health purposes, an effect greater than additive is considered as synergism. Biological mechanisms, if sufficiently understood may imply an alternative scale.

Although statistical methods have been developed for assessing effect modification, strict criteria for determining its presence have not been offered. Statistical significance alone is recognized to be an insufficient criterion (Greenland
1993). Additionally, inadequate statistical power often limits the assessment of
effect modification (Greenland 1983).

## Statistical Concepts

Statisticians have used the term "interaction" to refer to interdependence as
detected by a statistical approach or "model". Interaction, which is equivalent to
the epidemiologic concept of effect modification, has been typically assessed in a
regression framework using product terms of the risk factors of interest to test for
effect modification. For example, interaction between two risk factors, $x_1$ (for
example, smoking) and $x_2$ (for example, radon exposure) could be assessed using
the following model:

$$r(x_1, x_2) = r_0 (1 + \beta_1 x_1 + \beta_2 x_2 + \beta_3 x_1 x_2) \qquad (5)$$

In this linear model, the product or interaction term, $\beta_3 x_1 x_2$, estimates the joint contribution of the two agents to the risk. The model provides an estimate of the value of $\beta_3$ and a test of the statistical significance of $\beta_3$ for the null hypothesis: $\beta_3 = 0$. This modeling approach inherently assumes a mathematical scale on which the interaction is characterized, the usual choices being additive or multiplicative. Most often, primarily because of computational convenience, the multiplicative scale is used. Alternative approaches for assessing interaction have been described (Thomas 1981; Breslow and Storer 1985; Lubin and Gaffy 1988). These choices more flexibly estimate the combined effects of risk factors without imposing the rigidity of a particular scale. Imprecision may also limit estimates obtained from such modeling approaches.

## Implications for Interpreting Risk Estimates

These general considerations underscore the complexity of characterizing the joint effects of smoking and radon-progeny exposure using epidemiologic data, whether from cohort studies of uranium miners or case-control studies in the general population. In addition, there is no direct link between the biologic interaction of radon and smoking and the descriptive epidemiologic model. Limitations posed by imprecision and bias from measurement error and confounding further complicate the evaluation of the joint effects of radon and smoking.

## Characterizing the Burden of Radon-Related Lung Cancer

In describing the burden of disease, epidemiologists use a quantity referred to as the attributable risk (Rothman 1986). The attributable risk indicates the burden of disease that could be avoided if exposure to the agent of concern were fully prevented. One form of the attributable risk, the population attributable risk (AR) describes the proportion of disease in a population associated with exposure to an agent. For a factor, $x_1$, having associated relative risk $RR_1$, AR is calculated as below, where $I$ and $I_0$ are the disease rates in the population under current conditions and after effects of exposure are eliminated, respectively, and $P_1$ and $P_0$ are the probabilities of exposure and non-exposure, respectively:

$$AR(x)\frac{I - I_0}{I} =$$

$$\frac{P_1 I_1 + P_0 I_0 - I_0}{P_1 I_1 + P_0 I_0} = \tag{6}$$

$$\frac{P_1(RR_1 - 1)}{P_1(RR_1 - 1) + 1}$$

For diseases caused by multiple agents, the total burden of disease assessed individually may exceed the observed number of cases or 100%. For example, an estimate of radon-attributable lung cancer cases can be conceptualized as including those cases caused by radon in never smokers, those caused by radon in smokers, and those caused by radon and smoking in smokers. In the above formula, the attributable risk figure for smoking includes those cases caused by smoking alone and radon and smoking acting together; similarly, the attributable risk figure for radon includes those cases caused by radon alone and radon acting together with smoking. Combining the attributable risk estimates for smoking and radon counts the jointly determined cases twice. This subtlety of the attributable risk statistic is not universally appreciated and there is widespread misperception that the lung-cancer cases that could be caused by radon include only the 10 to 15 percent not directly attributed to smoking.

For two factors, $x_1$ and $x_2$, it is true that the sum of the individual exposure-specific AR estimates, $AR(x_1)$ and $AR(x_2)$, can exceed 100%. However, when evaluating two factors, these ARs are incorrectly determined by contamination of the referent groups, that is, the subgroup of individuals with $x_1 = 0$ includes individuals with $x_2 = 0/1$ and the subgroup of individuals with $x_2 = 0$ includes individuals with $x_1 = 0/1$.

For joint exposures to $x_1$ and $x_2$, AR is defined as:

$$AR(x_1, x_2) = \frac{P_{1,1}(RR_{1,1} - 1) + P_{1,0}(RR_{1,0} - 1) + P_{0,1}(RR_{0,1} - 1)}{P_{1,1}(RR_{1,1} - 1) + P_{1,0}(RR_{1,0} - 1) + P_{0,1}(RR_{0,1} - 1) + 1} \tag{7}$$

The AR for two exposures, for example smoking and radon, is the sum of components due to smoking in the absence of radon exposure, to radon exposure absent smoking, that is, in never-smokers, and to the combined effect of radon exposure and smoking. $AR(x_1,x_2)$, calculated with the above formula, cannot exceed 100%.

For a general exposure distribution f, for a casual factor x,

$$AR = \frac{I - I_0}{I}$$

$$= \frac{\int I(x)f(x)dx - I_0}{\int I(x)f(x)dx} \tag{8}$$

$$= \frac{\int RR_x f(x)dx - 1}{\int RR_x f(x)dx}$$

where $RR_x$ is the relative risk for exposure level x, relative to zero exposure.

The formulae for AR given above, assumes complete elimination of the exposure(s) from the population. However, complete elimination of exposure may not be practicable. Suppose there is a single exposure and $I^*$ is the disease rate in the population after exposure has been modified. The "effective" AR ($AR_E$), the amount of the disease burden that can be eliminated by changing exposure from distribution f to distribution g, is defined as:

$$AR_E = \frac{I - I^*}{I}$$

$$= \frac{\int I(x)\{f(x) - g(x)\}dx}{\int I(x)f(x)dx} \tag{9}$$

$$= \frac{\int RR_x\{f(x) - g(x)\}dx}{\int RR_x f(x)dx}$$

For example, for residential radon exposure, assume that f is the radon distribution in U.S. houses. Then, based on the BEIR IV model, the AR is about 10%, which represents the reduction of lung cancers that is estimated to be achievable if radon were eliminated in all houses. Complete elimination of radon is not feasible, however. If homes above 148 Bqm$^{-3}$ (4 pCiL$^{-1}$) were mitigated to levels below this concentration, then the $AR_E$ would be about 3% (Lubin and Boice 1989) .

## LUNG CANCER IN THE GENERAL POPULATION

### Trends of Lung-Cancer Occurrence Over the Century

For the United States, vital statistics document lung-cancer deaths from the late 1930s (Table C-3). These early mortality counts indicate less than 10,000

TABLE C-3   Numbers of lung-cancer deaths in the U.S. for selected years

| Year | Male | Female |
|---|---|---|
| 1937[a] | 5,244 | 1,784 |
| 1940[a] | 7,002 | 2,088 |
| 1950[b] | 13,981 | 3,162 |
| 1960[b] | 31,257 | 5,163 |
| 1970[c] | 52,801 | 12,367 |
| 1980[c] | 75,535 | 28,309 |
| 1989[c] | 89,052 | 48,089 |

[a]Cancer of the respiratory system.
[b]ICD 162 and 163.
[c]ICD 162.

deaths per year from lung cancer. However, lung cancer mortality rates quickly rose towards mid-century and by the 1940s, an epidemic of lung cancer was evident among U.S. males (Figure C-1). About 20 years later, a similarly sharp rise was evident in women (Figure C-2). These patterns of lung-cancer occurrence are consistent with patterns of smoking across the century (Figures C-3 and C-4) (Burns 1994; USDHHS 1991). Statistical models representing the relationship between lung-cancer occurrence and indicators of cigarette-smoking in the population confirm this visual impression (Samet 1995; USDHHS 1991).

The relative infrequency of lung cancer early in the century, in comparison with present rates, has been cited as evidence that indoor radon is not a significant cause of lung cancer (Yalow 1995). Proponents of this viewpoint argue that lung cancer would have been a more prominent cause of death before the presently dominant impact of cigarette-smoking, if the risks of indoor radon were in reality as high as estimated in currently applied risk models. Rates earlier in the century, particularly for women, have also been cited as indicative of the maximum number of lung-cancer cases that indoor radon might have caused, assuming that other causes of lung cancer had little impact (Yalow 1995). Similarly, rates of lung cancer in population groups considered to be largely never-smokers, such as Mormons, have also been considered to provide bounds for the numbers of lung-cancer cases caused by indoor radon (Yalow 1995). For example, Mormon

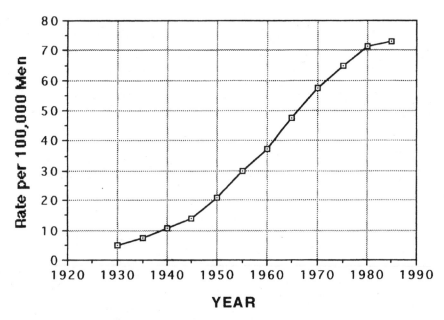

FIGURE C-1    Age-adjusted lung-cancer death rates for males. United States 1930-1986; adjusted to the age distribution of the 1970 population.

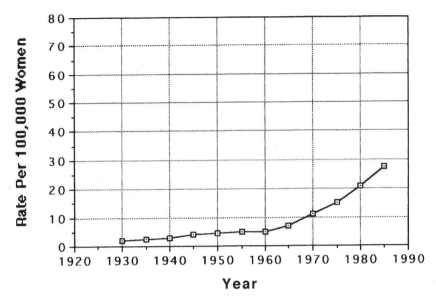

FIGURE C-2    Age-adjusted lung-cancer death rates for females; United States 1930-1986; adjusted to the age distribution of the 1970 U.S. Census population.

women in Utah had an annual lung-cancer incidence rate of approximately 4 per 100,000 for 1967-1975 (Lyon and others 1980).

The early lung-cancer counts and rates need to be interpreted with caution. There have been substantial changes in approaches to the diagnosis and management of lung cancer across the century that would be expected to impact on the apparent occurrence of this malignancy (Gilliland and Samet 1994). These changes include the evolution of diagnostic techniques, from reliance on tissue specimens obtained at surgery or autopsy to the less invasive techniques of fiberoptic bronchoscopy and needle aspiration and the non-invasive approach of sputum cytology. Improved imaging techniques, such as CT scanning, have sharpened the diagnosis of lung cancer as well. Finally, increasingly aggressive approaches to surgical removal of lung cancer have been introduced and improving post-operative care has extended the age range of patients undergoing surgery for diagnosis and management (Gilliland and Samet 1994). Additionally, earlier in the century, when tuberculosis was one of the most common causes of death, lung cancer may have been more often misdiagnosed as tuberculosis, absent tissue confirmation.

Thus, it is likely that lung-cancer statistics from earlier in the century underestimate the occurrence of lung cancer, as diagnosed with present techniques. Doll and Peto (1981) have argued that cancer mortality rates in persons under age 65 years of age are the most valid indicator of trends of cancer occurrence in

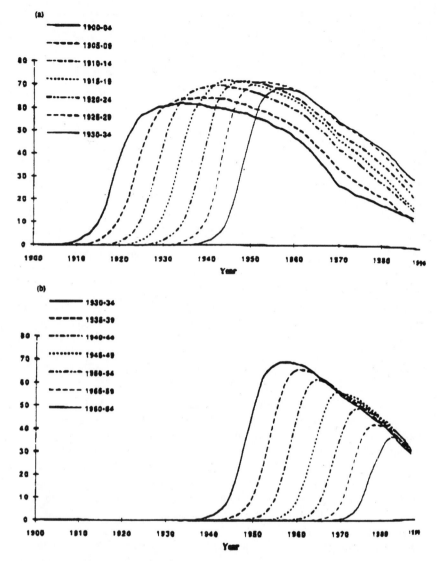

FIGURE C-3   White male smoking prevalence by year for 5-year birth cohorts; (a) 1900-1934; (b) 1934-1964.

populations because the intensity of diagnostic efforts would be most consistent in this younger age range. Analyses of data from the Surveillance, Epidemiology and End Results (SEER) Program of the National Cancer Institute indicate increasing histologic diagnoses of lung cancer in the elderly across the decades of the 1970s and 1980s and less reliance in making the diagnosis on a clinical basis alone (Gilliland and Samet 1994).

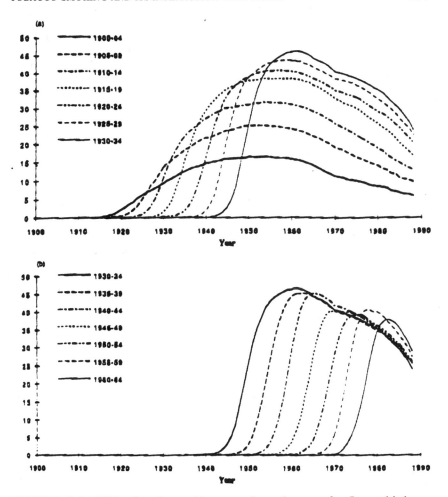

FIGURE C-4 White female smoking prevalence by year for 5-year birth cohorts; (a) 1900-1934; (b) 1934-1964.

These early figures on the occurrence of lung cancer were carefully scrutinized as the rising lung-cancer mortality was first identified and biases that may have artefactually increased lung-cancer mortality were considered (see, for example, Heady and Kennaway 1949; Macklin 1942; Fried 1931). These authors commented on the difficulty of making the diagnosis and the high-error rates of clinicians. A number of studies have been conducted across the century on the validity of death-certificate statements concerning cause of death in comparison to the medical record (Doll and Peto 1981), some of these studies have included information on certification of deaths as due to lung cancer. Anderson and colleagues (1989) reviewed the findings of five studies that provided information on the validity of clinical diagnoses of lung cancer (Table C-4). Surprisingly, the data did not indicate a trend of improving sensitivity and specificity across the period of the studies which extended from the 1940s through the 1970s. On the other hand, the studies provided little insight into the gains in validity anticipated with contemporary diagnostic methods such as fiberoptic bronchoscopy and CT scanning which have only been used during the last 20 years.

Others have documented lesser accuracy of lung-cancer diagnoses in the past. In autopsied cases at Los Angeles County General Hospital, 1933-1937, the diagnosis of lung cancer was incorrect in 28% of cases (Swartout and Webster 1940). For cases diagnosed at autopsy at Boston City Hospital, 1955-1965, incorrect diagnoses were reported for 49.1 % of cases (Bauer and Robbins 1972).

In view of the limitations of lung-cancer diagnoses before the advent of modern approaches, mortality rates from the early decades of the century, when tobacco-smoking had little impact on lung-cancer incidence, should be interpreted with caution. The extent of underdiagnosis, particularly among the elderly, is uncertain. Nevertheless, assuming that lung cancer was not underdiagnosed, the rates from the early decades of the century could be considered as an upper bound for the cases of lung cancer attributable to radon, absent contributions from other causes. Rates in never-smokers, such as the Mormon women, could be similarly interpreted.

This interpretation, however, is inconsistent with the understanding gained from the miner studies that radon exposure proportionally increases lung-cancer occurrence beyond background. For example, an overall lung-cancer mortality rate of 4 per 100,000 could arise from a relative risk of radon of 1.2 (assuming a binary exposure defined by the median) and a "radon-free" background rate of 3.6 per 100,000; the background rate would be 3.3 per 100,000 if the radon-associated relative risk were 1.5. Thus, lacking information on the determinants of background lung-cancer mortality rates earlier in the century, the rates can be interpreted only on the implausible assumption that there were no risk factors for lung cancer other than radon.

## Lung-Cancer Rates in Never-Smokers

Only limited information on lung-cancer occurrence in never-smokers is available as calculation of either incidence or mortality rates requires an estimate of the population of never-smokers at risk for lung cancer during some time interval (the denominator of a rate) and a count of the numbers of cases of lung cancer in never-smokers during the time period of interest (the numerator of a rate). The requisite data are not available from routine vital statistics and lung-cancer mortality rates in never-smokers have been primarily obtained by follow-up of large groups of persons participating in epidemiologic studies. Rates from

TABLE C-4  Accuracy of clinical diagnostics among autopsied persons dying of carcinoma of the bronchus and lung[a]

| Author(s) and Year(s) of Study | Clinical Diagnosis | Autopsy Diagnosis | | Total | Sensitivity/ Specificity | Clinical Accuracy | |
|---|---|---|---|---|---|---|---|
| | | Positive | Negative | | | Positive Diagnosis | Negative Diagnosis |
| Munck, 1940-1949 | Positive | 14 | 2 | 16 | 70.0 | 87.5 | 99.4 |
| | Negative | 6 | 978 | 984 | 99.8 | | |
| Heasman and Lipworth, 1959 | Positive | 399 | 72 | 471 | 74.7 | 84.7 | 99.2 |
| | Negative | 135 | 8,895 | 9,030 | 99.2 | | |
| Britton, 1970-1971 | Positive | 9 | 1 | 10 | 75.0 | 90.0 | 99.2 |
| | Negative | 3 | 370 | 373 | 99.7 | | |
| Cechner, and others, 1948-1973 | Positive | 260 | 38 | 298 | 70.0 | 87.2 | 99.2 |
| | Negative | 117 | 13,659 | 13,776 | 99.7 | | |
| Cameron and McGoogan, 1975-1977 | Positive | 88 | 15 | 103 | 65.7 | 85.4 | 95.6 |
| | Negative | 46 | 1,003 | 1,049 | 98.5 | | |

[a]Based on Table 13, Anderson and others, 1989.

groups like Mormons and certain Native American tribes can also be interpreted as providing an indication of lung-cancer occurrence among never-smokers.

The two large, longitudinal or cohort studies conducted by the American Cancer Society have provided mortality rates from lung cancer and other diseases in never-smokers. The populations in both of these studies, numbering about one million in each, were enrolled by American Cancer Society volunteers who identified the participants, obtained information on smoking and other risk factors for cancer, and periodically determined the vital status of the enrollees (USDHHS 1996). Participants in the first study, now referred to as Cancer Prevention Study I (CPS I), were followed between 1960 and 1972; participants in the second, CPS II, were enrolled in 1980 and follow-up continues. Lung-cancer rates for never smokers are shown in Table C-5 from the CPS I and CPS II studies. These rates should be interpreted with awareness of the relatively small numbers of lung-cancer deaths in specific strata of age and sex, in spite of the large number of participants in studies overall. With this constraint, several patterns are evident: the rates increase with age in both sexes and rates for men tend to be higher than for women. In older males, rates are somewhat higher in CPS II; a similar pattern is evident among older women. Ignoring competing causes of death, these mortality rates can be used to estimate a risk of lung-cancer death up to age 85 years of about one percent in men and 0.5 percent in women.

Rates for never-smokers have also been reported from the study of U.S. male veterans, originally started by Rogot and Murray (1980) (Table C-6) and from other studies (Samet 1988; Enstrom 1979). Rates in Mormon women (Lyon and others 1980) and southwestern Native American tribes (Wiggins and Becker 1993; Lyon and others 1980) with only a small proportion of smokers are similar to the rates in never-smokers from these cohort studies.

TABLE C-5   Lung-cancer rates per 100,000 in never-smokers (ACS CPS I and CPS II studies)

| Age | Female | | Male | |
|-----|--------|--------|------|------|
|     | CPS I  | CPS II | CPS I | CPS II |
| 30- | —    | —    | —    | —    |
| 35- | —    | —    | —    | —    |
| 40- | 4.0  | —    | 0.7  | —    |
| 45- | 6.0  | 6.0  | 4.3  | 1.9  |
| 50- | 5.7  | 5.5  | 5.1  | 5.8  |
| 55- | 13.6 | 5.3  | 6.2  | 7.2  |
| 60- | 21.0 | 11.6 | 12.0 | 12.3 |
| 65- | 23.1 | 21.5 | 13.4 | 16.7 |
| 70- | 29.7 | 34.9 | 15.9 | 30.5 |
| 75- | 31.0 | 52.0 | 24.9 | 32.5 |
| 80- | 67.5 | 89.2 | 43.4 | 57.6 |
| 85+ | 35.3 | 86.8 | 35.9 | 60.6 |

TABLE C-6    Lung cancer mortality rates (per 100,000) among never-smokers in the study of U.S. veterans

| Age | 1955-1959 | 1960-1964 | 1965-1969 |
|---|---|---|---|
| 40 - 44 | – | – | (103.5) |
| 45 - 49 | – | – | (80.6) |
| 50 - 54 | – | – | – |
| 55 - 59 | (12.0) | – | – |
| 60 - 64 | 11.2 | – | (48.0) |
| 65 - 69 | 25.1 | (10.7) | 43.5 |
| 70 - 74 | 139.9 | 16.9 | 38.2 |
| 75 - 79 | 39.9 | 40.5 | 47.2 |
| 80 - 84 | (37.8) | (15.0) | (20.6) |
| 85 - 89 | – | (200.6) | – |

## The Risks of Cigarette-Smoking

The lung-cancer mortality rates observed in never-smokers are markedly increased by cigarette smoking (Figures C-5 and C-6). Figures C-5 and C-6 provide the age-specific mortality rates in the two cohort studies conducted by the American Cancer Society for reported never-smokers and current-smokers on enrollment (Thun and others 1997). At all ages, current-smokers have markedly higher rates. Increases in mortality rates among smokers of both sexes are also evident, comparing the CPS II with CPS I. The extent of the increase varies strongly with number of cigarettes smoked daily and duration of cigarette-smoking (Table C-7). Cigarette-smoking increases the risk for each of the principal histologic types of lung cancer: squamous cell carcinoma, small cell carcinoma, adenocarcinoma, and large cell carcinoma (Wu-Williams and Samet 1994). A synergistic pattern of effect modification between radon and cigarette-smoking implies that the already high risks of lung cancer in cigarette smokers are augmented more than additively by the additional risk imposed by exposure to indoor radon.

## Lung-Cancer Numbers in Never-Smokers and Ever-Smokers

The total number of lung-cancer cases occurring annually in the United States can be estimated using the population-based incidence rates available from the SEER Program or from the number of deaths with lung cancer listed as the underlying cause. Estimates from incidence would only slightly exceed those based on mortality because of the high case-fatality rate of lung cancer. For 1994, these estimates were 172,000 for incidence and 153,000 for mortality (USDHHS 1995).

The Office on Smoking and Health of the Centers for Disease Control and Prevention annually reports estimates of the numbers of deaths from lung cancer

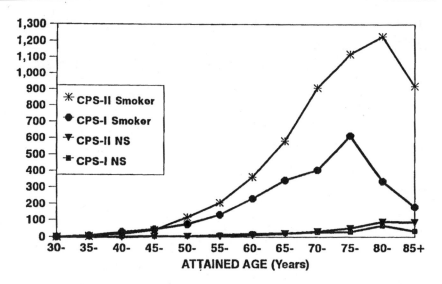

FIGURE C-5    Death rates for men from lung cancer by age and smoking status, CPS-I and CPS-II.

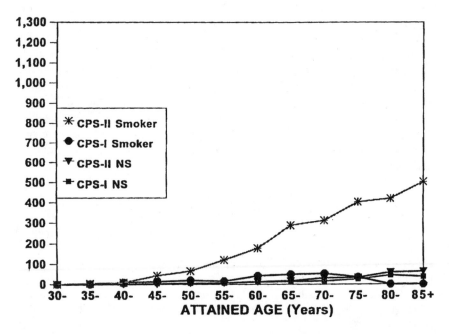

FIGURE C-6    Death rates for women from lung cancer by age and smoking status, CPS-I and CPS-II.

TABLE C-7    Standardized mortality ratios for lung cancer in women participants in CPS II[a]

| Duration Of Smoking | Cigarettes Per Day | | | | |
|---|---|---|---|---|---|
| (In Years) | 1-10 | 11-19 | 20 | 21-30 | 31+ |
| 21 - 30 | 2.9 | 6.7 | 13.6 | 18.4 | 18.9 |
| 31 - 40 | 7.9 | 19.2 | 19.2 | 26.5 | 25.3 |
| 41 - 70 | 10.0 | 17.0 | 25.1 | 34.3 | 38.8 |

[a]Source: Garfinkel and Stellman (1988).

and other diseases attributable to cigarette-smoking. These estimates are based on an attributable risk approach, implemented with the Smoking-Attributable Mortality, Morbidity, and Economic Cost (SAMMEC) software (SAMMEC 1992). SAMMEC applies attributable risk formulas to mortality data. For example, for 1990, 418,690 deaths in the U.S. were attributed to smoking, of which 116,920 were due to lung cancer (SAMMEC 1992). For lung cancer, the relative risk values assumed, versus never-smokers, are 22.4 and 9.4 for male current and former smokers, respectively, and 11.9 and 4.7 for females. These relative risk estimates were derived from the Cancer Prevention Study (CPS) II conducted by the American Cancer Society. They translate into attributable risk estimates of approximately 90% for males and somewhat less for females.

These attributable risk figures estimate the numbers of lung-cancer deaths due to smoking and not the actual number among ever-smokers, which is somewhat higher, as all cases among ever-smokers are not attributed to smoking. In recent population-based case-control studies, approximately 5 to 10% of cases have been in never-smokers (Table C-8). Using an estimate of approximately 160,000 lung-cancer deaths from 1995, the number of cases in never-smokers is estimated to range from about 8,000 to 16,000. In the analysis of attributable risk of lung cancer due to radon, 85% of the 149,000 lung cancer deaths in 1993 were assumed to have occurred in current or former smokers; the complementary figure for never-smokers, 15%, appears too high in view of the data from the case-control studies.

## EXPERIMENTAL MODELS

### In Vivo Approaches

The report of the BEIR IV committee (NRC 1988) reviewed the animal studies that included exposure to both radon progeny and cigarette smoke. The relevant studies included the experiments involving rats conducted by the Compagnie Generale des Matieres Nucleaires (COGEMA) in France and the experiments involving dogs conducted by Pacific Northwest Laboratory (PNL) in the United States. In summarizing the evidence, the report noted that the

TABLE C-8   Proportion of lung cancer in never-smokers:  selected U.S. case-control studies

| Reference | Location | Years | Sex | N | Percent Never-Smokers |
|---|---|---|---|---|---|
| Wynder and Graham, 1950 | United States; Various | 1948-49 (?) | M | 605 | 1.3 |
| Wynder and others 1956 | United States; Various | 1953-55 | F | 47 | 57.4 |
| Haenszel and others 1958 | United States; Various | 1955-57 | F | 157 | 48.4 |
| Wynder and others 1970 | United States; Various | 1966-69 | M | 270 | 3.3 |
| Stayner and Weyman 1983 | United States; Various | 1969-71 | M | 420 | 13.8 |
| Samet and others 1988 | New Mexico | 1980-82 | M | 356 | 2.5 |
|  |  |  | F | 232 | 8.6 |
| Wu and others 1985 | Los Angeles | 1981-82 | F | 220 | 14.1 |
| Schoenberg and others 1989 | New Jersey | 1982-83 | F | 994 | 10.0 |
| Ives and others 1988 | Houston | 1977-80 | F | 259 | 4.6 |

COGEMA experiments showed synergism if the exposure to cigarette smoke followed the exposure to radon progeny; synergism was not found if the smoke exposure preceded the radon-progeny exposure. In the PNL experiments, lung-tumor incidence was decreased when the animals were exposed to radon progeny and cigarette smoke simultaneously.

Since the BEIR IV Report, there have been several additional reports from the COGEMA and PNL groups (Monchaux and others 1994; Cross and others 1995). Cross has reviewed the newer studies (Cross 1994a,b; Cross and others 1995). The PNL group conducted initiation-promotion-initiation experiments with cigarette smoke and radon exposure (Cross and others 1995). These experiments involved various sequences of exposure to smoking and radon progeny as well as splitting the dose of radon progeny. Only preliminary findings have been reported to date and the findings are not yet available for lung tumors (Cross and others 1995). The findings of the COGEMA studies have been summarized recently (Table C-9) (Anonymous 1993; Marchaux and others 1994). The extent to which lung-cancer incidence was increased by cigarette-smoke exposure following radon exposure was shown to depend on the duration of exposure to smoke. Decreasing duration was associated with decreasing number of lung-cancer cases.

TABLE C-9   Incidence of lung carcinomas after combined exposure to radon and tobacco smoke according to the cumulative dose of radon and its progeny and to the cumulative exposure to tobacco smoke[a]

|  | Number of Rats | Number of Lung Carcinomas | Proportion (%) of Lung Carcinomas |
|---|---|---|---|
| Radon exposure only (40 WLM[b]) | 21 | 1 | 5 |
| Tobacco smoke (350 hr) after radon exposure (40 WLM) | 27 | 1 | 3.5 |
| Radon exposure only (200 WLM) | 63 | 7 | 11 |
| Tobacco smoke (350 hr) after radon exposure (200 WLM) | 75 | 16 | 21 |
| Radon exposure only (1600 WLM) | 208 | 81 | 39 |
| Tobacco smoke (350 hr) after radon exposure (1600 WLM) | 138 | 106 | 77 |
| Radon exposure only (1600 WLM) | 30 | 7 | 23 |
| Tobacco smoke (100 hr) after radon exposure (1600 WLM) | 35 | 11 | 31 |
| Radon exposure only (1600 WLM) | 74 | 22 | 30 |
| Tobacco smoke (60 hr) after radon exposure (1600 WLM) | 64 | 19 | 30 |
| Radon exposure only (1600 WLM) | 30 | 7 | 23 |
| Tobacco smoke (30 hr) after radon exposure (1600 WLM) | 35 | 1 | 3 |

[a]From Marchaux and others 1994, Table 2.
[b]WLM, working-level-month.

In spite of long-term research by two groups of investigators, the animal experiments on smoking and radon progeny do not supply strong evidence on the combined effects of the two exposures. The findings are inconsistent and dependent on the sequence of exposures. In the residential setting, exposure to smoking and to radon progeny occur essentially simultaneously throughout adulthood. Among the miners, smoking and radon exposure may take place simultaneously or radon exposure may begin before or after smoking has been started (Thomas and others 1994). The unique pattern of smoking by people, which has not been replicated in the animal experiments, is an additional barrier in extending the findings of the animal studies to humans.

# EPIDEMIOLOGIC APPROACHES

## Case-Control Studies in the General Population

The combined effect of radon and smoking can be assessed with data obtained using the case-control approach in the general population. However, such studies have been shown to have extremely limited power for distinguishing alternative patterns of interaction of potential interest, from both the biologic and public health perspectives (Thomas and others 1994; Lubin and others 1995c). Nevertheless, reports of several studies provide information on the combined effect of the two agents (Table C-10). Methods for assessing exposure to radon have varied among these studies with two using residential construction as a surrogate (Svensson and others 1989, Axelson and others 1988) and the remainder using measurements of radon concentration (Alavanja and others 1994; Pershagen and others 1994; Ruosteenoja 1991; Schoenberg and others 1990; Blot and others 1990). The analytic approach for assessing the combined effect of radon and smoking has also varied among the studies. Sample size limits interpretation of all of the studies and none of the individual studies have the number

TABLE C-10   Case-control studies of the combined effect of smoking and radon

| Reference | Study Location/Years | Total Cases/ Total Never- Smoking Cases | Findings |
|---|---|---|---|
| Axelson and others, 1998 | Sweden/1960-81 | 177 / 15 | Increased risk for non and occasional smokers vs. regular smokers in rural areas |
| Svensson and others, 1989 | Sweden/1983-85 | 210 / 35 | Greater risk for smokers vs. never-smokers |
| Blot and others, 1990 | China/1985-87 | 308 / 123 | Nonsignificantly greater risk in smokers (p = 0.15) |
| Schoenberg and others, 1990 | New Jersey/1982-83 | 433 / 61 | Exposure response strongest in light smokers; inverse in heaviest smokers |
| Ruosteenoja, 1991 | Finland/1980-85 | 238 / 4 | No pattern comparing lighter with heavier smokers |
| Pershagen and others, 1994 | Sweden/1980-84 | 1,360 / 178 | Higher excess relative risk in current smokers vs. never-smokers. Additivity rejected by data (p = 0.02) |

of participants needed to distinguish additive from multiplicative patterns of combined effects.

The findings of the studies are seemingly inconsistent with regard to the combined effect of smoking and radon. However, it is unlikely that there is significant heterogeneity among the studies, but direct comparison cannot be made because of the varying analytic methods. Data from the largest study, the Swedish study reported by Pershagen and others (1994) were not consistent with additivity when smoking and radon were considered as continuous variables. In this study, excess relative risk estimates per unit of radon exposure were greater for current smokers of less than 10 cigarettes per day (0.16 per 100 Bqm$^{-3}$) and 10 or more cigarettes per day (0.19 per 100 Bqm$^{-3}$) in comparison with never-smokers (0.07 per 100 Bqm$^{-3}$). However, neither the 0.16 or the 0.19 were statistically found to be significantly different from the 0.07.

## Epidemiologic Studies of Underground Miners

Studies of underground miners have been the principal source of epidemiologic information on the combined effect of smoking and radon. These studies have been limited for characterizing the joint effect of the two agents by the extent of the data available on the smoking habits of the miners, with some studies having no information and others having information obtained through surveys or at the time of medical examinations (Table C-11). Documentation of cigarette-smoking was also limited to the time of employment in most of these studies. To extend the smoking information across the entire time of smoking, case-control approaches have been used in studies of the Beaverlodge uranium miners (L'Abbe and others 1991) and Yunnan tin miners (Yao and others 1994). A separate cohort study of a group of 773 Ontario miners was recently reported (Finkelstein and Kusiak 1995); the analysis was limited by the small number of lung-cancer cases (N = 42) that occurred over the 18-year follow-up.

The report of the BEIR IV committee comprehensively reviewed the information on the combined effect of smoking and radon progeny in appendix VII. Since the publication of this report, the findings of several individual studies have been published that provide information on the combined effect of smoking and radon exposures; the study populations include Ontario uranium miners (Finkelstein and Kusiak 1995; L'Abbe and others 1991), New Mexico uranium miners (Samet and others 1991), and Chinese tin miners (Yao and others 1994; Qiao and others 1989). Additional analyses of data from the Colorado Plateau miners have also been reported: Moolgavkar and others (1993) applied a two-mutation model of carcinogenesis to these data and Thomas and others (1994) addressed the effect of the sequence in which exposures to radiation and smoking occurred (see below). The findings of the two-stage model did not indicate interaction between smoking and radon on the first- and second-stage mutation rates or on the rate of proliferation of initiated cells (see also Thomas and others 1993).

TABLE C-11  Summary of ascertainment and availability of tobacco use information[a]

| Study | Variables Available | Source Of Smoking Data | Comment |
|---|---|---|---|
| China | Smoker: yes/no | From 1976 lung cancer screening, ascertained at time of entry into cohort. | Data missing on 24% of subjects. There were 25 (out of 907) non-smoking lung cancer; 1 not exposed to radon. |
| Colorado | Cigarette use: Duration, rate, cessation | From medical examination between 1950-60 and a series of surveys between 1963-69 carried out at irregular intervals. | Smoking data unavailable after 1969. There were 25 (out of 294) non-smoking cases; all exposed to radon. |
| Newfoundland | Type of product, and duration, cessation, rate (cigarette) | From surveys conducted in 1960, 1966, 1970 and 1978. | Smoking data on 48% of cohort, including 25 cases. |
| Sweden | Type of product, amount, cessation | From 1972-73 survey of all active miners, 1977 survey of pensioners, and next-of-kin survey of deceased cases. | Data from 35% random sample of active miners in 1972, supplemented by subsequent surveys. |
| New Mexico | Cigarette use: Duration, rate, cessation | At time of physical examination at Grants Clinic. | Smoking data available through time of last physical examination. |
| Radium Hill | Status: Ever, never, unknown | From survey of all cohort members and next-of-kin, started in 1984. | Data on about half the cohort. One non-smoking lung cancer. |

[a]Source: Lubin and others 1994a, table 16.

The report of the pooled analysis of data from the studies of underground miners (Lubin and others 1994a) included analyses of the combined effects of smoking and radon within six of the cohorts individually along with combined analyses of the six data sets. Reports from the individual studies had also addressed the joint effects of smoking and radon but analytic approaches varied among the studies (Xuan and others 1993; Samet and others 1991; Hornung and Meinhardt 1987; Radford and St Clair Renard 1984). A uniform analytic strategy was applied to each of the data sets that compared additive and multiplicative models to a model fully parameterized in smoking and radon-progeny exposure. A geometric mixture model was also fit to the data and the maximum likelihood estimate of the mixing parameter, $\lambda$, was obtained. If $\lambda = 1$, the data are consistent with a multiplicative interaction while an estimate that $\lambda = 0$ indicates an additive interaction; values less than zero indicate subadditivity while values greater than unity imply supra-multiplicativity. The findings are reported in Table C-12.

Of the individual studies, only the studies of the Chinese tin miners and of the Colorado Plateau uranium miners have sufficient cases to provide insights into the pattern of interaction. The findings of the two studies were discrepant, with the Chinese data not fitting either the additive or the multiplicative model while the Colorado data were consistent with the multiplicative model. As in previous analyses, the estimate of $\lambda$ for the Colorado Plateau data indicated a submultiplicative interaction (Hornung and Meinhardt 1987; NRC 1988). The maximum likelihood fit for $\lambda$ to the Chinese cohort data suggested a sub-additive interaction or antagonism.

The six cohorts considered in the joint analysis included 2,798 workers with a history of never smoking; these men had experienced 64 lung-cancer cases during 50,493 person-years of follow-up. Risk increased with cumulative exposure to radon progeny (Table C-13). The estimate of excess relative risk (ERR)

TABLE C-12 Analyses of the combined effect of smoking and radon progeny exposure in six studies of underground miners[a]

| Study | p-value | | Mixture | |
| | Multiplicative | Additive | $\lambda$ | p-value[b] |
| --- | --- | --- | --- | --- |
| Chinese tin miners | 0.02 | 0.08 | −0.3 | 0.39 |
| Colorado uranium miners | 0.58 | 0.04 | 0.7 | 0.49 |
| Newfoundland fluorospan miners | 0.53 | 0.67 | −0.1 | 0.85 |
| New Mexico uranium miners | 0.15 | 0.11 | 0.4 | 0.16 |
| Radium Hill uranium miners[c] | — | — | — | — |
| Swedish iron miners | 0.43 | 0.31 | 0.3 | 0.38 |

[a]Based on Tables 17a, 18a, 19a, 20a, and 21a in Lubin and others, 1994a.
[b]Refers to fit of mixture model versus full model.
[c]Insufficient numbers for modeling.

per WLM was higher for never-smokers (ERR/WLM%=1.03, 95% CI 0.2%, 5.7%) than for smokers (ERR/WLM%=0.34, 95% CI 0.08%, 1.7%). This pattern is consistent with a sub-multiplicative interaction.

Further evidence on smoking and radon-progeny exposure in the Chinese tin miners has recently been reported from a population-based case-control study conducted by Yao and colleagues (Yao and others 1994) in Gejiu City, the site of the Yunnan Tin Corporation. This study included 460 cases, of whom 368 had been miners, and 1,043 controls. Tobacco is smoked by study participants as cigarettes or using water pipes or Chinese long-stem pipes; a mixed pattern of smoking was most common. In contrast to the cohort analysis of the Yunnan tin miners reported by Lubin and others (1994a), the case-control data were consistent with a multiplicative model, although the best-fitting model was intermediate between additive and multiplicative.

The joint effect of smoking and radon-progeny exposure could plausibly vary with the sequence of the two exposures. Thomas and others (1994) analyzed the Colorado Plateau data using a case-control approach to assess temporal modification of the radon progeny-smoking interaction. They characterized the temporal sequence of the two exposures as simultaneous, radon before smoking, and radon following smoking. The temporal sequence of the two exposures was a strong modifier of the interaction. Radon followed by smoking was associated with an essentially additive effect, while smoking followed by radon was associated with a more than multiplicative effect on a relative risk scale. Thomas and colleagues interpreted this finding as suggesting that smoking could act to promote radon-initiated cells. By contrast, in the case-control study conducted by Yao and others (1994) relative risk estimates were highest for those with overlapping exposures to smoking and radon.

TABLE C-13   Lung-cancer mortality rates and relative risks by cumulative WLM using data from all cohorts on 2,620 workers who never smoked

|  | Cumulative WLM | | | | | |
|---|---|---|---|---|---|---|
|  | <100 | 100-399 | 400-799 | 800-1599 | 1600 | Total |
| Cases | 6 | 11 | 25 | 13 | 9 | 64 |
| Person-years | 29,142 | 12,193 | 4,816 | 2,967 | 1,375 | 50,493 |
| Mean WLM | 22.0 | 213.5 | 556.0 | 1,153 | 2,332 | 248.5 |
| Rate x 1000 | 0.2 | 0.9 | 5.2 | 4.4 | 6.5 | 1.3 |
| RR[a] | 1.00 | 2.08 | 10.9 | 10.8 | 23.8 |  |
| 95% CI | 0.7 - 6.3 | 3.5 - 33.9 | 3.3 - 35.8 | 6.6 - 86.2 |  |  |

[a]RRs adjusted for cohort, previous occupational exposure, and attained age. Overall (ERR/WLM)% was 1.03 with 95% CI (0.2%, 5.7%); among smokers (ERR/WLM)% was 0.34 with 95% CI (0.08%, 1.7%).

## Populations Exposed to Low-LET Radiation

The combined effect of smoking and radiation exposure has been assessed in only a few populations exposed to low-LET radiation. These populations include the atomic-bomb survivors (Kopecky and others 1988; Prentice and others 1983), persons receiving therapeutic irradiation for breast cancer (Van Leeuwen and others 1995; Neugut and Murray 1994; Kaldor and others 1992), and workers with mixed exposure to external gamma radiation and internal emitters (Petersen and others 1990). The relevance of findings in populations exposed to low-LET radiation is uncertain for alpha radiation from radon progeny. Small sample size further constrains interpretation of the individual studies.

Of the cancers associated with radiation exposure in the atomic-bomb survivors, the relative risks for lung cancer are among the highest (Mabuchi and others 1992). A series of studies conducted by the Radiation Effects Research Foundation have explored the effect of smoking on lung cancer in the atomic-bomb survivors. Kopecky and others (1988) reported an analysis of the combined effects of smoking and radiation in a selected cohort with information available on smoking. A total of 351 cases occurred in a cohort of 29,332 exposed survivors. Poisson regression models were used to assess the effects of radiation exposure, using the T65 dosimetry, and smoking, with control for other factors including age at the time of the bombing. Using an additive model for the excess relative risk, Kopecky and others (1988) found that both radiation exposure and cigarette-smoking were determinants of lung-cancer risk; an interaction term for the two exposures was not statistically significant (p = 0.72). While Kopecky and others (1988) expressed a preference for the additive model based on these analyses, further analyses by the BEIR IV committee (NRC 1988) showed that the data were equally compatible with a multiplicative model.

Kopecky and others (1988) also explored the risk of radiation for the three principal histologic types of lung cancer in the study: adenocarcinoma, epidermoid carcinoma, and small cell carcinoma. Histologic classification was based on review of available materials. The excess relative risk was greatest for small cell carcinoma but the heterogeneity of radiation risk among the three histologic types was not statistically significant. Other recent investigations of lung cancer in the atomic-bomb survivors have addressed histologic patterns of lung cancer (Land and others 1993) and p53 mutations (Takeshima and others 1993) (see below).

Two studies have examined modification by cigarette-smoking of the risk of lung cancer following therapeutic irradiation. One of the studies involved follow-up of persons receiving radiation therapy for Hodgkin's disease and the other, persons treated for breast cancer. Another study examined risk of lung cancer in survivors of Hodgkin's disease in relation to radiation and smoking but the combined effect of the two agents was not considered (Kaldor and others 1992).

Neugat and Murray (1994) conducted a case-control study of Connecticut women with a second primary cancer following an initial diagnosis of breast cancer. The cases (N = 94) were women with lung cancer as the second primary whereas the controls (N = 598) had a second malignancy of a type not associated with smoking or radiation. The pattern of the increased risk associated with both smoking and radiation therapy for the initial breast cancer was consistent with a multiplicative interaction; however, the consistency of the data with different models was not formally assessed.

Van Leeuwen and others (1995) used a nested case-control design to assess lung-cancer risk in relation to radiation and smoking in a cohort of 1939 patients who had received treatment for Hodgkin's disease in The Netherlands. The 30 cases occurring during an 18-year follow-up were matched to 82 controls. Radiation doses to the region of the lung where the case developed cancer were estimated and information on smoking was obtained from multiple sources. There was a significantly greater increase in risk among smokers in relation to estimated radiation dose than among nonsmokers. However, in reviewing the findings, Boivin (1995) showed that the pattern of combined effects was consistent with additivity of the excess relative risks. This study is limited by the small number of lung-cancer cases and by the potential modifying effects of chemotherapy.

### "Signatures" of Radon-Progeny Exposure and Cigarette-Smoking

If the characteristics of lung cancers caused by radon progeny and cigarette-smoking were distinct, having specific "signatures", lung cancers could be separated by cause on this basis. Histologic type of lung cancer has been assessed in underground miners and in the general population as a potential indicator of the responsible causal agent (Churg 1994; NRC 1988). More recently, molecular markers of genetic change have also been examined as potential signatures (NRC 1994a).

Histologic type of lung cancer has not proven to be an indicator of the causes of lung cancer in the general population. Cigarette-smoking increases risk for each of the principal histologic types of lung cancer, although the exposure-response relationships differ among the types (Wu-Williams and Samet 1994); risks of small-cell and squamous-cell carcinoma rise most steeply with smoking. Adenocarcinoma predominates in never-smokers (Churg 1994). There is an unexplained trend of increasing adenocarcinoma in the United States and other countries over the last 20 years (Gilliland and Samet 1994).

Histologic type of lung cancer has also not proven to be a definitive indicator of lung cancer caused by radon progeny in underground miners exposed to radon (Churg 1994; NRC 1988). An appendix to the BEIR IV report provides an extensive review of the evidence published through approximately 1987. Early reports indicated a predominance of small-cell lung cancer in underground uranium miners in the Colorado Plateau, but later analyses showed that the risk of each type rose with level of radon exposure and on follow-up the proportion of small-cell lung cancer declined over time to the level observed in the general population (Saccomanno and others 1988). Other groups of miners studied cross-sectionally have shown a higher proportion of small-cell lung cancer than observed in the general population (NRC 1988), although squamous-cell carcinoma predominated in the Yunnan tin miners (Yao and others 1994). In the Yunnan tin miners, the dose-response relationship was steeper for small-cell carcinoma and adenocarcinoma, in comparison with squamous-cell carcinoma.

Land and colleagues (1993) compared the histopathology of lung-cancer cases in uranium miners (N = 92) and survivors of the atomic bombings of Hiroshima and Nagasaki (N = 108) and assessed the distributions of histopathology in relation to radiation exposures of the two groups. In both populations, radiation-associated lung cancers were more likely to be small-cell lung cancer and less likely to be adenocarcinoma. While there is evidence of increased small-cell lung cancer in miners exposed to radon progeny, the temporal pattern of histopathology is variable and there is no basis for using the distribution of lung-cancer histologic type in the general population to estimate the numbers of lung-cancer cases caused by radon.

The search for signatures has now moved to the level of genetic change (NRC 1994a). As reviewed elsewhere in this report, molecular and cellular markers are under active investigation, but specific changes indicating that a cancer has been caused by radon have not yet been identified.

## Previous Risk Models

As discussed above, in using a model to estimate the numbers of radon-attributable lung-cancer cases, an assumption is needed as to the pattern of combined effects of smoking and radon. In most, but not all, risk assessments, a multiplicative pattern of interaction between smoking and indoor radon has been assumed. The BEIR IV model assumed that the combined effect of the two agents was multiplicative, although the committee acknowledged the uncertainty of this assumption. The strongest evidence available to the committee was provided by the Colorado Plateau study. The most extensive analysis, that published by Whittemore and McMillan (1983), found that the data were not consistent with an additive model but were fit by a multiplicative model. A subsequent analysis by Hornung and Meinhardt (1987) and the BEIR IV committee's analysis found that a submultiplicative interaction was most consistent with the data from the Colorado Plateau cohort, although the data were consistent with a multiplicative interaction.

Both additive and multiplicative models had been published before the BEIR IV Report. The model developed by the National Council for Radiation Protection and Measurements (1984) was a modified 2-parameter additive model with the important modification being a term for reduction in risk with time since exposure. Publication 50 of the International Commission for Radiological Protection (1987) describes a multiplicative model, as does Publication 65, published in 1993. The Commission assumed a multiplicative projection model because such models were assumed ". . . to be more representative of the time distribution of the excess risk," while acknowledging the inconsistency of the evidence from the miners. The U.S. Environmental Protection Agency (1992b) also assumes a multiplicative interaction in making its risk projections.

### Passive Smoking and Lung Cancer

Because passive smoking is also a cause of lung cancer, there has also been interest in the combined effect of passive smoking and radon exposure on lung-cancer risk (Samet 1989). Interaction between passive smoking and lung cancer could occur at several levels. Smoking adds particulate matter to the air of a room, thereby increasing the attached fraction and concentration of radon progeny because plateout of unattached progeny onto surfaces is reduced. Tobacco smoke has been shown to increase radon progeny level but the reduction of the unattached fraction by tobacco smoke would be anticipated to reduce the dose to target cells.

The 1991 report of the National Research Council (NRC 1991) addressed the effect of passive smoking on dose of delivered alpha energy. The modeling compared delivered doses under average circumstances of smoking (unattached fraction of 0.03) and during the circumstance of exposure to average smoking passively (unattached fraction of 0.01). The aerosol generated by smoking was considered to have a larger aerodynamic diameter compared with that normally present (0.25 versus 0.15 μm). Smoking was projected to reduce the dose. In the extreme circumstance of the aerosol generated at the time of active smoking, the projected exposure-dose coefficient was only half that under normal conditions. There has not yet been formal epidemiologic investigation of the combined effect of passive smoking and radon and statistical power for assessing the joint effect would be anticipated to be extremely limited.

### CONCLUSIONS: IMPLICATIONS FOR THE BEIR VI REPORT

Tobacco-smoking, primarily in the form of smoking manufactured and hand-rolled tobacco, is the cause of most lung-cancer cases in the United States and many other developed countries. The trends of lung-cancer occurrence across the century are largely reflective of smoking patterns. Consequently, any risk assessment for indoor radon needs to address the effect of radon on never-smokers and ever-smokers of tobacco. The evidence from the epidemiologic studies of underground miners shows differing patterns of effect of radon-progeny exposure on never-smokers and ever-smokers. Although the data are not sufficiently abundant to describe this effect modification in great detail, the data indicate synergism between the two agents that is most consistent statistically with less than multiplicative interaction for describing the joint effect. The committee's risk projection models separately characterize the risks in ever-smokers and never-smokers. The miner data remain the principal basis for this separate characterization. The case-control studies in the general population have not proved informative on this issue and valid signatures at the histologic and molecular levels have not yet been identified.

Some have argued that indoor radon could not contribute substantially to lung cancer in the population because mortality rates from lung cancer in never-smokers are substantially lower than overall rates in the general population. By implication, all lung cancer is assumed to be caused by smoking, or preventable by smoking prevention and control. However, the synergism between radon exposure and smoking implies that the number of radon-attributable cases will vary with the background rate. From the public health perspective, it would be inappropriate to dismiss the large numbers of cases that occur in ever-smokers because of their high background rate. The historical record of lung-cancer rates across the century provides little insight into the significance of radon as a public health problem. The validity of the diagnosis has probably varied over time. Mortality rates from lung cancer earlier in this century and rates in never-smokers offer at most a biologically inappropriate upper bound for the numbers of lung cancers attributable to radon progeny absent cigarette-smoking.

# Appendix D

# Miner Studies

Data from epidemiologic studies of underground miners have played a central role in most efforts to evaluate radon risks, including those of the BEIR VI committee. The BEIR IV report reviewed epidemiologic studies of the following groups of underground miners: uranium miners in the Colorado Plateau, Czechoslovakia, Canada (Ontario, Port Radium, and Beaverlodge), and France; tin miners in Cornwall, United Kingdom, and China; iron and other miners in Sweden; fluorspar miners in Canada; and niobium miners in Norway. The BEIR IV committee based its risk model on analyses of combined data from 4 of these studies: Malmberget in Sweden, Colorado Plateau in the U.S., and Beaverlodge and Ontario in Canada.

Since publication of the BEIR IV report, additional information, including both updating of follow-up and new data analyses, has become available from several of the cohorts noted above, including the cohorts in Colorado, Czechoslovakia, France, China, Ontario, and Port Radium. Results from additional cohorts—New Mexico and Radium Hill in Australia—have also been published. These studies consistently demonstrate excess lung-cancer mortality compared with expected numbers from the general population and increasing risk with increasing exposure to radon progeny.

The recently published studies provide information on issues that could not be addressed adequately in the BEIR IV report. The study in China includes a large number of subjects exposed under the age of 20 and has provided an opportunity for examining the effect on risks of age at first exposure. The Chinese and Ontario studies have provided an opportunity to examine the effects of arsenic exposure. Six of the studies (Colorado, Sweden, China, New Mexico,

Newfoundland, and Radium Hill) include smoking data and have allowed a more effective examination of the combined effects of exposure to smoking and radon progeny. Additional years of follow-up and addition of new studies have also increased the information available on the modifying effect of age at risk, time since exposure, and time since last exposure. Finally, the studies now available include a wide range of exposure rates and have increased the information available on the modifying effect of exposure rate both within individual cohorts and in analyses of combined data from several cohorts.

Not only have new data become available, but informative statistical analyses of the data have been conducted. In most cases, investigators have modeled the hazard (age-specific risk) as a function of exposure and other variables, and this has allowed rigorous examination of the modifying effects of time-related variables, age at exposure, exposure rate, and smoking. That approach is comparable with that used by the BEIR IV committee, and many of the reports have compared findings from the study being evaluated with those predicted by the BEIR IV model. Special analyses addressing the combined effects of smoking and radon have been conducted by Thomas and others (1994), Yao and others (1994), and Moolgavkar and others (1993).

Most important, a working group of principal investigators, under the sponsorship of the U.S. National Cancer Institute (NCI), has analyzed combined data from 11 miner cohorts, including all available cohorts having at least 40 lung-cancer deaths and estimates of each participant's exposure to radon progeny (Lubin and others 1994a, 1995a). These analyses included 2,700 lung-cancer deaths in 68,000 miners, compared with 360 lung-cancer deaths in 22,200 miners included in the BEIR IV committee's analyses. Recent additional analyses of these data have provided a more-detailed examination of exposure-rate effects (Lubin and others 1995b) and of the low-exposure miner data (Lubin and others 1997).

The risk model recommended in the current report is based on analyses of data from the 11 cohorts evaluated by NCI, although for some cohorts data have been updated or modified in other ways. Although the committee chose to conduct its own analyses of these data to develop its risk model, it drew heavily on the extensive results included in the NCI report and papers noted above, particularly for addressing the modifying effects of such variables as smoking, exposure rate, age at first exposure, and time since exposure.

The purposes of this appendix are to describe the characteristics of the epidemiologic studies that were used to develop the committee's risk model and to summarize and discuss results of published analyses of data from these studies, particularly the extensive analyses conducted by NCI. Emphasis is given to information and analyses that have become available since publication of the BEIR IV report. The appendix is limited to lung-cancer risks; analyses of the miner data that address other health end points are discussed in chapter 4.

This appendix begins by providing an overall description of the characteristics of each of the 11 cohorts included in the combined analyses noted above and then summarizes and compares the cohorts with respect to their informativeness on several issues. It then discusses both the approach and results of statistical analyses, primarily the NCI combined analyses. Analyses reported by investigators for individual studies are not discussed unless they provide insights that are not available from the combined analyses. Discussion of the statistical analysis begins with an overview, which is followed by sections on specific topics. It concludes with an overall evaluation of the analyses and risk models developed in the NCI report. Analyses conducted specifically for the BEIR VI report are described in chapter 3 and appendix A, and are not discussed here. Appendix E provides a comprehensive description of the method for exposure estimation in each study.

## SUMMARY AND EVALUATION OF THE MINER COHORTS

The basic features of each of the 11 cohorts included in the NCI analyses are summarized in the text and tables of this section. The statistics presented in Tables D-1 through D-11 refer to the data used in analyses to develop the risk models for this report, which, for 7 of the cohorts, are very similar to data used in the NCI analyses, although in some cases changes in the approach used to allocate person-years resulted in minor changes from tables presented in the NCI report. For Czechoslovakia, exposures were re-evaluated and follow-up data were improved (Tomášek and Darby 1995) in the data set used for the BEIR VI analyses. Some modifications of the data were also made for China, Newfoundland, and France.

Tables D-1 through D-18 present brief descriptions of tabular information on the cohorts. For additional details on these cohorts, the reader is referred to Lubin and others (1994a) and to papers describing results from the individual studies. The cohorts are presented in descending order of number of lung-cancer deaths. In Tables D-1 through D-11, mean values are computed with person-years as weights.

### The China Cohort (Table D-1)

Tin mining in the Yunnan Province in southern China dates back almost 2000 years. The China cohort consists of about 17,000 employees of the Yunnan Tin Corporation (YTC), a large nonferrous-metals industry. YTC, which was formed in 1883 and nationalized in 1949, is the largest employer in Geiju City. The NCI became involved with the health-research unit of YTC and the Cancer Institute of the Chinese Academy of Medical Sciences in the study of the China cohort in 1985. The cohort has the largest number of lung-cancer deaths. Miners were exposed to a range of exposure rates, and exposure was relatively long, being second only to the study of Swedish miners. This is the only cohort with substantial numbers of workers exposed as children, and 735 of 980 lung cancers occurred in miners first exposed under age 20.

TABLE D-1   Summary of China cohort
_____

Study site:  Yunnan Province, China
Type of mine:  Tin
Recent references:  Xuan and others 1993; Yao and others 1994
Definition and identification of cohort:
  All participants in a 1976 occupational survey of both active and retired miners who
    worked in one of 5 divisions of YTC were included.  The survey included all employees
    within these 5 units, which were responsible for all underground mining activities.
Methods for follow-up and ascertainment of lung-cancer deaths:
  Vital status was determined from YTC medical, payroll, and retirement records.  Lung-
    cancer deaths were ascertained from a YTC-operated cancer registry.  Because YTC
    compensates lung-cancer cases and their families, ascertainment is considered to be
    complete.
Number of exposed miners in cohort: 13,649
Number of exposed person-years: 134,842
Average cumulative exposure:  286.0 WLM
Average duration of exposure: 12.9 years
Average exposure rate:  1.7 WL
Period of follow-up:  1976-1987
Average length of follow-up:  10.2 years
Average year of first exposure:  1955.6
Average age at first exposure:  18.8 years
Number of exposed lung-cancer deaths:
  Total: 936
  Cumulative exposure < 100 WLM: 72 (7.7%)
  Cumulative exposure < 50 WLM: 33 (3.5%)
  Average exposure rate < 0.5 WL: 9 (1.0%)
Other special characteristics of cohort:
  A substantial proportion were under age 20 at the start of exposure (73% of lung cancers
    occurred in those first exposed under age 20).  The exposure duration was much longer
    than the average of 5.7 for all cohorts.
Available data on smoking:
  Data on tobacco use, including cigarettes and bamboo water pipes, are available from 1976
    survey but missing for 24% of survey participants.  Information on duration and amount
    smoked is incomplete.
Available data on other mining exposures:
  Cumulative arsenic in mg/months/m$^3$—based on measurements from the 1950s.
  Miners were assumed exposed to arsenic 7 h/d and above-ground workers 8 h/d.
Results of NCI analyses:
  Estimated ERR/exposure = 0.16% after adjustment for arsenic exposure.
  No evidence of nonlinearity in dose-response relation.
  Significant decrease in ERR/exposure with increasing attained age, decreasing exposure
    duration, increasing average exposure rate, and increasing time since last exposure.
  Significant variation of ERR/exposure with categories of age at first exposure, but pattern
    was not consistent.
Special problems:
  Difficulties were encountered in linking cases with survey forms because of removal of
    forms from files, but this difficulty is thought to have been resolved satisfactorily.
Special analyses:
  Combined effects of smoking and radon exposure are addressed in analyses by Yao and
    others (1994).

# The Czech Cohort (Table D-2)

Metal mining in western Bohemia has been carried on since the 15th and 16th centuries, and the study of miners who began mining between 1948 and 1957 (know as the "S cohort") was initiated in 1970. The first results were published in 1972, making it one of the first miner studies to report findings; several updates have been published. As noted in Table D-2, most analyses have focused on the S cohort. The Czech cohort includes a wide range of exposure levels.

TABLE D-2    Summary of Czech cohort

Study site:  Western Bohemia, Czech Republic
Type of mine:  Uranium
Recent references:  Ševc and others 1988, 1993; Tomášek and others 1993, 1994a,b
Definition and identification of cohort:
  Most analyses, including NCI analyses, have focused on the "S cohort," which consists of uranium miners who started working in western Bohemia in 1948-1957. Ševc and others (1988) also evaluated additional workers, including miners who started working in middle Bohemia in 1968-1975, iron miners, and burnt-clay miners; these additional workers had much lower cumulative exposures than the early uranium miners and were not included in the NCI analyses.
Methods for follow-up and ascertainment of lung-cancer deaths:
  Both vital status and disease outcome were established from the population registry at the Ministry of Interior, examination of district death-registry records, oncologic notification records that were maintained by the Ministry of Health, and pathology records at district hospitals.
Number of exposed miners in cohort:  4,320
Number of exposed person-years:  102,650
Average cumulative exposure:  196.8 WLM
Average duration of exposure:  6.7 years
Average exposure rate:  2.8 WL
Period of follow-up:  1952-1990
Average length of follow-up:  25.2 years
Average year of first exposure:  1951.0
Average age at first exposure:  30.1 years
Number of exposed lung-cancer deaths:
  Total:  701
  Cumulative exposure < 100 WLM:  73 (10%)
  Cumulative exposure < 50 WLM:  11 (1.6%)
  Average exposure rate < 0.5 WL:  0 (0.0%)
Available data on smoking:
  No information is available on the cohort analyzed at NCI.
Available data on other mining exposures:
  No specific data are available, but Tomášek and others (1994a) note that miners were potentially exposed to arsenic and that miners who worked in the Jachymov mine were exposed to much higher levels of arsenic than those who worked in other mines. Data indicating how long mine workers were employed in various specific mines were available to Tomášek and others (1994a).

## TABLE D-2    Continued

Results of NCI analyses:
Estimated ERR/exposure = 0.34% (0.2-0.6%)
Some evidence of nonlinearity in dose-response relation with decrease in risks at highest cumulative exposures.
Significant decrease in ERR/exposure with increasing attained age, decreasing exposure duration, increasing average exposure rate, decreasing age at first exposure, and increasing time since last exposure.
Special studies:
Ševc and others (1988) reported increased incidence of basal cell carcinoma of skin.
Special analyses:
Tomášek and others (1993) examined mortality from cancer other than lung cancer (see chapter 4)
Additional results:
Ševc and others (1988) did not find evidence of significantly increased risks in the recent uranium miners (initially exposed 1968 and later), but risks were compatible with those obtained from the more highly exposed early workers. Statistically significant excess risks were observed in both iron and burnt-clay miners, and their experience was also comparable with that of the early uranium miners (S cohort). Ševc and others (1988) also conducted analyses of specific cell types.
Tomášek and others (1994a) found that the ERR/exposure was significantly larger for miners who worked more than 20% of their employment period in the Jachymov mine; arsenic exposure is a possible explanation for this finding. Tomášek and others also found no evidence of an exposure-rate effect in analyses that excluded miners who ever experienced exposure rates exceeding 10 WL; however, these analyses excluded large numbers of miners with modest average exposure rates, and lack of power might partially explain this finding.
Analyses by Tomášek and others (1994a) were based on exposure and follow-up data that had been improved over those used in earlier analyses, including the NCI analyses. The exposure revision involved correcting previous errors, estimating exposure for each month in each man's employment history, and taking account of exposures received in other mines. Improvements in follow-up, described by Tomášek and others (1994b), increased the numbers of persons who had died from lung cancer available for study. The overall ERR/exposure estimate based on the revised data was 0.61%, compared with 0.37% based on the earlier data.

## The Colorado Plateau Cohort (Table D-3)

The study of uranium miners in the Colorado plateau is one of the earliest studies to document excess lung cancer rigorously; initial findings were published in the 1960's. The study was established by the U.S. Public Health Service, and cohort follow-up is now conducted by the U.S. National Institute for Occupational Safety and Health. Mining in the Colorado plateau—including parts of Arizona, Colorado, New Mexico, and Utah—expanded rapidly in the late 1940s, but by 1970 many mines had closed. Exposure rates for this cohort were relatively high, and this cohort has both the highest mean exposure and highest

## TABLE D-3   Summary of the Colorado cohort

Study site:  Colorado Plateau, United States

Type of mine:  Uranium

Recent references:  Hornung and Meinhardt 1987; Moolgavkar and others 1993; Thomas and others 1994

Definition and identification of cohort:

 Workers in Arizona, Colorado, New Mexico, and Utah who had completed at least 1 month of underground uranium mining, who volunteered for at least one medical examination between 1950 and 1960, and who provided personal and occupational data of sufficient detail for follow-up and for exposure estimation.  Another 115 workers were also included but included only once in the joint analysis; these workers were retained for separate NCI analyses of the Colorado cohort, but included only once in joint analyses.  Most NCI analyses were restricted to the portion of the data with cumulative exposure less than 3,200 WLM, and the statistics below are also based on this restriction.

Methods for follow-up and ascertainment of lung-cancer deaths:

 Vital status was ascertained from records of mining companies, state vital-statistics offices, the U.S. Social Security Administration, the Internal Revenue Service, and the Veterans Administration; from the National Death Index; and by direct contact.  Cause of death was determined from state death certificates.

Number of exposed miners: 3,347

Number of exposed person-years: 79,556

Average cumulative exposure: 578.6 WLM

Average duration of exposure: 3.9 years

Average exposure rate: 11.7 WL

Period of follow-up:  1950-1990

Average length of follow-up: 26.3 years

Average year of first exposure:  1953.0

Average age at first exposure: 31.8 years

Number of exposed lung cancer deaths:

Total:

 Cumulative exposure < 3,200 WLM: 334

 Cumulative exposure < 100 WLM: 20 (6.0%)

 Cumulative exposure < 50 WLM: 13 (3.9%)

 Average exposure rate < 0.5 WL:  0 (0.0%)

Available data on smoking:

 Data were obtained from annual censuses of miners and from mail questionnaires on 1-4 occasions from 1950 to 1960, and at other times from 1963 to 1969.  Data have been updated more recently but results were unavailable.

Available data on other mining exposures:

 Information on number of years previously worked in hard-rock mines and estimates of total WLM received in these mines were available.

Results of NCI analyses:

 Estimated ERR/exposure = 0.42% (0.3-0.7%).

 Evidence of nonlinearity in dose-response relation when all data included; consistent with linearity when analyses were restricted to cumulative exposures not exceeding 3,200 WLM, as in NCI analyses.

 Significant decrease in ERR/exposure with increasing attained age, decreasing exposure duration, increasing average exposure rate, and increasing time since exposure.  No evidence of modification by age at first exposure.

TABLE D-3    Continued

Special analyses:
  Thomas and others (1994) used a case-control approach to investigate the joint effects of
  smoking and radon exposure. Their analyses also included consideration of the detailed
  exposure-rate history. Moolgavkar and others (1993) applied biologically based 2-stage
  clonal expansion model to the Colorado data.

average exposure rate (nearly the same as that of Port Radium) of the 11 cohorts evaluated here. The cohort contributes almost no direct information on effects of exposure at low exposure rates (< 1 WL).

## The Ontario Cohort (Table D-4)

The Ontario cohort used in the NCI analyses includes only uranium miners, although the complete cohort includes those engaged in other types of mining (gold, nickel, and copper) in the province of Ontario. Uranium mines began operation in 1955 and continued into the early 1960s. The cohort includes some workers who had previously worked in gold mines. The first findings of the Ontario study were published in 1981. This cohort has the largest number of workers, and makes the strongest contribution of the 11 cohorts to estimation of effects at lower cumulative exposures and exposure rates. The average exposure rate is one of the lowest, along with those of Sweden, Beaverlodge, Radium Hill, and France.

TABLE D-4    Summary of the Ontario cohort

Study site: Ontario, Canada
Type of mine: Uranium
Recent references: Muller and others 1984; Kusiak and others 1991, 1993.
Definition and identification of cohort:
  The cohort included miners who attended required miners' chest clinics from January 1
    1955, to December 31, 1984, and who had been employed for a minimum of 60 months
    in dusty jobs in a nonuranium mine or for a minimum of 2 weeks in a uranium mine.
    Miners who had also worked in mines outside Ontario, females, and men with missing
    dates of birth were excluded.
Methods for follow-up and ascertainment of lung-cancer deaths:
  Vital status and cause of death were determined for 1955-1986 by searching the Mortality
    Database of Statistics, Canada.
Number of exposed miners in cohort: 21,346
Number of exposed person-years: 300,608
Average cumulative exposure: 31.0 WLM
Average duration of exposure: 3.0 years
Average exposure rate: 0.9 WL
Period of follow-up: 1955-1986

## TABLE D-4   Continued

Average length of follow-up: 17.8 years
Average year of first exposure: 1963.8
Average age at first exposure: 26.4 years
Number of exposed lung-cancer deaths:
  Total: 285
  Cumulative exposure < 100 WLM: 225 (79%)
  Cumulative exposure < 50 WLM: 174 (61%)
  Average exposure rate < 0.5 WL: 100 (35%)
Available data on smoking:
  Data from several surveys are available. Beginning in 1976, each miner's smoking history
    was recorded when he made his annual visit to a chest clinic; in 1990-1991, a
    questionnaire was sent to all miners in Ontario whose addresses could be found.
Available data on other mining exposures:
  Estimates of total radon-progeny exposure and arsenic exposure from gold mining are
    available. Miners with known exposure to asbestos or radon-progeny outside Ontario
    were excluded, but many Ontario uranium miners had previously worked in gold mines.
Results of NCI analyses:
  Estimated ERR/exposure = 0.89% (0.5-1.5%)
  No evidence of nonlinearity in dose-response relation.
  Significant decrease in ERR/exposure with increasing average exposure rate. No evidence
    of effect modification by attained age, exposure duration, time since exposure, or age at
    first exposure.
Special analyses:
  A report by Muller and others (1984) included evaluation of risks of cancers other than
    lung cancer. A report by Kusiak and others (1993) included evaluation of smoking data
    and of specific cell types of lung cancer.

### The Newfoundland Cohort (Table D-5)

Underground mining of fluorspar in St. Lawrence, Newfoundland, began in 1936 and continued until 1978. The average exposure rate is high, and the cohort has the second highest mean WLM.

### The Swedish Cohort (Table D-6)

The Swedish cohort includes iron miners in Malmberget, Sweden, located above the Arctic circle. The study includes some who began mining as early as 1897. This is an older cohort with the longest mean duration of exposure of the 11 cohorts. The average exposure rate is the lowest of the 11 cohorts, and this cohort makes a strong contribution to the estimation of effects at lower exposure rates.

### The New Mexico Cohort (Table D-7)

New Mexico was a leading producer of uranium from the 1960s through the 1980s, and the New Mexico cohort is the most recently employed miner cohort in

TABLE D-5     Summary of the Newfoundland cohort

Study site: Newfoundland, Canada
Type of mine: Fluorspar
Recent references: Morrison and others 1988.
Definition and identification of the cohort:
  Men who were employed by one of 2 local mining companies from 1933 to 1978 and who
    had adequate personal identifying information were included.
Methods for follow-up and ascertainment of lung-cancer deaths:
  Vital status and cause of death were determined for the years 1950-1984 by searching the
    Mortality Database of Statistics, Canada.
Number of exposed miners: 1,751
Number of exposed person-years: 33,795
Average cumulative exposure: 388.4 WLM
Average duration of exposure: 4.8 years
Average exposure rate: 4.9 WL
Period of follow-up: 1950-1984
Average length of follow-up: 23.3 years
Average year of first exposure: 1954.1
Average age at first exposure: 27.5 years
Number of exposed lung-cancer deaths:
  Total: 112
  Cumulative exposure < 100 WLM: 18 (16%)
  Cumulative exposure < 50 WLM: 15 (13%)
  Average exposure rate < 0.5 WL: 8 (7.1%)
Available data on smoking:
  Information on smoking was obtained through several surveys (1960, 1966, 1970, 1978)
    and was available for 48% of the cohort.
Available data on other mining exposures: None.
Results of NCI analyses:
Estimated ERR/exposure = 0.76% (0.4-1.3%).
  Modest evidence of nonlinearity in dose-response relationship (p = 0.06); estimated power
    of exposure was 1.32.
  Significant decrease in ERR/exposure with increasing attained age, decreasing exposure
    duration, increasing average exposure rate, and increasing time since exposure. No
    evidence of modification by age at first exposure.

the United States. Exposures of this cohort were generally lower than those of the earlier Colorado plateau miners. The study of these miners was initiated by the University of New Mexico in 1977.

## The Beaverlodge Cohort (Table D-8)

The Beaverlodge uranium mine, in northern Saskatchewan, Canada, began operations in 1949 and closed in 1982. The mines were operated by Eldorado Resources Ltd., a government corporation. Exposure rates were low in this cohort, leading to the second lowest mean WLM.

## TABLE D-6  Summary of the Swedish cohort

Study site:  Malmberget area in northern Sweden

Type of mine:  Iron

Recent references:  Radford and St. Clair Renard 1984

Definition and identification of the cohort:

  Men who were born in 1880-1919, were alive on January 1, 1951, and worked
    underground in more than one calendar year in various mines in northern Sweden were
    included.  Company and union records were the principal sources used to identify
    workers; medical surveys and parish records were also used.

Methods for follow-up and ascertainment of lung-cancer deaths:

  Vital status and cause of death were determined by using each worker's Swedish personal
    identification number and parish records; this information is thought to be nearly
    complete.  Death certificates were used as source for cases; 70% of lung-cancer deaths
    confirmed by autopsy or thoracotomy.

Number of exposed miners: 1,294

Number of exposed person-years: 32,452

Average cumulative exposure: 80.6 WLM

Average duration of exposure: 18.2 years

Average exposure rate: 0.4 WL

Period of follow-up:  1951-1991

Average length of follow-up: 25.7 years

Average year of first exposure:  1934.1

Average age at first exposure: 27.4 years

Number of exposed lung-cancer deaths:

  Total: 79

  Cumulative exposure < 100 WLM: 36 (46%)

  Cumulative exposure < 50 WLM: 17 (22%)

  Average exposure rate < 0.5 WL: 78 (99%)

Available data on smoking:

  As a result of several surveys, there is information on smoking history of more than half
    the men living in 1970; such information is available for all lung-cancer deaths.

Available data on other mining exposures:

  None.  However, arsenic, chromium, and nickel were not present beyond trace amounts in
    the bedrock and were not detected in samples of mine air.  Diesel equipment was not
    introduced in the mines until 1960, after most of the lung-cancer deaths had left the
    mines or died.

Results of NCI analyses:

  Estimated ERR/exposure = 0.95% (0.1-4.1%).

  No evidence of nonlinearity in dose-response relation.

  Significant decrease in ERR/exposure with increasing attained age and increasing time
    since exposure.

  No evidence of significant modification by exposure duration, time since exposure, or age
    at first exposure.  Modification by exposure rate was of borderline statistical
    significance when treated quantitatively.

Special studies:

  The role of silicosis was investigated in a case-control study; the conclusion was that
    silicosis did not contribute to lung-cancer risk.

## TABLE D-7  Summary of the New Mexico cohort

Study site: Grants, New Mexico, United States

Type of mine: Uranium

Recent references: Samet and others 1989, 1991, 1994.

Definition and identification of the cohort:

Men who had undergone a mining-related physical examination at the Grants Clinic in Grants, New Mexico, and who had worked at least 1 year underground in New Mexico before December 31, 1976, were included. Personnel records were reviewed to document underground exposure.

Methods for follow-up and ascertainment of lung-cancer deaths:

Vital status was determined by searching New Mexico vital-statistics records, the New Mexico Tumor Registry, and state driver's license records; the cohort was also matched against records of the Social Security Administration and the U.S. National Death Index. Death certificates were obtained and the cause of death coded by one nosologist.

Number of exposed miners: 3,457

Number of exposed person-years: 46,800

Average cumulative exposure: 110.9 WLM

Average duration of exposure: 5.6 years

Average exposure rate: 1.6 WL

Period of follow-up: 1943-1985

Average length of follow-up: 17.0 years

Average year of first exposure: 1965.7

Average age at first exposure: 28.0 years

Number of exposed lung-cancer deaths:

Total: 68

Cumulative exposure < 100 WLM: 10 (15%)

Cumulative exposure < 50 WLM: 6 (8.8%)

Average exposure rate < 0.5 WL: 3 (4.4%)

Available data on smoking:

Grants Clinic records provided information on whether workers were current, former, or never-smokers.

Available data on other mining exposures:

Information on whether workers had previously worked in a hard-rock mine was available from Grants Clinic records.

Results of NCI analyses:

Estimated ERR/exposure = 1.72% (0.6-6.7%).

No evidence of nonlinearity in dose-response relation.

Significant decrease in ERR/exposure with attained age and decreasing exposure duration.

No evidence of significant modification by exposure rate or time since exposure.

Modification by age at first exposure was of borderline significance when treated quantitatively.

Special characteristics:

A high proportion (44%) of the study population was Hispanic or American Indian. Analyses were adjusted for ethnicity.

Special analyses:

Samet and others (1994) investigated the effect of exposure to silica; no evidence of such an association was found.

## TABLE D-8    Summary of the Beaverlodge cohort

Study site:  Saskatchewan, Canada

Type of mine:  Uranium

Recent references:  Howe and others 1986; L'Abbe and others 1991; Howe and Stager 1996

Definition and identification of the cohort:

Men who had ever worked at the Beaverlodge uranium mine during 1948-1980 were
included except those with no recorded birth year, incomplete job histories, evidence of
error in recorded birth date, or history of working for Eldorado Resources Ltd. at sites
other than Beaverlodge.

Methods for follow-up and ascertainment of lung-cancer deaths:

Vital status and cause of death were determined for 1950-1980 by searching the Mortality
Database of Statistics, Canada.

Number of exposed miners:  6,895

Number of exposed person-years:  67,080

Average cumulative exposure:  21.2 WLM

Average duration of exposure:  1.7 years

Average exposure rate:  1.3 WL

Period of follow-up:  1950-1980

Average length of follow-up:  14.0 years

Average year of first exposure:  1962.6

Average age at first exposure:  28.0 years

Number of exposed lung-cancer deaths:

Total: 56

Cumulative exposure < 100 WLM: 38 (68%)

Cumulative exposure < 50 WLM: 31 (55%)

Average exposure rate < 0.5 WL: 13 (23%)

Available data on smoking:

No information is available for analyses conducted at NCI.  Results from case-control
study suggest that tobacco use was unlikely to have biased the estimates of ERR/
exposure for this cohort.

Available data on other mining exposures:  None.

Results of NCI analyses:

Estimated ERR/exposure = 2.21% (0.9-5.6%)

No evidence of nonlinearity in dose-response relation.

Significant decrease in ERR/exposure with decreasing exposure duration and increasing
average exposure rate.

No evidence of significant modification by attained age, time since exposure, but evidence
of significant modification by time since last exposure, or age at first exposure.

Special studies:

L'Abbe and others (1991) conducted a case-control study to investigate the possibility of
confounding by smoking and other mining experience.  Little evidence of such
confounding was found.

Howe and Stager (1996) conducted a nested case-control study in which exposures were
re-estimated for 65 men who died of lung cancer and 126 matched controls.  The
revised estimates were based on a more-thorough review of individual employment
records that allowed subjects to be assigned to more-specific mine areas than in the
initial assessment and also used arithmetic means of measurements of area exposure
instead of the geometric means that had been used earlier.  Overall, the revised
cumulative exposures were about 60% higher than the original estimates.  In spite of
this increase, the estimated lung-cancer ERR per 100 WLM based on the revised
estimates was 3.26, higher than the value of 2.70 based on the original estimates.  This

## TABLE D-8    Continued

increase was attributed to a reduction in random exposure-measurement error. In addition, analyses based on the revised exposure estimates provided no evidence of an inverse exposure-rate effect, whereas earlier cohort analyses based on the original estimates provided strong evidence of such an effect.

### The Port Radium Cohort (Table D-9)

Like the Beaverlodge mine, the Port Radium uranium mine, in the Northwest Territories, Canada, was operated by Eldorado Resources Ltd. The mine was in operation from 1930 to 1960; because of its earlier dates of operation, miners were exposed to substantially higher exposure rates than those at Beaverlodge. The Port Radium and Colorado cohorts had nearly the same average exposure rate, which was the highest of the 11 cohorts.

## TABLE D-9    Summary of the Port Radium cohort

Study site:  Northwest Territories, Canada
Type of mine:  Uranium
Recent references:  Howe and others 1987
Definition and identification of the cohort:
  The cohort consisted of men who had worked in the mine since 1940 and who were known to be alive on January 1, 1945, except those with various missing or invalid data.
Methods for follow-up and ascertainment of lung-cancer deaths:
  Vital status and cause of death were determined for the years 1950-1980 by searching the Mortality Database of Statistics, Canada.
Number of exposed miners: 1,420
Number of exposed person-years: 30,454
Average cumulative exposure: 243.0 WLM
Average duration of exposure: 1.2 years
Average exposure rate: 14.9 WL
Period of follow-up: 1950-1980
Average length of follow-up: 25.3 years
Average year of first exposure: 1952.3
Average age at first exposure: 27.6 years
Number of exposed lung-cancer deaths:
  Total: 39
  Cumulative exposure < 100 WLM: 7 (18%)
  Cumulative exposure < 50 WLM: 2 (5.1%)
  Average exposure rate < 0.5 WL: 0 (0.0%)
Available data on smoking:  None
Available data on other mining exposures:  None
Results of NCI analyses:
  Estimated ERR/exposure = 0.19% (0.1-0.6%).
  No evidence of nonlinearity in dose-response relation.
  Significant decrease in ERR/exposure with increasing attained age, decreasing exposure duration, and time since last exposure.
  No evidence of significant modification by exposure rate or age at first exposure.

## The Radium Hill Cohort (Table D-10)

The Radium Hill mine, in a remote area of eastern South Australia, produced uranium ore for export to Great Britain and the United States in 1952-1961. The mine was owned and operated by the South Australian Department of Mines. A limitation of the study is that, because of emigration, 26% of the miners could not be traced. Exposure rates were low, and the study had the lowest mean WLM of the 11 cohort studies.

TABLE D-10    Summary of the Radium Hill cohort

Study site:  South Australia
Type of mine:  Uranium
Recent references:  Woodward and others 1991
Definition and identification of the cohort:
   Hourly workers employed at Radium Hill during its period of operation (1952-1961) were
      included.
Methods for follow-up and ascertainment of lung-cancer deaths:
   Vital status was determined by searching death records throughout Australia for the period
      1960-1987.  For earlier years (1952-1959), the search was restricted to South Australia.
      Other sources were also used to locate workers.  For the data used in the NCI analyses,
      26% of the cohort could not be traced.
Number of exposed miners: 1,457
Number of exposed person-years: 24,138
Average cumulative exposure: 7.6 WLM
Average duration of exposure: 1.1 years
Average exposure rate: 0.7 WL
Period of follow-up:  1948-1987
Average length of follow-up: 21.9 years
Average year of first exposure: 1956.0
Average age at first exposure: 29.2 years
Number of exposed lung-cancer deaths:
   Total: 31
   Cumulative exposure < 100 WLM: 30 (97%)
   Cumulative exposure < 50 WLM: 29 (94%)
   Average exposure rate < 0.5 WL: 16 (52%)
Available data on smoking:
   Smoking-status data (never-smoker, ever-smoker, or unknown) were available for about
      half the cohort from a 1984 survey of cohort members and their next of kin; collection
      of smoking information is continuing.
Available data on other mining exposures:  None
Results of NCI analyses:
   Estimated ERR/exposure = 5.06% (1.0-12.2%)
   No evidence of nonlinearity in dose-response relation.
   No evidence of significant modification by attained age, exposure duration, exposure rate,
      age at first exposure, or time since last exposure.
Special problems:
   A high proportion (26%) of subjects could not be traced.

## The French Cohort (Table D-11)

Uranium mining in France began in 1946, and the cohort includes miners in several uranium mines in various locations in France, some of which are still in operation. These mines are operated by a subsidiary of the Commissariat a l'Energie Atomique. Duration of exposure was relatively long in the French mines, but exposure rates were low.

TABLE D-11   Summary of the French cohort

Study site: Mines in 3 areas in the center of France and in an area near the western coast.
Type of mine: Uranium
Recent references: Tirmarche and others 1993
Definition and identification of the cohort:
  Men who had worked at least 2 years underground in uranium mines in France and who started worked in 1946-1972 were included. Thirty-nine foreign workers were excluded.
Methods for follow-up and ascertainment of lung-cancer deaths:
  Vital status was determined from several sources, including company medical records, national records, and local physicians and hospitals and by tracing workers through last known addresses. Cause of death was determined for 96% of workers who had been identified as deceased.
Number of exposed miners: 1,769
Number of exposed person-years: 29,172
Average cumulative exposure: 59.4 WLM
Average duration of exposure: 7.2 years
Average exposure rate: 0.8 WL
Period of follow-up: 1948-1986
Average length of follow-up: 24.7 years
Average year of first exposure: 1956.8
Average age at first exposure: 29.5 years
Number of exposed lung-cancer deaths:
  Total: 45
  Cumulative exposure < 100 WLM: 33 (73%)
  Cumulative exposure < 50 WLM: 22 (49%)
  Average exposure rate < 0.5 WL: 24 (53%)
Available data on smoking: None.
Available data on other mining exposures:
Available data indicate only whether workers had previous mining experience.
Results of NCI analyses:
  Estimated ERR/exposure = 0.36% (0.0-1.3%).
  No evidence of nonlinearity in dose-response relation.
  No evidence of significant modification by attained age, exposure duration, exposure rate, or time since last exposure. Significant modification by age at first exposure when treated quantitatively.

## Summary and Comparison of the Underground-Miner Cohorts

The informativeness of an epidemiologic study for quantifying the exposure-response relationship depends on several factors, including the size of the population, the number of cases of the disease of interest, and the size and distribution of the exposures. In general, studies with the greatest ranges of exposure and exposure rate are likely to be most informative with respect to investigating the shape of the exposure-response function and modifying effects of exposure rate. However, cohorts in which most subjects have been exposed primarily at lower exposure rates and to lower cumulative exposures might be more relevant for evaluating the effect of exposures at residential levels. Tables D-12 through D-15 provide a comparison of exposures and exposure rates among the cohorts and give special attention to describing cohorts in terms of the information available at lower exposures and exposure rates.

In addition to providing quantitative information on the dependence of risks on exposure and exposure rate, the miner data are used to address the modifying effects of attained age, time since exposure, and age at first exposure. Table D-16 provides summary information on the 11 miner cohorts, and Table D-17 provides

TABLE D-12    Number of exposed miners, person-years, lung cancer deaths, and mean WLM and WL

| Study | Number of workers | Number of person-years | Number of lung cancers | Mean $WLM^a$ | Mean duration | Mean $WL^a$ | Weighted mean $WL^b$ |
|---|---|---|---|---|---|---|---|
| China | 13,649 | 134,842 | 936 | 286.0 | 12.9 | 1.7 | 2.3 |
| Czechoslovakia | 4,320 | 102,650 | 701 | 196.8 | 6.7 | 2.8 | 4.2 |
| Colorado$^c$ | 3,347 | 79,556 | 334 | 578.6 | 3.9 | 11.7 | 17.0 |
| Ontario | 21,346 | 300,608 | 285 | 31.0 | 3.0 | 0.9 | 1.6 |
| | | | | | | | |
| Newfoundland | 1,751 | 33,795 | 112 | 388.4 | 4.8 | 4.9 | 12.2 |
| Sweden | 1,294 | 32,452 | 79 | 80.6 | 18.2 | 0.4 | 0.4 |
| New Mexico | 3,457 | 46,800 | 68 | 110.9 | 5.6 | 1.6 | 5.7 |
| Beaverlodge | 6,895 | 67,080 | 56 | 21.2 | 1.7 | 1.3 | 2.5 |
| | | | | | | | |
| Port Radium | 1,420 | 31,454 | 39 | 243.0 | 1.2 | 14.9 | 33.4 |
| Radium Hill | 1,457 | 24,138 | 31 | 7.6 | 1.1 | 0.7 | 1.0 |
| France | 1,769 | 39,172 | 45 | 59.4 | 7.2 | 0.8 | 2.6 |
| | | | | | | | |
| Total$^d$ | 60,606 | 888,906 | 2,674 | 164.4 | 5.7 | 2.9 | 10.8 |

$^a$Weighted by person-years; includes 5-year lag interval.
$^b$Weighted by WLM received at each exposure rate.
$^c$Exposure limited to < 3,200 WLM.
$^d$Totals adjusted for miners and lung cancers that were included in both the Colorado and New Mexico Studies.

TABLE D-13   Number of person-years, lung-cancer deaths, and mean WLM
and WL among exposed miners with cumulative exposure less than 100 WLM

| Study | Number of person-years | Number of lung cancers | Mean WLM[a] | Mean duration[a] | Mean WL[a] | Weighted mean WL[b] |
|---|---|---|---|---|---|---|
| China | 41,656 | 72 | 48.2 | 4.0 | 1.1 | 1.2 |
| Czechoslovakia | 27,100 | 73 | 60.8 | 4.4 | 1.3 | 1.0 |
| Colorado | 17,956 | 20 | 43.4 | 0.9 | 4.6 | 7.4 |
| Ontario | 276,838 | 225 | 20.4 | 2.7 | 0.8 | 0.9 |
| Newfoundland | 18,251 | 18 | 19.5 | 16.9 | 2.0 | 4.7 |
| Sweden | 21,836 | 36 | 45.7 | 12.7 | 0.3 | 0.4 |
| New Mexico | 28,403 | 10 | 32.2 | 3.8 | 0.8 | 1.3 |
| Beaverlodge | 59,305 | 38 | 14.1 | 1.6 | 1.1 | 2.0 |
| Port Radium | 16,858 | 7 | 29.0 | 0.8 | 5.4 | 8.4 |
| Radium Hill | 24,047 | 30 | 7.5 | 1.1 | 0.7 | 1.0 |
| France | 33,087 | 33 | 29.9 | 8.8 | 0.4 | 0.6 |
| Total[c] | 564,772 | 562 | 26.2 | 3.8 | 1.1 | 1.5 |

[a]Weighted by person-years; includes 5-year lag interval.
[b]Weighted by WLM received at each exposure rate.
[c]Totals adjusted for miners and lung cancers that were included in both the Colorado and New Mexico studies.

information on attained age, age at first exposure, and time since last exposure in a format suitable for comparison.

Table D-12 shows several measures of the size of the individual miner cohorts and of the magnitude of the exposures and exposure rates.  Although the number of exposed miners is greatest for the Ontario cohort, the China, Czechoslovakia, and Colorado cohorts all have more lung-cancer deaths because of their higher exposures.  Those 4 cohorts contribute 2,256 (84%) of the total of 2,674 lung-cancer deaths in exposed miners.  The highest average exposure rates (WL) are found in the Colorado, Newfoundland, and Port Radium cohorts, whereas the lowest average rates are found in Ontario, Sweden, Beaverlodge, Radium Hill, and France.  The average cumulative exposures (WLM) follow similar patterns, although China and Czechoslovakia also have relatively large WLM because miners in these countries were employed for relatively long periods.  Overall, the average duration of exposure is short, 5.7 years, with China, Czechoslovakia, Sweden, and France exhibiting somewhat longer durations.

Tables D-13 through D-15 provide similar information but limited to portions of the cohort exposed at levels more directly applicable to estimating risks at lower exposures and exposure rates.  Tables D-13 and D-14 show that Ontario contributes 225 (40%) of the total of 562 lung cancers in exposed miners with

TABLE D-14    Number of person-years, lung cancers, and mean WLM and
WL among exposed miners with cumulative exposure less than 50 WLM

| Study | Number of person-years | Number of lung cancers | Mean $WLM^a$ | Mean $duration^a$ | Mean $WL^a$ | Weighted mean $WL^b$ |
|---|---|---|---|---|---|---|
| China | 21,815 | 33 | 26.4 | 2.8 | 0.8 | 0.8 |
| Czechoslovakia | 9,182 | 11 | 27.4 | 2.5 | 1.0 | 0.7 |
| Colorado | 11,750 | 13 | 26.0 | 0.7 | 3.7 | 6.5 |
| Ontario | 244,785 | 174 | 13.9 | 2.2 | 0.7 | 0.5 |
| Newfoundland | 15,458 | 15 | 9.8 | 19.3 | 1.0 | 2.5 |
| Sweden | 12,625 | 17 | 25.7 | 9.3 | 0.3 | 0.3 |
| New Mexico | 20,504 | 6 | 16.6 | 3.4 | 0.5 | 0.8 |
| Beaverlodge | 55,717 | 31 | 10.5 | 1.4 | 1.0 | 1.5 |
| Port Radium | 12,127 | 2 | 15.5 | 0.8 | 3.2 | 5.3 |
| Radium Hill | 23,693 | 29 | 6.7 | 1.0 | 0.7 | 1.0 |
| France | 25,770 | 22 | 18.1 | 7.9 | 0.3 | 0.4 |
| $Total^c$ | 454,159 | 353 | 14.8 | 3.2 | 0.9 | 1.0 |

[a]Weighted by person-years; includes 5-year lag interval.
[b]Weighted by WLM received at each exposure rate.
[c]Totals adjusted for miners and lung cancers that were included in both the Colorado and New Mexico studies.

cumulative exposures less than 100 WLM, and 174 (49%) of the total of 353 lung cancer deaths in miners with cumulative exposures less than 50 WLM. Table D-15 shows that Ontario and Sweden contribute 178 (71%) of the total of 251 lung cancers in exposed miners with average exposures less the 0.5 WL; however, because of the longer duration of exposure in Sweden, the mean exposure among those in this low-exposure-rate category is much lower for Ontario (10.7 WLM) than for Sweden (79.4 WLM). With the committee's categorical exposure-rate model, the quantitative risk estimates for exposure at residential levels are based on the portion of the miner data with exposure rates less than 0.5 WL, although estimates of parameters indicating the effects of attained age and time since exposure were based on all miner data.

Table D-16 shows that the average age at first exposure is much lower in China than in any of the other cohorts. Table D-17 shows that there are no striking differences among the cohorts with regard to the distribution of cancers by attained age and time since last exposure although Sweden contributes little information to the under 55 age category. China contributes 735 (85%) of the 862 lung cancers among those first exposed under age 20 and all 54 lung cancers among those first exposed under age 10.

TABLE D-15   Number of person-years, lung cancers, and mean WLM and WL among exposed miners with average exposure rates less than 0.5 WL (zero exposure data omitted)

| Study | Number of person-years | Number of lung cancers | Mean WLM[a] | Mean duration[a] | Mean WL[a] | Weighted mean WL[b] |
|---|---|---|---|---|---|---|
| China | 1,106 | 9 | 35.6 | 9.6 | 0.31 | 0.38 |
| Czechoslovakia | 33 | 0 | —[c] | —[c] | —[c] | —[c] |
| Colorado | 291 | 0 | —[c] | —[c] | —[c] | —[c] |
| Ontario | 87,670 | 100 | 10.7 | 4.4 | 0.22 | 0.32 |
| Newfoundland | 7,362 | 8 | 8.4 | 4.1 | 0.16 | 0.28 |
| Sweden | 30,646 | 78 | 79.4 | 18.2 | 0.37 | 0.41 |
| New Mexico | 11,710 | 3 | 11.6 | 3.9 | 0.22 | 0.34 |
| Beaverlodge | 27,692 | 13 | 3.5 | 1.3 | 0.36 | 0.30 |
| Port Radium | 103 | 0 | —[b] | —[b] | —[b] | —[b] |
| Radium Hill | 8,781 | 16 | 7.9 | 1.7 | 0.39 | 0.41 |
| France | 23,720 | 24 | 25.2 | 7.6 | 0.25 | 0.36 |
| Total[d] | 198,720 | 251 | 21.1 | 6.3 | 0.27 | 0.39 |

[a]Weighted by person-years; includes 5-year lag interval.
[b]Weighted by WLM received at each exposure rate.
[c]There were no lung cancers and very few person-years in the Czechoslovakia, Colorado, and Port Radium cohorts with average exposure rates < 0.5 WL, so means are not presented for these cohorts.
[d]Totals adjusted for miners and lung cancers that were included in both the Colorado and New Mexico studies.

Figures D-1 and D-2 present further information comparing various aspects of exposure among the 11 cohorts. Because of the considerable variation in the numbers of miners and levels of exposure among the cohorts, it was necessary to use different scales for the ordinates. Figure D-1a contrasts the times and levels of exposure and clearly illustrates the generally high levels in Colorado, New-foundland, and Port Radium and the decreasing exposure levels in the 1950s and 1960s. The very constant estimated exposure levels in China and Sweden cohorts could reflect the use of natural ventilation in the early years of mining but could also reflect the need to rely on retrospective evaluation of exposures because of the absence of exposure measurements in the early years of mine operations. Figure D-1b clearly shows the large size and long duration of employment for the China cohort, the long duration of employment for the Sweden cohort, and the strong rise and fall in employment for the Czechoslovakia and Ontario cohorts. Figure D-1c shows the number of workers in follow-up by calendar year; these numbers increase as cohort members are added but decrease as deaths occur. Most of the miners still in follow-up in 1990 were in the Ontario and China cohorts.

TABLE D-16    Follow-up period, year first exposed, age first exposed, and years exposed

|  | | Mean among exposed | | |
|---|---|---|---|---|
| Study | Follow-up period | Length of follow-up | Year of first exposure | Age at first exposure |
| China | 1976-1987 | 10.2 | 1955.6 | 18.8 |
| Czechoslovakia | 1952-1990 | 25.2 | 1951.0 | 30.1 |
| Colorado[a] | 1950-1990 | 26.3 | 1953.0 | 31.8 |
| Ontario | 1955-1986 | 17.8 | 1963.8 | 26.4 |
| Newfoundland | 1950-1984 | 23.3 | 1954.1 | 27.5 |
| Sweden | 1951-1991 | 25.7 | 1934.1 | 27.4 |
| New Mexico | 1943-1985 | 17.0 | 1965.6 | 28.0 |
| Beaverlodge | 1950-1980 | 14.0 | 1962.6 | 28.0 |
| Port Radium | 1950-1980 | 25.3 | 1952.3 | 27.6 |
| Radium Hill | 1948-1987 | 21.9 | 1956.0 | 29.2 |
| France | 1948-1986 | 24.7 | 1956.8 | 29.5 |
| Total | 1943-1991 | 17.7 | 1954.0 | 25.2 |

[a]Exposures limited to <3,200 WLM.

Figure D-2 shows the number of miners in each of the cohorts by duration of exposure and illustrates that miners with long exposure durations were predominantly in China, Czechoslovakia, Ontario, and Sweden. This figure also illustrates the relatively short duration of employment for many of the workers in the Ontario and Beaverlodge cohorts.

## Data on Other Exposures

Some cohorts of underground miners have been exposed to dust, arsenic, silica, and diesel exhaust in addition to radon progeny. Because these exposures are not expected in conjunction with residential exposure, a concern is that the presence of such exposure in mines might have resulted in risks per WLM that overestimate risks in the residential setting. If relevant data are available, it is possible to adjust partially for such exposures; thus, in cohorts in which these exposures have occurred, the adequacy of available data is a factor in evaluating the usefulness of the cohort in estimating residential risks. However, because such data are probably never sufficiently detailed and accurate to permit certainty that adjustments are adequate, cohorts that are relatively free of such exposures probably have the least potential for bias from this source. Further discussion of these exposures is found in appendix F.

Smoking is of concern as a confounder and because the high prevalence of smoking among miners might have led to higher risks than would be expected in the general population. In addition, there is considerable interest in understand-

TABLE D-17 Number of lung cancers among exposed miners, by attained age, age at first exposure, and time since last exposure

| | Number of lung cancers by | | | | | | | | |
|---|---|---|---|---|---|---|---|---|---|
| | Attained age[a] | | | Age at first exposure | | | Time since last exposure | | |
| Study | < 55 | 55- | 65+ | < 20 | 20- | 30+ | < 5 | 5-15 | >15 |
| China | 348 | 484 | 148 | 735[b] | 131 | 70 | 121 | 293 | 522 |
| Czechoslovakia | 276 | 275 | 154 | 21 | 227 | 454 | 150 | 266 | 286 |
| Colorado[c] | 134 | 105 | 97 | 47 | 83 | 204 | 54 | 110 | 170 |
| Ontario | 101 | 125 | 65 | 21 | 125 | 139 | 46 | 92 | 147 |
| Newfoundland | 71 | 35 | 12 | 25 | 42 | 45 | 40 | 41 | 31 |
| Sweden | 5 | 22 | 52 | 10 | 51 | 18 | 25 | 19 | 35 |
| New Mexico | 30 | 28 | 11 | 4 | 15 | 49 | 33 | 23 | 12 |
| Beaverlodge | 25 | 25 | 15 | 0 | 1 | 55 | 19 | 21 | 16 |
| Port Radium | 20 | 22 | 15 | 1 | 4 | 35 | 4 | 12 | 24 |
| Radium Hill | 15 | 20 | 19 | 0 | 8 | 24 | 1 | 5 | 26 |
| France | 21 | 17 | 7 | 1 | 12 | 32 | 12 | 17 | 16 |
| Total[d] | 1,039 | 1,156 | 592 | 862 | 695 | 1,120 | 500 | 893 | 1,284 |

[a]Results for attained age include lung cancers in unexposed minors.
[b]54 of these lung cancers occurred in miners first exposed under age 10 (no others occurred in such miners).
[c]Exposures limited to < 3200 WLM.
[d]Totals adjusted for miners and lung cancers that were included in both the Colorado and New Mexico studies.

ing the modifying effects of smoking on risks associated with radon exposure. These issues are discussed in detail in appendix C. Tables D-1 to D-11 describe the available data on smoking for each of the cohorts and also note analyses related to smoking that have been conducted.

## RESULTS OF STATISTICAL ANALYSES OF DATA ON UNDERGROUND MINERS

### Radon and Lung Cancer Risk: A Joint Analysis of 11 Underground Miners Studies

Analyses based on combined data from 11 underground-miner cohorts have recently been conducted (Lubin and others 1994a) as a collaborative effort among the principal investigators. Those analyses provide the most comprehensive summary of available data on lung-cancer risks in miners. Although, as noted above, the BEIR VI committee chose to base its risk model on analyses specifically conducted for this purpose, the committee relied on the NCI analyses for

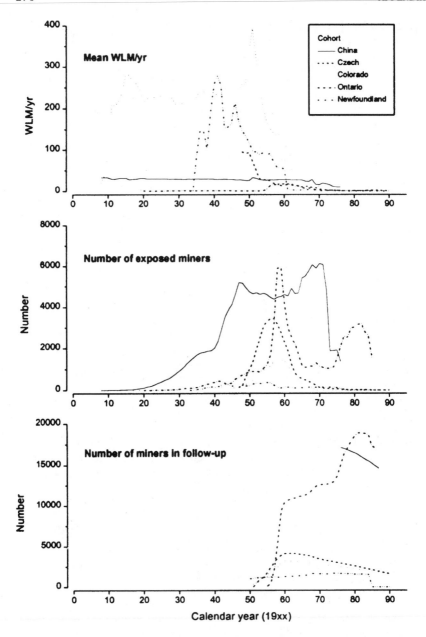

FIGURE D-1    Mean WLM/yr, number of exposed miners, and number of miners in each cohort by calendar year.

FIGURE D-1    Continued

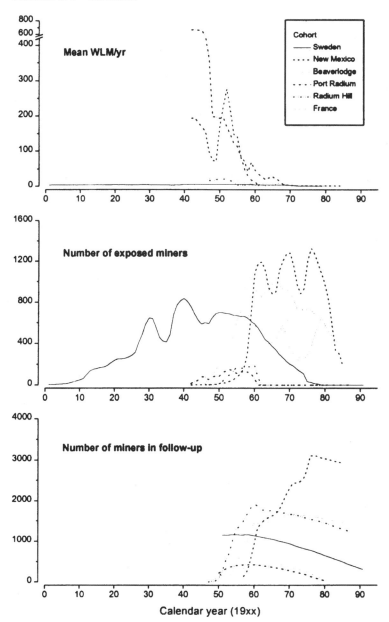

Calendar year (19xx)

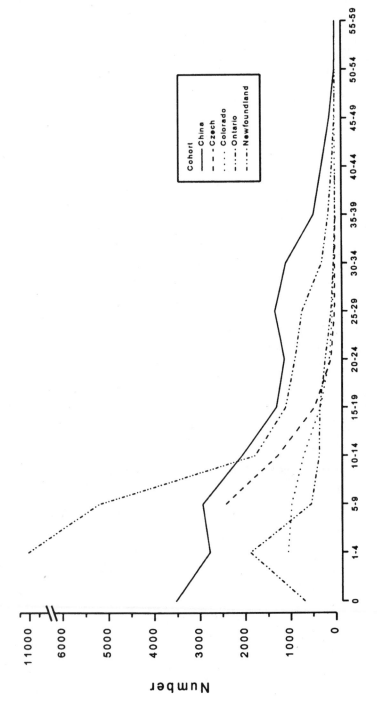

FIGURE D-2  Number of miners in each cohort by duration of radon exposure.

FIGURE D-2  Continued

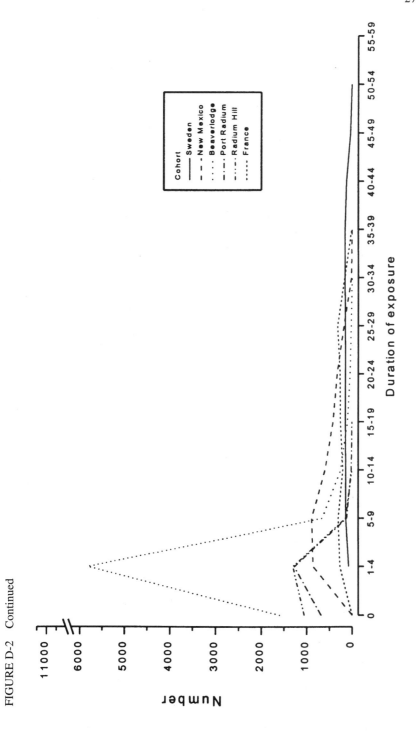

detailed exploration of several relevant issues. An objective of this appendix is to summarize and evaluate these combined analyses. For full detail, the reader is referred to the NCI report or one of the related publications (Lubin and others 1995a,b, 1997).

The general approach taken in the NCI analyses was to model the hazard (age-specific risk) as a function of exposure and other variables. Specifically, most analyses were based on linear relative-risk models in which the relative risk (RR) is written as $RR = 1 + \beta w$, where w represents cumulative exposure in WLM and $\beta$ is the excess relative risk (ERR) per exposure. Modification of risk by such factors as exposure rate was investigated by fitting categorical models of the form $RR = 1 + \beta_j w$, where j indexes categories of the modifying factor or by fitting models that treating such factors as quantitative variables, z. With the latter approach, both exponential models, $RR = 1 + w \exp(\gamma z)$, and power models, $RR = 1 + wz^\gamma$, were fitted.

Although time since *last* exposure was evaluated with the approach described above, time since exposure was handled by expressing the cumulative exposure (minus a 5-year lag interval) as the sum of components received in various time periods. For example $w_{5-14}$ was the cumulative exposure received 5-14 years before the age at risk being evaluated. Separate coefficients were then fitted for the different time components. This approach was similar to that used by the BEIR IV committee, although the extended follow-up of the cohorts allowed the inclusion of an additional category.

Analyses relied on internally based comparisons and did not make use of external vital statistics. With a relative-risk model, the risk of lung cancer is expressed as the product of baseline risk and relative risk. The baseline risk was modeled by including separate parameters for each 5-year age group. Some analyses allowed the baseline risk to depend on calendar-year period, although the final models did not do so. When data were available, baseline risks were adjusted for categories of arsenic exposure or for whether subjects had previous mining experience. For the Beaverlodge cohort, it was also found necessary to introduce a parameter expressing the difference in baseline risks in exposed and unexposed miners; and for New Mexico, analyses were adjusted for ethnicity. In joint analyses based on combined data from several cohorts, separate baseline-risk parameters were estimated for each cohort.

The NCI report first presents separate analyses of each of the 11 cohorts, starting with a simple model with a constant ERR/exposure, $RR = 1 + \beta w$. Results from those analyses are shown in the first column of Table D-18. Departures from linearity in each of the cohorts were then investigated by evaluating models of the form $RR = [1 + \beta w^\kappa] e^{\theta w}$. That was followed by investigation of the modifying effects of attained age, age at first exposure, exposure rate, exposure duration, time since last exposure, and time since exposure. The effects of those variables were investigated both individually and after the inclusion of other variables. Results for the individual cohorts based on exponential model are summarized in Table D-18. Special analyses addressing the effects of smoking and of other miner exposures were conducted for cohorts for which data on these factors were available. Those results are discussed in appendix A.

After presenting "parallel analyses" of each of the individual cohorts, the NCI report presents joint analyses based on data from all cohorts. Those joint analyses evaluated the modifying effects of all the variables noted above, both acting individually and in combination with other variables. Analyses also investigated homogeneity both of the overall magnitude of the risk estimate and of parameters describing effect modification.

On the basis of the analyses described above, evidence of an exposure-response relationship was found for all cohorts evaluated, and this relationship was found to be adequately described by a linear dose-response function (although for Colorado it was necessary to restrict exposures to those less than 3,200 WLM to achieve consistency with linearity). As in the BEIR IV committee analyses, the magnitude of ERR/exposure was found to depend on time since exposure and attained age. In addition, ERR/exposure was found to depend on exposure rate (or, alternatively, exposure duration), with an increase in risk with decreasing exposure rate (or increasing exposure duration). The magnitude of ERR/exposure was not found to depend on age at start of exposure. Modifying effects of the above variables were judged to be reasonably consistent among the 11 cohorts.

The overall magnitude of ERR/exposure varied substantially among the cohorts, ranging from 0.2 to 5.1 per 100 WLM, as is demonstrated in Table D-18. Even after adjustment for the modifying effects of time since exposure, attained age, and either exposure rate or exposure duration, evidence of heterogeneity remained. It was therefore necessary to include "between-cohort" variation in expressing the uncertainty in the risk estimates.

Four models for estimating risks of lung cancer resulting from exposure to radon and radon progeny were developed. Two of them were based on continuous treatment of modifying variables, and 2 were based on categorical treatment of these variables. The latter models were recommended as being more appropriate for estimating individual risk for both occupational and residential exposure. In addition, models based on both exposure rate and exposure duration were developed. The categorical models were very similar to those recommended by the current BEIR VI committee, although parameter values differed slightly because of the changes in the data described earlier in this appendix.

Lifetime risks were estimated for each of these models, with separate estimates for males and females and for ever-smokers and never-smokers. In these calculations, the same ERR was applied regardless of sex or smoking status (that is, a multiplicative model was assumed). Risk estimates were provided for both lifetime and occupational exposure to radon and for a range of exposure rates. In addition, attributable risks associated with an estimated exposure distribution for the United States were calculated. For the additional calculations, the ERR was

TABLE D-18 Estimates of ERR/exposure[a] (with 95% CI) and of various parameters[b] (with p-values) to quantify modifying effects

| Study | ERR/exposure,% | Modifying effect of [c] | | | |
| | | Attained age 60 | Rn progeny (WL) | Age at first exposure > 30 | Time since last exposure > 10 |
|---|---|---|---|---|---|
| China | 0.16 (0.1-0.2) | 0.92 (0.002) | 0.57 (<0.001) | 0.94 (0.01) | 0.86 (<0.001) |
| Czechoslovakia | 0.34 (0.2-0.6) | 0.93 (<0.001) | 0.77 (<0.001) | 1.05 (<0.001) | 0.93 (<0.001) |
| Colorado | 0.42 (0.3-0.7) | 0.92 (<0.001) | 0.94 (<0.001) | 1.00 (0.90) | 0.98 (0.02) |
| Ontario | 0.89 (0.5-1.5) | 0.99 (0.05) | 0.99 (0.37) | 0.99 (0.37) | 0.99 (0.37) |
| Newfoundland | 0.76 (0.4-1.3) | 0.92 (<0.001) | 0.94 (0.002) | 0.99 (0.44) | 0.94 (<0.001) |
| Sweden | 0.95 (0.1-4.1) | 0.86 (0.01) | 0.02 (0.07) | 0.96 (0.36) | 0.88 (0.02) |
| New Mexico | 1.72 (0.6-6.7) | 0.90 (0.02) | 0.94 (0.19) | 0.95 (0.05) | 0.97 (0.33) |
| Beaverlodge | 2.21 (0.9-5.6) | 0.99 (0.89) | 0.62 (<0.001) | 1.03 (0.34) | 0.87 (<0.001) |
| Port Radium | 0.19 (0.1-0.6) | 0.87 (0.01) | 0.98 (0.18) | 1.00 (0.92) | 0.90 (0.01) |
| Radium Rill | 5.06 (1.0-12.2) | 0.97 (0.67) | 0.46 (0.43) | 0.98 (0.55) | 1.02 (0.79) |
| France | 0.36 (0.0-1.2) | 0.89 (0.14) | 1.14 (0.59) | 0.55 (0.004) | 1.00 (0.96) |

[a] Estimates adjusted for age (all studies), other mine exposure (China, Colorado, Ontario, New Mexico and France), an indicator of Rn exposure (Beaverlodge), and ethnicity (New Mexico). Taken from table 5, Lubin and others (1994a).

[b] Estimates of proportional change (exp (γ)) in ERR/exposure based on model in which ERR is given by βw exp(γ x), where w is exposure in WLM and x is variable of interest. Taken from table 10, Lubin and others (1994a).

[c] p value for test of null hypothesis, γ = 0. Taken from table 10, Lubin and others (1994a).

estimated separately for ever-smokers and never-smokers by using data from the 6 cohorts for which data on smoking were available. Even though the data were reasonably consistent with a multiplicative interaction, the best-fitting model was submultiplicative, and, in separate analyses by smoking status, ERR for never-smokers was estimated to be about 3 times that for ever-smokers. Lubin and Steindorf (1995) discuss the issue of accounting for smoking in calculating risks and provide justification for the approach used for calculating attributable risks in the NCI report; this same approach was used for the committee's risk assessment, as described in chapter 3.

Lifetime risks based on those models were similar to the risks based on the comparable models developed by the BEIR VI committee. No attempt was made in the NCI report to evaluate uncertainty in the resulting estimates of lifetime risk. The analyses of the BEIR VI committee add greater attention to uncertainty, as discussed in chapter 3 and appendix A.

## Analyses Evaluating the Shape of the Exposure-Response Function

Current estimates of lung-cancer risks resulting from exposure to radon and radon progeny at the lower exposures encountered in the residential setting have been obtained through linear extrapolation from risk estimates derived from studies of underground miners. The risk models presented in the NCI report are also based on linear extrapolation, and this choice was made after evaluation of the shape of the exposure-response function both in individual cohorts and in the joint analyses.

To investigate departures from linearity on the multiplicative scale, models of the form $RR = [1 + \beta w]e^{\theta}$ (linear-exponential model) and $RR = [1 + \beta w^{\kappa}]$ (nonlinear model) were fitted to determine whether nonzero values for the parameters $\theta$ and $\kappa$ substantially improved the fit of the model. The analyses were conducted with the entire exposure range for each of the cohorts and were restricted to cumulative exposures less than 200 WLM. Only for the Colorado cohort was there clear evidence of nonlinearity; for this reason, analyses of this cohort were restricted to cumulative exposures under 3,200 WLM; with this restriction, tests for nonlinearity were no longer statistically significant. For other cohorts, the only instance where a p value less than 0.05 was achieved was for the linear-exponential model in the Czechoslovakian cohort; in this case, the p value for a test of $\gamma = 0$ was 0.03, whereas that for the nonlinear model was 0.07. With 11 cohorts and 2 models, a single p value under 0.05 could occur by chance.

Details on tests for nonlinearity are presented in the NCI report only for a simple model that did not include modifying effects of other variables. However, those tests were repeated with the variables included; the result was that only for Czechoslovakia was there any evidence of significant departure from linearity.

As discussed earlier, there is evidence that the exposure-rate effect depends on cumulative exposure. If that were the case, the shape of the exposure-response curve would necessarily be different for various fixed exposure rates; in particular, it could not be linear at all exposure rates.

### Analyses Addressing the Modifying Effects of Attained Age, Age at Exposure, Time Since Exposure, Time Since Last Exposure, and Exposure Rate

The BEIR IV committee found that relative risk depended on both attained age and time since exposure, and it included the modifying effects of these factors in its recommended risk model. Although the BEIR IV analyses indicated that exposure rate modified risk in the Colorado cohort, such evidence was not found in the other 3 cohorts, and the recommended model did not include modification by exposure rate. No evidence of modification by age at first exposure was found by the BEIR IV committee, but data on miners exposed under age 20 were sparse in the 4 cohorts evaluated (see Tables D-16 and D-17).

A difficulty in evaluating the effects of the variables considered in this section is that they are all strongly interrelated. As workers age, time since exposure or time since last exposure might also be increasing. It is also possible that smoking habits changed with time, and available smoking data were not adequate to evaluate how this might affect results. For many of the cohorts, exposure rates were much higher in the earlier calendar years than in later ones, and the higher rates might predominate in the longer time-since-exposure periods. In general, those dependences were addressed in the NCI report by evaluating the modifying effect of a specific variable after adjustment for other variables. However, it is not possible to be certain that those adjustments were adequate.

In describing the data over the range of the variables covered by them, it might not be important whether or not a particular variable exhibits a causal relationship or which variables are used to describe the observed pattern. However, in extrapolating to values outside the range of the data, the choice of variables and how they are used to model the data can be very important. For example, if the effect of other variables were incorrectly attributed to exposure rate, this could lead to erroneous estimates of risk at the relatively low rates encountered in residential exposures.

Interactions among the modifying variables evaluated or with cumulative exposure are also possible. For example, the effect of time since exposure might vary with age at exposure, or the effect of any of the variables might vary with exposure rate or with cumulative exposure. Such interactions were generally not investigated, and it is doubtful that the available data are adequate to do so effectively. However, joint analyses estimating the various modifying effects were conducted with various restrictions on cumulative WLM; in general, these analyses provided little evidence of important variation of modifying effects with cumulative WLM.

The NCI analyses on which the final models were based did not include adjustment of baseline risks for calendar year. Analyses based on a simple linear model without consideration of modifying effects of other variables were repeated with adjustment for calendar year; the results changed little.

## Attained Age

Like previous analyses by the BEIR IV committee, the NCI analyses indicated that ERR/exposure declined with attained age; that is, the increase with age in the excess lung-cancer risk attributable to radon was not as rapid as the background rate for the nonexposed.

Initial analyses examined the effect of attained age in each of the 11 cohorts, treating this as both a categorical and a continuous variable. With the categorical treatment, significant effects were seen for the China, Czechoslovakia, Colorado, Newfoundland, and New Mexico cohorts; with the quantitative treatment, Sweden and Port Radium could be added to this list indicated in Table D-18. For all cohorts, the estimated quantitative parameter indicated a decline in risk with increasing attained age. Joint analyses of the 11 cohorts indicated that the effect persisted with inclusion of time since exposure and of either exposure rate or duration and that there was no clear evidence of heterogeneity of this effect among cohorts.

## Age at First Exposure

Evaluation of age at exposure would require creating "windows" in a manner comparable with the treatment of time since exposure. As an alternative (and simpler) approach, age at first exposure was examined. For miners with short exposure durations, the 2 variables should be highly correlated, but this is not necessarily the case for miners with longer exposure durations. In the China cohort, which contributes many of the data on those exposed early in life, many miners were employed for longer periods.

Only for the China cohort were there substantial data for persons exposed under age 20. Even in that cohort, few workers were first exposed under age 10, and, exposure must have been skewed to the older end of the age range. Age-at-exposure effects that have been identified for low-LET exposure have been strongest in the very young (UNSCEAR 1994). Also, if the decline in risk with attained age and time since exposure applies to those exposed early in life, risk of radon-induced lung cancer would have become negligible by the ages when lung cancer usually occurs, and this would greatly limit the ability to detect an age-at-exposure effect. Whether the decline in risk with age and time depends on the age at initial exposure was not investigated, and it is unlikely that data were adequate to address this question adequately.

With a categorical treatment of age at first exposure, substantial improvement in the fit of the model was observed only for China and Czechoslovakia. With quantitative treatment (see Table D-18), New Mexico and France also demonstrated such improvement. For China, New Mexico, and France, the quantitative estimates indicated a decrease with increasing age at first exposure. For Czechoslovakia, the effect was in the opposite direction, but analyses by Tomášek and others (1994a), based on revised exposure and follow-up data, did not identify such an effect. No clear trend was present with the categorical treatment applied to all 11 cohorts and ERR/exposure for those under age 20 was generally similar to ERR/exposure for those age 20 and older. The effects of age at first exposure were re-evaluated with inclusion of other variables (time since exposure, attained age, and exposure rate or duration); details are not given, but this approach did not lead to modification of the decision not to include age at first exposure in the final model.

## Time Since Exposure and Time Since Last Exposure

In initial analyses of individual cohorts, NCI addressed the effects of both time since *last* exposure and time since exposure,[1] with both treated as categorical variables. Inclusion of either of the variables substantially improved the fit of the model ($p < 0.05$) in the China, Czechoslovakia, Colorado, Newfoundland, and Port Radium cohorts. For the Beaverlodge cohort, time since last exposure substantially improved the fit, but time since exposure did not, and, in general, the improvement in fit seemed to be stronger for time since last exposure. For cohorts with substantial improvement in fit, ERR/exposure was found to decrease with either decreasing time since last exposure or decreasing time since exposure. Limited power could have been the reason that effects were not demonstrated in the remaining cohorts; the Ontario cohort was the only large cohort in this category.

Analyses with time since last exposure treated as a quantitative variable were also conducted and were based on the exponential model (see Table D-18). Using this approach, all cohorts noted above showed a substantial improvement in fit; in addition, such an improvement was found for the Swedish cohort. In all

---

[1]To address the effects of time since exposure (as opposed to time since last exposure) on radon-induced lung cancer, separate parameters were fitted for each of 4 time-since-exposure windows. This model can be written $RR = 1 + \beta w^*$, where $w^* = \theta_1 w_{5-14} + \theta_2 w_{15-24} + \theta_3 w_{25-34} + \theta_4 w_{35+}$ and $\theta_1$ is set equal to 1, and where, for example, $w_{5-14}$ indicates the exposure received 5-14 years earlier. This is the same approach as used in the BEIR IV report, although BEIR IV combined the last 2 categories; the last 2 categories were also combined in the NCI analyses for some cohorts with insufficient data.

but Radium Hill and France, the estimated effect indicated a decline in risk with increasing time since last exposure.

Joint analyses of data from all 11 cohorts were based on time-since-exposure windows. Time since exposure was found to significantly improve the fit of the model even after inclusion of attained age and of either exposure rate or exposure duration. Joint analyses also included specific statistical tests to address the homogeneity of the time-since-last-exposure effect across cohorts after adjustment for other variables. Those analyses, which treated time since last exposure as a quantitative variable, provided evidence that the modifying effects of this variable varied significantly across cohorts. No analyses addressing lack of homogeneity of time since exposure effects are presented.

Although results based on both time since exposure and time since last exposure are presented, it is difficult to separate the effects of these 2 variables, because many miners had fairly short durations of exposure. Comparison of p values for the 2 approaches applied to individual cohorts suggests that time since last exposure improved the fit slightly more than time since exposure.

Another aspect of the time-since-exposure effect is the minimal latency period. To address that issue, analyses based on each of several lag periods were conducted. They indicated that the 5-year lag period was a reasonable choice.

## Exposure rate and duration of exposure

Considerable variation in exposure rate occurred among the 11 cohorts. The highest exposure rates generally occurred in early calendar-year periods, and they declined in later years. Exposure rate varied from < 0.3 WL to more than 30 WL, and exposure duration varied from < 1 year to more than 35 years.

It is important to note that exposure rates for individual miners must be inferred from average estimates of both WL and hours spent in the mine in a specific period. In most cohorts, these values were available only on an annual basis, and both were subject to measurement error. A particular concern is that earlier measurements, when exposure rates were largest, were generally subject to much greater errors than later measurements.

Analyses treating both exposure duration and average exposure rate as categorical variables were conducted for each of the 11 cohorts. For exposure duration, significant improvements in fit were found for all cohorts except Ontario, Sweden, Radium Hill, and France. For exposure rate, significant improvements were found for all cohorts except Sweden, New Mexico, Port Radium, Radium Hill, and France. The direction of these effects with quantitative treatment of exposure rate indicated a decrease in risk with increasing exposure rate in all cohorts except France (see Table D-18).

Because higher exposures were generally observed in earlier periods of mine operation, there was concern that the exposure-rate effect might represent a time-since-exposure effect. However, the exposure-rate effect persisted after adjustment for time since exposure.

Considerable effort was given to determining whether the effect of exposure rate was best described by using exposure rate directly or by using exposure duration and to determining whether a power or exponential model was more appropriate. The analyses did not provide a clear-cut answer, and final risk models used power models for both exposure rate and exposure duration. Analyses addressing homogeneity of the effect across cohorts provided evidence of lack of homogeneity for exposure rate with the exponential model but not for the other models evaluated (duration with the exponential model and both rate and duration with the power model).

Brenner and Hall (1990) have postulated that the inverse exposure-rate effect might be primarily a high-dose (or high-dose-rate) phenomenon. The NCI report presents ERR/exposure estimate by categories defined by both cumulative exposure and exposure rate. These suggest that the exposure-rate effect is strongest at the highest cumulative exposures, but data are inadequate to estimate the exposure-rate effect reliably at very low cumulative exposures.

Lubin and others (1995b) recently conducted further analyses of data from the 11-underground miner cohorts addressing the inverse exposure-rate effect. These analyses confirmed inverse exposure-rate effects in all cohorts but one. Separate measures of the exposure-rate effect (based on the power model) were estimated for each of 6 categories defined by cumulative exposure, and they indicated a lessening of the effect with decreasing exposure. Data in the lowest exposure category (< 50 WLM) were compatible with no inverse exposure-rate effect.

Most of the NCI analyses were based on either total duration of exposure or average exposure rate obtained as the total WLM divided by exposure duration. Although those variables were allowed to change as miners were followed, analyses did not take full account of the variation in exposure rate that might have occurred over a miner's employment period. For example, once the exposure was completed and the latency period had passed, a miner with a constant exposure rate of 4 WL would be treated similarly to a miner with an exposure rate of 7 WL for the first half of his exposure period and similarly to a miner with a rate of 1 WL for the second half. For miners with longer exposure periods, this approximation might not have been adequate. However, analyses were conducted that included separate estimates of the modifying effect of exposure rate (or exposure duration) for each of 4 time-since-exposure windows; these analyses yielded no indication that this treatment improved the fit over analyses based on a single average exposure rate.

Tomášek and others (1994a) and Thomas and others (1994) conducted analyses of data from the Czech and Colorado cohorts, respectively, in a way that took into account detailed exposure rate histories. They fitted a model based on $\Sigma_i w_i^\phi$, where $i$ indexes periods (months for the Czech cohort and years for the Colorado cohort), and $w_i$ indicates the exposure rate in period $i$. It can be shown that if the exposure rate were constant over the entire exposure period, then $\phi - 1$ would correspond to $\gamma$ in the power model described above that was applied in the NCI combined analyses. For both the Czech and Colorado data, $\phi - 1$ was estimated to be about –0.5. The comparable values estimated for those cohorts in the NCI analyses were similar: –0.66 for the Czech cohort and –0.78 for the Colorado cohort.

## Arsenic and Other Exposures

Two of the miner cohorts (China and Ontario) had quantitative data on arsenic exposure, and Ontario, Colorado, New Mexico, and France had data indicating whether miners had previous mining experience. Analyses were conducted to investigate the effect of those variables on lung-cancer risks after adjustment for radon WLM. Risks were found to increase with increasing arsenic exposure and to be larger for subjects with previous mining experience than for subjects without such experience. ERR/exposure for radon exposure was estimated both with and without adjustment for arsenic exposure or previous mining experience. For the China cohort, that reduced ERR/exposure from 0.61% to 0.16% but did not have a large effect on estimates from the other cohorts. There was no significant variation in ERR/exposure across categories of arsenic exposure or previous mining experience. It is noted that in all NCI analyses discussed thus far, the baseline risk was adjusted for arsenic and other exposures in cohorts for which data were available. It is possible, of course, that inadequate data or lack of data on such exposures could have biased results for any of the cohorts.

The effect of exposure to silica was investigated by Samet and others (1994) By examining whether the presence of silicosis, a fibrotic lung disease caused by silica, was associated with lung cancer in a case-control study of New Mexico underground uranium miners. No evidence of such an association was found, but data were too sparse to rule out the possibility that silica exposure could substantially bias lung-cancer risk estimates for miners. Radford and St. Clair Renard (1984) investigated the role of silicosis in a case-control study and found no evidence of association with lung-cancer risk.

## Overall Evaluation of Statistical Analyses Conducted
## Thus Far, with Emphasis on NCI Report

Overall, the NCI analyses provide a comprehensive summary of nearly all the relevant data on underground miners exposed to radon and radon progeny. The application of the same methods to all cohorts (parallel analyses) facilitate comparing results across cohorts, and combining data across cohorts (joint analyses) provides greater power for investigating various issues than would be available from any single cohort. The statistical methods are appropriate and in general provide an extremely thorough investigation of issues that are important for radon risk assessment. Careful attention is given to investigating the homogeneity of effects across cohorts and to determining which models provide the best description of the data.

Nevertheless, additional analyses might be desirable. For example, investigation of age at exposure was limited to consideration of age at first exposure, possibly because the use of "windows" for both age at exposure and time since exposure would have been too cumbersome with the use of Poisson regression. Analyses based on age at exposure would be desirable especially if substantial numbers of workers with very early ages at start of exposure continued to be exposed for many years.

The NCI analyses of exposure rate (or exposure duration) were limited primarily to consideration of average exposure rate and did not take account of detailed exposure histories. However, efforts by Tomášek and others (1994a) and by Thomas and others (1994) to use those histories more precisely yielded results that were similar to those obtained in the NCI analyses and based on the average exposure rate.

The NCI analyses did not take account of errors in the exposure measurements (including estimates of both exposure rate and duration, both of which are needed to estimate exposure). Those errors are generally thought to be largest for early periods of mine operations, when exposure rates were highest. In addition, there is considerable variation in the quality of exposure measurements across cohorts. In general, random error in exposure measurements tends to bias overall risk coefficients downward and might also distort the shape of the exposure-response curve. Because estimates of exposure in early years of mine operations are often subject to greater errors than estimates in more recent years, estimates of the modifying effects of exposure rate and possibly of other time-related factors might be exaggerated. Statistical methods are available for adjusting for exposure-measurement errors but tend to be difficult to use. Furthermore, the application of those methods requires that the error structure be specified, an extremely difficult undertaking, given the complexity of the structure and the lack of adequate data for quantifying many sources of error.

It is difficult to investigate the separate modifying effects of the variables evaluated in the NCI analyses, and this could have important implications for extrapolating to values outside the range of the data. A general difficulty is that we do not have adequate knowledge of the biologic rationale of patterns of risk associated with various factors. The lack of adequate biologic understanding has necessitated the descriptive approach taken in the NCI analyses and in analyses conducted to develop the BEIR VI risk models. However, despite its limitations, the descriptive approach provides extremely valuable information and very likely must serve, with appropriate caution, as the basis for developing risk estimates for both occupational and residential exposure.

# Appendix E

# Exposures of Miners to Radon Progeny

## INTRODUCTION

The epidemiologic studies of underground miners have become the principal basis for estimating the quantitative risk of lung cancer associated with exposure to radon progeny. As a result, there has been persistent concern about the extent of error in estimates of the exposures of the underground miners to radon progeny and of any resulting bias in risk estimates. The exposures of the underground miners have been estimated on the basis of incomplete information and ad hoc procedures have been used to complete gaps in the measurement data.

The BEIR IV committee recognized the potential for measurement error to affect risk estimates and consequently, the BEIR IV report provided extensive descriptions of the approaches used to estimate exposures in the four cohorts that served as the basis for the risk model. Since publication of the BEIR IV report, there has been increasing methodologic research on measurement error and its consequences, and on methods for adjusting for bias resulting from measurement error (Thomas and others 1993). The number of cohorts used to develop the BEIR-VI risk model has been increased from the four used by BEIR IV to 11, bringing a need to understand the varying approaches followed to assess exposures among the cohorts. Additionally, exposures have been re-estimated for several studies, giving insights into the potential magnitude of error in the studies.

The presence of errors in the exposure estimates for the miners has been widely recognized. Some investigators have addressed the problem by carrying out subgroup analyses. Risk patterns for miners considered as having the "better" exposure data were compared to risk patterns for miners with the "poorer" exposure data. Criteria for the identification of subgroups having better quality exposure data were based on surrogate indicators, such as calendar year of first employment, calendar year of follow-up, age at first employment, or exposure rate. Another approach was to limit analyses to miners with the more extensive measurement data. In more recent years, some mining companies have developed individual estimates of exposure, using increasingly extensive radon progeny measurement data and detailed information on exposure time and work location. This era of improved exposure assessment temporally corresponds to the lower radon progeny levels in modern mines and corresponding lower total exposure values (Jhm$^{-3}$ or WLM), compared with earlier years. In addition, relatively few miners started working after the improved exposure assessment procedures were put in place. Thus, information on risks to these more contemporary miners is still limited.

For a truly linear exposure-response relationship, it is widely recognized that misspecification of exposures tends to reduce the gradient of the trend, and to induce curvilinearity, from below. However, it is less well-recognized that misspecification of exposure does not *always* bias the exposure-response towards the null (Dosemeci and others 1990), although with the error patterns that prevail among the miner data sets, error would be unlikely to steepen the exposure-response relationship. With multiple exposure variables subject to error, for example, exposure to radon progeny, exposure to arsenic-containing dusts and cigarette-smoking, correlations among the variables could lead to either positive or negative bias in the radon progeny exposure-response relationship.

Interpretation of analyses of the miner data is further complicated by the relationships among radon progeny level, calendar year, and degree of error. High radon progeny levels generally occurred in the earliest years of operations of the mines and the high levels tended to occur during an era with limited numbers of measurements, incomplete coverage of work areas, and measurements of radon rather than radon progeny. In addition, work histories were usually less accurate in the early years of mining operations. Moreover, improvements in ventilation and reduction of radon levels were generally carried out over an extended period of time. Any attempt in analysis to designate specific years as "good" versus "bad" with regard to data quality is of necessity a substantial oversimplification.

Exposure estimation required work history information on time spent underground, which in most of the studies was obtained from company employment or medical history records. Thus, errors in the estimation of radon progeny exposure depended on the completeness and accuracy of these data. Exposure estimation was further compromised in some studies by lack of information on exposures that occurred outside the recorded work periods, and prior or subsequent to employment in the study mines. For example, some miners in the Colorado cohort were reported to mine on weekends, although the specific details on the amount of extra time and the exposed miners were lacking.

The impact of measurement error on radiation risk estimates was explored in the study of atomic-bomb survivors (Jablon 1971). Analyses were also conducted by Pierce and others (1990) using the Lifespan Study data. The intent was to adjust the linear dose-response estimate to account for random error in the dose estimates. The investigators assumed that errors in dose estimates were proportional to dose, and therefore worked on the logarithm scale for dose. Using a lognormal error distribution and a plausible estimate of error of about 30%, the authors found that the adjusted dose-response estimate was about 5 to 15% greater than if account was not taken of exposure error.

This appendix brings together information on exposure estimates made for the miners in the epidemiologic studies and related work on error in these estimates. One annex to the chapter describes the basis for the exposure estimates in each of the studies. An additional annex provides the proceedings of a workshop convened by the committee on exposure estimates; valuable insights were gained from participants who were extremely knowledgeable on the history of the U.S. and Canadian uranium mining industries and consequently the BEIR VI committee has included the proceedings in its report.

## ESTIMATION OF EXPOSURES OF MINERS TO RADON PROGENY

Published descriptions of the mines that are the basis of the epidemiologic studies indicate that the sources of radon included the ore being mined, air flowing into the areas where miners were working, and radon-containing water in the mines (NRC 1988; Lubin and others 1994a). Under the circumstances of mine operation, it is likely that concentrations of radon and progeny varied spatially and temporally within a particular mine, although little data have been published that document such variation. In the New Mexico mines, for example, information presented at the committee's workshop on dosimetry documented extensive variation in concentrations of radon progeny across various locations within mines in Ambrosia Lake, New Mexico (Table E-1). Thus, exposure estimates for individual miners would be ideally based on either a personal dosimeter, as used for low-LET occupational exposures, or on detailed information on concentrations at all locations in mines where participants in the studies received exposure (SENES 1989). For the participants, information would be needed on the locations where time was spent, the duration of time spent in the locations, and the concentrations in the locations when the miners were present. A miner might have spent time in a number of different locations during a typical working day including the stope (the area where mining actually takes place), the haulage way leading to the stope, and perhaps a separate lunch area. Personal dosimeters for radon progeny have not been developed until recently and their usage has been

TABLE E-1    Measurement data for individual mines in New Mexico

| Mine | Dates | Location | Working Level (WL) |
|---|---|---|---|
| A | 5/5-13/60 | Room 6-1, SE Area | 6.8 |
| | | Room 9-1, SE Area | 15.2 |
| | | Station | 10.0 |
| | | Room 9-1, SE Area | $5.0^a$ |
| | | Room 6-1, SE Area | $4.6^a$ |
| | | Room 8-1, SE Area | $6.7^a$ |
| | | 1 Left-15, NW Area | $0.1^a$ |
| | | Station | $2.4^a$ |
| | 8/5/60 | Station | 20.2 |
| | | 13 West | 3.9 |
| | 8/10-11/60 | Station | 10.6 |
| | | 11 West, N End | 0.9 |
| | | 11 West, N End | 1.1 |
| | | Shaft Collar | 15.0 |
| | | Station | $9.6^a$ |
| | | Shaft Collar | $10.8^a$ |
| B | 2/19/60 | Near old magazine | 37.0 |
| | | Slusher station | 25.0 |
| | | Ventilation course | 16.0 |
| | | Station | 20.0 |
| | 9/29/60 | Breakthrough to drift | 26.0 |
| | | Breakthrough to drift | 27.0 |
| | | Drift | 1.5 |
| | | Top of fresh air incline | 2.0 |
| | | Station | 23.5 |
| | 6/15-16/61 | Shop | 0.5 |
| | | Station | 0.4 |
| | | South area | 10.5 |
| | | Drift | 10.3 |
| | | South area | $9.0^a$ |
| | | Drift | $6.2^a$ |
| | | Station | $3.4^a$ |

$^a$Measured after intervention.

limited, and detailed information on concentrations within mines and time spent in various locations has not been available to most of the epidemiologic studies.

In the epidemiologic studies, these ideal approaches have been replaced by various, pragmatically determined strategies for exposure estimation that draw on measurements made for regulatory and research purposes and extend the measurements using interpolation and extrapolation to complete gaps for mines in particular years. Additionally, missing information for mines in the earliest years of some of the studies was completed by either expert judgment or by re-creation of operating conditions. The numbers of measurements made also varies widely across the different studies included in the pooled analysis and within studies the numbers of measurements tend to be greater in the later years of operation of the mines, when the exposures were generally lowest.

A pilot case-control study of lead-210 levels in the skull and lung-cancer risk in the Chinese tin miners points to the promise of this technique for exposure estimation. In this study, exposures are estimated for 19 miners with lung cancer and 141 age-matched controls. Lead-210 levels were measured in the skull and lung-cancer risk was estimated in relationship to lead-210 level estimated for the time of last radon exposure. There was a gradient of risk with exposure. This technique is also applied in an exploratory fashion with former uranium miners in New Mexico (Lauer and others 1993).

## PREVIOUS WORK ON ERRORS IN EXPOSURE ESTIMATES IN STUDIES OF MINERS

### Cohort Study of Colorado Plateau Uranium Miners

Epidemiologists conducting research on lung cancer in radon-exposed miners have long been aware of the potential for measurement error. The study of Colorado Plateau uranium miners conducted by the U.S. Public Health Service involved approximately 4,000 men, both whites and Native Americans, who worked in thousands of mines scattered across remote regions of four states. Measurements were not made regularly in all of the mines and no measurements were made in some of the mines. Consequently, the investigators based exposures on the actual measurements and on interpolated and extrapolated concentrations and on "guesstimates". For mines with missing data, concentrations were imputed from other mines in a hierarchy that began at the level of mining district and moved to the state level (see description in this chapter).

A descriptive evaluation of the consequences of using this approach was published in Joint Monograph No. 1 of the National Institute for Occupational Safety and Health and the National Institute for Environmental Health Sciences (Lundin and others 1971). The coefficient of variation was provided by year and by average concentration value ($Jm^{-3}$ or WL) for the small number of mines with at least five or more samples in a particular year. The coefficient of variation tended to be less than 100 percent and not to vary over time.

To assess error associated with their interpolation procedure for completing gaps in measurements of radon progeny concentrations, the investigators used data for the mines having the longest period of continuous radon progeny measurements to compare estimated with actual values. Comparisons were made for mines during three periods, 1952-1954, 1956-1960, and 1960-1968. Lundin and others (1971) compared actual values to estimates based on averaging adjacent years. This analysis showed that the error was greatest for the earliest time period; because of the general trend of declining progeny concentrations, backward projection tended to underestimate and forward projection to overestimate.

A further analysis was directed at the imputation of missing values based on mining locality. Four of the largest uranium mining areas were selected for this analysis. Individual mines were omitted from the calculation of the average for the locality and the average for the locality was compared to the actual measurements. The differences tended to be large for the earliest years and declined substantially by 1968. The direction of bias tended to be positive; the investigators reported that during the four years considered, 55 to 70 percent of the estimates exceeded mine averages based on measurements.

This report also examined the potential for distortion of the results by errors in exposure assessment. Standardized mortality ratios (SMR) for lung-cancer deaths for miners who had 25% or more of their cumulative exposure based on measurement data (called exposures of the "highest quality", 1,325 miners including 20 lung-cancer cases, from a total of 3,325 miners and 70 lung-cancer cases) were compared with SMRs for miners with and without previous hardrock mining experience, which was used as a surrogate for previous radon progeny exposure. While this analysis of previous hardrock mining experience does not directly address exposure error and the quality of the exposure estimates, exposures for miners with previous mining are likely estimated with substantial error. Results of the comparison of the highest quality data and mines with and without previous hardrock experience are displayed in Figure E-1. In this informal analysis, there was no difference in the exposure-response for the high quality exposure data and either those with or without previous hardrock mining experience.

The 1987 National Institute for Occupational Safety and Health document, *A Recommended Standard for Occupational Exposure to Radon Progeny in Underground Mines* (NIOSH 1987), includes a systematic attempt to quantify the errors associated with the exposure (Jhm$^{-3}$ or WLM) estimates for participants in the Colorado Plateau Study. Following the approach of Joint Monograph No. 1 (Lundin and others 1971), four sources of error were identified: 1) actual measurements; 2) interpolation or extrapolation in time; 3) geographic area estimation; and 4) estimates for years prior to 1950 when measurements were largely unavailable. The estimated coefficients of variation for each of these sources of error were 1.13, 1.21, 1.49, and 1.86, respectively. A pooled estimate of 1.37 was calculated as a weighted average. Using this figure, an average coefficient of variation of 0.97 was estimated for the cumulative exposure received by the miners. The report comments that if these errors are lognormally distributed then there would be minimal bias in the preferred power function model for the exposure-response relationship of lung-cancer risk with exposure to radon progeny. The possible effects of systematic errors were not considered.

At a 1994 BEIR VI committee workshop, Dr. Duncan Thomas presented preliminary results from an analysis of exposure errors using a method for inferring annual dose rates directly from miners' exposure histories, allowing for gaps

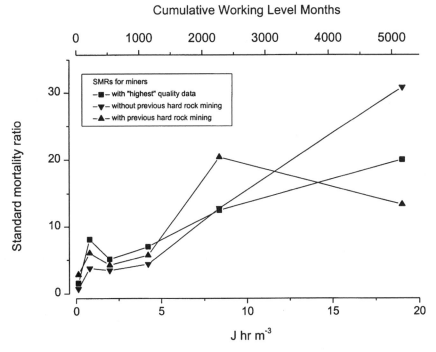

FIGURE E-1   Standardized mortality ratios for Colorado Plateau uranium miners study by cumulative radon progeny exposure and quality of exposure data (Lundin and others 1971).

in exposure. Data from the 1987 follow-up of the Colorado miners were used. Unlike the Colorado data set used in the committee's analysis of pooled miner data, Thomas did not limit exposures to $< 11.20$ Jhm$^{-3}$ ($< 3,200$ WLM). Several assumptions were made that represented a simplified characterization of the true error structure (Table E-2).

Thomas considered the effects of exposure error on curvilinearity, and the modification of the exposure-response relationship by time since exposure and exposure rate. Table E-3 shows that there was significant curvilinearity in the exposure-response trend if no adjustment was made for exposure errors ($\chi^2_1 = 5.22$, $p = 0.02$). With adjustment for error, the deviation from the linear exposure-response model was not significant ($\chi^2_1 = 1.94$, $p = 0.16$). For the linear model and the linear-exponential model, the magnitude of the exposure-response trend increased when adjustment was made for exposure errors. Further, the estimate of $\beta$ for the linear model with error adjustment was about the same magnitude as the unadjusted linear-exponential model, suggesting that the curvilinearity was largely the result of exposure errors.

TABLE E-2    Assumptions made by D. Thomas in analysis of exposure errors in the Colorado Plateau uranium miners study presented to the committee

- Classical error model applied to annual exposure rates
- True exposure rate in WLM/yr (x) has a Weibull distribution
- Observed exposure rate z=xu, where the error distribution (u) is log-normal with mean 0 and standard deviation:
  - CV = 1.00 before 1960, 0.50 after 1960 (large error)
  - CV = 0.50 before 1960, 0.25 after 1960 (small error)
- True exposure x integrates over time to produce cumulative exposure X.
- Relative risk (RR) models:
  - Curvilinearity: $RR = 1 + \beta x_e^{-\alpha X}$
  - Time since exposure[a]: $RR = 1 + \beta(x_{5\text{-}9} + \theta_2 x_{10\text{-}19} + \theta_3 x_{20+})$
  - Exposure-rate:

    Model 1: $RR = 1 + \int_0^t \beta x(u) x(u)^\gamma du$
    Model 2[b]: $RR = 1 + \beta x_e^{-\gamma \overline{x}}$

[a]$x_{5\text{-}9}$, $x_{10\text{-}19}$, and $x_{20+}$ denote true exposures accumulated 5-9, 10-19, and 20 years and more prior to age at risk.
[b]$\overline{x}$ is average exposure-rate while exposed.

The consequences of adjusting for exposure error on the effects of time since exposure and exposure rate were also considered by Thomas.  Results indicated that adjustment for errors had little effect on the pattern of risk with time since exposure, but adjustment for errors reduced the effects of exposure rate (Tables E-4 and E-5).

The analyses suggested that exposure errors were likely to have had their greatest influence on the modeling of the inverse exposure-rate effect.  However, the precise magnitude of the influence of the adjustment for errors was uncertain, due to limitations of the analysis.  For example, exposure rates were inferred, rather than computed directly from mine records and measurement data, and the assumption that errors were independent across individuals can be questioned, since the same concentration value $Jhm^{-3}$ (WL) would often be given for all

TABLE E-3    Parameter estimates and change in deviance for modeling curvilinearity in the exposure-response for the Colorado data.[a]  Adjusted estimates account for error in exposures.  Presented to the committee by D. Thomas

|  | $\beta x 100$ | $\alpha$ | Change in deviance |
|---|---|---|---|
| Unadjusted | 0.40 |  | — |
|  | 0.69 | −0.011 | 5.2 |
| Adjusted[b] | 0.73 |  | — |
|  | 1.08 | −0.012 | 1.9 |

[a]Model of the form: $RR = 1 + \beta x_e^{-\alpha X}$ where $x$ is cumulative exposure in WLM.
[b]Assumed error: CV=1.00 before 1960 and CV=0.50 after 1960.

TABLE E-4   Parameter estimates and change in deviance for modeling time since exposure in the exposure-response for the Colorado data.[a]   Adjusted estimates account for error in exposures. Presented to the committee by D. Thomas

|  | $\beta \times 100$ | $\theta_2$ | $\theta_3$ | Change in deviance |
|---|---|---|---|---|
| Unadjusted | 1.26 | 0.65 | 0.13 | 19.2 |
| Adjusted[b] | 1.92 | 0.80 | 0.15 | 23.8 |

[a]Model of the form: $RR = 1 + \beta(X_{5-9} + \theta_2 X_{10-19} + \theta_3 X_{20+})$ where $X_{5-9}$, $X_{10-19}$, and $X_{20+}$ are cumulative WLM exposures 5-9, 10-19, and 20 years and more prior to age at risk.
[b]Assumed error: CV = 1.00 before 1960 and CV = 0.50 after 1960.

workers in a particular mine and workers' movements among mines or among companies were not independent. Finally, in his presentation Thomas indicated that the Weibull distribution provided a poor fit to the exposure rate data.

### Cohort Study of New Mexico Uranium Miners

Samet and others (1986b) addressed several issues in the estimation of exposures for the New Mexico uranium miners. For selected years, relatively large numbers of measurements had been made in the large mines of Ambrosia Lake. During several day visits to the mines, the mine inspectors sometimes made over 100 measurements. Samet and colleagues addressed the most appropriate parameter for summarizing the measurements. In New Mexico, the inspectors from the state calculated a person-weighted index, termed the "Total Mine Index" which weighted the measurements by the approximate number of workers associated with the measurement by the inspector. Using the Total Mine Index as the standard for comparison in a regression analysis, Samet and colleagues compared the performance of the arithmetic mean and various trimmed means and the median in estimating the Total Mine Index. The arithmetic mean was found to be

TABLE E-5   Parameter estimates and change in deviance for modeling the effect of exposure rate on the exposure-response for the Colorado data.[a]   Adjusted estimates account for error in exposures. Presented to the committee by D. Thomas

|  | $\beta \times 100$ | $\alpha$ | Change in deviance |
|---|---|---|---|
| Unadjusted | 0.40 |  | — |
|  | 1.89 | −0.35 | 8.0 |
| Adjusted[b] | 0.73 |  | — |
|  | 1.06 | −0.21 | 2.0 |

[a]Model of the form: $RR = 1 + \int_0^t \beta x(u) x(u)^\gamma du$ where x is exposure rate in WLM/yr.
[b]Assumed error: CV = 1.00 before 1960 and CV = 0.50 after 1960.

an unbiased predictor of the Total Mine Index, suggesting that measurements had been made in approximate proportion to the distribution of the exposed miners within the mine. The median was lower than the mean and the Total Mine Index, as anticipated for the skewed distribution of the measurements.

## Cohort Study Of Czechoslovakian Uranium Miners

Estimation of the "true" exposures for all miners in the epidemiologic studies is not currently feasible. However, improving the overall quality of study data reduces the effects of errors. Recent analysis of the study of Czech uranium miners (sometimes referred to as the West Bohemian uranium miners) have resulted in two rather different estimates of the ERR/exposure. An analysis by Tomášek and others (1994a) estimated the ERR/exposure in WLM as 0.0064 (95 percent CI 0.004, 0.011), while in the pooled analysis of 11 cohorts, Lubin and others (1994a) estimated the ERR/exposure in WLM for the Czech study as 0.0034 (95 percent CI 0.002,0.006). The former estimate was based on using male Czechoslovakian lung-cancer mortality rates as an external referent population, while the latter estimated was based on an internal referent population (Tomášek and Darby 1995).

Tomášek and Darby (1995) showed that methodologic differences did not explain the different estimates of ERR/exposure, but that the principal difference in the two estimates was the result of improvements in two aspects of data quality. First, follow-up information was re-examined, using additional sources, such as pension offices, and local inquiries. These efforts resulted in an increase in the number of deaths and a reduction in the number of miners who were lost to follow-up. Following the reevaluation, the number of lung-cancer deaths increased from 661 to 705. However, the added number of lung-cancer deaths had minimal impact on the ERR/exposure estimate (Tomášek and Darby 1995). Second, exposures for all miners were re-assessed, correcting arithmetic and transcription errors and, for some men, accounted for exposures at other uranium mines which previously had not been included. Revised exposures were recomputed for each miners for each month of employment.

Improving the exposure data nearly doubled the estimate of ERR/exposure. It should be noted that the Czech data, along with the Colorado data, exhibited curvilinearity in the exposure-response in the pooled miner analysis. Unlike the Colorado data (see below), the exposure-response relationship in the Czech data continued to exhibit significant curvilinearity after the exposure reassessment (Tomášek and Darby 1995).

## Cohort Study of Beaverlodge Uranium Miners

Exposures have also been recalculated for the Beaverlodge, Canada, cohort of uranium miners for those included in a case-control study of lung cancer, smoking, and radon-progeny exposure (L'Abbe and others 1991). The new exposures replaced the median by the arithmetic mean for the assignment of values to mines and company records were used to place the participants in more specific areas within mines rather than simply using mine-wide averages (Howe and Stager 1996). The new exposure estimates were higher than the previous ones: 0.28 Jhm$^{-3}$ (81.3 WLM) versus 0.18 Jhm$^{-3}$ (50.6 WLM). The new and old values were moderately correlated (r = 0.66). There were few major changes in individual estimates.

The estimated excess relative risks were calculated for the cohort study and the case-control study with original exposure estimates and for the case-control study with the new exposure estimates. The values of ERR/exposure were 0.01/100 Jhm$^{-3}$ (2.63/100 WLM) for the cohort study, 0.01/100 Jhm$^{-3}$ (2.70/100 WLM) for the case-control study with the original exposure estimates, and 1/100 Jhm$^{-3}$ (3.25/100 WLM) for the case-control study with the new exposure estimates. The authors attributed the increase in the estimate of excess relative risk with the new estimates to reduction in measurement error which more than compensated for the increase in average exposure.

### General Assessments Of Exposure Error In Miner Studies

In 1989 the effects of exposure error were considered for the Czechoslovakia, Colorado, Ontario, Sweden, Beaverlodge, and Port Radium studies in a study conducted by SENES Consultants Limited in Canada (SENES 1989). This report described several sources of uncertainty in the miner studies and was one of the earliest efforts to address the consequences of error analytically. The focus was on adjusting the exposure-response parameter for the excess RR model and for the absolute excess risk model. The approach was limited, because the authors did not have access to data on individual miners and utilized only published RRs within categories of exposure (Jhm$^{-3}$ or WLM). No attempt could be made to evaluate patterns of error for different mining periods within a study. For the excess RR model, the authors estimated that the most likely range for the ERR/exposure parameter for this group of studies was 0.009 to 0.00005 per Jhm$^{-3}$ (0.005 to 0.015 per WLM).

In the pooled analysis, Lubin and others (1994a) considered the impact of errors in exposure only in the most general way and only within the context of the modification of the exposure-response relationship by exposure rate and exposure duration. Exposure error was considered greatest in the earliest years of mining, years in which exposure rates were at their highest. Exposure error would therefore have tended to attenuate the effects of high exposures, and potentially induced an observed inverse exposure-rate pattern.

To assess the contribution of exposure error to the inverse exposure-rate effect, the authors analyzed RR patterns within categories of several variables which were considered as indicators of the magnitude of exposure error, for example, year of first exposure, attained age, years since last exposure, and total cumulative exposure. Figure E-2 shows the result of one of these analyses. Observed RRs, adjusted for cumulative exposure (Jhm$^{-3}$ or WLM) and other factors, are plotted by categories of calendar year of first exposure from the pooled analysis of 11 underground miner studies (Lubin and others 1994a). Also, shown are log-linear models fit to the observed RRs. The overall levels of the RRs were affected by the small numbers of lung-cancer cases in the referent category <0.1 WL (6, 5 and 8 cases in the <1945, 1945-54, and ≥ 1955 year of first exposure categories, respectively), but the figure shows a similar pattern of declining RRs with WL, suggesting that it was unlikely that the inverse exposure-rate effects were entirely the result of exposure errors.

In a report to the Canadian Atomic Energy Control Board, Howe and Armstrong presented a sensitivity analysis of the potential impact of measurement error on risk of lung cancer using hypothetical data, but based on the pooled analysis of 11 cohorts of underground miners (Howe and Armstrong 1994). The

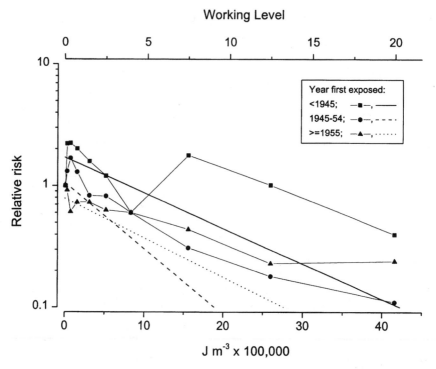

FIGURE E-2   Relative risks by categories of exposure rate and fitted trends by calendar year first exposure for data from analysis of underground miner studies (Lubin and others 1994a).   Relative risks adjusted for attained age, cohort, ethnicity, and cumulative exposure.

error structure was assumed a combination of Berkson and classical errors, with no systematic error in exposures. The measurement errors were assumed to have a Berkson component, that is errors arising from the assignment of an overall mine or area mean to a group of miners. Bias in risk estimates occurs because the true mine or area mean for the group of miners was unknown and must be estimated from measurement samples. The estimation of the area mean exposure rate introduced a classical error structure, that is, error was independent of the true exposure (Howe and Armstrong 1994).

In the analysis, exposure time in working months was assumed as known without error. While some error in individual employment records was inevitable, the assumption was considered a reasonable approximation, particularly in comparison with the amount of error introduced by exposure rate. There was additional uncertainty about the amount of time miners spent at their underground work sites and in travelways, but again this source of error was considered of less consequence for biasing risk modeling (Howe and Armstrong 1994).

The authors presented results from two simulation studies. One simulation, which closely mimicked average data in actual studies, generated matched case-control data with mean numbers of cases from 300 to 835, and equal numbers of controls. Exposure rate was assumed log-normally distributed with geometric mean 5.0 WL and standard deviation (of log WL) of 0.5, 1.0, and 2.0. Three schemes for the number of measurements per year were used: (1) one measurement per year for all years of the study; (2) one measurement per year increasing linearly to 100 measurements per year over the years of the study; and (3) 100 measurements per year in all years. For true data generated using a simple linear excess RR model, $RR = 1 + \beta(WLM)(WL)^{\alpha}$ with true $\beta$ equal to 0.010 and true $\alpha = 0.00$, bias in the estimate of $\beta$ was minimal under measurement schemes 2 and 3 (Table E-6). For these schemes, there was some attenuation of the exposure-response relationship only with a standard deviation of log(WL) of 2.0. With schemes 2 and 3, a significant exposure rate effect was induced only with the extreme standard deviation. In contrast, with a single measurement per year, there was induced curvilinearity when the standard deviation was 1.0 or greater.

Due to computational limitations, a second simulation using simplified cohort data was carried out, in which exposure did not overlap follow-up. The simulation assumed that all exposure was accumulated by age 30 years, and that follow-up started at age 35. Subgroups of the cohort were given exposure rates 1, . . . , 20 WL and follow-up continued for 20-50 years. In each simulated data set, there was 14,000 subjects and cumulative exposure ranged from 0.35 to 7.00 $Jhm^{-3}$ (100 to 2,000 WLM). True RR again followed the simple linear excess RR model, with true $\beta = 0.010$. The same three standard deviations for log(WL) and three schemes for measurements per year were used. Again, attenuation of the exposure-response was observed, particularly with the scheme of one WL measurement per year.

TABLE E-6   Effect of measurement error in exposure on the excess relative risk estimate in a linear excess relative risk model,[a] based on computer simulation of case-control data with parameters $\beta = 1.00$ and $\alpha = 0.00$. Table adapted from Howe and Armstrong (1994)

| WL measurement scheme[b] | Standard deviation of log(WL) | Median ($\beta$) | Median ($\alpha$) |
|---|---|---|---|
| 1-1 | 0.5 | 0.90 | −0.075 |
|  | 1.0 | 0.70 | −0.41 |
|  | 2.0 | 0.26 | −0.79 |
| 1-100 | 0.5 | 1.10 | 0.021 |
|  | 1.0 | 1.10 | −0.07 |
|  | 2.0 | 0.54 | −0.32 |
| 100-100 | 0.5 | 1.01 | 0.026 |
|  | 1.0 | 1.00 | 0.019 |
|  | 2.0 | 0.77 | −0.22 |

[a]Data simulated under the model: $RR = 1 + \beta(WLM)(WL)^{\alpha}$
[b]Measurement schemes for WL:
    1-1:  one measurement per year for all (25) exposure years;
    1-100: one measurement per year increasing linearly to 100 measurements per year over exposure years;
    100-100: 100 measurements per year in exposure years.

Simulations also addressed the effect of measurement error on estimating the effects of time since exposure, age at risk, and exposure rate. With 100 measurements per year (scheme 3), there was little effect on the patterns of risk with any of these factors, although as previously observed, there was some bias in the main risk effect of exposure. With scheme 3, there was some induced bias when the standard deviation of the error was large 2.0. With measurement scheme 2, some bias was induced, but only in the extreme category ($\geq 25$ years since exposure, $\geq 65$ years of age, or exposure rate $\geq 15$ WL). With a single measurement per year (scheme 1), there was marked bias.

Howe and Armstrong identified two key elements for assessing the effects of measurement error: variation of the true exposure rate, that is, the degree that the true exposure rate for individuals differed from the true mean exposure rate; and the number of measurements used to estimate the true mean exposure rate.

The authors concluded that "measurement error leads to (a) reduction in the main effects coefficient, that is, that for cumulative exposure; (b) increasing downward bias in risk estimates with increasing time since exposure; (c) increasing downward bias in risk estimates with increasing age at risk and (d) increasing downward bias with increasing exposure rate." They further conclude that "biases are likely to be negligible if estimates are based on 100 or more samples per year and if the standard deviation of log(exposure rate) has a value of 1.0 or less." And "these conditions are met for the majority of the miners' cohort studies, and hence, although there may well be a contribution by measurement error to these apparent effects within some of the studies, overall it appears that measurement error of itself does not account for the existence of these effects."

## SUMMARY

There was marked variation among the cohorts in the approaches used to estimate exposures and in the extent of data available (see Annex to this chapter). All exposure estimates are subject to measurement error. Within cohorts, the degree of measurement error likely depends on the calendar years during which exposures were incurred; across cohorts, there is likely a varying impact of measurement error. Some work has addressed the consequences of error, both in individual cohorts and more generally. These analyses show that error would generally blunt exposure-response relationships. Time-dependent errors pose an additional constraint in interpreting time-dependence of effect, such as the inverse dose-rate effect. The substantial variation in methods for exposure assessment among the cohorts undoubtedly contributes to the heterogeneity of risk estimates from the individual studies.

# E-ANNEX 1

# Exposures to Miner Cohorts: Review of Estimates for the Studies

## Colorado Plateau Uranium Miners

### Introduction

Uranium mining in the Colorado Plateau expanded rapidly in the post-World War II period to include more than 200 mines by 1950 (see Time Line E-1). The start of an industry and the boom times did not lead to orderly administration and record keeping, c.f. Czechoslovakia and Ontario below. Moreover, some of the miners who worked in the mines during the post-war uranium boom had previously worked the same ore bodies for radium and vanadium without any accounting of exposure to radon progeny. Most of the early mines were small and depended on natural ventilation so that ambient temperature change was the driving force for exchange of the mines' air with outside air. Until 1967, mining operations were regulated only by the states where mining was taking place, even though all ore was sold to the Atomic Energy Commission. There was no requirement in place for measurement of exposure and there was not a federal standard for exposure to radon progeny. Consequently adequate ventilation practices were not uniformly introduced from the outset and the extent of radon measurement was initially quite limited. As a result, estimates of cumulative exposures to uranium miners on the Colorado Plateau were largely based on various estimation procedures rather than direct measurements relating to a particular mine shaft or even the mine where a given worker was exposed.

The history of radon exposures to the miners was described by Holaday (1969) and the approaches followed by the U.S. Public Health Service for esti-

# Time Line E-1: Colorado Plateau Uranium Miners

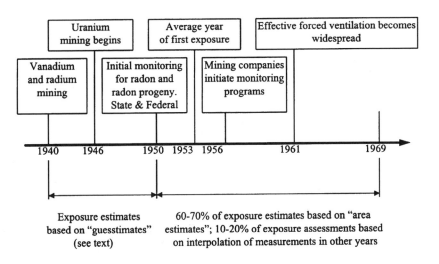

mating exposures of individual participants in the epidemiologic study of Colorado uranium miners are described in National Institute for Occupational Safety
and Health-National Institute for Environmental Sciences Joint Monograph No. 1
(Lundin and others 1971). A 1968 report of the Federal Radiation Council
addressed the accuracy of the exposure estimates. SENES Consultants Limited
of Ontario, Canada, has prepared a report entitled "Preliminary feasibility study
into the re-evaluation of exposure data for the Colorado Plateau uranium miner
cohort study" (SENES 1995). This report provides an extensive description of
the calculation of the WLM values for the epidemiologic study and gives insights
into the sources of variability and error in the estimates.

## Estimation of WLM

The following description is taken largely from the 1971 monograph
authored by Lundin and colleagues. The U.S. Public Health Service began
surveying for radon in uranium mines in 1949. In 1950 they were joined by the
Colorado State Department of Health and in 1951 by the U.S. Bureau of Mines
for mines on Indian reservations. Coverage was far from complete; 1949 "a few
measurements," 1950 "relatively few mines," 1951 "but again coverage was
incomplete," (Lundin and others 1971). By 1952 an effort was made to survey
all operating mines and radon progeny were sampled in 157 mines. This sampling may have examined most of the larger mines, but government records

indicate that over 450 mines shipped ore in 1951. Mining companies introduced radon surveys in 1956 and the state programs continued through 1960. Both company and state-sampling efforts were made in work areas for information purposes, not for control purposes, and "are considered to be representative of the areas in the mines in which miners were exposed" (Lundin and others 1971). This early data base is of primary importance in considering the adequacy and precision of miner's exposure estimates as utilized in epidemiology assessments of risks due to radon since a large portion of the cumulative exposure occurred in the 1950's.

By 1960, exposure levels had dropped precipitously in anticipation of Colorado's adoption of a 10 WL shutdown level in 1961. However, regulatory control probably reduced the validity of the measurements in mines for epidemiologic purposes. As outlined in Joint Monograph No. 1, the most complete description of the Colorado Plateau miner data (Lundin and others 1971) "Most radon daughter measurements available from Colorado, Utah, and Wyoming after 1960 were made by mine inspectors who measured air samples primarily for control purposes." This may have led to bias in the estimated exposures. As noted by Lundin and others (Lundin and others 1971), "Proportionately more measurements were made in sections of mines having high levels which tended to yield radon-progeny values greater than would have been obtained by sampling all work areas with equal frequency." In addition more measurements were concentrated in mines having high levels of radon. The U.S. P.H.S. investigators who assembled the data base for estimating cumulative exposures chose to exclude company measurements made after 1960 on the grounds that they might have been "minimized to avoid regulatory action." The aim was clearly "to assure a consistent direction of bias, that is, over estimation of radon daughter levels" (Lundin and others 1971).

Even though the number of radon-progeny measurements increased during the 1960's, the number per mine increased only slowly from about six in 1960 to almost 12 in 1968 (Figure E Annex 1-1). Measurements of radon progeny in a particular mine were never extensive and, more importantly, were not made on even a once per year basis in the majority of mines. Only 341 miners, about 10% in the Colorado Plateau miner cohort, had their exposure assignments based on measured radon-progeny concentration. For the majority of the miners, information on measured levels was combined with estimates made using a variety of methods as described by Lundin and others (1971).

Many of the uranium miners were also employed as hardrock miners or previous to 1950 some had mined the same ore bodies, where uranium was found, for radium, vanadium etc., particularly in the Urivan Mineral Belt in Colorado. In the epidemiologic study, hardrock miners were assigned an exposure level of 1 WL for mining that occurred before 1935, 0.5 WL for 1935 through 1939, and 0.3 WL for later years (Lundin and others 1971). No information is given as to the basis of these estimates but a statement is included in Joint Monograph No. 1

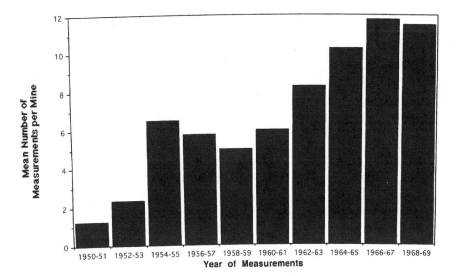

FIGURE E ANNEX 1-1   Frequency of radon-progeny measurements on the Colorado Plateau in two-year intervals 1950-1969.  Source: Presentation to the committee of analysis of the data tapes for the Colorado Plateau miners by Duncan Thomas and Dan Stram, September 1995.

(Lundin and others 1971) which indicates these estimates were thought to have been too high and that the average exposure was less.

A re-evaluation of a sample of the Colorado Plateau cohort for exposure during hardrock mining is described in Monograph 1.  This reassessment indicates that the tabulation of hard rock mining duration was subject to error and that misclassification of exposure was fairly common for that portion of a cohort member's work experience.  For example, for a sample of 101 cases and 202 controls, misclassification was only about 10% for cumulative exposures of less than 20 WLM but 50 % or more at higher levels.  Nevertheless, hard rock mining may be a relatively unimportant source of exposure compared to the mining of uranium-bearing ores for which exposure levels were often much higher than 1 WLM.

Because relatively few mines were initially monitored for radon or radon progeny, exposure estimates in uranium mines that occurred before 1951 were referred to as "guesstimates" in Joint Monograph No. 1 (Lundin and others 1971). According to that report, "guesstimates" were made on the basis of knowledge concerning ore bodies, ventilation practices, emission rates from different types of ores, and such radon or radon progeny measurements as were performed in 1951 and 1952.

For mining that occurred after 1950, three other methods were used to estimate exposure levels. By far the most common was a process called area average
estimation. This consisted of using the available, albeit often sparse, measured
values to estimate concentrations in a given locality to obtain an "area average."
In order to reduce sampling variability for these area averages it was required that
three or more mines and ten or more samples had to be available for a locality in
a year, otherwise the locality was assigned the average for the district in which it
was located (Lundin and others 1971). If sufficient data for a district were not
available, a state average was used or, in a few cases for which state data were
insufficient, data for the state of Colorado were used. The degree to which area
estimates were used to obtain exposure estimates is not often appreciated. Area
estimates account for most of the exposure assignments throughout the study
period of the Colorado Plateau cohort (Figure E Annex 1-2) Monograph 1
implies that when an individual mine was thought to differ appreciably from
others in the same locality due to its ore quality or mining practices, guesstimation
was substituted for an area average.

To complete gaps in the measurements in calculating individual WLM estimates, a system of extrapolation, interpolation, and expert judgment was used to
estimate the exposure in mines monitored less frequently than once a year. For
mines with actual measurements at least once every five years, working-level
estimates were obtained by interpolation, that is, averaging the measured values

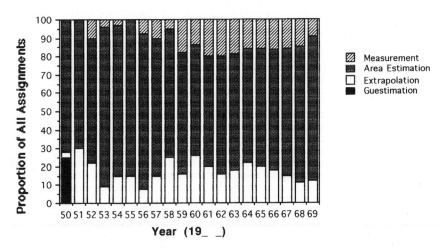

FIGURE E ANNEX 1-2   Bases for the assignment of exposure estimates by
calendar year 1950-1969. Source: Presentation to the committee of analysis of
the data tapes for the Colorado Plateau miners by Duncan Thomas and Dan
Stram, September 1995.

(Lundin and others 1971). Approximately 20% of the exposure assignments were made using this method (Figure E Annex 1-2). An assessment of this extrapolation procedure described in Monograph No. 1 indicates that it tended to overestimate exposures in the early years of mining but became more valid in the 1960s as information from more frequent measurements became available.

For mines with yearly monitoring information available, the measured concentration was used to assign a worker's cumulative exposure in a given year. Table IV-3 in the BEIR IV Report (NRC 1988) indicates that the number of measurements per year per mine surveyed was usually between 10 and 20 after 1959 so that the measured values provide a reasonably stable estimate of the average working levels in those areas monitored. However, the average number per mine was somewhat less, 8-9 as illustrated in Figure E Annex 1-1. Although nearly 43,000 measurements were obtained (Lundin and others 1971), there were about 2,500 mines and measured concentrations were not a frequent method of exposure assignment. Figure E Annex 1-2 indicates that from 1959 to 1969 only 10-20% of the exposure assignments in a given year were based on direct measurement of radon progeny concentration and that even fewer were made on such direct information prior to 1959, when exposure levels were, on the whole, much higher.

## Assessment of Errors in the WLM Estimates

A comparison of exposure estimates in relation to calendar year is given in Figure E Annex 1-3 for each assignment method. Except for 1950, estimates based on the extrapolation procedures are in reasonable agreement with those based on direct measurement while area average estimates tend to be somewhat greater than obtained by other methods. This may in part be due to measurements having been made more frequently in large mines having more employees and because of larger capital investment in better ventilation.

The degree of variation in exposures among workers in a given mine was not well characterized. Before 1960 mechanical ventilation was not commonly used and a near equilibrium between radon and progeny was probably the rule under conditions of convective ventilation as indicated by the early data described by Holaday (1969). There appears to be no information on aerosol size distribution or even the unattached fraction in early mines. Even though diesel power was not common, compressed air or electricity was used to operate equipment including ore cars; dust was plentiful from drilling and hauling operations so that it is likely that the unattached fraction was low.

An extensive study of air quality in nine uranium mines was carried out by the AEC Health and Safety Laboratory (HASL), now the DoD Environmental Monitoring Laboratory, in 1967-1968. Mines were selected by the U.S. Bureau of Mines to represent a cross section of the uranium mining industry (Breslin and others 1969). This investigation was in response to the concerns expressed at the

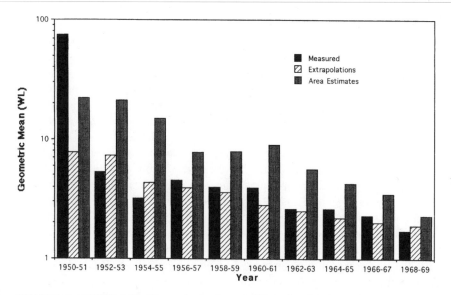

FIGURE E ANNEX 1-3   Comparison of mean WL estimation by various methods in two year intervals 1950-1969.  Source: Presentation to the committee of analysis of the data tapes for the Colorado Plateau miners by Duncan Thomas and Dan Stram, September 1995.

Joint Committee on Atomic Energy hearings in 1967 in which the validity of exposure and early risk estimates of increased lung cancer in miners were questioned.  A particular point in question was "the extreme variation of atmospheric characteristics within a mine and among mines"; the HASL study was directed at exploring this question (Breslin and others 1969).

The nine mines studied ranged in size from having two to 112 workers.  Ore production varied from 150 to 11,000 tons per month.  Mechanical ventilation rates varied from 5,600 to 100,000 cu. ft. per minute.  Given this range of conditions, atmospheric conditions were surprisingly uniform, giving some credence to the validity of the estimation methods described above.  In most of the mines the variation in radon-progeny concentration at different times and locations was only occasionally as large as a factor of two and 80% of the time had a coefficient of variation of 30% or less.  The average WL ratio (pCi progeny to pCi radon) averaged 0.23 with a geometric standard deviation of 1.6 and showed limited variation with the absolute level of radon progeny.  Equilibrium values F were also in a narrow range: about two-thirds were between 0.20 and 0.30; mode 0.25.  Polonium-214 was most often at 16% of the equilibrium value, range 0.09 to 0.49.

Simultaneous measurements of radon progeny were made at various locations in stopes (mining chambers) and in drifts (tunnels).  While drifts showed greater variation than stopes, as indicated in Figure E Annex 1-4, the report's authors indicated that sampling location was not critical within a radius of 10 to 20 feet of the miners' location and that breathing-zone sampling was unnecessary. Similarly, the HASL study indicated that differences between various mining operations, for example, drilling, mucking, etc., had little effect on the measured working level (Figure E Annex 1-5). No measurements were taken immediately after blasting but such areas would not have been occupied because of other safety considerations. While the HASL study does indicate that mine-wide averaging probably provides a useful measure of worker exposure in the mines studies, this is probably less accurate for the high exposures which occurred before the introduction of mechanical ventilation in U.S. uranium mines.

The recent report from SENES Consultants Limited provides additional relevant information. Tables for several mines demonstrate substantial variation in WL values within a mine during a single visit by an inspector, typically one day. For example, U.S. Bureau of Mines data for one Utah mine in 1968 showed variation from 0.4 to 5.4 WL across the mine (Table-E Annex 1-1).

The Public Health Service investigators used self-reported mining history as the basis for estimating time spent underground in specific mines. This information was collected both retrospectively and prospectively during the annual miner censuses. The possibility of error in these histories has been acknowledged. The SENES report provides a series of case descriptions documenting inconsistencies in these histories and gives a compilation of exposure estimates for 78 miners for whom exposures have been calculated for both the epidemiologic study and for other purposes. Substantial variation is evident in these

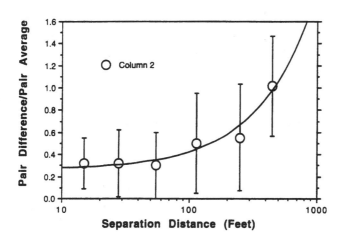

FIGURE E ANNEX 1-4 Variation of radon concentration with distance in ventilated uranium mine drifts on the Colorado Plateau (Breslin and others 1969).

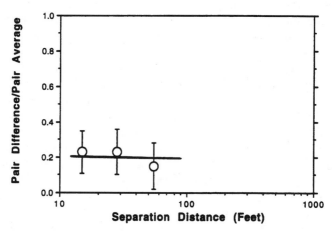

FIGURE E ANNEX 1-5   Variation of radon concentration with distance in ventilated uranium mine stopes on the Colorado Plateau (Breslin and others 1969).

estimates, largely reflecting various discrepancies in the alternative work histories used for the purpose of estimating the exposure.

## New Mexico Uranium Miners

Large-scale uranium mining began in the early 1950s (see Time Line E-2) with the opening of the Jackpile mine, an open-pit mine.  By the late 1950s, a number of large mines were operating at Ambrosia Lake and the Churchrock mining district became active in the late 1970s. The industry continued operating into the early 1990s, longer than in other U.S. locations, so that miners working after 1968 have individual exposure records (work location estimates and estimates of exposure) for this period of employment. These were calculated based on area measurements and work locations.  For the most part, post-1968 employment was in very large industrial operations with state of the art ventilation. Mean annual exposures in 1968 were about 3.8 WLM and declined to 1.2 WLM or less by 1972 (Samet and others 1986b).  Earlier exposures were not estimated as accurately, although the State Health Department and the State Mine Inspector had implemented active measurement programs by the late 1950s.  The state implemented a progressively more stringent series of shut-down concentrations. As for the Colorado Plateau miners (see above), median annual exposures were considerably larger during the earlier years of the industry, about 30 WLM in the 1960's.  Some members of the New Mexico cohort, who had also mined in the Colorado Plateau, had annual exposures as high as 300 WLM or more (Samet and others 1991).

# Time Line E-2: New Mexico Uranium Miners

| | | Average year of first exposure | | | |
|---|---|---|---|---|---|
| Large-scale uranium mining | Extensive ventilation to reduce radon introduced | | Individual exposure records based on extensive monitoring | Underground mining declines |

```
   1952              1961    1965   1968                    1990
```

Exposures estimated on basis of a person-weighted mine-wide average and limited radon and radon measurements available

State of the art exposure data

Investigators directed substantial effort at tracing employment histories for the purpose of estimating the cumulative exposures for those employed before exposure estimates were individualized (Samet and others 1991). The miners' underground employment and exposures in specific mines were traced by examining company personnel records and self-reported work histories taken at the time of periodic medical examinations. Estimated exposures for miners who had worked underground on the Colorado Plateau were supplied by the USPHS (Lundin and others 1971; Samet and others 1991). Contributions to the total mean exposure from various information sources are listed in Table E Annex 1-2 (Samet and others 1991).

With the notable exception of those members of the work force employed on the Colorado Plateau, this cohort probably has maintained the most extensively documented exposure estimates. In this regard, it should be noted that the state of New Mexico had more extensive and more frequent monitoring for radon then was common elsewhere in the early 1950's when exposures were very high (Lundin and others 1971). From 1957 to 1967 exposure estimates are based on 20,086 measurements taken during 1,886 visits. Most annual exposures were relatively low during this period, mean 4-5 WLM per year, so that this cohort has a large sub-cohort of miners exposed at low rates and relatively low cumulative exposures.

## Beaverlodge Uranium Miners

The BEIR-IV report also includes a description of the exposure estimates for this cohort (NRC 1988). Exploratory uranium mining at Beaverlodge,

TABLE E ANNEX 1-1    U.S. Bureau of Mines February 1968 survey at North Alice Mine, Utah[a]

| No. of Men | Location, Operation | Estimated Average Full Shift Exposure to Radon Daughters[b] (WL[c]) |
|---|---|---|
| 2 men, night shift | 416 NE from 360 NW; mining | 0.5 |
| 2 men, day shift | 236 from 325 S; mining | 1.7 |
| 1 man night shift | | |
| 1 man day shift | 240 W inclind station and hoist | 1.0 |
| 1 man night shift | | |
| 1 man day shift | 248 SE from 225 N; mining | 5.4 |
| 2 men, night shift | 242 S from 190 W; mining | 3.6 |
| 1 man day shift | | |
| 1 man night shift | 240 W incline to main incline; tramming | 2.2 |
| 2 men, night shift | 100 S area; mining | 0.6 |
| 2 men, day shift | 147 N from 130 E; mining | 2.8 |
| 3 men, day shift | | |
| 2 men, night shift | 128 S from 145 E; mining | 1.5 |
| 1 man day shift | | |
| 1 man night shift | main incline; trip rider | 0.4 |
| 1 man day shift | all areas; electrician | 1.4 |
| 2 men | all areas; mechanics | 1.6 |
| 5 men | all areas; shift bosses | 1.4 |
| 3 men | all areas; staff | 1.0 |
| 3 men | all areas; bratticemen | 2.6 |

[a]This table is from a February 1968 report on a Radiation Survey prepared by U.S. Bureau of Mines, obtained from SENES 1995.

[b]Average Levels are estimated from information gained by questioning the miners about where there time is spent and weighing the radon daughter concentrations in each place by the time spent in that place.

[c]NIOSH database: 1967 WL is 1.3, based on 39 measurements; 1968 WL is 3.3, based on 120 measurements.

Saskatchewan started in 1949 and commercial production began with a greatly expanded labor force in 1953 (see Time Line E-3). Radon monitoring was carried out in 1954 and 1956 but only sporadically until the end of 1961. A number of radon-progeny measurements were also made at this time but monitoring was mostly for radon and viewed as a check on ventilation rather than as a tool for exposure control. Nevertheless, the frequency of radon-progeny measurements increased and by 1961 exposure records were maintained for all full-time underground employees. These records listed each worker's occupancy time at each work place on a daily basis. In 1970 worker's exposure records were estimated retrospectively to 1 November 1966 and in 1971 part-time underground workers were included in the exposure assessment (SENES 1989).

TABLE E ANNEX 1-2

| Source of Information on Underground Employment | Contributions To Mean Cumulative Exposure |
|---|---|
| Work outside New Mexico | 24.9 WLM |
| New Mexico Employment records | 59.5 WLM |
| Self reported work histories | 11.8 WLM |
| Company individual records (1967 and later) | 10.0 WLM |
| Other (1967 and later) | 5.2 WLM |

Two assessments of lung-cancer risk observed in Beaverlodge miners have been made by Howe and colleagues (Howe and others 1986; Howe and Stager 1996) using two related but differing exposure estimates. The first of these estimates was prepared by Frost (1983) who, observing a wide dispersal in the recorded concentrations in a given year, assigned the median of this quasi log-normal distribution as the best measure of exposure. Although, it was possible to assign work locations for service personnel, for miners, mine-wide medians were used in the cohort study reported by Howe and colleagues (1986).

A reassessment of the Beaverlodge exposure estimates was carried by SENES Consultants, Ltd. at the direction of the Atomic Energy Control Board (SENES 1989, 1991). This included a painstaking reconstruction of mining activity and its correlation with exposure information for the Beaverlodge mining complex. The revised exposure estimates were used by Howe as the basis for a recent case-

## Time Line E-3: Beaverlodge Uranium Miners

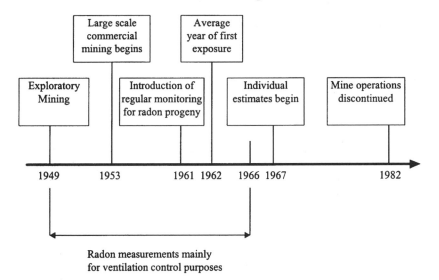

control analysis of lung-cancer mortality (Howe and Stager 1996), which used cases and controls from a previous analysis that used the original exposure estimates (L'Abbe and others 1991). In the new analysis, averages rather than medians of the individual measurement data were used to estimate exposure levels in a given location. There was also a systematic consideration of the locations where mining took place at a given period of time and, in many cases, individual miners could be assigned to a given mine face, as recorded in bonus-pay information, removing in some measure the radon error inherent in using mine-wide averages.

The effects of these changes is illustrative of what an improved exposure assessment can accomplish. The exposure estimates for each of the miners in the case-control study were compared to those used in the original cohort and case-control studies. In general, the more recent exposure estimates were considerably higher than the original estimates, the mean exposure increasing from 50.6 WLM to 81.3 WLM. There is evidence that the new estimates reduced exposure misclassification. Table E Annex 1-3 compares the cumulative exposure estimates used in the original cohort study to the newer estimates for the case control study.

Because of the wide intervals of grouped exposures, most workers remained in the same exposure category even though their estimated exposures were on the average considerable larger in the revised exposure estimates. However, there was a decrease in the number of workers receiving low exposures and a corresponding increase at higher levels. For example, the number of workers in the 200+ WLM group increased from 10 to 15 (Table E Annex 1-3). Because there is less misclassification of estimated exposures, the slope of the regression of risk on cumulative exposure is increased even though the estimated exposures increased. The original estimate of the excess relative risk from the case-control was an excess relative risk of 2.70% per 100 WLM while the revised exposure assessment was 3.25% (Howe and Stager 1996).

TABLE E ANNEX 1-3   Number of miners in each exposure category

| Cumulative Exposure WLM | Cohort Study | Case Control Study |
| --- | --- | --- |
| 0 | 43 | 42 |
| 1-24 | 90 | 80 |
| 25-49 | 15 | 18 |
| 50-99 | 15 | 17 |
| 100-199 | 18 | 19 |
| 200+ | 10 | 15 |

### Ontario Uranium Miners

Uranium mining in Ontario, Canada, started in 1953, somewhat later then in the United States and was conducted in relatively few mines in comparison to the United States (see Time Line E-4). A 12 WLM annual limit was adopted in 1954 with a concomitant decrease in annual exposure thereafter. Radon measurement and radon control programs were instituted within two years of the start of mining. Except for exposures occurring before 1958, exposure estimates are largely based on actual measurements (SENES 1989). However, some Ontario uranium miners had worked earlier as gold miners and were exposed to both radon progeny and arsenic in those operations. These miners had an estimated average cumulative exposure 2 WLM due to gold mining (Kusiak and others 1991) compared to an average of 30 WLM in uranium mines (Kusiak and others 1993). Even for those with gold mining experience, the approximated exposures from gold mining are only a minor portion of the total exposure.

The BEIR IV report provides a complete description of the exposure estimates for their cohort (NRC 1988). The radon-progeny measurement program was extensive: 131,000 measurements in 15 mines. Exposures were estimated using different methods for 1967 and earlier years and for 1968 and later years for which WLM estimates made by the companies were used. For 1957-1967, WLM were calculated by combining WL data with work histories. Two separate sets of estimates were derived for these years: the "standard" or lower WL values were

## Time Line E-4: Ontario Uranium Miners

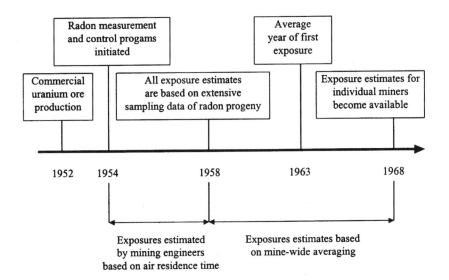

the averages of the four quarterly averages or three four-month averages for a particular year while the "special" or upper WL values were a weighted average of the four highest quarterly measurements or the three highest four-month measurements in headings, stoops, raises, and travel ways. The differences between these two sets of values varied by mine and by year, with the special values being up to four times as high as the standard values. The investigators considered that the true exposures were bounded by the two sets of values. For its analysis, the BEIR IV committee used the WLM values based on the standard WL values. Some estimation of exposures for the earliest years of the industry, before 1954, required extrapolation from measured values, taking into account such factors as ventilation. These years included the highest exposures and consequently 22 percent of the total WLM accumulated by the cohort was based on extrapolation of measured values.

### Port Radium Uranium Miners

The approach for exposure estimation for the Port Radium miners is well documented in a 1996 report by SENES Consultants Limited (1996b). Underground uranium and/or pitchblende mining at Port Radium started in 1932 and continued, with a two-year interruption, until 1960 (see Time Line E-5). Because records of employment before 1940 were not available, exposures occurring before that date have not been accounted for (Howe and others 1987, SENES

## Time Line E-5: Port Radium Uranium Miners

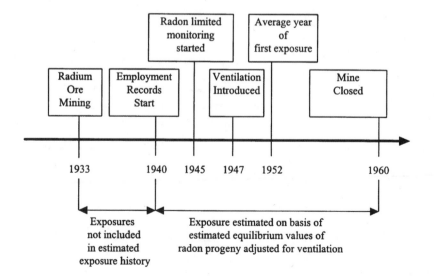

1989). Exposures occurring after 1939 have been estimated on the basis of rather sparse monitoring data for radon. Between 1945 and 1957, 261 radon measurements were made in seven years of this period, with from nine to 71 samples in an individual year. A few radon-progeny measurements were made and only three paired measurements of radon and progeny were obtained.

Absent information on concentrations of radon progeny, the equilibrium between radon and its progeny was estimated on the basis of knowledge of mine operations and by analogy from the Beaverlodge mine and the radon concentrations were then converted to concentrations of progeny. Although ventilation was introduced in 1947, it apparently was used in limited fashion during the winter season because of the cold. Consequently, the approach to estimating the WL values considered seasonal variation in equilibrium factor. The SENES report provides detailed documentation of the assumptions made in estimating the WL values from the radon measurements.

Reported radon concentrations were extremely variable ranging from 50-300,000 $pCiL^{-1}$ and it is thought that before ventilation was introduced, some exposures could have been as high as 1,000 WLM per year (SENES 1989). Unfortunately, such large annual exposures could not be assigned to the involved workers unless the exposures took place after 1940 because of the missing work-history information before 1940.

The potential limitations of the exposure data were acknowledged in the initial report on the findings of the epidemiologic study of Port Radium miners (Howe and others 1987). SENES Consultants Limited (1996b) have recently re-estimated exposures to radon progeny for 171 miners included in a case-control study (see Table E Annex 1-4). For the 171 miners, employment histories were reconstructed and used with revised WL estimates to calculate WLM. Substan-

TABLE E ANNEX 1-4   Summary of differences between the SENES reevaluation of miner exposures and epidemiology exposures used by Howe and others (1987). Data from 171 miners[a]

| Months Worked | Months Worked | WLM |
|---|---|---|
| Mean difference | −1.64[b] | −5.2[d] |
| Maximum | 66.93 | 2908 |
| 75th percentile | 0.26 | 69 |
| Median | −0.23[c] | 0[d] |
| 25th percentile | −1.97 | −12 |
| Inter quartile range | 2.24 | 80 |
| Minimum | −38.04 | 2348 |

[a]Based on SENES 1989, table 4.1.
[b]Mean difference significantly different at 5% level.
[c]Median difference significantly different from 0 at 1% level.
[d]Median difference and mean difference *not* significantly different from 0.

tial differences were found in individual estimates, although the mean WLM values were comparable for the two sets of estimates. There were large differences in employment duration for some of the men and changes in estimated exposure as large as 2900 WLM were found. The report comments on key sources of uncertainty in the exposure estimates for the Port Radium miners: incomplete employment histories for other employment; lack of employment information for years before 1940 when exposures were extremely high; and the numerous assumptions made in calculating the WL values from the radon measurements. Finally, the report also indicates that Port Radium ores contained "significant concentrations of various elements including for example arsenic, nickel, and cobalt."

### Czechoslovakia Uranium Miners

A cohort of miners in Czechoslovakia, who started uranium mining between 1948 and 1957, has been described extensively in the literature by Sevc and Placek (1976), Sevc and others (1988), Kunz and others (1978, 1979), and more recently by Tomášek and others (1993, 1994a,b). This cohort is often designated as group S by the Czech authors (see Time Line E-6). Compared to miner studies in other countries, exposure information for group S is among the most extensive. Measurement of radon and other potentially hazardous materials had become routine in Czech mines before 1948 so that estimation was not necessary for periods of employment during which radon measurements were not made. Even so, exposure estimates for radon-progeny exposures prior to 1961 are based on radon concentrations as concentrations of progeny were not measured.

## Time Line E-6: Czechoslovakian Uranium Miners

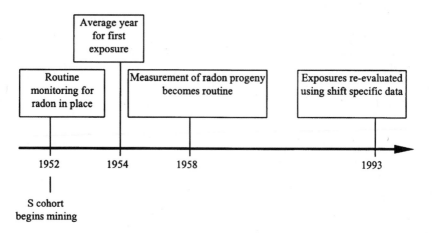

These radon measurements were, however, extensive, the annual number varying from 100 to 700 per shaft (Ševc 1993). Estimates of WLM for years before 1961, during which radon measurements only were made, were based on F values taken during periods of ventilation failure which occurred in 1969 and 1973. These data indicated an F value of 0.86 for the earliest years 1948-1952. From 1961-1969 the mean number of annual radon-progeny measurements recorded per shaft was 952, about 3 per work day.

In 1953, natural ventilation was augmented by mechanical means and F values decreased to an estimated average value of 0.55 in 1953-1959 and to 0.36 thereafter. Czech investigators estimate the coefficient of variation in converting radon levels to working levels as 28%. The fraction of unattached plutonium-214 has been estimated by Czech investigators as 0.1 (Hamilton and others 1990).

Estimated exposures of these miners have recently been reevaluated (Tomášek and others 1994a,b). The principal change appears to have been a more thorough investigation of workers' employment histories to take account of prior mining experience and the assignment of exposure for each month based on the particular shaft in which the miner worked. In any event, the newest account given below is more complete than those published previously. Additional background can be found in a report of a WHO-sponsored trip to Czechoslovakia made in 1988 by L.D. Hamilton, L.W. Swent, and D. B. Chambers (published by SENES Consultants Limited—see Hamilton and others 1990).

"During 1949-1963 about 39,000 measurements of radon gas were made in the 19 mine shafts in Jachymov and Horoni Slavkov in which the men were employed. Some men also worked, particularly after the closure of most shafts at Jachymov and Horoni Slakov in 1963, at other Czechoslovak uranium mines, and substantial numbers of measurements were also made in these mines. An initial review of exposure estimates used in previous reports found a considerable number of errors, and for some miners a part of their employment histories had not been taken into account. Therefore, exposure estimates have been completely revised for the present analysis, based on a review of all available information. The radon-gas measurements were converted into estimates in terms of working levels using equilibrium factors based on radon-progeny measurements made after 1960, and on data collected during two accidents in the uranium mines at Zadni Chodov in 1969 and at Pribram in 1973, when mechanical ventilation was stopped for at least a month. An estimate of each man's exposure in each month in terms of working months (WLM) was calculated from the time he spent in each mine shaft in conjunction with the year- and shaft-specific WL estimates. Men worked 6 days per week with 1-month of holiday each year. For most men underground work was assumed to last 8 hours per day, but for geologists, safety and ventilation technicians, and emergency workers, it was estimated that 70% of working time was spent underground, while for managers 50% was estimated. About 300 men were involved in exploratory work, which was normally carried out in shallow shafts near the surface. Explicit radon measurements are not available for this work, but exposures are thought to have been low, and are estimated at 3.3 WLM per year" (Tomášek and others 1994a).

With regard to the magnitude of exposure error, it is of interest that less than 5% of the new exposure estimates differed by as much as 50% from those used in previous analyses of these data. The mean cumulative exposure in the new evaluation is 219 WLM compared with 227 WLM in the older work (Ševc 1993). Presumably, random error as well as systematic biases were reduced in the re-evaluation. Using a simple model in which the estimated relative risk is linearly related to the cumulative exposure yields a relative risk of 0.37%/WLM (95% C.I. = 0.18-0.55) with the old exposure estimates. With the new ones the estimated relative risk is 0.61%/WLM (95% C.I. = 0.29-0.8) (Tomášek and others 1994), providing evidence that the dose-response was flattened by errors in the original exposure estimates.

### French Uranium Miners

Uranium mining in France started in 1946 with exploratory operations that continued through 1948 when extensive commercial operations commenced (Tirmarche and others 1984, 1993) (see Time Line E-7). The first reported radon sampling occurred in 1953 when 40 measurements were taken, an average of 10 per mine. Large-scale radon monitoring began in 1956 when forced ventilation was introduced. Exposures prior to this date have been estimated retrospectively by an expert group which considered mine characteristics and type and duration of work. For this early period, before forced ventilation, exposures were relatively high and varied substantially between individuals. The estimated median annual exposure from 1947 through 1955 was 11 WLM and varied for the 3rd quartile up to 55 WLM per year.

Exposures declined rapidly after forced ventilation was introduced, median exposures averaging about 3 WLM per year from 1956 to 1975 with a further decline to 1 WLM by the early 1980's. About half of the French miners started their underground employment before 1956 but most of their person years of exposure occurred after monitoring became comprehensive, ventilation improved, and exposures were relatively low. Nevertheless, for a large number of the French miners, a major portion of their cumulative exposure was based on estimations by experts for the period when ventilation was poor and routine monitoring lacking. After sufficient monitoring data became available, worker-exposure assessments were individualized to some extent by considering the type and location of work performed. Recently, personal dosimeters, using track-etch dosimeters, have provided direct information on individual miners. This has allowed a comparison of exposure based on area monitoring in 1982 with direct personal measurements in 1983. The comparison indicated that annual exposures based on area monitoring and work locations were, on the average, underestimated by almost 30% (Bernhard and others 1984; Piechowski and others 1981). Con-

# Time Line E-7: French Uranium Miners

```
                    ┌─────────────┐
                    │ Initial radon│
                    │  sampling   │
                    └──────┬──────┘
                           │
┌───────────┐ ┌───────────┐│┌─────────────────────┐  ┌─────────────────────┐
│Exploratory│ │Commercial ││││ Average year of first│  │Routine use of personal│
│  mining   │ │  mining   ││││    exposure. Routine │  │dosimeters for exposure│
└─────┬─────┘ └─────┬─────┘││monitoring of radon progeny│  │    to progeny       │
      │             │    │ │└─────────────────────┘  └──────────┬──────────┘
      │             │    │ │                                    │
──────┼─────────────┼────┼─┼────────────────────────────────────┼──────────────►
      │             │    │ │                                    │
    1946          1948  1953 1956                              1983
```

|  Exposure estimated on                    Exposure estimated for  |
| basis of radon measurements          individuals based on systematic |
| mostly in travelways and exhaust              monitoring |
| points. Equilibrium calculated from |
| estimated air residence time. |

sidering the year-to-year variation in the true levels and miner location, as well as the accuracy of the personal dosimeters, the reported difference may not be indicative of a significant bias in the exposure estimates.

## Radium Hill Uranium Miners

The Radium Hill mine began operations in Australia in 1952 and radon monitoring began two years later (see Time Line E-8). In estimating exposures, exposure levels in the prior years were assumed to be the same as in early 1954 (Woodward and others 1991). A total of 56 samples were collected by 1 April 1955. Early radon concentrations were low (estimated 1.8 WL) even before forced ventilation was introduced and declined substantially thereafter—range 0.10-0.55 WL. Apparently only radon concentrations were measured; WL concentrations were estimated by means of a calculated equilibrium factor based on ventilation rates and air volumes at various locations but such methods do not account for plate out and recirculation of progeny.

Enough radon measurements were made to allow exposure estimates by job category for work after 1 April 1955. The estimated exposures for workers show an exponential distribution with a median exposure of 3 WLM; a mean of 7 WLM and a 3rd quartile limit of 7.4 WLM. A few heavily exposed workers received about 80 WLM (Woodward and others 1991).

# Time Line E-8: Radium Hill Uranium Miners

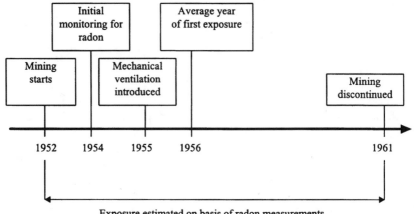

Exposure estimated on basis of radon measurements
mostly in travelways and exhaust points. Equilibrium
calculated from estimated air residence time.

## Chinese Tin Miners

Exposure assessments of Chinese miners employed by the Yunnan Tin Corporation are largely retrospective as no measurements of either radon or radon
progeny were made prior to 1972 when mechanical ventilation was introduced
(see Time Line E-9). Exposure estimates for two periods prior to 1972 reflect
changes in the mining industry that occurred after nationalization in 1949. Before nationalization, mining was conducted in small mines with back hauling
performed manually, often by children (Xuan and others 1993). To estimate
exposures under these conditions, 117 measurements were made in 13 local mine
pits that had been in operation before the large-scale expansion of the tin mines
that started in 1953.

Exposure estimates for miners employed between 1953 and 1972 were based
on 413 measurements obtained in the 1990's by recreating conditions in tunnels
and galleries in original areas or in similarly configured areas in nearby mines
that used techniques similar to those in the index year (Xuan and others 1993).
Evidently, there was little change in radon progeny concentration in the larger,
post 1953 mines. The reported average mean WL before 1950 was $2.3 \pm 0.8$ and
$2.2 \pm 1.2$ thereafter. Mechanical ventilation was introduced rather slowly with
priority given to new tunnels. Working levels decreased moderately in 1971-
1975 to $1.7 \pm 1.1$, to $1.2 \pm 0.8$ in 1980 and $0.9 \pm 0.3$ in 1985 (Xiang-Zhen and
others 1993). Exposure estimates from experience since 1972 have been based
on over 26,000 measurements of radon progeny.

# Time Line E-9: China Tin Workers, Yanna Province

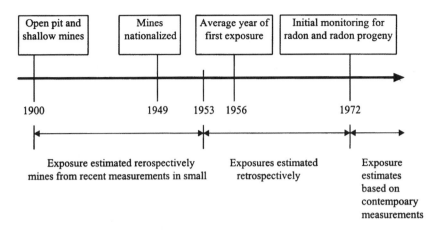

Little information is available on other characteristics of the occupational environment. Equilibrium factors, F, were measured in 1977-1978 and found to vary from 20% to 84% with a mean of 72% in "small pit operations" and of 62% in "larger tunnels" (Lubin and others 1990b). Evidently the mean of 72% refers to conditions prior to 1953 and the mean of 62% to the expanded operations. In any event the mines were very dusty by contemporary standards. Airborne dust was first measured in the 1950's and maximum levels were between 20 and 192 mg/m³. Wet drilling was introduced in the late 1950's and became widespread in 1964 when dust levels fell to about 6.2 mg/m³ (Xiang-Zhen and others 1993). Given these levels of dust, it is probable that equilibrium levels remained rather high throughout and that the unattached fraction was small.

For the epidemiological studies, workers were assumed to be exposed to radon progeny seven hours per day. For exposures occurring after 1972, estimates of exposures were adjusted by the worker's job title to take account of those exposed intermittently (Lubin and others 1990b). Exposure to arsenic in airborne dust was also accounted for in these studies and shows a large decrease over time from 0.4 mg/m³ in the mid 1950's to .01 mg/m³ in 1985. The radon exposure estimates for the Yunnan tin miners are not very well documented but, given the apparently uniform level of exposure throughout the period of miner employment, about a factor of 2, the estimates may be less subject to errors in estimation then for those uranium mines where exposure levels varied over time by factors of ten or more due to changes in ventilation practices. Arsenic exposure, on the other hand, did decrease appreciably as wet drilling became standard practice.

### Newfoundland Fluorspar Miners

Underground mining for fluorspar started in 1936 in Newfoundland, Canada, and continued for more than three decades before monitoring for radon and radon progeny was initiated in 1960 (see Time Line E-10). Radon levels were found to be highly variable, range 0-190 WL (Morrison and others 1988). Ventilation was immediately introduced and in 1960 levels declined to an average of 0.5 WL (1960-1967) and then to an average of 0.17 WL (1969-1978) (Morrison and others 1988). Therefore, an average worker for the entire period of radon control would have accumulated about 30 WL as opposed to an estimated average exposure for all cohort members of 382.8 WL.

Exposure estimates for epidemiological purposes have been developed by Dory and Cockill (1984) and are described in SENES 1989. Exposure estimates for the period before 1961 were based on maps of the various mines, reports by mine inspectors, and workers' recollections. Apparently experience gained when the mines were monitored was also taken into account as well as the entry of water, the source of the radon, into the mines. Eventually a computational model was developed to simulate the annual radon progeny concentration in each mine so as to yield "average workplace concentration for high, medium, and low areas" (SENES 1989). For epidemiological purposes, workers have on the basis of their jobs been assigned to approximate areas of radon concentration, high, medium, etc. for a particular year and cumulative exposures estimated on this basis.

For exposures occurring after 1960, worker job records and monitoring data have been used to assign individual exposure estimates. As noted above, this period of employment is likely to be relatively unimportant for risk estimation because of the relatively small exposures. It is impossible to estimate the accu-

## Time Line E-10: New Foundland Flurospar Miners

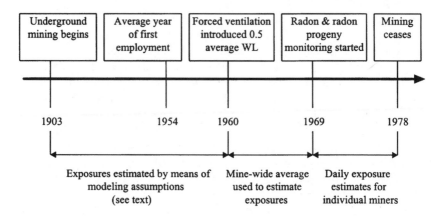

racy of the exposure estimates for before 1960. Radon concentrations in mine water varied form 300 to 1300 pCiL⁻¹ in 1960 (Report of the Royal Commision 1969). Under such conditions, actual exposures must have been highly variable and in spite of the thorough assessment performed by Corkhill and Dory, the estimated cumulative exposures are guesstimates c.f. Colorado Plateau miners.

### Swedish Iron Miners

The BEIR IV report provides an extensive description of exposure estimates for this cohort (NRC 1988). Additional description can be found in a report submitted to the BEIR IV committe, "Comments to the U.S. Mine Safety and Health Administration for the American Mining Congress", prepared by L.W. Swent and D.B. Chambers. This report describes a visit to the mine by Swent and Chambers and their discussion with mine personnel. Much of the material is described in the 1989 report by SENES Consultants Limited (1989).

Exposure estimates for the Swedish iron miners are primarily retrospective. The cohort includes those who started work earlier but for the cohort as a whole the average year of first exposure was 1934 (see Time Line E-11). The first extensive measurements of radon in these mines were not made until 1968; radon-progeny measurements were initiated somewhat later. Most of the mine radon came from water seepage and there is limited evidence that this source was relatively constant in strength from 1915 to 1972. Comparison of radon measurements in water taken in 1915 with data from 1972 and 1975 indicated constant groundwater concentration of radon. Exposures have been estimated using the assumption that levels of radon progeny were constant until forced ventilation was introduced in 1972 (Radford and St. Clair Renard 1984). This assumption

## Time Line E-11: Swedish Iron Miners in the Malmberget Area

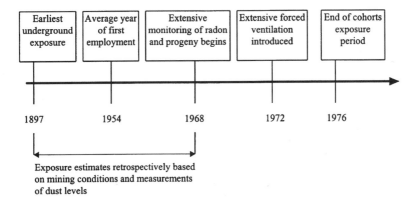

Exposure estimates retrospectively based
on mining conditions and measurements
of dust levels

was supported by consideration of the pattern of natural convection and by data on quartz dust concentrations that extended to the 1930s. On the basis of their visit, Swent and Chambers have questioned the assumption of stable ventilation and suggest that the estimates of exposure were low by a factor of two or more due to air recirculation and changing ventilation conditions as the mines became deeper (SENES 1989). Because there are no measurement data for years before 1968, it is a matter of speculation as to how much exposures varied with time. It is likely, however, that the estimated error of the exposure estimated initially made by Radford and St.Clair Renard (1984), about 30%, was unduly optimistic.

# E-ANNEX 2

# Workshop on Uncertainty in Estimating Exposures to Radon Progeny in Studies of Underground Miners

## INTRODUCTION

The epidemiologic studies of underground miners have been the principal basis for estimating the risk of indoor radon (NCRP 1984; NRC 1988; USEPA 1992c).

To estimate exposure to radon progeny, information is needed on the concentrations to which the miners have been exposed as well as the time spent at these concentrations. There are now 12 studies that include estimates of exposure to radon progeny (Lubin and others 1995b; Darby and others 1995). The exposures received by the miners in these studies began as long ago as the end of the nineteenth century in the case of the Malmberget iron miners (Radford and St. Clair Renard 1984) and are continuing for some of the more contemporary groups, such as the Chinese tin miners (Qiao and others 1989) and the French uranium miners (Tirmarche and others 1993). The information available for estimating exposures varied among the cohorts and even within the cohorts by time period. The measurements tended to be more sparse during the initial years of mining operations, the same years during which exposures were generally highest. In some of the studies (Chinese tin miners, Czechoslovakian uranium miners, Colorado Plateau uranium miners, Ontario uranium miners, and Radium Hill uranium miners, and French uranium miners) concentrations of radon progeny were not measured in the early years of operations and it was necessary to estimate WL based on radon measurements, assuming a value for the equilibrium of radon with its progeny (Lubin and others 1994a). Information was also used on mining practices and measurements were made in the Chinese tin mines based on re-created mine conditions of earlier years. Personnel records were generally used to document time underground, although the detail of the information available also varied among studies. Gaps in the information available for estimating exposures of the underground miners in these epidemiologic studies are an acknowledged source of exposure misclassification with attendant implications for uncertainty in the risk estimates derived from these studies.

The consequences for risk estimation of the errors in exposure estimates have been of concern to the BEIR VI committee. The committee recognizes that exposure misclassification is inevitable in the epidemiologic studies of miners. However, techniques are becoming available to account for these errors in estimating exposure-response relationships (Thomas and others 1994). Statistical approaches to measurement error, with the specific application to the underground miner studies, was a topic of a one-half day workshop held in 1994 by the BEIR VI committee.

To further address the issue of measurement error and its consequences of risk estimation, the BEIR VI committee convened a second workshop on January 23 and 24, 1995. The workshop was designed to bring together geologists and mining and ventilation engineers who had worked on exposure issues in the underground mines with statisticians and epidemiologists who are now contending with the measurement-error problem. The workshop's goal was to obtain additional documentation on approaches followed to assess exposures and to obtain historical insights that might lead to better quantification of the errors in the exposure estimates. The committee also invited statisticians engaged in investigation of approaches for correcting for the effect of measurement error on risk estimates.

The workshop participants selected were appropriate for these objectives. William Chenoweth, now a consulting geologist in Grand Junction, Colorado, worked for many years for the Atomic Energy Commission. He has written extensively on the history of the various uranium mining districts in the United States. James Cleveland, an engineer, worked for Kerr-McGee Corporation in New Mexico for many years, directing ventilation and safety for the Ambrosia Lake operations of Kerr McGee, by far the largest uranium producer in the Grants Mineral Belt. Andreas George, from the Environmental Measurement Laboratory of the U.S. Department of Energy, made measurements of the attached and unattached fractions of radon progeny during the late 1960s and early 1970s; these data represent the only published, historical information on the distribution of progeny between attached and unattached fractions. Douglas Chambers from SENES Consultants Limited in Ottawa, Canada, has had long-standing interest in the assessment of exposures to radon progeny and the consequences of error for risk estimation. Neal Nelson from the U.S. Environmental Protection Agency, also a participant, has followed the exposure assessment issue closely for many years.

The practical experience of these participants was complemented by the statistical expertise of Dan Stram and Duncan Thomas, both from the University of Southern California. Stram and Thomas have been applying new statistical methods for errors-in-variables to data from the Colorado Plateau miners. Their models make adjustments to risk estimates for measurement error. The workshop was also attended by BEIR VI committee members and Jonathan Samet, the committee's chair, presented descriptive analyses of exposure data for the New Mexico cohort of uranium miners.

The workshop provided valuable and not-well-documented information concerning the U.S. and Canadian uranium mining industries. Consequently, the committee decided to publish portions of the workshop proceedings as an appendix to its report. We include the historical material presented and not the discussion of ongoing work on measurement error by Stram and Thomas.

## Workshop Introduction

The workshop began with a welcome and introductions by Jonathan Samet and Evan Douple, the study director. Ethel Gilbert, the workshop chair, reviewed the goals of the workshop and called attention to questions to be considered throughout the workshop. Roger McClellan reminded participants to focus on "uncertainties," not just on "errors." Philip Hopke asked that participants try to recall details regarding the underground mining procedures, especially environmental conditions and the methods used for measuring radon and radon progeny. He asked participants to review those details that have not been previously published or documented. These committee members provided a reminder that the committee needs to characterize the uncertainties that affect different levels of exposure received by the miners.

## William Chenoweth (Grand Junction, Colorado)

Chenoweth joined the Atomic Energy Commission (AEC) in 1953 as a geologist. His work initially involved the vanadium mines of the Colorado Plateau. Vanadium mining was initiated in the mid-1930s; he noted that the mines were not ventilated. He then reviewed the uranium mining operations across the years 1947-1981; approximately two-thirds of the mines were in the Colorado Plateau region. The peaks of uranium mining production were in the 1960's and the 1980's on the Colorado Plateau (Moab, Monticello, and Uravan areas and the Grants Mineral Belt). The peak number of mines in operation was around 750, a total reached in the mid-1950s. The AEC stopped buying ore in 1962. However, it continued to purchase concentrates (yellowcake) from the mills through 1970.

Most uranium ore was mined from sandstone. Mining involves drilling, blasting, and mucking, the extraction of the ore. The needs for mining include compressed air and water; powder, blasting caps, and fuses; and a shovel or machine to load the ore. Typically blasting was done at night and the miners would return the next morning without ventilation operative in the mine. In the early mines, ventilation was used primarily to reduce blasting fumes. Chenoweth explained that the typical cut was made with a burn of multiple fuses. The miners used wheelbarrows in remote areas and the larger mines used compressed-air locomotives. There was no ventilation when the 1952 measurements were made. There were incentives for miners to be productive since the AEC needed the uranium.

Chenoweth showed the structure of a larger Grants mine with a central ventilation shaft. He also showed a map of the Uravan area near the Colorado-Utah border. Brenner asked whether some of the vanadium miners beginning in 1936 were also uranium miners later. The answer was "yes." The same mines were mined back in the 1920's for radium; there was a radium industry from 1910 through 1923. Douglas Chambers stressed that many of the miners worked their own mines on weekends and others were vanadium miners before they worked in uranium mines. Krewski asked whether there were houses in the area. The answer was "no," as the mines tended to be in very remote locations with just a few ranches nearby.

The mines were generally 400 to 800 feet below the surface, although some were as much as 2000 feet deep. Chenoweth provided definitions of various mining terms: a "stope," the site of mining, is an area at the end of a "drift," or passage; a "raise" is a passage up between drifts whereas a "winze" is a passage down between drifts. A map of one early mine showed that there was no ventilation as the rooms were interconnected and the open rooms had randomly placed pillars. The Deremo mine was the largest that Union Carbide mined in the Uravan area. Pictures taken by the U.S. Bureau of Mines in 1953-1955 do not show ventilation; one photo showed a fan that was not in use. Chenoweth surmised that it may have been used for venting blasting powder. Between 1948 and 1956, smaller mines probably had less ventilation. Ventilation rates are affected by temperature and recirculation of radon progeny could have been a problem.

There were various types of operations [U.S. Vanadium (USV) contractors, large mining companies like Walter Duncan, Climax Uranium, Kerr-McGee, Union Carbide, small mining companies, Vanadium Corporation of American (VCA) mines, VCA leasors, small independent operators, and one- or two-man operations]. Some of these mines probably had adequate ventilation, that is, about 500 cubic ft per minute, enough to keep oxygen concentration at about 20%. Early miners did not live in the mines but typically lived in boarding houses; they worked 12-hour shifts and worked 30 days and then spent four or five days in town.

Chenoweth presented a summary of the key events in the history of uranium mine ventilation. Diesel-powered trucks and loaders were introduced in the 1960s. Respirators were issued, but they were not necessarily worn. A Bureau of Mines report (IC07908) states that the Climax Uranium Co. Was the first to use rubber-tired diesel equipment in their mines in the Uranan area in the early 1950s.

Union Carbide adopted extensive air sampling in 1957; it began using diesel in 1958. A shut-down level of 10 WL was adopted in 1958. The U.S. Department of Labor proposed a 3.6 WLM exposure limit in 1969.

Samet asked whether there is any information on arsenic in these mines. Chenoweth explained that a few years ago the Bureau of Land Management asked for a sampling for trace metals in mine dumps. Arsenic was a constituent of some uranium ores. With respect to the question of how many total miners were employed, Chenoweth pointed out that 780 Native American miners were included in the Public Health Service's epidemiologic study, but the Navajo Tribe has registered approximately 2,500 former miners through its Office of Navajo Uranium Workers. There were about 75 miners in 1948 and 5,000 in 1960. Frank Lundin had advised Neal Nelson that 15,000 miners were identified in the Public Health Service surveys, beyond those included in the epidemiologic study (U.S. Bureau of Mines). The current number of miners worldwide is uncertain but underground uranium mining is no longer carried out in the U.S. Some of the current sites of active uranium mining include Canada, China, and Russia. Uranium production is down substantially since the early 1980s and there is now an oversupply of uranium. Countries producing uranium in 1995 were:

- Canada        32% (of world total)
- Australia      11%
- Niger          9%
- USA            7%
- Russia         6%
- Uzbekistan     6%
- Kazakstan      6%

- Namibia        6%
- South Africa   4%
- France         3%
- China          2%
- Gabon          2%
- Other          6%

(From the International Atomic Energy Agency)

### James Cleveland (Edmond, Oklahoma)

From 1960 until 1985, Cleveland was with Kerr-McGee. He provided a historical perspective on uranium mining in New Mexico. Uranium mining in New Mexico began with small mines on the Navajo reservation, which operated from the 1948 to 1968. Most Kerr-McGee mines on the reservation were small, operated by less than 20 people and usually by only two or three. The Shiprock uranium mill operated by Kerr-McGee was closed in 1963. The Grants mineral belt, the principal site of uranium mining in the state, stretches 200 miles from the Arizona border to the city of Albuquerque; it is about 25 to 50 miles wide. The Ambrosia Lake and the Jackpill-Paguate districts were the first large-scale production mining areas in New Mexico. In 1980, Grants-area mines employed 4,500 to 5,000 people.

Cleveland described a four-day meeting in 1961 between Archer and Wagoner, from the Public Health Service, and the industry. A warning concerning the health effects of radon and radon progeny was presented. Three aspects of exposure control were addressed: 1) instrumentation for making measurements, then lacking and very temperamental; 2) ventilation; and 3) record-keeping.

1) Instrumentation for measurements initially involved off-site equipment. Evacuated flasks were opened underground and then shipped to the Massachusetts Institute of Technology or elsewhere for measurement of radon. The Kusnetz method was then developed for measuring radon progeny. The Working Level (WL) unit of concentration and the Working Level Month (WLM) unit of exposure were established by the U.S. Public Health Service. A joint survey with state health department officials and mine inspectors found high levels of radon. The findings were discussed with the industry in 1961. The Junod instrument was used universally until 1967 to measure WLs, but this instrument was very sensitive to environmental conditions. Gas- and battery-powered air samplers were used. The samples were frequently four-hours old before the measurements were made. A probe was then developed for alpha counting and used underground. One could then measure for 40 to 90 minutes using the Kuznetz method. Next, Eberline Instruments developed an instant WL meter which could provide readings after a 2.5 minute count. Several groups developed other instruments for measuring radon. Area monitors were then designed that would give warnings. Cleveland estimated accuracy of the early instruments as plus/minus about 50%.

A discussion followed about the accuracy of the measurements. Andreas George indicated that the uncertainties associated with the measurements were in the range of 50%. Duncan Thomas asked what was reported—the highs, the lows or an average? Douglas Chambers indicated that if there is a bias it is probably small, but with a substantial uncertainty. Lubin asked a series of questions about the measurements: Where were the readings taken? The answer was "in the worker's area in a stope." Within a room the concentration tended to not vary substantially if the air was not very mobile. How much difference was there between measurements? The answer was the variation depended on where the miners were working, and was possibly up to 4 WLM. In the Public Health Service study, Archer assembled all of the measurements and then averaged them for a particular mine. That average was then multiplied by the number of months worked in the mine to obtain WLM.

2) Ventilation was increased in 1962 in most of the mines and the mining companies started to develop and maintain records. New Mexico adopted a 10 WL cease and desist level. Up to that time, ventilation had been natural or used gas-powered fans. The large mines in Ambrosia Lake had 30-inch diameter fans. The major purpose of the fans was to remove powder smoke which contained nitrogen oxides. Simple ventilation was used before 1961. Specifications called for 500 cubic feet per man per minute and, if diesel was used, greater ventilation was called for. State officials had the authority to shut-down mines if ventilation

was not adequate. Eventually, large-diameter vent holes, up to 72-inch diameter, were installed to increase ventilation. The mines needed parallel ventilation to deliver fresh air to all parts of the mines. Secondary protection measures, such as filters, respiratory protection (a simple two-canister and air supply unit, or a self-supply unit, depending on levels), and electrostatic precipitators, were proscribed and the latter were used in isolated instances for short periods of time.

3) Records of measurements taken once or twice a quarter were initiated by the companies in 1962. The records used the worker's time in travel, time in the stope, and in other locations, and assigned WLM values to each worker. In 1972, the work week was assumed to include 40 hours. When levels were particularly high, measurements were sometimes taken more than once per week. Total WL-hours for the day were then expanded to the work in a week and the exposure was developed for the month. From 1967 on, records became quite complete. Before 1962, however, records were generally of poor quality. Exposures were grossly underestimated in the early days of the industry, when the work week was longer than 40 hours. Ambrosia Lake mines worked a 48-hour week until 1966.

Hopke asked whether there was smoking underground. Smoking was banned in Ambrosia Lake in about 1975. Standards for exposure to radon progeny were not in place until 1972 when MESA adopted a 4 WLM standard.

## Andreas George (New York, New York)

George addressed sources of variation in the measurements made in mines. Sometimes measurements were not made in proximity to the miners. George provided tables showing some typical measurement values and the range of concentrations measured over three days in various locations in the mines (Note: the mines were not named). In the table (Table E Annex 2-1), the arrow indicates either downcast or upcast ventilation by its direction. The WL values relate to specific locations in the mines but not to individual miners. The measurements were made as often as every 30 minutes. The variation in WL values within a stope was generally not large. Ziemer asked whether there were WL values for samples taken at similar locations but at different times of the year. The answer was "no."

Methods were developed for measuring the unattached fraction and applied in the mining environment. George recalled that measurements were made in four different mines for one week. Polonium-218 was measured to determine the uncombined fraction, which was almost related to the unattached fraction (less than 0.2 %). The uncombined fraction tends to be inversely related to particle concentrations. In the Beaverlodge uranium mines, diesel was used infrequently and the vehicles were primarily powered by electricity. Blasting dust remained in the mine for long time intervals. Ventilation ducts were occasionally damaged during blasting.

## TABLE E ANNEX 2-1

| Date | Mine | Location | Activity | T(°F) |
|---|---|---|---|---|
| 9/26-28/67 | A | 1-stope | none (dry) | 57 |
| 9/26-28/67 | A | 2-stope | track laying | 50 |
| 9/26-28/67 | A | 3-drift | blasting - ore hauling | 50 |
| 9/29-10/2/67 | B | 1-drift | none (wet) | 52 |
| 9/29-10/2/67 | B | 2-stope | slushing/mucking | 53 |
| 9/29-10/2/67 | B | 3-stope | drilling/slushing | 53 |
| 10/3-4/67 | C | 1-stope | mucking/slushing (wet) | 52 |
| 10/3-4/67 | C | 2-stope | blasting/slushing | 52 |
| 10/3-4/67 | C | 3-drift | none | 51 |
| 11/2/67 | D | 1-drift | mucking ore (dry) | 57 |
| 11/2-4/67 | D | 2-drift | drilling/charging | 55 |
| 11/3-4/67 | D | 3-drift | hauling ore | 55 |
| 11/2-4/67 | D | 4-stope | drilling/mucking | 58 |
| 11/6-8/67 | E | 1-drift | drilling/slushing (dry) | 50 |
| 11/6-8/67 | E | 2-drift | hauling ore | 50 |
| 11/6-8/67 | E | 3-drift | drilling/charging/slushing | 50 |
| 11/10-15/67 | F | 1-stope | drilling/mucking (wet) | 55 |
| 11 | F | 2-stope | drilling/mucking (wet) | 60 |
| 11/10, 11/13/67 | F | 3-stope | drilling/mucking (wet) | 59 |
| 11 | F | 4-stope | drilling/mucking (wet) | 59 |
| 11/14, 11/15/67 | F | 5-drift | slushing/mucking (wet) | 49 |
| 11/10, 11/13/67 | F | 6-stope | drilling/mucking (wet) | 53 |
| 11/15/67 | F | 7-drift | drilling/hauling (wet) | 60 |
| 1/24-26/68 | G | 1-drift | drilling/mucking/hauling | 63 |
| 1/24-26/68 | G | 2-drift | drilling/mucking/hauling | 53 |
| 1/24, 1/25/68 | G | 3-heading | mucking (dry) | 67 |
| 1/26/68 | G | 4-heading | drilling/charging | 69 |
| 2/1-2/68 | H | 1-drift | drilling/slushing | 34 |
| 1/30-31/68 | H | 2-drift | slushing/mucking | 39 |
| 1/30, 2/1/68 | H | 3-drift | drilling/slushing/hauling | 45 |
| 2/1-2/68 | H | 4-drift | drilling/slushing | 44 |
| 2/2/68 | H | 5-drift | none | 40 |
| 1/30, 3/1/68 | H | 6-stope | drilling/slushing/mucking | 42 |
| 2/6, 7/68 | I | 1-drift | near shaft/ore hauling | 55 |
| 2/5/68 | I | 2-drift | blasting/slushing | 55 |
| 2/5-7/68 | I | 3-heading | drilling/slushing | 62 |
| 2/5-7/68 | I | 4-cross-cut | | 60 |

| RH(%) | (Ft$^3$/min) | Ventilation (pCiL$^{-1}$) | Radon WL | Mines |
|---|---|---|---|---|
| 79 | 5,500 ↑ | ND | 0.67-2.8 | Beaver |
| 93 | 2,200-4,400 | ND | 1.4-4.5 | Mesa, |
| 93 | 4,500-5,900 | ND | 1.9-4.5 | Colorado |
| | | | | |
| 94 | 22,000-34,000 ↓ | ND | 0.95-1.4 | Beaver |
| 95 | 200 | ND | 2.1-2.4 | Mesa, |
| 97 | 1,000 | ND | 2.1-2.3 | Colorado |
| | | | | |
| 96 | 2,000 ↑ | ND | 5.0-5.5 | Uravan, |
| 96 | 3,500 | ND | 3.8-4.1 | Colorado |
| 97 | 13,000 | ND | (3.8) | |
| | | | | |
| 47 | 27,000 ↓ | (410) | (0.69) | Uravan, |
| 78 | 900 | 190-380 | 0.66-1.13 | Colorado |
| 73 | 3,000 | (260) | (1.1) | |
| 71 | 1,000 | 410-1000 | 0.41-0.78 | |
| | | | | |
| 81 | 500 ↓ | 460-1000 | 1.27-3.10 | |
| 63 | 3,000 | ND | 1.15-2.0 | |
| 82 | 3,000 (on/off) | 180-270 | 0.36-0.67 | |
| | | | | |
| 57 | 1,500-3,000 ↓ | (430) | 0.35-2.13 | Uranum, |
| 84 | none | (490) | (1.28) | Colorado |
| 95 | none | (360) | (2.36) | |
| 92 | none | (340) | (1.74) | |
| 66 | fresh air 5,000-9,000 | 88-110 | 0.22-0.27 | |
| 92 | 7,000-9,000 | 180-220 | 0.42-0.46 | |
| 94 | ND | (540) | (1.28) | |
| | | | | |
| 81 | 14,000 ↑ | 380-420 | 0.8-1.0 | Ambrosia Lake, NM |
| 80 | 8,000 | 1350-1790 | 3.1-5.10 | |
| 90 | (convection) | 1900-2300 | (1.70) | |
| 92 | ND | (680) | | |
| | | | | |
| 57 | 26,000-29,000 ↑ | 330-370 | 0.26-0.43 | Ambrosia Lake, NM |
| 66 | 4,000 | 830-1010 | 1.04-1.08 | |
| 71 | 3,000-6,000 | 780-1150 | 1.4-2.1 | |
| 60 | 6,000 | 770-920 | 1.28-1.60 | |
| 62 | 15,000 (exhaust) | (870) | (2.13) | |
| 67 | 3,000 | 670-960 | 1.10-1.37 | |
| | | | | |
| 84 | (fresh air) 66,000-72,000 ↓ | 84-190 | 1.19-0.23 | Ambrosia Lake, NM |
| 84 | 3,000 | (160) | (0.26) | |
| 94 | 3,000-5,000 | 640-900 | 2.2-2.7 | |
| 95 | 1,000-2,000 | 1020-1440 | 3.0-3.8 | |

Measurements were made using diffusion batteries developed in his (George's) laboratory. A bimodal distribution of alpha activity was found in some mines, similar to more recent information from homes. The 5-6 nm size was critical because this size is relevant for the bronchial deposition. Classical, bimodal, or unimodal (diesel mines) distributions were obtained for activity-weighted sizes. Five nm is the critical size for the tracheal deposition. NCRP 1978 gives a number of about 1.7, which should be compared to 4 in the publication "Summary of dose conversion factors from reanalysis of New Mexico uranium mine particle-size data." Most of the progeny are attached to the larger particles while the smaller particles are deposited in the lung. It is realistic to assume two-fold variation in measurements from mine to mine.

### Douglas Chambers (Ottawa, Canada)

Chambers discussed the characteristics of several mines, beginning with the Newfoundland fluorspar mines. The Black Duck Mine opened in 1933 and there was not forced ventilation until the 1950s. For the Newfoundland fluorspar mines, uncertainties are quite high for data before 1967, perhaps as high as 300-fold. The miners smoked heavily and it was a very dusty environment. Chambers was not aware of other contaminants in the mines, such as arsenic. The reconstruction of exposures for these mines has been difficult and the approaches used have been as much as can reasonably be done.

The original client for uranium from the Ontario mines was the U.S. AEC. Two uranium mines were operational in Ontario in 1955; in 1958 there were 15 mines. Between 1955 and 1981, 131,000 radon-daughter measurements were made over 141 mine-years of operation, averaging 929 measurements per mine-year. Two sets of exposures were calculated, "standard" and "special" (see 1988 NRC BEIR IV Report for a description of these two sets of exposures). Exposure to aluminum powder was used in an attempt to prevent silicosis. The Canadian report on health and safety in mines is a potential source of information (Canadian Task Force on the Periodic Health Examination 1990).

A comparison was made between past and present exposure conditions. Pre-1958 conditions included radon progeny at 0.3 to 1.4 WL and mineral dust at approximately 1 to 9 mg/m$^3$ and for 1990 conditions at 0.05 to 0.3 WL and 0.05 to 1 mg/m$^3$, respectively. The lung-cancer experience of Ontario uranium miners with and without gold-mining history was provided. The observed to expected ratio was greater for those with gold-mining experience.

Chambers also provided information concerning the Port Radium uranium mine. In the Port Radium mine, the ventilation was frequently turned off in the winter months. Arsenic may have been 6-7% of the pitchblende ore. Exposures to radon progeny were estimated based on retrospective reconstruction. The ore grade was high and yielded high WLs (50-100 WL). The original estimates were done by Frost with Eldorado Nuclear. Chambers indicated that he has all of the records for the Port Radium mine.

Chambers also discussed the Beaverlodge mines. There has recently been a reassessment of exposures for some of the miners in the epidemiologic cohort of Beaverlodge mines. He noted that at Beaverlodge, while the person running the drill would be expected to get the highest dose, this was not necessarily the case. In 1990, the Atomic Energy Control Board asked if exposures could be reconstructed from the records. The Schwartzwalter Mine in Colorado is similar to the Beaverlodge Mines.

Chambers then discussed the estimation of WLM for Beaverlodge miners. Exposure was estimated by year and type of workplace. Men tended to migrate from one mine to another in the Beaverlodge area and this was not accounted for in the exposure reconstructions. There were six or seven mine areas and nine work-type categories. The reconstructed exposures were compared to the original estimates used in the epidemiologic report of Howe and colleagues. The correlation was strong, although the original exposure estimates tended to be less than the revised exposure estimates. Means rather than medians of individual measurements were used in the new estimates. A summary of observations in the cohort was provided. Many of the miners lived in homes built on uranium-containing foundations. A positive correlation was found between WLMs and konimeter data for particle counts. The category of a miner at first work was a factor (miners needed previous work experience).

Finally, Chambers provided some remarks concerning the Colorado Plateau study. In the Colorado mines, exposures were deliberately overestimated (see Lundin and others 1971). For other hardrock mining, exposures at concentrations of 1.0, 0.5, or 0.3 WL were assigned. A crude trace of work histories of 29 people was presented from Archer, 1966. The uncertainties were related to the location of the miners and their work histories. Studies in the past have not sufficiently considered the uncertainties in the exposures. The exposures after 1969 were probably not that significant, particularly in comparison to earlier years.

## Jonathan Samet (Baltimore, Maryland)

Samet began by recommending a book, *Uranium Frenzy* by Raye C. Ringholz (University of New Mexico Press), for a review of the early history of the uranium-mining industry. He reviewed the measurements made in the New Mexico mines. Measurements were first made in the 1950s and by 1961-1967, WL measurements were being made routinely by the State Mine Inspector, the State Health Department, and the industry. He showed examples of data sheets for the years through 1967. Person-weighted totals of the individual measurements were made. These were referred to as Total Mine Indexes; the data sheets also included measurements by the type of area. Samet showed the scatter in the data and the substantial variation in the measurements made within individual visits. Jim Cleveland mentioned that inspections by the State Agencies were unannounced. Attention was not paid to ventilation if the inspection results were satisfactory. Exposures in the Colorado Plateau were grossly underestimated by small companies.

## Discussion Of Exposure Estimates

Samet reminded the participants that time-dependent errors affect the exposure estimates with implications for the committee's modeling. Hopke commented that there is ample qualitative information on errors, but it is not clear what should be done quantitatively. Brenner indicated that the uncertainty appears to be more of a problem in the small mines. Can mine size be incorporated on the basis of the total number of miners? Gilbert mentioned that the committee may want to develop a questionnaire. Can we remove certain cohorts for which the uncertainty was largest? Can we rank cohorts? Can we obtain additional data? Chambers mentioned that the presence of other factors in the dust should be considered. Lubin mentioned that data on arsenic are available for the China and Ontario cohorts. The Swedish study has some information on silica and the Czech data have some additional information as well. Chambers mentioned that he had inquired about this issue (other contaminants) in the mines that he visited. Do higher levels of exposure to radon progeny entail higher exposures to other agents? Has there been consideration of parallel analyses of entire cohorts versus the group with exposure to uranium alone. Ziemer asked "How consistent are the ore forms"? There appears to be a difference from location to location. Bill Chenoweth has a report (PP-320) describing measurements made by the U.S. Geological Survey of trace metals in different mines. There are data available from Union Carbide in Grand Junction from retired ventilation engineers and state records (by engineers such as Vern Bishop, Bob Beverly, and Ben Kilgore). Umetro Minerals Corp. is now doing restoration. Lubin asked—How common was it that workers worked at their own mines on weekends and holidays? Hopke asked—What do we mean by "other hardrock"? Gold and silver mines were common, as opposed to vanadium (or copper mines). The Port Radium mine was reopened as a silver mine (Chambers). The participants were reminded that AMSA keeps records on other metals in mines (Nelson).

What fraction of mines used diesel? Jim Cleveland answered that Kerr-McGee was almost all electric until the last few years; United Nuclear was mostly diesel. The Ambrosia Lake mines were mostly diesel. Was the use of diesel always accompanied by better ventilation? Yes. Good ventilation was needed for the diesel. Diesel was used in the '60's. Typically the miners blasted at noon and at 8:00 pm. They could not leave the lunchroom until 30 minutes after blasting. There was no diesel used until 1958 on the Colorado Plateau. By 1971, diesel use was as much as 90% and reached 100% by 1980. What were other sources of particles in the air of the mines? Drilling and diesel were the major causes. The drilling did not fracture sandstone sand grains and the rock was very wet (20% moisture) at Ambrosia Lake. Slushing created some dust. Colorado mines were mostly dry mines with substantial dust. Mucking, pushing ore out the haulway, dropping it down chutes, and like activities, all created dust. In the late 1970s and early 1980s, gravel was hauled in for building good roadways. Was the mineralogy about the same throughout the Colorado mines? No, the mines differed in vanadium and other metals.

# Appendix F

# Exposures Other Than Radon in Underground Mines

## OVERVIEW

Underground miners are exposed to a number of agents, in addition to radon progeny which may adversely affect the lung. Several of these agents are known or suspect carcinogens (arsenic, diesel exhaust, and silica), and some may cause airways inflammation (blasting fumes and diesel exhaust). Silica exposure causes silicosis and several investigations have assessed modification of the effect of radon progeny by the presence of this fibrotic disorder.

These exposures of miners, in addition to radon progeny, are a source of uncertainty in extending risk estimates based on the epidemiologic studies of miners to the general population. Inflammatory changes in the epithelium might non-specifically affect the risk of lung cancer from radon progeny and the additional exposure to other carcinogens might alter the risk of radon progeny as well. These other exposures were considered in the BEIR IV report (NRC 1988) and subsequently in the radon dose panel report (NRC 1991).

In this appendix, we update the earlier reviews for exposure to arsenic, silica, and diesel exhaust. Information on exposures of the miners to the agents is limited and only a few studies provide human information on arsenic and silica. None of the studies have direct information on exposure to diesel exhaust. The limited data available on these exposures are summarized by cohort in appendix D. Use of diesel engines in U.S. mines is described in the workshop summary that is part of appendix E annex 2. The more general topic of interactions between agents is addressed in appendix C in considering the combined effect of cigarette smoking and radon.

## Arsenic

Although evidence in experimental animal studies of the carcinogenicity of arsenic is limited, there is substantial evidence that inorganic arsenic is a carcinogen in humans (Blot and Fraumeni 1994; IARC 1987). Neubauer (1947) reports that arsenic was suspect as being carcinogenic as early as 1879 as a result of high lung-cancer rates in German miners (Bates and others 1992; Furst 1983). The ingestion of arsenic in drinking water and in pharmaceuticals has been associated with a number of disease outcomes, such as liver angiosarcoma and meningioma, and cancers of the skin, bladder, kidney, and colon, as well as black-foot disease (IARC 1987). Studies have also clearly shown that inhaled arsenic (arsenic trioxide) is a human lung carcinogen (IARC 1987). The principal concern for this committee is the role of exposure to airborne arsenic in mine dusts as a primary risk factor for lung cancer, and how its presence might affect the evaluation of the relationship between radon-progeny exposure and lung cancer.

Occupational studies have been the main source of data on the effects of exposure to arsenic and risk of lung cancer. These studies have included workers manufacturing and using arsenical-containing pesticides (Hill and Faning 1948; Roth 1957; Ott and others 1974; Mabuchi and others 1985), smelter workers and underground miners (for summaries, see Blot and Fraumeni 1994 and IARC 1987). Although the majority of occupational studies of arsenic exposure have been conducted in smelter workers, an increased risk of lung cancer with arsenic exposure has been observed in several studies of miner populations (Taylor and others 1989; Kusiak and others 1993; Xuan and others 1993; Enterline and others 1987; Simonato and others 1994). However, among the studies of miners, only the investigations of Chinese tin miners (Xuan and others 1993) and Ontario uranium miners (Kusiak and others 1993) have included a quantitative evaluation of arsenic and of the joint association of arsenic and radon-progeny exposure.

Although studies have consistently shown an increasing risk of lung cancer with greater cumulative exposure to arsenic, there have been few detailed analyses of the shape of the dose-response curve for arsenic exposure. The analysis by Enterline and others (1995) and a meta-analysis of published studies (Hertz-Picciotto and Smith 1993) suggested a curvilinear relationship with a decrease in the excess relative risk per unit exposure as exposure increases, that is, the exposure-response curve was concave from below. Analyses of the Ontario miners (Kusiak and others 1993) and Chinese miners (Lubin, communication to the committee) showed a similar concave relationship, even after adjustment for radon-progeny exposure.

The distribution of histological types of lung cancer in arsenic-exposed populations has not been extensively studied. There have been several small investigations, with little consistency in their finding. Based on 25 cases, Newman and others (1976) reported a higher proportion of poorly differentiated epidemiod carcinoma, while Wichs and others (1981) studied 42 smelter workers and 42 matched controls and found an excess of adenocarcinomas. In contrast, in a larger study of 93 lung-cancer cases highly exposed to arsenic and 136 referent lung-cancer cases, Pershagen and others (1987) found no variation in the histo-logical distribution of lung-cancer cases when data were classified by a measure of arsenic exposure. The distributions of histological type in underground miners have been reported, but are potentially confounded by smoking and radon-progeny exposure.

Mathematical models, based on the Armitage-Doll multistage theory for carcinogenesis (Armitage and Doll 1961), were applied to data on lung cancer from two studies of copper-smelter workers in Tacoma (Mazumdar and others 1989) and in Montana (Brown and Chu 1983). Both analyses drew similar conclusions, namely, arsenic exposure acts primarily as a late-stage carcinogen, but that the possibility of an early-stage effect cannot be ruled out. However, one limitation of both analyses was the inability to directly incorporate cigarette-smoking into the modeling, a factor which is thought to act as both an early- and late-stage carcinogen.

In the miner pooled analysis by Lubin and others (1994a), adjustment for arsenic exposure reduced the ERR/WLM in the Chinese miners from 0.61% to 0.16%. Interpretation of the reductions is hampered by the high correlation coefficient, 0.48, between cumulative radon-progeny exposure and arsenic expo-sure among jointly exposed miners. This suggests that the best estimate of the ERR/WLM for the radon progeny exposure-lung cancer relationship lies between 0.0061 and 0.0016. In the Ontario data, adjustment for arsenic exposure reduced the ERR/WLM from 0.0093 to 0.0084. The correlation coefficient between radon-progeny exposure and arsenic exposure was 0.02. After adjustment for arsenic exposure as a primary risk factor, the ERR/WLM did not vary signifi-cantly with level of arsenic exposure in either study (Lubin and others 1994a). This pattern is consistent with a multiplicative association between radon-progeny exposure and arsenic exposure. However, interpretation of these results is ham-pered by differences in definition of the arsenic-exposure measure, which was percent arsenic in rock multiplied by duration of exposure for the Ontario study and duration of arsenic exposures ($mgm^{-3}y$) for the China study. The evidence appears to suggest a greater than additive (synergistic) association for the com-bined relative risks for cigarette use and airborne arsenic exposure (Hertz-Picciotto and others 1992). In miner populations, the joint association of the three factors, radon progeny, arsenic, and smoking, has not been evaluated.

## Silica

Silica, a ubiquitous exposure in many types of underground mining, is of particular interest in that it not only causes silicosis but also has been identified as a suspect human carcinogen by the International Agency for Research on Cancer (IARC 1987). In classifying crystalline silica as carcinogenic, IARC indicated that evidence of silica was sufficient in animals while limited in humans. For a detailed review of silica and lung cancer, see Goldsmith and Samet (1994). Abelson (1991) has identified silica in mines as one of the key factors contributing to uncertainty in the use of radon-associated lung cancers for miners to estimate population risks for radon. Silica might modify the risk of radon directly as an additional carcinogenic exposure or indirectly by causing fibrosis and airways damage.

With regard to this possible indirect mechanism, there have been several studies on respiratory disease patients that suggest a significant association between obstructive lung function and lung cancer (Davis 1976; Skillrud and others 1986; Tockman and others 1987). Similar findings have been reported for pneumoconiotic workers by Harber and others (1986) and by Carta and others (1991). Carta and others (1994) suggest that "airways obstruction may be an independent risk factor for bronchogenic carcinoma." Accordingly, they studied the lung-cancer mortality in relation to airways obstruction among Sardinian metal miners exposed to silica and low levels of radon progeny. In one of the two mines studied, the quartz concentration in the respirable dust was between 0.2% and 2.0% while the radon exposures averaged 0.07 $Jm^{-3}$ (0.13 WL) with the maximum cumulative exposure in the 0.28 to 0.42 $Jhm^{-3}$ (80-120 WLM) range. In the second mine, the silica levels were much greater, ranging from 6.5% to 29%, while the radon levels were lower than in the first mine. The cohort included some 1,741 miners and a total of 25,842.5 person-years of exposure. Lung function tests, chest radiographs, and smoking histories were available for all subjects entering the cohort. A total of seventeen subjects from the first mine and seven from the second died of lung cancer. The standardized mortality ratio (SMR) for lung cancer was higher for the first mine. Furthermore, among miners with initial spirometric airways obstruction, those in the first mine showed the highest risk. Carta and others concluded that crystalline silica as such does not affect lung-cancer mortality. They further suggest that impaired pulmonary function may be an independent predictor of lung cancer and may be an important risk factor because of enhancement of residence times for inhaled carcinogens.

An important investigation that considered silica dust and silicosis as risk factors for lung cancer in underground miners was reported by Radford and St. Clair Renard (1984). They conducted a case-control study of silicosis in Swedish iron miners involving 50 lung-cancer cases in deceased miners and 100 controls matched on age, year mining began, and duration of time mining. Both the severity of silicosis and the frequency of radiographic evidence of silicosis were comparable for the cases and the controls, indicating no effect of this disease on lung-cancer risk.

Epidemiological evidence of increased lung-cancer risk in silicotic patients has been reported by Koskela and others (1990) as well as by Chiyotani and others (1990) and Merlo and others (1990). However, there have been a number of studies that present conflicting results on lung-cancer risks for workers with and without silicosis exposed to dust that contained silica. These studies include: Hessel and others (1990), Meijers and others (1990), Ng and others (1990), Ahlman and others (1991), Amandus and Costello (1991), Carta and others (1991), Chen and others (1991), Chia and others (1991), Hnizdo and Sluis-Cremer (1991), Kusiak and others (1991), and McLaughlin and others (1992). Generally these studies demonstrate no clear dose-response relationship for silica exposure even though an overall association between lung cancer and the presence of silicosis was observed in some of the studies.

Samet and coworkers (1994) conducted a case-control study in the cohort of underground uranium miners in New Mexico to assess the presence of radiographic silicosis as a risk factor for lung cancer. This is one of the cohorts included in the pooled data set. The presence of silicosis as determined by chest radiographs taken at or near the beginning of employment was determined for 65 lung-cancer cases and 216 controls. Data on the individual exposures to silica were not available, but there are data available that demonstrate the presence of silica in mines in the region of the study. Also, silicosis is well documented in underground uranium miners in the southwestern states. The study showed that the presence of silicosis was not associated with lung-cancer risk after adjustment was made for cumulative exposure to radon. These investigators recognized that the findings were limited by the small number of subjects, but they were able to conclude nonetheless that there was a lack of association of silicosis with lung cancer. They stated that "silica exposure should not be regarded as a major uncertainty in extrapolating radon risk estimates from miners to the general population."

Finkelstein (1995) examined the presence of radiographic silicosis as a lung-cancer risk-factor in miners from the Ontario Silicosis Surveillance Database. In contrast to the findings of Samet and others (1994), he found that silicosis was a highly significant risk factor for lung cancer. Accordingly, he concluded that the radon lung-cancer risk decreased if an adjustment for the presence of silicosis was made. However, Archer (1996) has criticized Finkelstein's conclusion on the basis that early lung cancer is very difficult to discern from radiographs of individuals whose lungs contain fibrotic abnormalities. Archer states that it is likely that at the time they were admitted into the study the silicotics in Finkelstein's cohort had more undetected cancers than did the controls. Archer also criticized Finkelstein's assumption that radon exposures for the non-uranium miners was zero.

Recently, Enderle and Friedrich (1995) published a review of the exposure conditions and the health consequences for the East German uranium miners in the Saxony and Thuringia regions. They point out that in the 1946 to 1955 period working conditions were extremely poor and the miners were exposed not only to radon progeny, but also to very high dust levels, and to toxic chemicals or elements including arsenic and crystalline silica. They offer no direct evidence relating silica and lung cancer for these miners, but they do cite a study by Melhorn (1992) that reports a high rate of bronchial carcinoma occurring in miners with known silicosis. They also cite the work by Tockman and Samet (1994) who describe silicosis as a risk factor for lung cancer.

Goldsmith and coworkers (1995) have also shown that, in addition to having increased mortality from nonmalignant respiratory diseases and from tuberculosis, silicotics have a significantly elevated risk of death from cancers of the trachea, bronchus, and lung.

### Diesel Engine Exhaust and Fumes

Exposure to diesel is also relevant to extrapolation of risks from miners to the population. Some uranium-mining operations used diesel engine-powered equipment resulting in the exposure of miners to diesel exhaust. As will be discussed below, the diesel soot particles are readily respirable. They are carbonaceous particles and have associated hydrocarbons some of which are mutagenic and also carcinogenic. This raises the potential for the diesel soot to be carcinogenic, and further raises the possibility that diesel exhaust may induce lung cancer. In turn, this raises the possibility for diesel exhaust to be a confounding factor in evaluating the lung-cancer risks of exposure to radon.

In this section, the evidence is reviewed for diesel exhaust causing lung cancer. This is followed by a discussion of the possible role of diesel exhaust as a causative factor in lung cancers observed in uranium miners.

The diesel engine, patented by Rudolph Diesel in 1892, has found wide use in commerce, including use in mining operations and in railroad locomotives. The dieselization of railroads occurred principally after World War II, reached its midpoint in 1952, and by 1959, approximately 95% of the locomotives in the United States were diesel powered (U.S. Department of Labor 1972).

Concern for health effects of exposure to diesel exhaust has existed for some time. This concern relates to the readily inhalable size of diesel soot particles, 0.1 to 0.5 $\mu g$ (Cheng and others 1984), giving concern for the development of lung cancer. This concern is heightened by an awareness that a significant portion, typically 10 to 15%, of the diesel soot particles by weight consist of organic compounds readily extractable by organic solvents (Johnson 1988). The extracted material includes many polycyclic aromatic hydrocarbons including many that are mutagenic and some that are carcinogenic (Schuetzle and Jensen 1985; Schuetzle and Lewtas 1986). Kotin and others (1955) demonstrated that organic solvent extracts of diesel soot were carcinogenic when painted on mouse skin.

The prospect for increased use of diesel engines in light-duty vehicles in the late 1970s increased concern for the cancer risks of inhalation exposure to diesel soot. This concern stimulated the conduct of epidemiological investigations, bioassays in laboratory animals, and a wide range of mechanistic studies at all levels of biological organization from cells to populations of mammals.

The epidemiological studies have been recently reviewed by Cohen and Higgins (1995) and Nauss and others (1995) in a special report prepared by the

Health Effects Institute (1995). Two figures from that report provide a summary of the currently available data on lung-cancer risks evaluated in railroad workers (Figure F-1), and truck drivers (Figure F-2). From these figures, it is clear that the relative risk of lung cancer measured in the various studies is only elevated significantly if at all in a few studies. A major confounder in these studies, as is usually the case, is cigarette-smoking which is a dominant causative factor in lung cancer. This is illustrated by considering Table F-1 taken from Garshick and others (1987). The slightly elevated lung-cancer risk (odds ratio = 1.41, 95% CE = 1.06, 1.88) contrasts sharply with the substantial risk measured for cigarette-smoking. Cigarette-smoking risk increased with amount of cigarette smoking and age to an odds ratio of 9.14, 59% CE = 6.11, 13.70 for cases age greater than 65 years and >50 pack-years of cigarette smoking. Crump and others (1991) reanalyzed the data used by Garshick and others (1987) as well as additional data on the same population and was unable to discern an exposure-related increase in lung-cancer risk.

In the late 1980s, results of a number of well-conducted laboratory animal bioassays of diesel exhaust became available. These results have been extensively reviewed (Mauderly 1992; Health Effects Institute 1995; McClellan 1987). The results, summarized in Figure F-3 taken from the HEI report (1995), clearly indicate that long-term high-level exposure to diesel exhaust increases an excess of lung cancer in rats. Mice and Syrian hamsters similarly exposed have yielded negative or equivocal results. An excellent example of this contrasting result is apparent from the studies of rats (increased lung cancer) and mice (no increase in

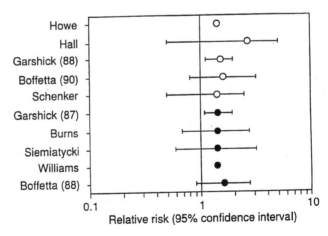

FIGURE F-1   Lung cancer and exposure to diesel exhaust in railroad workers. ● = RR adjusted for cigarette-smoking; ○ = RR not adjusted for cigarette-smoking. For two studies (adapted from Nauss and The HEI Diesel Working Group, 1995).

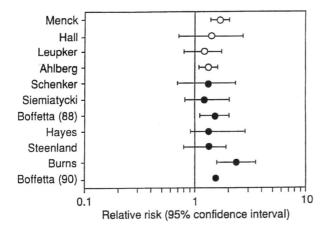

FIGURE F-2   Lung cancer and exposure to diesel exhaust in truck drivers. ● = RR adjusted for cigarette smoking; ○ = RR not adjusted for cigarette smoking. For the study by Williams (William and others 1977), CLs were not reported and could not be calculated. For the Steenland study (Steenland and others 1992), the data were gathered from the union reports of long-haul truckers; for the 1988 Boffetta study (Boffetta and others 1988), the data were self-reported by diesel truck drivers; and for the Siemiatycki study (Siemiatycki 1991), they were self-reported by heavy-duty truck drivers (personal communication).

lung cancer) reported by Mauderly and others (1996) to diesel exhaust from the same source.

As the significance of the diesel exhaust rat lung-cancer findings was discussed, it was noted that chronic inhalation exposure of other particulate materials (lacking in capability to directly damage DNA) also caused an increase in lung cancer in rats (Vostal 1986). This raised questions as to the mechanisms by which diesel exhaust and these other materials might be acting. It was speculated that the effects of these materials might be related to their ability, when inhaled at high concentrations, to overload lung-clearance mechanisms and cause chronic inflammation and, ultimately, lung cancer (Vostal 1986; Morrow 1988; McClellan 1990).

To test this hypothesis, studies were conducted in which rats were chronically exposed to carbon black particles, which were relatively devoid of mutagenic organic compounds. Two major laboratories found that carbon black had about the same effectiveness as diesel exhaust in producing lung cancer in rats. Recently, Driscoll and others (1996) and Oberdörster (1996) have shown that exposure to high concentrations of carbon black produced persistent pulmonary inflammation, and an increase in mutations in lung epithelial cells. These results provide a plausible mechanism for the pathogenesis of the particle-induced lung cancer in rats. This is illustrated schematically in Figure F-4.

APPENDIX F

TABLE F-1    Regression results using diesel exhaust exposure as a single continuous variable (diesel-years) adjusted for cigarette-smoking and asbestos exposure

| Exposure Category | Odds Coefficient | Ratio | 95% Cl | p Values |
|---|---|---|---|---|
| Case age ≤64 | | | | |
| Diesel-years | 0.01719 | 1.41[a] | 1.06, 1.88 | 0.02 |
| Asbestos, Y/N | 0.18111 | 1.20 | 0.87, 9.65 | 0.27 |
| ≤50 pack-years[b] | 1.19196 | 3.29 | 1.57, 6.93 | <0.01 |
| >50 pack-years[b] | 1.73606 | 5.68 | 2.73, 11.80 | <0.01 |
| Pack-years missing[b] | 1.37975 | 3.97 | 1.86, 8.51 | <0.01 |
| Case age ≥ 65 | | | | |
| Diesel-years | –0.00461 | 0.91[a] | 0.71, 1.17 | 0.47 |
| Asbestos, Y/N | –0.01807 | 0.98 | 0.81, 1.20 | 0.86 |
| ≤50 pack-years[b] | 1.47641 | 4.38 | 2.90, 6.60 | <0.01 |
| >50 pack-years[b] | 2.21321 | 9.14 | 6.11, 13.70 | <0.01 |
| Pack-years missing[b] | 1.35379 | 3.87 | 2.56, 5.84 | <0.01 |

[a]Calculated on the basis of 20 years of exposure.
[b]Reference category of zero pack-years (never-smokers).
From Garshick and others (1987).

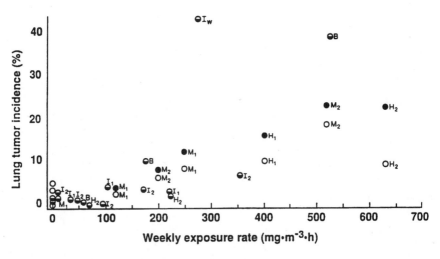

FIGURE F-3   The relation between rat lung-tumor incidence and exposure rates for diesel exhaust particulate matter.  Data point code is: B = Brightwell and others 1989; $H_1$ = Heinrich and others 1995; $I_1$ = Ishinishi and others 1986; (exhaust from 1.8-L engine); $I_2$ = Ishinishi and others 1986 (exhaust from 11-L engine); $I_w$ = Iwai and others 1986; $M_1$ = Mauderly and others 1987; $M_2$ = Mauderly and others 1994. ● = Includes lesions identified by the investigator as "benign squamous tumors"; ○ = excludes these lesions.

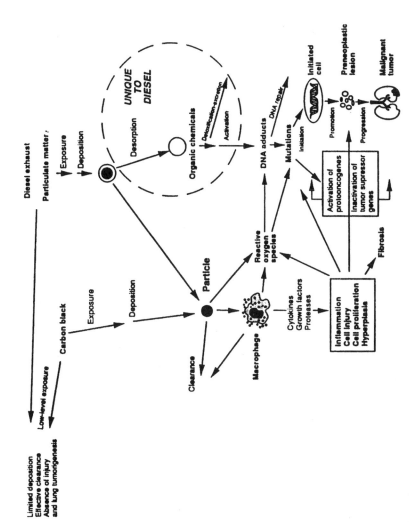

FIGURE F-4  Schematic representation of the pathogenesis of lung cancer in rats with prolonged exposure of high concentrations of diesel exhaust or carbon black particles.  From Health Effects Institute (HEI 1995).

The recent cancer results from rats exposed to diesel exhaust raises questions as to the appropriateness of their use for defining the carcinogenic risk of diesel exhaust to humans (McClellan 1996). Thus, the human lung-cancer risk of diesel exhaust exposure should be based exclusively on the epidemiological data reviewed earlier.

With the above information as background, the potential interaction between radon and diesel exhaust can be considered. In the absence of either epidemiological studies of radon-exposed individuals that have included characterization of diesel exhaust exposure as a confounder or laboratory animal studies with controlled exposure to radon and diesel exhaust, it is only possible to speculate on the potential combined effects of radon and diesel exhaust. Some insight may be gained by considering the data in Table F-1 from the study of railroad workers exposed to diesel exhaust. The results of this study can be interpreted as identifying diesel exhaust as a potential low-potency carcinogen. The characterization of diesel exhaust as a low-potency carcinogen is made with reference to cigarette-smoking that was substantially more potent than diesel exhaust depending on the extent of smoking. It is unlikely that the uranium miners had smoking histories substantially different from the railroad workers. Compared to diesel exhaust, radon exposure may also be classed as a high-potency risk factor. It can be argued that in an exposure environment involving the two high-potency risk factors, radon and cigarette-smoking, the addition of a low-potency risk factor, diesel exhaust, would be unlikely to affect the combined risk from the two high-potency risk factors.

Muscat and Wynder (1995) conducted a case-control study to determine the effects of exposure to diesel engine exhaust and fumes. The subjects were truck drivers, mine workers, firefighters, and railroad workers, and included 235 male hospital patients with laryngeal cancer. These investigators showed that diesel engine exhaust is unrelated to laryngeal cancer risk. They offered no suggestion that lung cancer would be directly related to diesel fume exposures.

## Mycotoxins

Recently there has been speculation about the possible role of mycotoxins in the production of lung cancers. In a letter to the editor of *Lancet*, Venitt and Biggs (1994) suggested that exposure of uranium miners to mycotoxins such as sterigmatocystin could account for the mutations in p53 at codon 249 that had been reported by Taylor and coworkers (1994). Taylor and others had suggested that the codon 249 mutation may be a marker for radon-induced lung cancer, but Venitt and Biggs point out that the gross damage caused by $\alpha$ particles would be expected to produce gross damage to the DNA rather than a precise mutation at a specific codon. Likewise, Hei and others (1994b) suggest that although a point mutation could be induced by $\alpha$ particles, complete loss of the p53 gene would be more likely. Bartsch and others (1995) also assert that a radon-induced hotspot mutation would be surprising since one would expect mainly random DNA strand breaks. They screened for the presence of the codon 249 mutation in lung cancers from the Saxony, Germany uranium miners and found that none of the 50 lung tumors analyzed showed the hotspot mutation. Lo and others (1995) also raise the possibility that the results found by Taylor and colleagues could be related to mycotoxins.

Hypotheses concerning a possible role of mycotoxins are presently speculative and not supported by any observational data. Information on exposures is completely lacking.

Sram and coworkers (1993) have reported that Czech uranium miners are exposed to chemical mutagens as well as radon. They found molds in throat swabs from 27% of the miners studied as compared to only 5% in controls. Various mycotoxins were found in the swabs, including sterigmatocystin, a bisfuranoid mycotoxin that is structurally related to aflatoxins. Sterigmatocystin is reported by Gopalakrishnan and others (1992) to be a potent carcinogen and mutagen that produces squamous carcinomas and adenocarcinomas in animal lungs.

## SUMMARY

Exposures other than radon progeny sustained by underground miners could plausibly modify the lung-cancer risk associated with exposure to radon progeny. The relevant data for assessing such modification in the miner cohorts are scant. Uncontrolled arsenic exposure may be a source of positive bias, as shown for example in the Chinese tin miners. The role of silica has not been directly assessed; the scant epidemiologic evidence indicates that the presence of silicosis is not a strong modification of the risk of radon. Diesel exhaust, present in some of the more recent miners, was also probably not a strong modifier of the risk of radon progeny.

# Appendix G

# Epidemiologic Studies in the Indoor Environment

This appendix examines the epidemiologic evidence of an association between indoor radon-progeny exposure and lung cancer. Although data from indoor-radon studies are not yet sufficient to develop a general risk-assessment model or to estimate precisely the magnitude of risk posed by radon in houses, the data do support a small increase in lung-cancer risk due to indoor radon exposure and are consistent both with the extrapolation of lung-cancer risk using miner-based models and with relative risks among miners with cumulative exposures similar to exposures that might be experienced by long-term residents in houses that exceed the Environmental Protection Agency (EPA) action level. However, there are sufficient uncertainties in current epidemiologic studies that the residential data alone do not conclusively support a definable excess lung-cancer risk associated with radon-progeny exposure.

Ecologic studies and analytic case-control studies are the 2 types of epidemiologic studies that have considered the issue. In an ecologic study, regional rates of lung cancer are related to a measure of regional radon concentration. The measures of radon concentration are regional mean radon concentrations obtained from direct measurement in a small number of houses and purported correlates of indoor-radon concentration, such as geologic formations and housing characteristics. In an analytic case-control study, data are obtained directly from lung-cancer cases and controls, or their surrogates, through personal interviews. Radon-progeny exposure is estimated for each person and is based on either direct data from indoor-radon measurement or surrogate measures, such as housing type.

The committee concludes that only analytic case-control studies that rely on direct measurement of radon in houses are useful for evaluating the risk of lung cancer posed by indoor-radon exposure. In contrast, because of the inability to control for confounding at the level of the individual, limitations in the use of a few radon measurements to represent exposures for an entire region, and the large risk associated with cigarette-smoking (an excess lung-cancer risk of 1,000-2,000%, with an estimated 20-30% for indoor radon-progeny exposure), the committee believes that ecologic studies of indoor-radon exposure and lung cancer are essentially noninformative and shed little light on the association of indoor radon-progeny exposure and lung cancer.

In this appendix, we review the sources of exposure to radon progeny in the general population and the epidemiologic studies of indoor exposure, and we consider the results of the epidemiologic studies and their design limitations.

## SOURCES OF ENVIRONMENTAL RADON EXPOSURE AMONG NONMINERS

The principal source of radon-progeny exposure in buildings is emanation from soil and rock below ground. In a few special situations, well water or building materials can contribute substantially but they make relatively small contributions to the overall dose (NCRP 1984).

The most-complete survey of radon concentrations in U.S. dwellings, the National Residential Radon Study (NRRS), was performed by EPA in 125 counties in 50 states (Marcinowski and others 1994). The arithmetic mean radon concentration was 46.3 $Bqm^{-3}$ (geometric standard deviation, 3.11); the EPA action level is 148 $Bqm^{-3}$. Figure 1-4 from Marcinowski and others (1994) shows that single-family homes have a slightly higher radon concentration (arithmetic mean 54.0 $Bqm^{-3}$; geometric standard deviation, 2.97) than all dwellings, whereas multi-family units (defined as attached single-family dwellings, townhouses, apartments, duplexes, and condominiums) have a markedly lower mean radon concentration (arithmetic mean, 24.1 $Bqm^{-3}$; geometric standard deviation, 3.23). In the survey, 6.1% of houses exceeded the EPA action level; this confirms earlier estimates, which were based on smaller studies compiled by Nero and others (1986), but it is much lower than estimates from commercial test vendors of 19% (Cohen and others 1984; Cohen and Gromicko 1988) and 23% (Alter and Oswald 1987). Presumably, the latter higher estimates result from a selection bias of homeowners who suspect they have a radon problem and not from biased measurement. Also in agreement with the earlier findings of Nero and others (1986), the NRRS found that the distribution of radon concentrations in residences could be satisfactorily represented by a lognormal distribution.

The NRRS also confirmed earlier studies that indicated that basements have higher average radon concentrations than higher floors. Their data showed that average radon concentrations in first-floor rooms were about 40% of those in basements, average second-floor room concentrations were about 90% of those of first floors, and average third-floor or higher room concentrations were about 84% of those of second floors.

There is no evidence that average radon concentrations in U.S. dwellings are significantly different from those in other nontropical countries. The most-comprehensive international survey was compiled in the 1988 UN Scientific Committee on Effects of Atomic Radiation (UNSCEAR) report (UNSCEAR 1988), which analyzed worldwide radon surveys. The population-weighted arithmetic mean in temperate and high-latitude countries was estimated to be about 50 Bqm$^{-3}$, which was combined with a "guesstimate" of about 20 Bqm$^{-3}$ in tropical countries to yield a worldwide population-weighted arithmetic mean of about 40 Bqm$^{-3}$.

Within a given dwelling or building, the radon concentration is determined essentially by the ratio (Scott 1992) of the average radon concentration in soil gas near foundation openings ($C_{soil}$) to the airflow resistance of soil around the house foundation ($I_{soil}$). That ratio is often called the soil-radon potential (SRP), and much effort has been devoted to its characterization. $C_{soil}$ depends on depth, soil radium concentration, and water content (Rogers and Nielson 1991), and it is not well correlated with radium or uranium concentrations in the underlying bedrock (the poor correlation makes SRP predictions based on gross geologic considerations rather inaccurate). $I_{soil}$ depends essentially on the basement dimensions and the soil permeability to air, which varies widely from 10 to 18 m$^2$ for well-graded gravel, from 10 to 11 m$^2$ for sand and gravel, and from 10 to 15 m$^2$ for clay (Tanner 1990).

The various methods that have been used for estimating SRP have been reviewed by Yokel (1989). However, as the number of screening measurements of radon concentrations in homes has increased dramatically in the last decade, the need for predictive methods has correspondingly decreased, and radon-prone areas can be identified simply by analyzing the home measurements (Scott 1992, 1993). For example, the International Commission on Radiological Protection (ICRP 1993) has proposed that areas in which more than 1% of buildings have radon concentrations that are more than 10 times the national average might be designated as radon-prone.

In contrast with earlier expectations (for example, Rundo and others 1979; Cohen and Gromicko 1988), it now appears that house design and weatherproofing (both determinants of the rate of exchange of indoor with outdoor air) are not strong determinants of domestic radon concentration. For example, in a study of 2,000 British homes, Gunby and others (1993) found that house design, building-material type, and amount of weatherproofing together accounted for less than 5% of the observed variation in radon concentration; the implication is that, on the average, SRP is the dominant determinant.

The occupancy factor of schools and workplaces is about 2/7 that of homes (ICRP 1993). Thus, radon in schools and workplaces is likely to be an important contributor to the overall dose. ICRP (1993) has suggested that action levels in these buildings should be the reciprocal of that fraction, or 7/2, as high as in dwellings (and should be weighted by the relevant dose-conversion coefficients, if applicable), but this suggestion has not been implemented.

Relatively few studies of radon concentrations in the workplace have been undertaken (Cohen and others 1984; Saccomanno and others 1986; Turk and others 1986; Westin and others 1991). In general, concentrations have been lower than those in local dwellings, presumably because of the larger number of floors and the greater ventilation rates in workplaces (Nero and others 1988).

Radon concentrations in schools are of particular interest because of the possible variation in radon susceptibility of children, compared with adults (Probart 1989). During 1990 and 1991, EPA undertook a randomized national survey of radon concentrations in U.S. public schools (the National Schools Radon Survey, or NSRS, EPA 1993). Of a random sample consisting of 927 public schools, about 19% had at least 1 classroom with a radon concentration above 148 $Bqm^{-3}$, and 2.7% of all schoolrooms had concentrations above 148 $Bqm^{-3}$. The NSRS, however, is not useful for estimating the contribution of schools to total radon-progeny exposure of children, teachers, and others from time in schools, in that measurements were conducted continuously, including on weekends and vacations, and thus failed to account for the intermittent occupancy of schools by students, teachers, and others. Typically, heating, ventilating, and air-conditioning systems would not be in use when children and teachers were away from school, but the increased concentrations during those times would not contribute to personal exposure.

## ECOLOGIC STUDIES

In ecologic studies, data are considered at the group level, rather than the individual level, as in other epidemiologic studies (Morgenstern 1995). Ecologic studies typically use existing information, such as vital-statistics data, and are therefore relatively easy to perform. Ecologic studies have been useful for generating hypothesis about environmental exposures and disease but have been used less to characterize risks. Because recognized limitations in data and approximations of the form of the regression model, ecologic analyses are not generally useful for confirmatory purposes, such as risk estimation and hypothesis-testing. Piantadosi and others (1988) present several examples from a national health and nutrition survey that showed that ecologic regression coefficients based on aggregated data are larger and smaller than regression coefficients based on individual data and have opposite signs. In the case of radon risk, limitations of ecologic studies are particularly serious because of the presence of smoking, which constitutes an overarching risk factor for lung cancer; as noted above, cigarette-smoking causes a 1,000-2,000% excess risk of lung cancer.

Soon after the potential hazard of indoor radon was first identified, a number of ecologic studies were performed and reported. The findings of these studies were mixed in their support of the hypothesized lung-cancer risk associated with indoor radon (Stidley and Samet 1993). One particularly large on-going study of lung-cancer mortality by county in the United States, reported in several papers by Cohen (1990, 1995) and Cohen and Colditz (1990), even shows an unanticipated inverse association between lung-cancer mortality by county and estimated average radon exposure of residents of the counties. Stidley and Samet (1993) reviewed the ecologic approach to indoor radon and lung cancer and considered 15 studies published as of 1992. This chapter extends their review and considers the general utility of information from ecologic studies.

## The Ecologic Study Design

In ecologic studies, the relation between exposure (radon) and disease (lung cancer) is assessed by examining the association between a measure of disease occurrence (generally the age-adjusted lung-cancer mortality) in a group of people, usually those residing in a defined geographic unit, and the extent of exposure estimated for the group. Ecologic studies have proved feasible for developing hypotheses for further testing in studies at the individual level. For example, an ecologic association between breast-cancer mortality in a number of western countries and estimates of average fat consumption led to the hypothesis that higher intakes of fat increase breast-cancer risk (Armstrong and Doll 1975). Ecologic studies may also be useful if there is general homogeneity of exposure in a population. Typically, the ecologic design has not been used to assess risks associated with exposures at the individual level.

Morgenstern (1995) has classified ecologic studies by method of exposure measurement and method of grouping. Studies are termed *exploratory* if the primary exposure of interest is not measured and the data are analyzed to identify patterns that could lead to more-specific hypotheses. *Analytic* studies incorporate the exposure of interest; studies of radon have been of this type. With regard to the method of grouping, the groups in a study may come from multiple locations (multiple-group design), from multiple periods (time-trend design), or from multiple locations and periods (mixed design). The multiple-group design has been the principal ecologic approach to the study of indoor radon and lung cancer; in the application of this approach to indoor radon, lung-cancer mortality is compared across geographic groups assumed to have a range of associated exposures to indoor radon. Stidley and Samet (1993) further classified the studies of indoor radon and lung cancer according to the primary analytic approach—comparison of disease rates in different groups classified by radon exposure or regression of disease rates on a continuous estimate of radon exposure for the group.

## Ecologic Studies of Radon and Lung Cancer

In their 1993 publication, Stidley and Samet (1993) summarized 15 ecologic studies on lung cancer and residential radon exposure. Through 1995, 4 additional studies were reported, including a study of lung-cancer mortality in U.S. counties by Cohen (1995). Details of those studies are provided in Tables G-1 through G-4: their approach to estimation of radon exposure in Table G-1, their outcome variables and controlled covariates in Table G-2, their handling of smoking in Table G-3, and their findings in Table G-4. Following the approach of Stidley and Samet (1993), we have broadly grouped the studies as "comparison" or "regression" studies on the basis of the primary analytic approach for assessing the effect of the radon exposure measure. In the comparison studies, disease rates and mortality are compared in 2 or more groups; in the regression studies, the outcome measure is modeled as a function of exposure.

Diverse approaches have been used to estimate the exposures of the groups (Table G-1). In the comparison studies, exposure rankings have been assigned to the groups on the basis of geology, measurements, or other factors. In most of the regression studies, data on indoor radon concentrations from population-based surveys or from less formally developed samples were used to assign quantitative exposures to geographic units. Background gamma radiation and radon concentration in well water were also used as surrogates.

The outcome measure in the studies was either the age-adjusted incidence or mortality from lung cancer (Table G-2). The extent to which other factors were considered in the analyses was variable. Analyses were done separately by sex, with adjustment for sex, or with restriction to one sex. Socioeconomic factors and urbanization were incorporated in some studies.

A number of the studies included measures of smoking by the members of the analytic groups (Table G-3); these measures were based on cigarette-sales information and smoking surveys.

The finding of the studies vary widely, from positive and statistically significant associations between radon-exposure measures and lung-cancer rates to negative and statistically significant associations (Table G-4). A number of studies showed no evidence of association. The studies reported by Cohen have been particularly prominent because of the large number of U.S. counties included in the analyses and the strong negative association between estimated county-average radon exposure and lung-cancer mortality. We have cited two, representative reports based on Cohen's analyses, including the most recent report (1995).

In the most-recent report by Cohen (1995), data from 1,601 counties, representing most of the U.S. population, were used. Radon exposures were assigned to the counties by combining data from 3 sources: measurements made by the University of Pittsburgh, measurements made by EPA, and measurements compiled by individual states. Potential confounding by smoking was addressed by extending 1985 data on statewide prevalence to the county level with adjustment for the degree of urbanization of the county. The potential for confounding by sociodemographic factors or their correlates was explored by stratification on levels of 54 variables. Confounding by geography was assessed by stratification,

TABLE G-1    Characteristics of radon-exposure measures in 19 ecologic studies
of lung cancer and indoor radon

| Study | Exposure Measure |
|---|---|
| *Comparison studies*<br>Archer (1987) | Proportion of county covered by Reading Prong granites |
| Bean and others (1982a,b) | Radium concentration in municipal well-water supply |
| Dousset and Jammet (1985) | Two regions differing by a factor of 3-4 in indoor radon concentrations |
| Fleischer (1981) | Proximity of phosphate mines, deposits, or processing plants |
| Fleischer (1986) | Proportion of county within Reading Prong |
| Forastiere and others (1985) | Characteristics of soil |
| Hofmann and others (1985) | Adjacent areas varying by radon and thoron concentrations |
| Vonstille and Sacarello (1990) | Indoor radon |
| Ennemoser and others (1994) | Indoor radon |
| Neuberger and others (1994) | Indoor radon |
| *Regression studies*<br>Cohen (1993) | Indoor radon concentration |

| Group measure | Exposure grouping and number of groups |
| --- | --- |
| Location | Reading Prong (7 counties)<br>Border (9 counties)<br>Control (17 counties) |
| Averages based on average of 9 measurements per town taken between 1958 and 1979 | 0-74 $Bqm^{-3}$ (2-5 $pCiL^{-1}$) (10 towns)<br>74-185 $Bqm^{-3}$ (2-5 $pCiL^{-1}$) (9 towns)<br>>185 $Bqm^{-3}$ (2-5 $pCiL^{-1}$) (9 towns) |
| Location and radon | Limousin (high)<br>Poitou-Charentes (control) |
| Location | Counties with phosphate mines (25), deposits, or processing plants (total of 316 counties) |
| Location | Mostly within (3 counties)<br>Less than half within (10 counties)<br>Adjacent counties (138 counties adjacent to counties with mines) |
| Lithology | Volcanic (27 municipalities)<br>Nonvolcanic (4 municipalities)<br>(Total population for both groups <200,000) |
| Location, radon, and thoron | High background (0.38 WLM/yr) (764,696 person-yr)<br>Control (0.16 WLM/yr) (777,482 person/yr) |
| Averages for U.S. Geological Survey quadrangles based on statewide survey of 6,500 homes commissioned in 1985 | High (99 Zip-code areas)<br>Low (1,983 Zip-code areas)<br>None (918 Zip-code areas) |
| Location | Austria Alp area (mean, 4,121 $Bqm^{-3}$ in 178 homes), remainder of Tyrol |
| Location | <8$pCiL^{-1}$, 8-10 $pCiL^{-1}$, >10 $pCiL^{-1}$ |
| Geometric means based on several surveys (some nonrandom) of homes: 39,000 measurements in living areas and 29,000 measurements in basements | 411 countries |

## TABLE G-1 Continued

| Study | Exposure Measure |
|-------|------------------|
| Edling and others (1982) | Background gamma radiation |
| Haynes (1988) | Indoor radon concentration |
| Hess and others (1983) | Radon concentration in well water |
| Letourneau and others (1983) | Indoor radon concentration |
| Ruosteenoja (1991) | Average annual indoor radon concentration |
| Stranden (1987) | Indoor radon concentration |
| Magnus and others (1994) | Indoor radon concentration |
| Cohen (1995) | Indoor radon concentration |

Source: Samet and Stidley 1993.

| Group measure | Exposure grouping and number of groups |
|---|---|
| Averages based on county random sample of 1,500 homes; published in 1987 | 24 counties |
| Averages based on county survey of 2,309 homes; published in 1987 | 55 counties |
| Averages weighted by proportion of rock type; survey of 2,000 wells | 16 counties |
| Geometric means from survey of 14,000 homes conducted in summers of 1978-1980 | 18 cities (total population about 11,000,000) |
| Geometric means based on nonrandom survey of average of 120 homes per municipality conducted by end of 1985 | 18 municipalities (total population of 59,000 males in 1980) |
| Averages based on nonrandom sample of 20 houses in each municipality during heating season, 2 locations per house | 75 municipalities |
| Average based on national sample of 7,500 homes | 427 municipalities |
| Average based on 3 data sets | 1,601 counties |

**TABLE G-2** Outcome and controlled variables in 19 ecologic studies of lung cancer and indoor radon

| Study | Outcome variable | Controlled variable | Comments on controlled variables |
|---|---|---|---|
| *Comparison studies* | | | |
| Archer (1987) | Lung-cancer mortality, 1950-1979 | Age | Rates standardized to 1970 U.S. census population. |
| | | Sex | Both sexs combined; analyses by sex gave similar results. |
| | | Ethnicity | Analysis restricted to Caucasians. |
| | | Socioeconomic | Groups "similar." |
| | | Urbanization | Counties with large cities omitted; groups "similar." |
| | | Population growth | No adjustment, but rates differed for groups. |
| Bean and others (1982a,b) | Lung-cancer incidence, 1969-1978 (1972 excluded) | Age | Rates standardized to 1970 Iowa age distribution. |
| | | Sex | Analyses done by sex. |
| | | Smoking | Not included in model, but groups checked for similarity. |
| | | Socioeconomic | Included in regression model. |
| | | Urbanization | Included towns had 1970 population of 1,000-10,000; towns categorized by size. |
| | | Water characteristics | Used as exclusion criteria or included in regression model. |
| Dousset and Jammet (1985) | Lung-cancer mortality, 1968-1975 | Age | Rates standardized to 1968 population. |
| | | Sex | Analyses done by sex. |
| | | Smoking | Similar average tobacco consumption. |
| Fleischer (1981) | Lung-cancer incidence, 1950-1969 | Age | Rates standardized to the 1960 U.S. population. |
| | | Sex | Analyses done by sex. |
| | | Ethnicity | Main analysis restricted to Caucasians. |
| | | Urbanization | Analysis included stratification by population size. |

| Study | Outcome | Variable | Comments |
|---|---|---|---|
| Fleischer (1986) | Lung-cancer incidence, 1950-1969 | Age | Rates standardized to the 1960 U.S. population. |
| | | Sex | Analyses done by sex. |
| | | Ethnicity | Analysis restricted to Caucasians. |
| Forastiere and others (1985) | Lung-cancer mortality, 1969-1978 | Age | Analyses done by age groups or used age-adjusted rates; only those 35-74 yr old included. |
| | | Sex | Analyses done by sex or used sex-adjusted rates. |
| | | Smoking | Stratified by per capita yearly cigarette sales. |
| | | Urbanization | Largest town excluded; remaining municipalities stratified by population size. |
| Hofmann and others (1985) | Lung-cancer mortality, 1970-1983 | Age | Age-adjusted rates. |
| | | Sex | Sex-adjusted rates. |
| | | Smoking | Neighboring groups; assumed similar; women did not smoke. |
| | | Socioeconomic | Neighboring groups; assumed similar. |
| | | Urbanization | Both groups rural. |
| | | Mobility | Stable populations. |
| Vonstille and Sacarello (1990) | Percentage of serious illnesses that were malignant neoplasms | Age | Age-adjusted to 1985 Florida population. |
| | | Sex | Analyses done by sex. |
| | | Socioeconomic and mobility | Limited one analyses to the lower class in an attempt to reduce effect of mobility. |
| Ennemoser and others (1994) | Lung-cancer mortality, 1970-1991 | Age | Age-adjusted rates. |
| | | Sex | Sex-adjusted rates |
| Neuberger and others (1994) | Lung-cancer incidence, 1973-1990 | Age | Age-adjusted rates. |
| | | Sex | Women only |
| *Regression studies* | | | |
| Cohen (1993) | Lung-cancer mortality, 1950-1969 | Age | Age-adjusted rates. |
| | | Sex | Analyses done by sex. |
| | | Ethnicity | Analysis restricted to Caucasians. |

TABLE G-2  Continued

| Study | Outcome variable | Controlled variable | Comments on controlled variables |
|---|---|---|---|
| | | Smoking | Included in regression model. |
| | | Socioeconomic | Several variables included in regression models. |
| | | Urbanization | Several variables included in regression models. |
| | | Mobility | In some analyses, radon exposures were adjusted to account for mobility; Blacks omitted to reduce effect of mobility. |
| Edling and others (1982) | Lung-cancer mortality, 1969-1978 | Age | Rates standardized to 1974 Swedish population >40 yr; restricted to people >40 yr. |
| | | Sex | Analyses done by sex. |
| Haynes (1988) | Standardized mortality ratios for lung-cancer mortality, 1980-1983 | Age | Age adjustment attempted through standardized rate ratios. |
| | | Sex | Analyses done by sex. |
| | | Smoking | Included in regression model. |
| | | Socioeconomic | Included in regression model. |
| | | Urbanization | Population density included in regression model. |
| | | Diet | Vitamin A consumption included in regression model. |
| Hess and others (1983) | Lung-cancer mortality, 1950-1969 | Age | Age-adjusted rates from National Cancer Institute. |
| | | Sex | Analyses done by sex. |
| | | Smoking | Concluded that smoking did not account for observed differences. |
| | | Urbanization | Concluded that population density did not account for observed differences. |
| | | Mobility and growth | Acknowledged that mobility would diminish an effect, but observed that state population has been stable in 1900s. |

369

| Study | Health outcome | Covariate | Comments |
|---|---|---|---|
| Letourneau and others (1983) | Lung-cancer mortality, 1966-1979 | Age | Rates age-adjusted to 1971 Canadian population; restricted to those 45-79 yr old. |
| | | Sex | Analyses done by sex. |
| | | Smoking | Included in regression model. |
| | | Socioeconomic | Correlated with rates. |
| | | Mobility | Study restricted to people >45 yr old to restrict effect of mobility, but mobility still high. |
| Ruosteenoja (1991) | Lung-cancer incidence, 1973-1982 | Age | Rates age-adjusted to world standard population. |
| | | Sex | Study restricted to males. |
| | | Smoking | Included in regression model. |
| | | Urbanization | All groups rural. |
| | | Mobility | Stable population. |
| Stranden (1987) | Lung-cancer incidence, 1966-1985 | Age | Age-adjusted rates. |
| | | Sex | Analyses done by sex. |
| | | Smoking | Included in regression model. |
| | | Urbanization | Oslo, Norway excluded. |
| | | House characteristics | Examined, but not controlled. |
| Magnus and others (1994) | Lung-cancer incidence, 1979-1988 | Age | Age-adjusted rates. |
| | | Sex | Analyses done by sex. |
| | | Smoking | Included in regression model. |
| | | Asbestos | Included in regression model. |
| Cohen (1995) | Lung-cancer mortality, 1970-1979 | Age | Age-adjusted rates. |
| | | Sex | Sex-specific analyses. |
| | | Smoking | Included in regression model. |
| | | Socioeconomic | Included in regression model. |

Source: Samet and Stidley 1993.

TABLE G-3   Adjustments for cigarette-smoking in 19 ecologic studies of lung cancer and indoor radon

| Study | Adjustment for cigarette-smoking |
| --- | --- |
| *Comparison studies* | |
| Archer (1987) | No adjustment, but concluded that average smoking behavior should not differ significantly among groups. |
| Bean and others (1982a,b) | By examining lung-cancer rates in neighboring towns, they concluded that neighboring towns were similar to each other with respect to smoking behavior; from analysis of controls in National Collaborative Case-Control Study, they concluded that smoking rates were lower in counties with study town with high radium concentration in water than in counties with "low"-radium town. |
| Dousset and Jammet (1985) | No adjustment, because groups did not differ in average tobacco consumption. |
| Fleischer (1981) | No adjustment, but noted that average smoking rates differ only slightly among states. |
| Fleischer (1986) | No adjustment. |
| Forastiere and others (11985) | Stratified by per capita yearly cigarette sales from 1971 survey (thus, used current smoking behavior). |
| Hofmann and others (16) | No adjustment, because groups were assumed to be similar and women generally did not smoke. |
| Vonstille and Sacarello (1990) | No adjustment. |
| Ennemoser and others (1994) | No adjustment. |
| Neuberger and others (1994) | Smoking as of 1960 estimated from data from case-control study |

*Regression studies*

| | |
|---|---|
| Cohen (1993) | Average state cigarette sales and information from state tax collections were included in regression model restricted to state averages. |
| Edling and others (1982) | No adjustment. |
| Haynes (1988) | Regression model included average weekly household expenditure on cigarettes for 1961-1963; information obtained from 1962 Ministry of Labor report, from about 20 yr before lung-cancer deaths. |
| Hess and others (1983) | No variable included in regression model, but concluded that smoking did not account observed differences in lung-cancer mortality rates. |
| Letourneau and others (1983) | Regression model included percentage of people >45 yr old who were current smokers or ex-smokers; information obtained from Canadian Labour Force Surveys in 1977, 1979, and 1981, so current smoking behavior was considered. |
| Ruosteenoja (1991) | Regression model included percentage of smokers; information obtained from recent smoking survey of men 19-70-yr old in each municipality. |
| Stranden (1987) | Regression model included average number of cigarettes smoked/d; information obtained from a 1964-1965 study, 1-21 yr before considered lung-cancer cases. |
| Magnus and others (1994) | Smoking data from 1964-1965 mailed survey. |
| Cohen (1995) | 1985 smoking survey by state, adjusted for time trend. |

Source: Samet and Stidley 1993.

TABLE G-4  Findings of 19 ecologic studies of lung cancer and indoor radon

| Study | Location of study | Findings |
|---|---|---|
| *Comparison studies* | | |
| Archer (1987) | Reading Prong and | Increase in lung-cancer mortality for counties containing Reading Prong (p < 0.01), with increase from annual rate of 23.9 per 100,000 (95% CI[a], 23.4-24.4) for control to 31.3 per 100,000 (95% CI, 30.5-32.1) for Reading Prong counties (rate ratio of 1.3[b]) |
| Bean and others (1982a,b) | Towns in Iowa | Lung-cancer incidence for males increased with increasing average radium concentration in water (p<.002); relative risk of 1.68 for males with exposure >185 Bqm$^{-3}$ (5 pCiL$^{-1}$) to those with exposure <74 Bqm$^{-3}$ (2 pCiL$^{-1}$). Relative risk of 1.45 for females not statistically significant. |
| Dousset and Jammet (1985) | Two regions in France | No difference in lung-cancer mortality between 2 exposure groups (rate ratios, 0.97 for males and 1.00 for females).[b] |
| Fleischer (1981) | U.S. counties | More counties than expected with high lung-cancer rates in group with phosphate deposits or processing plants under assumption of no association between phosphate and lung-cancer rates (p < 0.0001). |
| Fleischer (1986) | Reading Prong counties | More counties than expected with high lung-cancer rates in group mostly within Reading Prong under hypothesis of no geographic association with lung-cancer rates (p = 0.017 for males and p = 0.038 for females). |
| Forastiere and others (1985) | Towns in central Italy | Nonsignificant increase in lung-cancer mortality in volcanic area over nonvolcanic area; standardized rate ratio of 1.22 for males (p = 0.22) with 95% CI, 0.89-1.68 and standardized rate ratio of 1.24 for females (p = 0.37) with 95% CI, 1.77-1.98.[b] |
| Hofmann and others (1985) | Two adjacent areas in China | No association between lung-cancer mortality and radon exposure; 2.7 deaths per 100,000 in high-exposure group and 2.9 per 100,000 in control (rate ratio, 0.93).[b] |
| Vonstille and Sacarello (1990) | Florida | No difference in percentage of total serious illnesses that were malignant neoplasms among 3 exposure groups, with 3.6%, 5.4%, and 3.9% for males in no-exposure, low-exposure, and high-exposure groups, respectively; percentage for females were lower and with no difference. |
| Ennemoser and others (1994) | Tyrol, Austria | SMR for lung cancer increased for high-radon region vs. Tyrol total; SMR = 6.2; 95% confidence interval, 4.4-8.4. |
| Neuberger and others (1994) | Counties in Iowa | Effect of radon with smoking, radon, and histologic type. |

373

*Regression studies*

| Study | Location | Findings |
|---|---|---|
| Cohen (1993) | US counties | Negative association between lung-cancer mortality and average indoor radon concentration were as follows: -0.45 (37 Bqm$^{-3}$)$^{-1}$ (1 pCiL$^{-1}$) per 100,000 (95% CI, -0.57 to -0.33) for females; -3.38 (95% CI, -4.03 to -2.73) for males.[b] |
| Edling and others (1982) | Counties in Sweden | Positive association between lung-cancer mortality and average background gamma-radiation exposure; correlation, 0.46 (p = 0.12) for males and 0.55 (p = 0.03) for females. |
| Haynes (1988) | Counties in Great Britain | Negative association between lung-cancer mortality and average indoor radon concentration; partial correlation, -0.20 (p < 0.01) for males and -0.16 (p < 0.01) for females after adjustment for population density, social class, smoking, and diet. |
| Hess and others (1983) | Counties in Maine | Positive association between lung-cancer mortality and average radon concentration in water; correlation 0.65 (p < 0.01) for females and 0.46 (p < 0.10) for males. |
| Letourneau and others (1983) | Cities in Canada | No significant association between lung-cancer mortality and average indoor radon exposure; correlation -0.34 for males and 0.13 for females; after adjustment for smoking, estimates of $\beta_{IE}$ in Model IIa were -2.7 (95% CI, -12. to -7.5) and 0.9 (95% CI, -1.4 to -3.2) for males and females, respectively. |
| Ruosteenoja (1991) | Municipalities in Finland | No significant association between lung-cancer incidence and average indoor radon concentration; adjusted for smoking, relative risk, 1.08 for 100 Bqm$^{-3}$ (95% CI, 0.92-1.27); weighted correlation, 0.36 (p = 0.14). |
| Stranden (1987) | Cities in Norway | Positive association between lung-cancer incidence and average radon exposure; 95% CI for lifetime relative risk, 0.001-0.003 (Bqm$^{-3}$)$^{-1}$ radon. |
| Magnus and others (1994) | Municipalities in Norway | No overall association with radon; significant increase for small-cell carcinoma in women. |
| Cohen (1995) | U.S. counties | Negative association between lung-cancer mortality and average indoor radon concentration; smoking-adjusted coefficients, -7.3 per pCiL$^{-1}$ per 100,000 for men and -8.3 per pCiL$^{-1}$ per 100,000 for women. |

[a]CI = confidence interval/
[b]Some numeric results were calculated from information provided in the articles.
Source: Samet and Stidley 1993.

and the sensitivity of the findings to outliers was examined. There was a strong negative association between 1970-1979 lung-cancer mortality and the county-average radon concentrations; the association could not be explained by confounding. In interpreting this finding, Cohen proposes that the negative association implies failure of the linear, non-threshold theory for carcinogenesis from inhaled radon decay products.

## Limitations of the Ecologic Design for Investigating Indoor Radon and Lung Cancer

Methodologic limitations of the ecologic design have received extensive treatment in recent publications in the epidemiologic literature (Greenland and Morgenstern 1989; Brenner and others 1992; Greenland and Robins 1994a; Morgenstern 1995). Morganstern's (1995) review provides a framework for considering the limitations of ecologic studies of indoor radon and lung cancer. He notes that the goal of an epidemiologic study might be to draw biologic inferences about individual risks or ecologic inferences about group rates. In the ecologic studies of radon and lung cancer, the goal is to make inferences about the radon-associated lung-cancer risk of individuals, so there is a potential for cross-level bias as observations made at the group level (such as, the county level) are extended to individuals (for example, the county residents). The extension of quantitative risk estimates from ecologic studies to the individual level is also problematic. Estimated risks depend strongly on the choice of model form (Morgenstern 1995). Control of confounding might be accomplished by regression modeling (which includes stratification) or standardization, as typically done for age. However, in the context of ecologic studies, regression modeling for control of confounding might be unsuccessful unless a series of conditions are met with regard to associations among predictors and disease rates (Greenland and others 1989; Morgenstern 1995). Standardization for control of confounding might be unsuccessful unless all predictors are mutually standardized for the same confounders—a condition that requires data on joint distributions that might not be available (Morgenstern 1995). Control of confounding can be further compromised by misclassification and misspecification (Brenner and others 1992; Morgenstern 1995). Effect modification, that is, interaction effects—further complicates interpretation of ecologic estimates of risk.

Ecologic bias has long been known to be a principal limitation of the ecologic study design. This bias refers to the difference between associations at the group and individual levels (Morgenstern 1995). Ecologic bias has been given quantitative definition (Greenland and others 1989). Greenland and Morgenstern (1989) have described sources of ecologic bias in using linear regression to estimate the exposure effect; these sources include biases acting within a group, confounding by group, and effect modification by group. Other forms of bias can affect ecologic studies, including inadequate control of confounding, model misspecification, and misclassification. Morgenstern (1995) also lists a lack of adequate data, temporal ambiguity of the exposure-disease relationship, collinearity of predictor variables within groups, and migration across groups.

In their 1993 review, Stidley and Samet (1993) specifically addressed limitations of ecologic studies of indoor radon and lung cancer, covering measurement error and model misspecification. Each of the 15 studies was reviewed for 14 potential limitations in those 3 broad categories. All studies were found to have multiple limitations.

Stidley and Samet noted 5 sources of measurement error: use of current exposure to represent the biologically relevant period of past exposure, the inherent error of the measurement devices, use of an indirect measure of indoor concentrations as an index of indoor radon exposure, use of sample rather than total-population information, and estimation of individual exposure by a group indicator, the ecologic fallacy.

Within a region, radon concentrations for houses are extremely variable (Piantadosi and others 1988) and estimates of regional mean concentrations are usually derived from relatively few measurements. Radon concentrations are derived from measurements in houses, which are occupied an average of 60-70% of the time; the remaining 30-40% of people's time spent in other houses, in workplaces, outside, and so on. Time spent in unmeasured areas increases exposure uncertainty. And current measurements might not accurately reflect radon exposures of individuals over the last 30 yr or more. The effects of exposure errors can bias results in many ways; trends can increase or decrease or even reverse direction.

With regard to the fourth source of measurement error (use of sample data), Stidley and Samet showed that there are substantial probabilities of misclassifying counties or other geographic units as to exposure if only a few measurements are available. In an expanded analysis, Stidley and Samet (1994) assessed the impact of measurement error on the estimated effect of radon and on the standard error of the regression coefficient describing the ecologic relationship between radon and lung-cancer risk. They found that the effect of radon and the standard error of the effect estimate were underestimated because of measurement error; the degree of bias was greater for smaller samples. The underestimation of the standard error would tend to overstate significance levels for tests on regression coefficients.

Model misspecification refers to a biologically incorrect formulation of the relationship between radon exposure and lung-cancer risk. Possible errors in model specificaton include omission of confounders (such as, age and smoking) or effect modifiers (such as smoking), the use of inappropriate functional forms (such as linear rather than loglinear increase in risk posed by an exposure if the latter were correct), and the use of an incorrect form of the model (such as an additive model for the joint association of radon exposure and smoking rather than a supra-additive model). Effect modification at the individual level is an intractable problem at the ecologic level (Stidley and Samet 1993, 1994).

Miner data clearly indicate that the relationship of lung-cancer occurrence in a person, to cumulative exposure, is not simply linear, and that the joint relationship of radon and smoking is not additive. The "best" models indicate that the regression relationship depends on cumulative radon progeny exposure and on attained age, time since exposure, and exposure rate, although at the concentrations of radon found in homes exposure rate might be of less importance. In addition, effects of smoking are greater than additive. Thus, for analyses of aggregated data, the model for *age-specific* rates is not a simple linear regression in exposure and smoking. Ecologic regressions typically fit linear models to *age-adjusted* rates and to estimates of radon concentration and smoking. The dependence of the radon-progeny exposure effects on the various factors implies that comparisons can be made only among regions that have the same population profile for age and past radon exposures, including exposure rates, or these factors must be age-standardized to a common population. Furthermore, regions must have a similar joint distribution for radon-progeny exposure and smoking.

When the exposure-response relationship is linear, there are no group-level effects, and regressor variables are measured without error, population cumulative exposure can be used to obtain an unbiased estimate of the exposure-response parameter. However, that simplification, which theoretically might arise when the true model for individuals is linear, does not apply for radon-progeny exposure and lung cancer. The complexity of the risk model at the individual level (exposure-response effects with age-specific, time-since-exposure and exposure-rate variations and multiplicative or submultiplicative effects of smoking) does not lend itself to a simply linear approximation for aggregated data and guarantees that a linear model for age-adjusted disease rates is misspecified.

Statistical power of published reports was considered in the 1993 review of Stidley and Samet (1993). They found power to be limited for the reported studies; given the expected magnitude of effect of radon on lung-cancer risk at typical indoor concentrations, inadequate statistical power can lead to the incorrect conclusion that there is no association.

Stidley and Samet (1994) and Greenland and Robins (1994a) have further considered limitations of ecologic studies of radon and lung cancer. Using simulation, Stidley and Samet (1994) assessed the sensitivity of the ecologic design to confounding by cigarette-smoking. The average estimate of the effect of radon was negative when the correlation between radon and smoking was between -0.17 and -1.00 (Figure G-1). In an additional series of simulations, they explored the consequences of model misspecification, assessing the findings of a simple linear-regression model when the underlying model is nonlinear. They showed that age-dependent risks and smoking-specific risks can be incorrectly estimated by simple regression methods.

Greenland and Robins (1994a) considered biases that affect ecologic studies, using a number of examples based on investigating radon and lung cancer. They provide an informative example based on a multiplicative relationship between

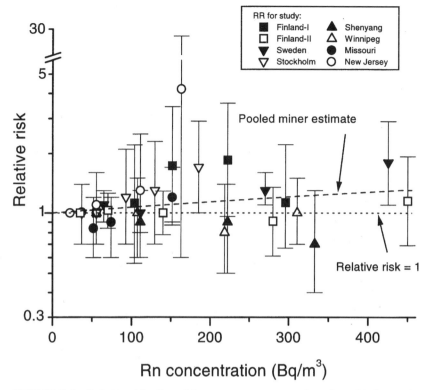

FIGURE G-1   Relative risks from 8 lung-cancer case-control studies of indoor radon. Dashed line, extrapolation of risk from miners (Lubin and others 1994); dotted line; relative risk of 1.

lung cancer and radon level and smoking that mimics the negative exposure-response results of Cohen (1995). The example is based on the following: lung-cancer occurrence is positively associated with radon concentration and cigarette-smoking rate, with the relationship linear in radon level and jointly multiplicative in radon concentration and cigarettes smoked per day; proportions of never-smokers, 1-pack/d smokers and 2-pack/d smokers vary by region; smoking rates vary by region (smoking rate is higher in regions with lower proportions of ever-smokers), so regional mean smoking rate is independent of region; and radon concentration is uniform within region but varies by region and is negatively correlated with the proportion of ever-smokers. Even though the "true" relationship specifies an increasing risk with radon concentration, and the ecologic regression of lung-cancer rates on mean regional smoking rate shows a positive exposure-response relationship, the regression of lung-cancer rates on regional radon concentration shows a *negative* exposure-response relationship.

The example does not prove that confounding from smoking is causing the negative regression in Cohen's analysis, but it shows that results of an ecologic regression can be affected by a risk factor that is confounding at the level of the individual, but not at the level of region.

That example and others in the same paper make it clear that conditions for confounding at the individual and ecologic levels are distinct and that regression methods might not fully control confounding. Greenland and Robbins also note misconceptions concerning ecologic regression: the incorrect assumption that a linear model should approximate the true model because of Taylor's theorem; failure to recognize that nonlinearities at the individual level can lead to ecologic bias; an incorrect belief that important departures from linearity in the individual-level model will be detected by a test of fit of the ecologic linear model; an incorrect belief that having a large number of analytic groups, such as, regions, will ensure a random relationship between exposure and covariates; and an incorrect assumption that for ecologic bias to be present, region must be a confounder on the individual level with control of other factors.

Cohen (1994), in responding to Greenland and Robins, dismissed those limitations as not applicable to ecologic studies of radon and lung cancer. He argued that his ecologic study has at least 4 advantages over an individual study: a large number of data points, the small degree of uncertainty affecting county mortality, the availability of "good" data on many potential confounding factors, and the size and diversity of the population being studied. However, Cohen's response did not specifically address the inherent limitations of ecologic studies and ecologic regression, as detailed by Stidley and Samet (1993) and Greenland and Robins (1994a). Greenland and Robins (1994b), in responding to Cohen, found little merit in his responses and disagreed with a principal assertion of his: that the ecologic fallacy does not affect a test of "linear-nonthreshold theory."

Piantadosi (1994), commenting on the exchange, suggested that Cohen's findings do "more to discredit the analysis than the theory." He elaborated: "The result of Cohen's analysis will seem biologically implausible to many investigators although it is probably theoretically possible at the individual level. Many epidemiologists will likely attribute the discrepancy between theory and result more to deficiencies in ecologic analyses than to failure of the dose-response theory. . . Most of us would not be willing to discard a useful theory on the basis of such a test." Like Greenland and Robins, Piantadosi is concerned by the limitations of ecologic analyses and the inability to determine whether bias is present and to estimate the direction and magnitude of its effect.

Uncontrolled confounding by smoking remains an explanation for the negative association between radon and lung cancer reported by Cohen. Stidley and Samet (1994) noted that there might be confounding in Cohen's analysis reflecting the higher concentrations of radon in western states, where smoking tends to be lower than elsewhere in the country. Gilbert (1994) further noted that other smoking-related cancers are also negatively associated with radon concentration in Cohen's data, possibly providing further evidence of confounding.

## Conclusions

Although a number of ecologic studies have been published since the BEIR IV report, the present committee did not find the new evidence to be informative concerning the risks posed by radon. The finding of a statistically significant negative association between radon and lung cancer in Cohen's analysis of lung-cancer mortality in the United States was considered to have resulted from inherent limitations of the ecologic method. That analysis has been widely cited as weighing against any risk of lung cancer at typical indoor radon concentrations (White and others 1992; Marcinowski and others 1994). The finding was considered to be an inappropriate basis for concluding that indoor radon is not a potential cause of lung cancer. We also note that the case-control studies reported to date, although limited in statistical power, have not yielded evidence of a negative association between exposure to radon progeny and lung-cancer risk. The ecologic studies were also not considered to be an appropriate basis for quantitatively estimating lung-cancer risk associated with radon exposure. Ecologic regression coefficients can be biased, and extensive individual-level data are available for estimating risk.

## CASE-CONTROL STUDIES

The most-direct evidence of health consequences of radon-progeny exposure in homes is offered by case-control studies, in which characteristics of lung-cancer patients are compared with those of control subjects who do not have the disease. After age, smoking, and other factors are accounted for, if residential radon causes lung cancer, it would be expected that the mean of a measure of exposure of cases would exceed the mean of controls, given proper assessment of statistical sampling variation. Exposure measures are usually based on a surrogate thought to be correlated with exposure, such as type of home construction, or on a more-direct correlate, such as measured radon in current and past homes.

Although straightforward in principle, case-control studies of residential radon are burdened with several limitations. These are discussed later in this chapter and include in particular an inability to measure radon in current and all past homes and thereby create an accurate measure of exposure, the lack of an estimate of radon exposure outside the home, and the relatively small relative risk (RR) that is expected even for long-term residents of higher-radon homes, which are not common. A small RR implies that mean exposures of cases and controls differ by only a small amount, thus limiting study power. The detection of an excess risk of lung cancer is potentially complicated also by an inability to control completely for other lung-cancer risk factors, particularly cigarette-smoking, which has an RR of 10-20.

In this section, we review case-control studies, first those which use surrogate measures of exposure and then those in which direct measurements of radon concentration in homes were used.

## Studies Using Surrogate Measures of Exposure

Many of the earliest studies of the effects of residential radon exposure relied on surrogate measures of exposure, such as housing style, for example, the presence or absence of a basement, the type of construction materials, or the characteristics of the local geology. Investigators often supplemented their observational data on houses with direct measurements of radon concentration to validate their "exposure" classifications. Table G-5 adapted from Samet (1989), lists the principal studies that used surrogate measures as the primary source of exposure classification. In those studies, measured concentrations were generally positively correlated with housing characteristics thought to be related to high indoor radon concentrations. For example, in several studies conducted in Sweden, radon measurements were related with their classification scheme whereby wood houses without basements on normal ground were classified as low-radon houses; wood houses on alum shale (known to have high radiation emanation rates), stone houses with basements, and stone houses without basements on alum shale were classified as high-radon houses; and the remaindor were classified as moderate-radon houses. But housing type was not always directly related to radon concentration. In the Stockholm County study by Svensson and others (1989), which was supplemented by direct radon measurements in houses and reported by Pershagen and others (1992), measured houses with ground contact classified by "type of ground" had geometric means of 99, 108 and 153 $Bqm^{-3}$ for low, moderate and high categories, respectively.

Results of the studies varied, but, the overall pattern of RRs suggests a positive association between the surrogate measure of radon concentration and lung-cancer risk, with an RR for the high-radon houses about twice that for the low-radon houses. When data were available, results were not materially affected by controlling for smoking. Because the links between the radon potential of a house and actual radon concentration and between radon concentration and individual exposure are uncertain, it is difficult to interpret the RRs in relation to extrapolations from miners or to studies in which radon concentrations in houses were measured.

The agreement among the studies that used surrogate measures complicates their interpretability, in that the results appear more consistent than do results of studies in which indoor radon was measured directly. Classification of indoor radon concentrations based on housing type or building materials might be expected to be less accurate and to have greater random and systematic errors in

TABLE G-5 Epidemiologic studies of residential redon-progeny exposure and lung cancer with surrogate measures of exposure

| Location: Reference | Study Design | Subjects | Exposure | Results[a] |
|---|---|---|---|---|
| Southern Sweden: Axelson and others 1979 | Case-control | 37 cases deceased in 1965-1977 and 178 controls deceased at same time as cases, excluding cancers; rural residents only | Residence type: wood without basement, "mixed," or stone with basement | RR=1.8 [95% CI (1.0,3.2)] for stone and mixed, compared with wood |
| Oeland, Sweden: Edling and others 1984 | Case-control | 23 cases deceased in 1960-1978 and 202 deceased controls | Residence type: wood without basement on normal ground, "mixed," or wood on alum shale, stone with basement and stone without basement on alum shale; 1 mo measurements in some homes | RRs of 1.2 [90% CI (0.5,3.1)] and 4.3 [90% CI (1.7,10.6)] for intermediate- and high- vs. low-exposure categories; p for trend <0.01. |
| Southern Sweden: Edling and others 1984 | Case-control | 23 cases and 202 controls | Measurement with α-sensitive film | RR increased for highest- vs. lowest-exposure categories. |
| Northern Sweden: Pershagen and others 1984 | Case-control (matched pairs) | 15 never-smoker and 15 ever-smoker case-control pairs | Construction characteristics | Estimated mean exposure significantly higher for smoking cases than controls; exposure not different for nonsmokers. |
| Sweden: Pershagen and others 1984 | Case-control (matched pairs) | 11 never-smoker and 12 ever-smoker case-control pairs | Construction characteristics | Estimated mean exposures similar for cases and controls, regardless of smoking status. |

TABLE G-5 Continued

| Location: Reference | Study Design | Subjects | Exposure | Results[a] |
|---|---|---|---|---|
| Northern Sweden: Damber and Larsson 1987 | Case-control | 589 male cases, 582 deceased controls, and 453 living controls | Residence type: wood or nonwood | RR not increased with or without smoking adjustment; RR increased for those never employed in non-lung-cancer-related occupations. |
| Stockholm, Sweden: Svensson and others 1987 | Case-control | 292 female cases diagnosed in 1972-1980 and 584 controls, resident in Stockholm for at least 28 of prior 30 yr | Geology and living near ground level; grab-sample measurements in some homes | RR=2.2 [95% CI (1.2,4.0)]; exposure-response trend not significant. |
| Southern Sweden: Axelson and others 1988 | Case-control | 177 cases deceased 1960-1981 and who lived in same house at least 30 yr before death and 677 controls deceased in same years of noncancer causes | Residence type: wood without basement on normal ground, "mixed," or wood on alum shale, stone with basement and stone without basement on alum shale; 2 mo measurements in some homes | RR=1.8 [90% CI (1.0,3.3)] for nonsmokers and light smokers in rural areas; no association for smokers in rural areas or for urban residents. |

| Stockholm County, Sweden[b]: Svensson and others, 1989 | Case-control | 187 female cases in 1983-1985, 160 "hospital" controls with suspect lung cancer found not to have the disease, and 177 population-based controls | Geology and living near ground level; 2-w radon measurements during the heating season in a sample of homes | RRs of 1.8 [95% CI (1.2,2.9)] and 1.7 [95% CI (0.9,3.3)] for intermediate- and high- vs. low-exposure categories; p for trend 0.03. Slight variation of risk by cell type or smoking status; steeper RR trend at highest ages. |
| --- | --- | --- | --- | --- |
| Washington County, Maryland, U.S.A.: Simpson and Comstock 1983 | Cohort | 298 cases over 12-yr period | Housing characteristics | No association of incidence with housing characteristics. |

[a]Parentheses provide 95% confidence interval for RR.
[b]Initial study, which was later expanded to include indoor radon measurements and reported by Pershagen and others 1992.
Source: adapted from Samet (1989).

exposure assessment; as a result, there would likely be greater attenuation of the observed association and greater variability in the outcomes of the independent studies. However, one should not necessarily conclude that studies using surrogate measures will be more misclassified than those using actual measurements, because their error structures differ fundamentally. There is a possibility that such might have some advantages. The ideal study would include both measurements and surrogates, and both should be included in an analysis incorporating measurement errors.

Results of these studies should therefore be interpreted cautiously. In addition, several of the studies had few lung-cancer cases, that often precluded subgroup analyses, which would permit evaluation of both internal consistency and consistency among studies.

## Studies with Direct Measurements of Indoor Radon

Potentially, the most important source of direct information on the consequences of exposure to indoor radon is epidemiologic studies in which long-term measurements of radon concentration were carried out. Several such studies have been done, and they are described below. All studies used a case-control design, in which estimates of radon or radon-progeny exposure of lung-cancer subjects are compared with estimates of exposure of controls selected from the same target population as the cases taking account of other factors that might influence the comparison—such as age, smoking status, and sex—are accounted for.

Case-control studies that incorporate direct measurement of indoor radon concentrations have several advantages over ecologic studies and over case-control studies that use surrogate exposure measures. Such case-control studies must be viewed as generally having greater validity for the identification and ultimately the quantification of an excess risk of lung cancer. In contrast with an ecologic study, a case-control study offers a well-defined target population, and outcome status is assessed unambiguously. Direct, long-term measurement of radon in houses permits estimation of exposures specific to individuals, thereby reducing exposure errors, compared with ecologic studies and studies that define exposure in terms of house type. Direct measurement data permit the reconstruction of historical exposure profiles and the evaluation of biologically plausible exposure periods. With direct measurement data, it is possible to evaluate the consequences of missing data and the effects of various imputation approaches. In addition, measurements from track-etch devices are generally more comparable across countries than are crude classifications by house type and so allow more valid comparison among studies from different countries. Thus, results of ecologic studies are considered noninformative, results of case-control studies that use surrogate exposure measures are provocative, and case-control studies with direct, long-term radon measurement offer the best opportunity for identifying an excess risk of lung cancer associated with indoor radon.

### New Jersey Case-Control Study of Females

*Study subjects.* The radon component of this study was an add-on to a continuing lung-cancer case-control study of females in New Jersey (Schoenberg and others 1990; Klotz and others 1993). The original case group consisted of 1,306 female residents with histologically confirmed lung cancer diagnosed from August 1982 through September 1983 (see Table G-6). They were identified from hospital pathology records and from the New Jersey State Cancer Registry and death-certificate files. Data were collected on 994 cases (76%) from 532 in-person interviews and 462 next-of-kin interviews.

Controls were selected from New Jersey drivers-license file, on those under age 65 and from Health Care Financing Administration file, on those 65 and over. For cases with next-of-kin respondents, controls were selected at random from death certificates, excluding deaths from respiratory disease. Controls were individually matched to cases by race, age, and, for deceased cases, closest date of death. A total of 1,449 controls were identified, and interview data were obtained on 995 women (69%).

Houses to be measured were defined in the study in 2 phases (Table G-7). In phase I, a single "index" residence per subject was chosen in which the subject lived for at least 10 yr in the period 10-30 yr before diagnosis or selection. In phase II, the residence criteria were broadened, to add subjects to the radon component of the study and houses for subjects selected in phase I. The eligibility period for the index residence was extended to cover the period 5-30 yr before diagnosis, and the study enrolled all houses in which a subject resided for 4 yr or more in the 6 New Jersey counties with high average radon concentrations, or for 7 yr or more in the rest of the state. Twelve subjects were excluded because their eligible residences represented less than 9 yr of coverage in the exposure-time window. Of the 994 cases and 995 controls with completed interviews in the original study, 661 cases (66%) and 667 controls (67%) had residences that were eligible under the expanded phase II criteria.

*Data collection.* Subjects or surrogate respondents were interviewed by trained interviewers. Study subjects provided 53% of the interviews, spouses 17%, and other next-of-kin 29%. Information was obtained on lifetime smoking history, smoking by other household members, lifetime residential and occupational histories, and consumption of food high in vitamin A. Information on specific addresses of past residences was collected several years after the original interview through telephone contacts. During eligible residencies, information was obtained on house characteristics, including heat circulation and modifications to the structure or to heating and ventilation.

*Methods of radon measurement.* Long-term $\alpha$-track detectors were deployed

TABLE G-6    Summary of results of New Jersey female case-control study

| Factor | Comment |
|---|---|
| Principal references | Schoenberg and others 1990, 1992. |
| Design | Case-control study in females. |
| Study subjects | *Cases:* Cases were selected from 1,306 histologically confirmed lung cancers in females diagnosed in August 1982 through September 1983 throughout the state. In original study, 994 women were interviewed: 532 subjects and 462 next-of-kin. Cases for the radon analysis were further restricted by measurement protocol Phases I and II.<br>*Controls:* For living cases, controls were selected randomly from New Jersey driver's license files (age < 65 yr) or Health Care Financing Administration files (age ≥ 65 yr). For next-of-kin cases, controls were selected randomly from death certificates that did not mention respiratory disease. Controls were matched by race, age, and, for deceased cases, date of death.<br>*Subjects in radon study:* Phase I included subjects who had lived in a single residence ≥ 10 yr in the period 10-30 yr before diagnosis or selection; phase II included subjects who had lived in one or more residences in the period 5-30 yr before. Phase II added subjects to phase I and added houses. Subjects restricted to those with 9 yr of coverage. Totals of 480 cases and 442 controls were included. |
| Lung-cancer histology | 480 cases: squamous 25.8%; small cell 29.8%; adenocarcinoma 21.9%; other 22.5%. |
| Rn-measurement protocol | *Measurements:* 1-yr α track in living area (76%), 1-yr α track in basement (5.4%), basement and upstairs charcoal canister (6.5%), upstairs charcoal canister (1.4%). Canister below minimal detectable concentrations assigned MDC concentration. Apartments above the 2nd floor assigned 0.4 $pCiL^{-1}$. Regressions linked basement and canister measurements to long-term values for living areas.<br>*Missing:* Under phase II eligibility, 74% of cases and 72% of controls had measurements. |
| Rn measurements | *Mean:* Cases, 0.7 $pCiL^{-1}$; controls 0.7 $pCiL^{-1}$.<br>*Median:* Cases, 0.5 $pCiL^{-1}$; controls 0.5 $CiL^{-1}$. |
| Rn-exposure estimation | *Exposure time window:* 5-30 yr before the date of case diagnosis or control selection.<br>*Coverage:* Median 22 yr for cases and for controls; 35% of subjects fully covered.<br>*Imputation for gaps:* None for TWA radon exposure; for cumulative exposure, 0.6 $pCiL^{-1}$ (22 $Bqm^{-3}$) was assigned for missing intervals. |
| Results | *Overall:* For categories <1, 1-1.9, 2-3.9, and ≥ 4 $pCiL^{-1}$, RRs were 1.0, 1.2, 1.1, and 8.7, with p value for 1-sided test of linear trend 0.04. Only 5 cases and 1 control in highest category. |

## TABLE G-6  Continued

| Factor | Comment |
|--------|---------|
|  | *Histology:* Increasing RRs with pCiL⁻1 only for "other" cell types; no trend with other histologic types (Lubin and others 1994). |
|  | *Smoking:* For RR with $pCiL^{-1}$ no trend in never-smokers, increasing RRs for <15 and 15-24 cigarettes/d, and decreasing trend for ≥ 25 cigarettes/d. |

TABLE G-7   Distribution of 944 cases and 955 controls in original New Jersey lung-cancer case-control study by status in radon substudy for Phase I only and Phase I and II

|  | No. (%) Subjects | | | |
|--------|------------------|--|------------------|--|
|  | Phase I only | | Phase II only | |
| Status | Cases | Controls | Cases | Controls |
| Included in radon study[a] | 433  (44) | 402  (40) | 480 (48) | 442  (44) |
| No address-specific information[b] | 140  (14) | 126  (13) | 168 (17) | 152  (15) |
| No address met residence criterion[c] | 253  (25) | 256  (26) | 165 (17) | 176  (18) |
| No radon testing at index address[d] | 168  (17) | 211  (21) | 181 (18) | 255  (23) |

[a]Index residence(s) tested for radon with α-track detectors or charcoal canisters. If index residence was apartments above the 2nd floor, radon exposure assumed to be < 1 pCiL⁻¹. Seven cases and 5 controls with complete measurements in phase I or II were excluded because they represented 8 yr or less of 25 yr exposure history.
[b]Respondent refused further contact after initial interview, respondent lost to follow-up, respondent refused address-specific information, or inadequate address-specific information.
[c]Subject did not meet residence criterion for inclusion in phase I (phase I and phase II).
[d]Index residence demolished, refusal by current resident, or no contact with current resident.
Source: Schoenberg and others 1992.

for 1-yr. In each dwelling, 1 detector was placed in the living area, usually the master bedroom, and another in the lowest habitable level, usually the basement. In addition, 4-d screening measurements were made with the house closed, during the heating season, with charcoal canister detectors. The screening measurements were used primarily as a backup if long-term measurements could not be completed and to identify homes that required immediate mitigation. The radon concentration used for the house was based on the nonbasement primary-living-area α-track measurement when available (76% of houses). When it was unavailable, the nonbasement radon concentrations were estimated from other measurements in descending order of priority: basement α-track (5.4%), basement charcoal canister with upstairs canister (6.5%), and upstairs charcoal canister (1.4%). The estimates were derived from regression equations based on complete sets of measurements which also took into account the heating system (forced air versus other).  Canister readings below the minimal detectable concentration (MDC) were assigned the MDC value.  Apartments above the 2nd floor (10.6%) were assigned a value of 14.8 Bqm$^{-3}$ (0.40 pCiL$^{-1}$).  Usable measurements were obtained for 480 cases and 422 controls or, 74% and 72%, respectively, of those eligible under the Phase II criteria.

For analysis, 2 measures of exposure were developed.  The time-weighted average (TWA) radon concentration was the mean concentration for all houses measured weighted by the years of residence in the exposure time window of 5-30 yr.  Cumulative radon exposure was computed as the product of residence time within the 5-30 yr and  measured radon concentration.  Within the 5-30 yr, unmeasured houses were assigned a radon concentration of 22.2 Bqm$^{-3}$ (0.6 pCiL$^{-1}$), the median concentration of all phase I control houses.

*Results.*  On the basis of phase II data, the mean residence time within the 5-30 yr was 22 yr for cases and for controls; 35% of the subjects had radon measurements for all their homes in the exposure period.  The houses in the New Jersey study had the lowest radon concentration of any of the current case-control studies;  most of the TWA concentrations were below 37 Bqm$^{-3}$ (1 pCiL$^{-1}$).  The median radon concentration was 18.5 Bqm$^{-3}$ (0.5 pCiL$^{-1}$) and was the same for cases and controls.

Table G-8 shows the RRs for categories of TWA radon concentration adjusted for cigarettes per day, cessation of smoking, age, occupation, type of respondent, and interaction of type of respondent with cigarettes per day (Schoenberg and others 1992).  RRs were flat and increased only in the highest category, 148 Bqm$^{-3}$ (4.0 pCiL$^{-1}$), which included 5 cases and 1 control.  The p-value for linear trend was significant at p = 0.05 but was based on a 1-sided, rather than the traditional 2-sided, test of the null hypothesis.  For this study, it should be pointed out that 90% CIs were used for RR rather than the more

TABLE G-8   Distribution of cases and controls and adjusted odds ratios[a] (OR) and confidence intervals (CI) by time weighted average (TWA) radon concentration for the New Jersey case-control study of females

| | TWA Rn concentration, pCiL$^{-1}$ | | | | | |
| | <1 | 1-1.9 | 2-3.9 | ≥ 4 | Total | P for trend |
|---|---|---|---|---|---|---|
| No. Cases | 384 | 72 | 19 | 5 | 480 | — |
| No. Controls | 360 | 69 | 12 | 1 | 442 | — |
| OR[a] | 1.0 | 1.2 | 1.1 | 8.7 | — | 0.05[b] |
| 90% CI | | (0.8,1.7) | (0.6,2.3) | (1.3,57.8) | — | — |

[a]Adjusted for lifetime average cigarettes/d, years since smoking cessation, age, occupation, respondent type, and interaction between respondent type and cigarettes/d.
[b]One-sided test of linear trend
Source: Schoenberg and others 1992.

conventional 95% CIs. Results for cumulative radon exposure were similar to those for TWA radon concentration (Schoenberg and others 1992).

## Shenyang China Case-Control Study of Females

*Study subjects.* Like the New Jersey study, the radon component of this study was an add-on to an existing lung-cancer case-control study of woman in Shenyang China (Xu and others 1989; Blot and others 1990). Potential cases were female residents of Shenyang, who were 30-69 yr old and were listed in the Shenyang Cancer Registry with primary lung cancer in September 1985 to September 1987 (Table G-9). In the full study, 75% of the diagnoses for the female lung-cancer cases were based on pathologic or cytologic material; histologic information was available on 73% of all female cases (Xu and others 1989).

A population-based, age-matched control group of women was selected from the Shenyang general population by using the system of area administrative units and neighborhood population lists. Controls were randomly selected in 5-year age groups to reflect the age distribution of the cases.

The radon component of the study was initiated 6 months after the start of the original study and, because of budgetary reasons, ended before completion of case acquisition in the full study. A total of 397 cases and 391 control subjects had detectors placed in their houses, representing 95% and 99% of eligible cases and controls, respectively.

*Data collection.* Trained nurses sought personal interviews with the subjects, except for those who were too ill or deceased. Participation rates were 95% for cases and 97% for controls. For most patients, the time between diagnosis and interview was less than 1 month. A structured questionnaire was used in an interview to inquire about smoking by the subject and other household members, occupation, prior medical conditions, residential history, and housing characteristics, such as indoor air pollution. A time-weighted index of lifetime air-pollution exposure was determined from housing characteristics, including type of heating, fuel used for cooking, and whether cooking facilities were in a separate kitchen or combined with living room or bedroom (Xu and others 1989).

*Methods of radon measurement.* Two $\alpha$-track detectors were placed for 1 yr in the current residence of each case and control; 1 detector was usually in the living room and 1 in the bedroom. Nearly all homes were single-story buildings. For persons who lived in the current house less than 5 years, a prior Shenyang residence was tested, provided that it was accessible and the subject had lived there at least 5 years. Detectors were collected for 308 cases (78%) and 356 controls (91%).

The maximum of the 2 measurements were used in analysis. Among the paired measurements, the correlation was 0.52, 77% were within 74 Bqm$^{-3}$ (2 pCiL$^{-1}$), and 78% of the ratios of the 2 measurements were less than a factor of 2.

*Results.* Among the subjects, the median number of reported residences was 3, the median residence time in the last home was 24 yr, and 76% lived in the

TABLE G-9    Summary of results for Shenyang, China female case-control
study

| Factor | Comment |
|---|---|
| Principal reference | Blot and others 1990. |
| Design | Case-control study of females. |
| Study subjects | *Cases*: Cases included all female residents of Shenyang, China, aged 30-69 yr with primary lung cancer diagnosed in September1985 to September 1987 and listed in the Shenyang Cancer Registry. All case diagnoses were reviewed.<br>*Controls*: Controls were randomly selected in 5-yr age groups from the general population.<br>*Subjects in radon study*: For the radon component, ascertainment was delayed 6 mo but included all subjects. A total of 308 cases and 356 controls had radon measurements. |
| Lung-cancer histology | 308 cases: squamous, 23.4%; small cell, 12.7%; adenocarcinoma, 30.8%; other or unknown 31.1%. |
| Rn-measurement protocol | *Measurements*: 1-yr $\alpha$-track detectors in living room and in bedroom of current home. For those who had lived for < 5 yr in the current home, 1-yr $\alpha$-track detectors were placed in the previous residence if it was in Shenyang and accessible and subject had lived there $\geq$ 5 yr.<br>*Missing*: Among those eligible, 79% of cases and 91% of controls had measurements. |
| Rn measurements | *Median*: Cases, 2.8 pCiL$^{-1}$; controls, 2.9 pCiL$^{-1}$. |
| Rn-exposure estimation | *Exposure-time window*: 5-30 yr before case diagnosis or control selection.<br>*Coverage*: Median residence in last home was 24 yr, and 76% lived in measured home $\geq$ 10 yr.<br>*Imputation for gaps*: None; analyzed only measured radon concentration. |
| Results | *Overall*: For categories <2, 2-3.9, 4-7.9, and $\geq$ 8 pCiL$^{-1}$, RRs were 1.0, 0.9, 0.9 and 0.7 and 1.0, 0.7, 1.2, and 0.7 when analyses were restricted to subjects who lived $\geq$ 25 yr in their last residence.<br>*Histology*: RRs for small cell for pCiL$^{-1}$ categories were 1.0, 1.2, 1.7, and 1.4, but with nonsignificant trend.<br>*Smoking*: Little evidence of a trend in RRs with pCiL$^{-1}$ in any smoking category.<br>*Subgroup analyses*: RR patterns were the same within levels of an index of indoor air pollution or after adjustment. |

TABLE G-10  Distribution of cases and controls and adjusted odds ratios[a] (OR) and confidence intervals (CI) by radon concentration in Shenyang, China, case-control study of females

| | Radon concentration (pCiL$^{-1}$) | | | | | |
| | < 2 | 2-3.9 | 4.0-7.9 | ≥ 8.0 | Total | P for trend |
|---|---|---|---|---|---|---|
| Cases | 91 | 131 | 60 | 26 | 308 | |
| Controls | 95 | 148 | 77 | 36 | 356 | |
| OR | 1.0 | 0.9 | 0.9 | 0.7 | | n.s. |
| 95% CI | | (0.6,1.3) | (0.5,1.4) | (0.4,1.3) | | |

[a]ORs adjusted for age, education, smoking status, and an index of indoor air pollution.
Source: Blot and others 1990.

measured home for 10 or more. On the average, subjects lived 66% of their adult lives in the measured home (Blot and others 1990). Using the maximum of the 2 radon measurements, the medians were 103.6 Bqm$^{-3}$ (2.8 pCiL$^{-1}$) for cases and 107.3 Bqm$^{-3}$ (2.9 pCiL$^{-1}$) for controls.

Categories of radon concentration ranged from < 74 Bqm$^{-3}$ (2 pCiL$^{-1}$) to 296 Bqm$^{-3}$ (8.0 pCiL$^{-1}$) (Table G-10). The RR for lung cancer adjusted for age, education, smoking status, and an index of indoor air pollution showed no increase with increasing radon concentration.

In the Shenyang data, cigarette-smoking and indoor air pollution were found to be significant risk factors for lung cancer for males and females (Xu and others 1989). Among females, 55% of cases and 35% of controls smoked cigarettes. The risk was over 9 times as high in women who smoked more than 1 pack/d for at least 40 yr, as in never-smokers. For females in the radon component of the study, the RR pattern with radon concentration was similar, that is, it showed no increase in never-smokers, light smokers and heavy smokers.

Indoor air pollution was found to increase lung-cancer risk by a factor of 2-3, depending on the variable analyzed. The greatest risks were associated with the use of a coal-burning *kang* (a brick bed under which heated smoke is passed through pipes before venting to the outside through a chimney or other opening) or cooking in the same room as the sleeping quarters.

An air-pollution index was developed to incorporate the type of heating for the home, the type of cooking fuel, and whether the kitchen and the bedroom were the same room. For females, no positive association was found with radon concentration for low or high categories of the indoor air-pollution index (Blot and others 1990).

### Stockholm, Sweden Case-Control Study of Females

*Study subjects.* The methods of the Swedish investigation have been described (Pershagen and others 1992) and are summarized in Table G-11. Women suspected of having lung cancer on admission in 1983-1985 to the 3 clinical departments of pulmonary medicine and the only department of thoracic surgery in Stockholm County were interviewed. Those later diagnosed as having lung cancer (210) were classified as cases.

Two control groups were selected. Hospital controls consisted of women suspected of having lung cancer who were later found not to have it (191), and population-based controls (209) were obtained from Stockholm County population registers. Results were reported for both control groups combined.

*Data collection.* Subjects were interviewed by physicians using a structured questionnaire. For cases and hospital controls, information was obtained on admission. Population controls were interviewed in visits or by telephone. Information

TABLE G-11   Summary of results of Stockholm female case-control study

| Factor | Comment |
|---|---|
| Principal reference | Pershagen and others 1992. |
| Design | Case-control study of females. |
| Study subjects | *Cases*: Cases (210) included females admitted to the 3 pulmonary departments and the 1 thoracic-medicine department in Stockholm County in September 1983 to December 1985. |
| | *Controls*: Two controls were selected: "hospital" controls (191) included females suspected to have had lung cancer but found not to, and population controls (209) selected randomly from County population registers. |
| | *Subjects in radon study*: For the radon analysis, 31 women (5%) could not be measured, leaving 201 cases and 378 controls with radon measurements. |
| Lung-cancer histology | 201 cases: squamous, 26.9%; small cell, 25.4%; adenocarcinoma, 34.4%; other 13.4%. |
| Rn-measurement protocol | *Measurements*: 1-yr $\alpha$-track detectors in living room and in bedroom (85.1%) or thermoluminescence detector (TLD) for 1 wk in living room followed by 1 wk in bedroom (14.9%) in all homes occupied 2 yr or more since 1945. TLD values were then adjusted empirically to link with $\alpha$-track measurements. |
| | *Missing*: 2,118 homes fulfilled criteria for measurement; $\alpha$-track detectors retrieved from 1,339 homes (63%) and TLD from 234 homes (11%). |
| Rn measurements | *Median*: Cases, 3.1 pCiL$^{-1}$; controls, 2.9 pCiL$^{-1}$ |
| | *Mean*: Cases, 3.6 pCiL$^{-1}$; controls, 3.7 pCiL$^{-1}$ (Lubin and others 1994). |
| Rn-exposure estimation | *Exposure time window*: From 1945 to 5 yr before interview. |
| | *Coverage*: 26.3 yr and 25.3 yr of residence corresponding to 78% and 77% of the time window. |
| | *Imputation for gaps*: None for TWA radon concentration; unclear for cumulative exposure-some analyses set missing to zero, and some replaced missing with estimates based on housing characteristics. |

was obtained on smoking, exposure to environmental tobacco smoke, occupational history, and consumption of foods rich in vitamins A and C. Also obtained was a history of all residences in which the subject lived for 2 yr or more since birth or arrival in Sweden. The residential history included information on type of house, building material, and year of construction. Data from parish registries on past residences were used to verify and supplement the residential histories.

*Methods of radon measurement.* Measurements were sought for all dwellings where the subject resided for 2 years or more between 1945 and the end of the observation period, 1983-1985. For cases and hospital controls, the end of the exposure observation period was 5 years before the date of the study interview; for the population controls, it was 5 years before the interview of the corresponding case. Of the 2,118 residences so identified, no measurements could be made in 27.4%—in 11.2% because the house no longer existed, in 4.4% because the house was abroad, in 3.2% because the current owner refused, and in 8.6% for various other reasons (Pershagen and others 1992).

Year-long radon-concentration measurements were made in 1,339 dwellings with 2 a-track detectors: one in the living room and the other in the bedroom. In 234 dwellings (15%), measurements were made for 2 weeks during the winter with thermoluminescence detectors (TLDs) designed by the Swedish Institute of Radiation Protection. A TLD was placed in the living room for 1 week, then moved to the bedroom for another week. The 2 methods gave readings that had correlations above 0.8, although the TLD values were higher on the average, reflecting decreased ventilation in the colder months and the greater likelihood of their placement in areas of high-radon ground emanation (Svennson and others 1989). For analyses, TLD values were adjusted empirically to reflect $\alpha$-track detector concentrations (Svensson and others 1988). The radon concentration assigned to a house was either the mean of the 2 $\alpha$-track readings or the adjusted TLD reading.

Two measures of exposure were developed for analysis. The TWA radon concentration was the mean concentration for all houses measured weighted by the years of residency in the exposure window from 1945 to 5 years before enrollment. Cumulative radon exposure was computed as the product of residency time in the exposure window and measured radon concentration. The handling of missing measurements in the calculation of cumulative radon exposure was unclear. It appears that missing measurements were sometime set to zero and that "in certain analyses, missing radon measurements were replaced by estimates based on dwellings actually measured and information from the interview questionnaire on type of house, building material, and year of construction" (Pershagen and others 1992). Specific details were not provided.

*Results.* For cases and controls, the mean times covered by measurement data were 26.3 years and 25.3 years and represented 78% and 77% of the relevant period, respectively. For the subjects, median radon concentrations were 114.70 Bqm$^{-3}$ (3.1 pCiL$^{-1}$) for cases and 107.3 Bqm$^{-3}$ (2.9 pCiL$^{-1}$) for controls.

Results show a significant increase in RRs with increasing TWA radon concentration (Table G-12). The p value for trend was 0.05. As indicated by Pershagen and others and expanded on in the pooled analysis by Lubin and others (1994a), the significance of the test for trend depended on the cut points and on the quantitative value used. The p value of 0.05 computed by Pershagen and others (1992) used the median for each category as the quantitative trend variable, whereas Lubin and others computed the p value as 0.46 by using the continuous value for radon concentration. The former approach minimized the impact of extreme values; the latter approach eliminates the arbitrariness of categorization. The trend of increasing RR was reduced when adjusted for occupancy or when exposure 15 years and more before was given half the weight in line with results of miner studies. These differences highlight the need to interpret the Stockholm results with caution.

Because of the small number of cases, the trends in the RR with level of radon concentration were probably statistically homogeneous by histological type, although no formal assessment was done. However, the gradient of increase appeared greater for squamous cell and small cell carcinomas (Pershagen and others 1992).

Similarly, there was no statistical evaluation of the joint association of radon concentration and smoking status; however, the trend appeared slightly greater in never-smokers than in ever-smokers.

TABLE G-12   Distribution of cases and controls and adjusted odds ratios[a] (OR) and confidence intervals (CI) by radon concentration for the Stockholm case-control study of females

|  | Radon concentration (pCiL$^{-1}$) | | | | | |
|  | < 2 | 2-2.9 | 3.0-4.0 | ≥ 4.1 | Total | P for trend |
|---|---|---|---|---|---|---|
| Cases | 43 | 59 | 38 | 61 | 201 | |
| Controls | 89 | 113 | 76 | 100 | 378 | |
| OR | 1.0 | 1.2 | 1.3 | 1.7 | | 0.05[b] |
| 95% CI | | (0.7,2.1) | (0.7,2.3) | (1.0,2.9) | | |
| OR[c] | 1.0 | 1.5 | 1.6 | 1.5 | | 0.19 |
| 95% CI | | (1.0,2.4) | (0.9,2.7) | (0.6,3,4) | | |
| OR[d] | 1.0 | 1.4 | 1.2 | 1.3 | | 0.65 |
| 95% CI | | (0.9,2.3) | (0.7,2.1) | (0.6,3.1) | | |

[a]ORs adjusted for age, smoking, and municipality of residence.
[b]For test of linear trend using category means, P = 0.05; using continuous exposure, P = 0.46.
[c]Exposure adjusted for occupancy.
[d]Exposure adjusted for BEIR IV weighting, exposures 5-15 yr before given full weight, exposures ≥ 15 yr before given 0.5 weight
Source: Pershagen and others 1992.

## Swedish National Case-Control Study

*Study subjects.* This study, the largest to date, relied on various national data files for the identification of subjects for the study (Pershagen and others 1994). The study is summarized in Table G-13. The study base was defined as all subjects 35-74 years old who had lived in any of 109 municipalities in Sweden at some time from January 1980 through December 1984 and who had been living in Sweden on January 1, 1947. The municipalities were selected to include areas suspected of having homes with high and low radon concentrations on the basis of measurement data or geologic and other information. Municipalities with mining activities and the large cities of Stockholm, Göteborg and Malmö were not included.

Using Swedish Cancer Registry files, cases included subjects diagnosed with primary lung cancer in 1980-1984. All 650 women and a radon sample of 850 men (about 40% of all men with lung cancer) were identified. After excluding those not in the study base, a total of 1,360 cases were enrolled 586 females and 774 males.

Two control groups were defined by using the population registers of Statistics Sweden. One control series consisted of a radon sample of women frequency-matched in 5-year age categories and calendar year of residence to the case group and included 1,424 subjects—730 women and 694 men. A second control group was selected by matching on age and calendar year and on vital status. Deceased controls were ascertained from the Swedish Cause of Death Registry, excluding subjects who had died of smoking-related diseases (cancer of the mouth, esophagus, liver, pancreas, larynx, or uterine cervix or bladder; ischemic heart disease; aortic aneurysm; cirrhosis of the liver; chronic bronchitis and emphysema; gastric ulcer; violent causes; or intoxication). In the second control group, there were a total of 1,423 subjects—650 women and 773 men.

At the time of selection on December 31, 1986, about 90% of the cases and of the second control group had died; about 9% of subjects in the first control group had died.

*Data collection.* All subjects or their next of kin were mailed a standardized questionnaire. Information was collected on smoking habits of the subject, and their spouses and parents and on lifetime occupational and residential histories since 1947. Residential history included information on type of house, building material, heating system, and time spent at home. For incomplete questionnaires or nonrespondents, telephone interviews were attempted. Data from parish registries on past residences were used to supplement residential histories from questionnaires.

*Methods of radon measurement.* Measurements were sought for all dwellings where a subject resided for 2 years or more between 1947 and 3 years before the end of the observation period, defined as year of diagnosis for the case and calendar year of selection for the controls. Of a total of 13,392 residences,

TABLE G-13   Summary of results of Swedish national case-control study

| Factor | Comment |
|---|---|
| Principal reference | Pershagen and others 1994. |
| Design | Case-control study of females and males. |
| Study subjects | *Cases*: A total of 1,500 subjects 35-74 yr old with primary lung cancer diagnosed in January 1980 to December 1984 were selected from the Swedish Cancer Registry, including all 650 females and 850 males. After various exclusions, 586 females and 774 males remained. |
| | *Controls*: Two controls were selected: 1 control group (730 females and 694 males) derived from a randomly selected sample from population registers, frequency matched on age to the cases; and 1 control group (650 females and 773 males) similarly selected and matched by vital status against the Swedish Cause of Death Registry. |
| | *Subjects in radon study*: For the radon analysis, measurements were not obtained on 27.4% of the homes. A total of 1,281 cases and 2,576 controls were included. |
| Lung-cancer histology | 1,281 cases: squamous, 33.1%; small cell, 23.1%; adenocarcinoma, 26.9%; other or unknown 16.8%. |
| Rn-measurement protocol | *Measurements*: 3-mo $\alpha$-track detectors in living room and in bedroom in all homes occupied 2 or more years since 1947. |
| | *Missing*: 12,394 homes fulfilled criteria for measurement; $\alpha$-track detectors were retrieved from 8,992 homes (73%). |
| Rn measurements | *Median*: 1.5 pCiL$^{-1}$. |
| Rn-exposure estimation | *Exposure time window*: From 1947 to 3 yr before end of followup diagnosis for cases or matched date of selection for controls. |
| | *Coverage*: 23.5 yr and 23.0 yr of residence corresponding to 72% and 71% of the exposure-time window. |
| | *Imputation for gaps*: None for TWA radon concentration; unclear for cumulative exposure—some analyses set missing to median concentration, and some replaced missing with estimates based on housing characteristics. |
| Results | *Overall*: For categories < 1.4, 1.4-2.1, 2.2-3.8, 3.8-10.8, and $\geq$ 10.8 pCiL$^{-1}$, RRs were 1.0, 1.2, 1.0, 1.3, and 1.8, with a P value for test of trend < 0.05. |
| | *Histology*: RR trends showed no difference by cell type. |
| | *Smoking*: No difference in RR trend greatest by smoking status, in contrast with the authors' view. |
| | *Subgroup analyses*: RR trend occurred only for subgroup that reportedly did not sleep near an open window; no trend was observed in those who sleep near an open window. |

addresses could not be identified for 7.5%. Of the remaining 12,394 residences, 27.4% could not be measured, usually because they no longer existed or because they were used only as summer houses (Pershagen and others 1994). In all, 73% of identified homes (8,992) were measured.

Three-month radon measurements were made during the heating season— from October 1 to April 30—with α-track detectors, which were processed by the Swedish Radiation Protection Institute. Two detectors were used: one in the living room and the other in the bedroom. For analyses, the mean of the 2 values was assigned to the residence. The authors estimate that the winter measurements might be 10-20% higher than yearly values, although the basis for this estimate is not provided.

Cumulative radon exposure since 1947 was estimated by multiplying the measured radon concentration concentration and the length of residency in each home. For each subject, TWA radon concentration was calculated by dividing cumulative radon exposure by the total residential time covered by radon measurements. Missing measurement time was not included; in effect the concentration and duration during those times were zero. For some analyses, imputation of missing measurement data was accomplished by replacing the missing data with the median radon concentration for all subjects or values adjusted to reflect residential characteristics (Pershagen and others 1994).

*Results.* Results were presented for males and females combined. There were a totals of 1,281 cases and 2,576 controls.

Radon measurements covered 23.5 years and 23.0 years of the exposure time for cases and controls, representing 72% and 71% of the intended period, respectively. For individuals, the median TWA radon concentration was 55.5 Bqm$^{-3}$ (1.5 pCiL$^{-1}$).

Results show a significant increase in RR with increasing radon concentration (Table G-14). The RR patterns appeared similar by cell type. For the 5 categories shown in Table G-14 RRs were: 1.0, 1.2, 1.3, 1.5, and 1.7 for squa-

TABLE G-14 Distribution of cases and controls (males and females combined) and adjusted odds ratios[a] (OR) and confidence intervals (CI) by radon concentration for the Swedish national case-control study

| | Radon concentration (pCiL$^{-1}$) | | | | | | |
|---|---|---|---|---|---|---|---|
| | <1.4 | 1.4-2.1 | 2.2-3.8 | 3.8-10.8 | ≥$^3$10.8 | Total | P for trend |
| Cases | 452 | 268 | 272 | 246 | 43 | 1,281 | |
| Controls | 952 | 561 | 568 | 436 | 59 | 2,576 | |
| OR | 1.0 | 1.1 | 1.0 | 1.3 | 1.8 | | <0.05 |
| 95% CI | | (0.9,1.3) | (0.8,1.3) | (1.1,1.6) | (1.1,2.9) | | |

[a]ORs adjusted for age, occupation, sex, smoking status, and urban compared with nonurban residence.

Souce: Pershagen and others 1992.

mous cell carcinomas; 1.0, 0.9 1.1, 1.2, and 2.8 for small cell carcinomas; and 1.0, 1.1, 1.0, 1.4, and 2.3 for adenocarcinomas. Differences in RRs for the highest category could have arisen by chance, in as much as the category included only 11, 15, and 12 squamous, small cell, and adenocarcinoma cases, respectively.

RR patterns were also similar by smoking status. For categories of radon concentration shown in Table G-14, RRs were: 1.0, 1.1, 1.0, 1.5, and 1.2 for never-smokers; 1.0, 0.9, 1.2, 1.7, and 0.4 for ex-smokers; 1.0, 1.0, 1.0, 1.2, and 4.0 for current smokers consuming fewer than 10 cigarettes/d; and 1.0 0.9, 0.9, 1.2, and 2.6 for current smokers consuming at least 10 cigarettes/d. RRs in the highest radon concentration category were based on 5, 1, 12, and 16 lung-cancer cases for never-smokers, ex-smokers, and current smokers of fewer than 10 and at least 10 cigarettes/d, respectively.

Pershagen and others found that the RR trend with radon concentration increased for those who reportedly sleep with their bedroom windows closed, but the trend disappeared for subjects who reported sleeping next to an open window. Those patterns of risk are difficult to interpret. Sleeping next to an open window is not itself a risk factor for lung cancer. Furthermore, the radon concentration of a bedroom with an open window will be reflected in a reduced radon measurement. Subjects or next of kin were interviewed about sleeping practices many years after disease occurrence. The relationship between the radon measurement, the current practice of sleeping with an open window, and whether the case or control subject slept with an open window at the time of enrollment is uncertain, particularly in homes that no longer were occupied by the subjects or their spouses. Effects of errors in exposure estimation might also play a role in the observed RR patterns. Control data suggest that in this age group about 70% of the population sleep with closed windows. Among subjects who sleep with an open window, measurements in homes, particularly former homes, are more likely to occur with owners who sleep with closed windows, thereby adding to error in exposure estimation and obscuring exposure-response effects. Among subjects who sleep with windows closed, measurements in former homes are more likely to occur with closed windows; however, for owners who sleep with open windows, there is a systematic (nondifferential) under estimation of exposure, a condition that can induce an increase in the trend of the exposure-response relation (Dosemeci and others 1990). No data on sleeping next to an open window were obtained at the time of the radon measurement. Further conclusions regarding sleeping with an open window are problematic, in that other studies have not considered the issue.

### Winnipeg Case-Control Study

*Study subjects.* This study, summarized in Table G-15, was a case-control study of lung cancer in males and females in Winnipeg, Canada. In Létourneau

TABLE G-15   Summary of results of Winnipeg, Canada, case-control study

| Factor | Comment |
|---|---|
| Principal reference | Létourneau and others 1994. |
| Design | Case-control study of males and females. |
| Study subjects | *Cases*: Cases included all residents of Winnipeg, Canada 35-80 yr old with histologically confirmed, primary lung cancer diagnosed in September 1983 to September 1990 and listed with the provincial cancer-incidence registry.<br><br>*Controls*: Controls were randomly selected from the Winnipeg telephone directory and individually matched on age within 5 yr and sex.<br><br>*Subjects in radon study*: A total of 759 pairs were assembled. After exclusion for misdiagnosis or improper control selection, 738 case-control pairs were enrolled. 257 cases and 78 controls had proxy interviews. |
| Lung-cancer histology | 738 cases: squamous, 31.4%; small cell, 15.9%; adenocarcinoma, 32.9%; other 19.8%. |
| Rn-measurement protocol | *Measurements*: Two sequential 6-mo α-track detectors in the bedroom and two in the basement of up to 3 homes in the Winnipeg metropolitan area. For apartments, only bedroom measurements were made. Yearly values were taken as the mean of the 2 measurements.<br><br>*Missing*: Subjects had a mean of 5 homes in the Winnipeg area; attempts were made to measure 3 homes. 7,318 homes were eligible, and 4,448 were measured (61%). |
| Rn measurements | *Mean:* For bedrooms: cases, 3.1 pCiL$^{-1}$; controls, 3.4 pCiL$^{-1}$. For basements: cases, 5.1 pCiL$^{-1}$; controls, 5.6 pCiL$^{-1}$. |
| Rn-exposure estimation | *Exposure time window*: Two windows defined: 5-30 yr and 5-15 yr before date of case diagnosis or control selection.<br><br>*Coverage*: About 67% of 5 to 30 yr window and 80% of 5 to 15 yr window.<br><br>*Imputation for gaps*: For cumulative exposure, used mean concentration for Winnipeg (3.3 in the living area and 5.3 in the basement). |
| Results | *Overall*: For categories (estimated from cumulative exposure) <1.9, 1.9-3.9, 3.9-7.8, and ≥ 7.8 pCiL$^{-1}$, RRs were 1.0, 1.0, 0.8, and 1.0 for the 5 to 30 yr window and 1.0, 1.0, 0.8, and 1.0 for the 5 to 15 yr window.<br><br>*Histology*: RRs similar and show no increased risk with exposure by cell type.<br><br>*Smoking*: Smoking patterns were used only for adjustment; no evaluation of effect modification of radon RRs was conducted.<br><br>*Subgroup analyses*: Data on occupational exposures were used only for adjustment; no evaluation of effect modification of radon RR was conducted. |

and others (1994), cases were histologically confirmed primary lung-cancer cases diagnosed between the ages of 35 and 80 years and, listed in the provincial cancer-incidence registry maintained by the Manitoba Cancer Treatment and Research Foundation for 1983-1990. All patients were residing in Winnipeg at the time of diagnosis. Controls were individually matched to cases on age within 5 years and on sex and were identified through the Winnipeg telephone directory.

A total of 759 matched pairs were initially identified. When cases that did not have primary lung cancer or had improperly matched controls were excluded, a total of 738 pairs remained for analysis.

*Data collection.* Information from in-person interviews was collected on demographic characteristics, education, and smoking practices and on detailed residential history. The questionnaire also incorporated a detailed occupational history, including information on specific job exposures.

*Methods of radon measurement.* There was a mean of 9 homes per subject, of which 5 were in the Winnipeg metropolitan area. It was not clear whether these means reflect lifetime residency or residencies in an exposure-time window. Radon was measured in 3 of these homes, although it was not clear precisely what criteria were used to select homes. The authors identified a total of 7,745 homes to be measured. This number was reduced for homes that had been occupied for less than 1 year (6%), for refusals (11%), for homes that no longer existed, were commercial institutions, or could not be located (24%), or where the dosimeter was lost or damaged (2%). Radon measurements were obtained for 4,448 homes (57%).

Year-long monitoring of the bedroom and, if there was one, the basement was undertaken. The basement was selected to provide the maximal possible residential exposure. Year-long monitoring was achieved through the sequential placement of two 6 mo detectors.

The detectors were developed and calibrated in house by laboratories of the Bureau of Radiation and Medical Devices in the Department of National Health and Welfare. Although the in-house calibration might affect comparisons with other studies in overall mean radon concentration, it should have no effect on the evaluation of trends in the exposure-response relation.

Two exposure windows were defined: 5-30 yr and 5-15 yr before the date of enrollment in the study. For estimation of cumulative radon exposure, imputation of missing measurements used the mean concentration in living areas [122.10 $Bqm^{-3}$ (3.3 $pCiL^{-1}$)] or in basements [196.10 $Bqm^{-3}$ (5.3 $pCiL^{-1}$)].

*Results.* Available radon measurements covered about 67% and 80% of the exposure windows of 5-30 yr and 5-15 yr, respectively. For cases and controls mean radon concentrations were 114.7 $Bqm^{-3}$ (3.1 $pCiL^{-1}$) and 125.8 $Bqm^{-3}$ (3.4 $pCiL^{-1}$) for bedrooms and 188.7 $Bqm^{-3}$ (5.1 $pCiL^{-1}$) and 207.2 $Bqm^{-3}$ (5.6 $pCiL^{-1}$) for basements, respectively.

Preliminary analysis revealed that cases were significantly less educated than controls and somewhat less likely to be born in Canada, although almost

TABLE G-16   Distribution of cases and controls and adjusted odds ratios[a] (OR) and confidence intervals (CI) by radon concentration in Winnipeg, Canada, case-control study (concentration level estimated from cumulative exposure)

| | Radon concentration ($pCiL^{-1}$) | | | | | |
|---|---|---|---|---|---|---|
| | < 1.9 | 1.9-3.9 | 3.9-7.8 | ≥ 7.8 | Total | P for trend |
| Cases | 92 | 488 | 118 | 40 | 738 | |
| Controls | 84 | 453 | 153 | 48 | 738 | |
| OR | 1.0 | 1.0 | 0.8 | 1.0 | | n.s. |
| 95% CI | | (0.6,1.5) | (0.5,1.4) | (0.7,1.5) | | |

[a]ORs adjusted for education and smoking with analyses matched on age and sex.
Source: Létourneau and others 1994.

80% of the subjects were born in Canada.  All analyses were adjusted for education and smoking status; age and sex were adjusted through the study matching. Table G-16 shows that there was no trend in the RRs with increasing TWA radon concentration in the bedrooms.  Similar results hold for basement measurements. The results were similar when cases were restricted by histologic type.

**Missouri Case-Control Study of Nonsmoking Females**

*Study subjects.*   The Missouri study was a population-based case-control study of white non-smoking woman, defined as lifelong never-smokers or former smokers who ceased 15 yr or more before interview (Alavanja and others 1994). The study is summarized in Table G-17.  Among former smokers, the median time since smoking cessation was 24 yr.  Cases were women 30-84 yr old with primary lung cancer who were reported to the Missouri Cancer Registry from June 1, 1986, to June 1, 1991.  After exclusion of ineligible cases, interviews were completed on 618 cases.  Radon measurements were obtained for 538 cases (83%); measurements were not obtained for 80 cases because of refusal, homes that were out of state or destroyed, or other reasons.  Although all cases were confirmed when diagnosis was reported to the registry, a separate panel of experts was established to review available slides; 409 of the 538 cases (76%) were reviewed.

A population-based control sample of white nonsmoking women was randomly selected by using Missouri state driver's license files (age 30-64 yr) or files of the Health Care Financing Administration (age 65-84 yr).  The controls were selected to match the age distribution of cases in 5-yr categories.  Of the 1,527 controls who satisfied enrollment criteria, 1,402 (92%) agreed to an interview and 1,183 (77%) had at least one valid year-long α-track measurement.

*Data collection.*   An initial telephone questionnaire was used to screen eligible subjects.  If a subject agreed to participate, a telephone-interview survey

TABLE G-17   Summary of results of Missouri case-control study of female never-smokers

| Factor | Comment |
| --- | --- |
| Principal reference | Alavanja and others 1994. |
| Design | Case-control study of female never-smokers and long-term former smokers. |
| Study subjects | *Cases*: 618 women 30-84 yr old with primary lung cancer listed with the Missouri Cancer Registry an June 1, 1986, to June 1, 1991, who never smoked or were long-term former smokers. |
| | *Controls*: Population-based controls (1,527) selected from state drivers-license files or files of the Health Care Finance Administration, frequency matched by age. |
| | *Subjects in radon study*: After refusals and other exclusions, 538 cases (87%) and 1,183 controls (78%) had at least 1 home in the 5-30 yr before enrollment measured for radon. |
| Lung-cancer histology | A histologic review, separate from registry notification, was conducted.  From 409 cases, there were 262 adenocarcinomas (53.5%); other cell types were not reported. |
| Rn-measurement protocol | *Measurements*: 1-yr $\alpha$-track detectors in kitchen and in bedroom in all homes in Missouri occupied 1 yr or more from 5-30 yr before date of enrollment. |
| | *Missing*: Radon measurements available for 74% of identified dwellings. |
| Rn measurements | *Mean and median*: Case and control values were the same, 1.8 $\mathrm{pCiL}^{-1}$ (mean) and 1.4 $\mathrm{pCiL}^{-1}$ (median).  About 7% had homes above 4 $\mathrm{pCiL}^{-1}$. |
| Rn-exposure estimation | *Exposure-time window*: 5-30 yr before to case incidence or control interview. |
| | *Coverage*: Mean 20 yr of residence corresponding to 78% of the exposure-time window. |
| | *Imputation for gaps*: None for TWA radon concentration; missing values for cases and controls replaced with means all cases and controls, respectively. |
| Results | *Overall*: For quintile categories < 0.8, 0.8-1.2, 1.2-1.7, 1.7-2.5, and $\geq$ 2.5 $\mathrm{pCiL}^{-1}$, RRs were 1.0, 1.0, 0.8, 0.9, and 1.2, P value for trend, 0.99 with continuous value for test and 0.19 with category means. |
| | *Histology*: RR showed a suggestive trend with adenocarcinoma cell type; P value for trend was 0.31 with continuous and 0.04 with categoric radon values. |
| | *Smoking*: No difference in RR trend with radon concentration for never-smokers or former smokers. |
| | *Subgroup analyses*: Suggestive RR trend (P = 0.06) for data restricted to in-person interview.  Since measured quantity, reason for differences uncertain. |

obtained information on demographic factors, occupational history, lifetime passive smoking, previous active smoking, diet, and previous diseases, and a detailed residential history.

*Methods of radon measurement.* For each subject, radon was measured in all homes in the state of Missouri occupied for at least 1 yr during the 30 yr before enrollment. One year-long $\alpha$-track detector was placed in the bedroom and one in the kitchen. Every 3 mo, subject's homes were checked to see whether the dosimeters were still in place. Quality control procedures—blind inclusion of blank and spiked dosimeters and duplicate detectors—suggested excellent validity in the measurement protocol. A small subsample of 3-mo winter measurements had a mean value twice the year-long readings.

The time-weighted radon concentration was computed for each subject by using all available measurement data; gaps in the exposure window were ignored. Cumulative radon exposure was estimated for an exposure window of 5-30 yr before lung-cancer incidence for cases and before interview for controls. Missing values for cases or controls were set to the mean radon concentration data for cases or controls, respectively.

Questionnaire data revealed that the subject occupancy factor was 84%. No special adjustment for occupancy was carried out.

*Results.* An average of about 20 yr of occupancy in the exposure period of 5-30 yr was covered by measurement data, 78% of the residency time. Mean and median radon concentrations were the same for cases and for controls: 66.6 Bqm$^{-3}$ (1.8 pCiL$^{-1}$) and 51.8 Bqm$^{-3}$ (1.4 pCiL$^{-1}$), respectively.

RR results were presented by categories defined by quintiles; the mean for the highest radon-concentration category was 151.7 Bqm$^{-3}$ (4.1 pCiL$^{-1}$). Table G-18 shows no increase in age-adjusted RR with increasing radon concentration. The RRs were adjusted only for age, but the pattern was unaffected by further adjustment of RRs for previous smoking, pack-years of smoking, previous lung disease, education, or intake of saturated fat.

TABLE G-18   Distribution of cases and controls (males and females combined) and adjusted odds ratios[a] (OR) and confidence intervals (CI) by quintiles of radon concentration in Missouri case-control study of female never-smokers

| | Radon concentration (pCiL$^{-1}$) | | | | | | |
| | < 0.8 | 0.8-1.2 | 1.2-1.7 | 1.7-2.5 | ≥ 2.5 | Total | P for trend |
|---|---|---|---|---|---|---|---|
| Cases | 112 | 112 | 93 | 99 | 122 | 538 | |
| Controls | 233 | 242 | 233 | 252 | 223 | 1,183 | |
| OR | 1.0 | 1.0 | 0.8 | 0.9 | 1.2 | | 0.19 |
| 95% CI | | (0.7,1.4) | (0.6,1.2) | (0.6,1.2) | (0.9,1.7) | | |

[a]ORs adjusted for age.
Source: Alavanja and others in review.

There was no RR trend with exposure within age categories or smoking status. Limiting cases to the 262 women with an adenocarcinoma cell type resulted in a suggestive RR trend. RRs for the 5 radon-concentration categories of Table G-18 were 1.0, 1.4, 1.1, 1.2, and 1.7; the p value for the test of linear trend was 0.04 when mean values for each category were used and 0.31 when continuous radon concentration.

### Finnish Case-Control Study (Finland I)

*Study subjects.* A population-based case-control study (denoted Finland I) of lung cancer in men was conducted in southern Finland in 19 municipalities. As of 1980, about 65,000 males lived in these areas (Ruosteenoja 1991). The study is summarized in Table G-19. Cases consisted of lung cancers diagnosed in men in the designated municipalities in 1980-1985. In the period 1980-1982, cases were identified from the Finnish Cancer Registry; the more-recent cases were accrued directly from the records of hospital that diagnose and treat lung cancer. A total of 291 cases were available for study.

Controls were a random sample of all men in the Finnish Population Registry files who were living in the designed area on January 1, 1980. Controls were frequency-matched to the age profile of the cases. Controls were then sent a mail questionnaire to obtain information on tobacco use. From the returned questionnaires (91%), controls were further selected to match the smoking proportions of the cases; 10% never-smokers, 10% ex-smokers who quit before 1979, and 80% current smokers or recent ex-smokers. A total of 495 controls were enrolled into the study: 50 never-smokers, 50 ex-smokers, and 395 current smokers.

*Data collection.* In-person interviews were conducted for cases and controls or, if they were deceased, with their next of kin. Information was collected on residential history, house type, smoking, education, and occupation.

*Methods of radon measurements.* The Finnish Centre for Radiation and Nuclear Safety conducted measurements of indoor radon concentrations in all dwellings occupied for 1 yr or more since 1950; $\alpha$-track detectors were placed for 2 mo in the winter between November 1, 1986 and April 30, 1987.

For analysis, an exposure window was defined as the 25-yr period from 1950 to 1975. For homes that could not be measured, a regression equation developed by Mäkeläinen and others (1987) that accounted for housing type and other factors was applied to estimate radon concentration. For dwellings higher than the ground floor, radon concentration was assigned the value 51.8 $Bqm^{-3}$ (1.4 $pCiL^{-1}$). Two radon measures were calculated: the TWA radon concentration based on available measurements and a TWA radon concentration for the entire 25-yr period based on measured and estimated concentrations.

*Results.* A total of 238 cases and 434 controls were available for analysis after exclusion for nonresponse to interview or inability to locate the subject. At the time of interview, 88% of cases and 21% of controls were deceased.

TABLE G-19    Summary of results of the Finnish case-control study

| Factor | Comment (1991) |
| --- | --- |
| Principal references | Ruosteenoja 1991, Ruosteenoja and others 1996. |
| Design | Case-control study of males. |
| Study subjects | *Cases*: 238 males with primary lung cancer diagnosed in 19 municipalities in Finland 1980-1985. For 1980-82 cases obtained from the Finnish Cancer Registry; from 1983-1985 cases from records of treatment hospitals. |
| | *Controls*: Population-based sample of mean living in 19 municipalities on January 1, 1980, frequency matched by age category. With information on smoking from a mail questionnaire, a random sample of 10 never-smokers, 10 ex-smokers and 395 current smokers was selected to serve as controls. |
| Lung-cancer histology | 238 cases: 91 (38.2%) squamous cell, 61 (25.6%) small cell, 18 (7.6%) adenocarcinomas, and 68 (28.6%) other or unknown. |
| Rn-measurement protocol | *Measurements*: 2-mo α-track detectors in the living room or bedroom in all homes occupied 1 yr or more 1950-1975. |
| | *Missing*: Radon measurements available for 50% of identified dwellings; and at least 1 measurement available for 76% of subjects. |
| Rn measurements | *Mean or median*: Not provided, but quintile cut-points indicate that 40% and 20% have concentrations above 4.7 and 7.4 $pCiL^{-3}$, respectively. |
| Rn-exposure estimation | *Exposure-time window*: 25 yr between 195 to 75 or 5-10 y prior to case incidence or control interview. |
| | *Coverage*: Mean 20 yr of residence corresponding to 78% of the exposure-time window. |
| | *Imputation for gaps*: For TWA radon concentration, missing values were estimated on basis of regression of housing type, municipality and other factors. |
| Results | *Overall*: For quintile categories < 2.2, 2.2-3.4, 3.4-4.7, 4.7-7.4, and ≥ 7.4 $pCiL^{-1}$, RRs were 1.0, 1.1, 1.7, 1.9, and 1.1; and the P value for trend was not significant. |
| | *Smoking*: Little effect of adjustment in pattern of RRs with radon concentration. |

Subjects resided in a total of 1,393 homes during the 1950-1975 period and indoor radon measurements were conducted in 696 homes (50%). Radon measurements in at least one house were obtained for 171 cases (72%) and 342 controls (79%); in this subgroup, the mean residency time covered by measurement data was about 20 yr or about 80% of the exposure window.

It was not entirely clear, but there seemed to be little difference in results between TWA radon-concentration measures. The distribution of cases and controls by category concentration was not provided by the author; however,

RRs, adjusted for age and smoking, for 5 categories based on quintiles were 1.0, 1.1, 1.7, 1.9, and 1.1 (Ruosteenoja 1991).

### Finnish Case-Control Study (Finland II)

*Study subjects.* Subjects for this case-control study, denoted Finland II, were selected from a subset of records of the Finnish Population Registry on persons living in the same single-family house (called the index dwelling) from January 1, 1967, or earlier to the end of 1985 (Auvinen and others 1996). Cases eligible for the study consisted of all persons with lung cancer diagnosed from January 1, 1986, to March 31, 1992, that were listed with the Finnish Cancer Registry. A total of 1,973 cases were identified. One control for each case was selected, matched by birth year and sex. The control had to be alive at the time of the diagnosis of the case. The investigators selected additional controls when possible, and a total of 2,885 controls were identified. There were 1,644 (83%) deceased cases and 326 (11%) deceased controls. The study is summarized in Table G-20.

*Data collection.* In September 1992, a mail questionnaire was sent to subjects or their next of kin. Information was obtained on residential history, smoking habits and occupational exposures. Information was also obtained on the daily number of hours spent indoors in the 1960s and 1970s. Response rates for the mail questionnaire were 55% for cases and 54% for controls.

*Methods of radon measurement.* One-year measurements with track-etch devices were undertaken for all subjects for their index dwelling. In the winter of 1992-1993, residents were mailed a detector and instructed to place it in a bedroom or living room. The detectors were returned the next winter. Houses were excluded for a number of reasons, the most common being the building of new houses on the same locations as index dwellings, uncertain dates of construction, measurements of less than 150 days, uninhabitation, and extensive renovation. About 20% of houses were excluded for those reasons.

The authors estimated the relative precision of the 1-yr measurements as ±20% for concentrations below 50 Bqm$^{-3}$ and ±15% for concentrations above 400 Bqm$^{-3}$, with a systematic bias of less than 10%.

*Results.* After exclusions for incomplete questionnaires and missing radon measurements, data were available on 1,055 cases and 1,544 controls. The study had originally been designed as an individually matched study, so results were presented only for analyses restricted to 517 case-control pairs. However, the authors state that unmatched results based on all available data were similar to the results from the matched analysis.

The mean radon concentrations were 103 Bqm$^{-3}$ for cases and 96 Bqm$^{-3}$ for controls; the median was 67 for both groups. The median occupancy times were similar, 11.5 h/d for cases and 11.7 h/d for controls.

Median residence times in the index house were 37 yr for cases and 35 yr for

TABLE G-20   Summary of results of Finland-II case-control study

| Factor | Comment (1991) |
| --- | --- |
| Principal reference | Auvinen and others 1996. |
| Design | Case-control study of males and females. |
| Study-subjects | Subjects were selected from the Finnish Population Registry of persons living in the same single-family house from January 1, 1967, or earlier until the end of 1985. *Cases*: All lung cancers diagnosed from January 1, 1986, to March 31, 1992, listed with the Finnish Cancer Registry; 1,973 cases were identified. *Controls*: For each case, subject was matched by birth year and sex and alive at the time of the diagnosis of the case. Additional controls were selected when possible, and a total of 2,885 controls were identified. For the matched analysis, 517 pairs were available. |
| Lung-cancer histology | Histologic or cytologic confirmation was available on 92% of cases. In the final series, the distribution was 36% squamous, 14% small cell, 13% adenocarcinoma, 9% other, and 28% undefined. Sex distribution was not provided. |
| Rn-measurement protocol | *Measurements*: One α-track detector was mailed to each subject with instructions for it to be placed in the bedroom or the living room. *Missing*: About 20% of houses (subjects) were excluded due because of missing information or other problems with the index dwelling. |
| Rn measurements | *Mean*: Cases, 103 Bqm$^{-3}$; and controls, 96 Bqm$^{-3}$. *Median*: 67 Bqm$^{-3}$ for cases and controls. |
| Rn-exposure estimation | *Exposure-time window*: Defined by study design; medians for years in the index house, 38 yr for cases and 35 yr for controls; median occupancy times, 11.5 hr/d for cases and 11.7 hr/d for controls. |
| Results | *Overall*: For categories < 50, 50-99, 100-199, 200-399, and ≥ 400 Bqm$^{-3}$, adjusted RRs (and 95% CIs) were 1.0, 1.03 (0.8-1.3), 1.00 (0.8-1.3), 0.91 (0.6-1.4), and 1.14 (0.7-1.9). RRs similar when radon levels weighted by occupancy. *Histology*: RR similar by cell type. *Smoking*: RR patterns were similar within smoking categories. |

controls, and 70% of cases and 76% of controls lived more than 30 yr in the index house. Although the design indicated a minimum of 19 yr of residency in the index house, registry information appears not to reflect actual residency. A total of 26 cases (5%) and 36 controls (7%) had less than 16 yr residency in the index house.

For indoor-radon categories of less than 50, 50-99, 100-199, 200-399, and ≥400 Bqm$^{-3}$, RRs (and 95% confidence intervals) adjusted for age, sex, and smoking were 1.0, 1.03 (0.8-1.3), 1.00 (0.8-1.3), 0.91 (0.6-1.4), and 1.14 (0.7-

1.9). RRs were similar when radon concentrations were weighted by occupancy or when adjusted for occupational asbestos exposure.

RRs were also similar when data were analyzed by histologic type of lung cancer.

### Israeli Case-Control Study

*Study subjects.* This was a small hospital-based case-control study at the Rambam Medical Center in Israel, summarized in Table G-21. Subjects were consecutive patients with primary lung cancer seen in the oncology ward in 1985-

TABLE G-21   Summary of results of Israeli case-control study

| Factor | Comment (1991) |
|---|---|
| Principal reference | Biberman and others 1993. |
| Design | Hospital-based case-control study. |
| Study subjects | *Cases*: Two case groups defined on the basis of consecutive patients with primary lung cancer at an oncology ward of the Rambam Medical Center in 1985-1989: 35 cases with small-cell carcinoma (SCC) (including ever-smokers and never-smokers) and 26 cases with non-small-cell carcinoma (16 adenocarcinomas) who were never-smokers (NS). Cases must have live in Israel for at least 10 yr before diagnosis.<br>*Controls*: Patients without lung cancer matched by sex and 5-yr age group who were admitted to the same hospital immediately after case admission and lived in Israel 10 yr or more.<br>*Subjects in radon study*: After exclusions and refusals, 52 cases and 43 controls were eligible; however, only 35 matched pairs (20 SCC pairs and 15 NS pairs) were available for analysis. |
| Rn measurement protocol | *Measurements*: $\alpha$-track detectors placed for an average of 9 mo from June or July 1990 through April 1991. |
| Rn measurements | *Median*: For cases and controls, 1.09 and 0.9 pCiL$^{-1}$ for SCC pairs and 0.9 and 1.07 pCiL$^{-1}$ for NS pairs, respectively. Differences were not statistically significant. Overall mean concentration was 1.0 pCiL$^{-1}$. |
| Rn exposure estimation | *Exposure-time window*: None defined; measurement in current house only.<br>*Coverage*: 28 (80%) cases and 19 (54%) controls lived 20 yr or more in measured house; 15 (43%) cases and 13 (37%) controls lived 30 yr or more in measured house. |
| Results | *Overall*: No significant differences in median radon concentrations between cases and controls. RR for $\geq 1$ pCiL$^{-1}$ compared with $<1$ pCiL$^{-1}$ was 1.5 with 90% CI (0.4,5.4) for SCC pairs and 0.5 with 90% CI (0.1,2.2) for NS pairs. |

1989 (Biberman and others 1993). Two case groups were defined: 35 patients with small cell carcinoma (denoted the SCC group), including ever-smokers and never-smokers, and 26 patients with non-small cell carcinoma who were never-smokers (denoted the NS group and including 16 patients with adenocarcinoma). All subjects had to have lived in Israel 10 yr or more before diagnosis.

Controls were matched by sex and age group within 5 yr from admissions to the same hospital immediately after case admission. Controls were also limited to those who had lived 10 yr or more in Israel before admission.

A total of 52 cases and 43 controls were eligible; however, after refusals or an inability to obtain radon measurements, a total of 35 pairs were available for analysis (20 SCC pairs and 15 NS pairs).

*Data collection.* Personal interviews with subjects or next of kin yielded information on residential, occupational, and smoking histories.

*Methods of radon measurement.* One α-track detector was placed in the bedroom of each subject in June-July 1990 and collected starting in April 1991. The detectors remained in place for a mean of 9 mo (Biberman and others 1993). Of the 70 dwellings, 33 (47%) were single-story or ground-floor units of multi-unit apartments. The geometric mean was 37 $Bqm^{-3}$ (1.0 $pCiL^{-1}$), and the range was 7.4-262.7 $Bqm^{-3}$ (0.2-7.1 $pCiL^{-1}$). Seven measurements were at or above 74 $Bqm^{-3}$ (2 $pCiL^{-1}$).

*Results.* In the SCC group, 17 subjects were males and 3 females; 19 subjects and 14 controls were ever-smokers. In the NS group, 4 subjects were males and 11 females; no subjects were smokers, and 2 controls were ever-smokers.

There was no significant difference in radon concentration between cases and controls for the SCC or NS groups. For cases and controls, median concentrations were 40.33 and 33.3 $Bqm^{-3}$ (1.09 and 0.9 $pCiL^{-1}$) for the SCC pairs and 32.93 to 39.59 $Bqm^{-3}$ (0.89 and 1.07 $pCiL^{-1}$) for the NS pairs. After adjustment for pack-years of cigarette use, the RR for 37 $Bqm^{-3}$ (1.0 $pCiL^{-1}$) compared with 37 $Bqm^{-3}$ (1.0 $pCiL^{-1}$) was 1.5 with a 90% CI of 0.4-5.4 for the SCC pairs and 0.5 with a 90% CI of 0.1-2.2 for the NS pairs. RRs were significantly increased for long-term residency on a ground floor for both case groups, but the interpretation of this result is clouded by other possible case-control differences for which no adjustment could be made.

## Port Hope Case-Control Study

One of the earliest case-control studies to estimate indoor radon from direct measurements is summarized in Table G-22. This study was conducted in Port Hope, a town of about 10,000 residents on the north shore of Lake Ontario. In 1932, mining operations included processing of ore and the recovery of radium; after 1939, operations shifted to the production of uranium. Disposal of residue from the operations occurred on the plant site and in other designated areas. In 1953, modification of the operations resulted in use of demolition rubble and

TABLE G-22    Summary of results for the Port Hope case-control study

| Factor | Comment (1991) |
|---|---|
| Principal reference | Lees and others 1987. |
| Design | Case-control study of males and females living in Port Hope, Ontario. |
| Study subjects | *Cases*: 27 lung-cancer cases diagnosed in 1969-1979, in persons who lived 7 yr or more in Port Hope, and were never employed at the uranium-refining plant. |
|  | *Controls*: 49 subjects matched on sex and date of birth, who lived 7 yr or more in Port Hope, with at least 1 of these years during the 7-yr period before the date of diagnosis of the matched case. One dead and 1 live control were matched to each deceased case, and 2 live controls were matched to each live case. |
| Lung cancer histology | Cell types included 11 squamous cell carcinomas, 6 adenocarcinoma, and 11 unknown. |
| Rn progeny measurement protocol | *Measurements*: Precise protocol was not provided, but apparently measurements were of WL. |
| WL measurements | *Mean or median*: Not provided. |
| Rn progeny exposure estimation | *Exposure-time window*: Estimates of exposure based on all homes occupied in Port Hope since 1933. Residences outside Port Hope area were ignored. Exposure estimates in WLM and adjusted by a background exposure of 0.229 WLM/yr. On basis of WLM distributions, it was estimated that for cases and controls mean WLMs were 2.7 and 0.5, including 33% and 49% with "zero" WLM exposure (below estimated background exposure), respectively; among exposed, means were 4.1 and 1.0 WLM. |
| Results | Overall RR of 1.55 with 95% CI (0.6,4.1) and with adjustment for smoking RR of 2.36 with 95% CI (0.8,7.1). |

reclaimed building materials throughout the town for various construction purposes (Lees and others 1987).

*Study subjects.* Subjects were defined as persons who died of lung cancer in 1969-1979 and who lived in Port Hope for 7 yr or more before the year of diagnosis. Cases were identified through the Provincial Cancer Registry and by contacting local physicians.

For each case, 2 controls, matched on sex and year of birth, were selected from among persons who had lived in Port Hope for 7 yr or more, with at least 1 yr during the 7-yr period before the date of diagnosis of the matched case.

Persons were excluded if they had worked in the uranium-refining plant or if they did not meet residency requirements. After exclusions, 27 lung-cancer cases and 57 matched controls were studied.

*Data collection.* Data were collected by personal interviews with subjects or next of kin, including information on residential, occupational, and smoking histories.

*Methods of radon measurement.* The protocol used to measure radon or radon progeny in homes was not given in Lees and others (1987). However, the authors indicate that in 1976 a "complete survey of the town was undertaken to delineate radiation contaminated areas and measure radon levels." It was not stated what was measured, but for the analysis exposure of each subject was estimated in cumulative $Jhm^{-3}$ (WLM), on the basis of all houses occupied in Port Hope from 1933. Estimated values were adjusted to exclude background exposures, on the basis of an estimate $0.0008015$ $Jhm^{-3}/yr$ ($0.229$ WLM/yr) and the assumption that a "worker" and a "nonworker" spend 60% and 85% of the year, respectively, inside the house.

*Results.* After adjustment of exposures, 67% of cases and 51% of controls had "nonzero" exposures. Among the exposed, mean exposures were $0.01435$ and $0.0035$ $Jhm^{-3}$ (4.1 and 1.0 WLM) for cases and controls, respectively. The matched RR estimate for exposure was 1.55 (95% CI, 0.6-4.1) without adjustment for smoking status and 2.36 (95% CI, 0.8-7.1) with adjustment for smoking status.

### Case-Control Studies of Indoor Radon in Progress

Extrapolations from miner studies suggest that lung-cancer risk posed by indoor radon exposure may be a potentially important public-health problem. There have been substantial interest in obtaining direct evidence of harmful effects of indoor radon to validate miner-based extrapolations and to identify an upper bound of the risk. Few studies have been published, and difficult design issues remain in the conduct of the studies (Lubin and others 1990a, 1995c; Stidley and Samet 1993). To expand the base of information on the consequences of indoor radon exposure, several case-control studies are under way. Table G-23 lists these studies, which total some 13,000 lung-cancer cases.

## SUMMARY OF STUDIES OF LUNG CANCER AND INDOOR RADON

In this section, we provide an overall perspective from the various studies of residential radon, including results of a meta-analysis of current indoor-radon studies.

### Ecologic Studies

Ecologic studies are limited by the inability to estimate relevant exposures, by the presence of an extremely strong risk factor for lung cancer (cigarette-smoking), and by the intrinsic confounding arising from regression analyses that use summary data and model misspecification We conclude that ecologic studies are noninformative for estimating risks posed by exposure to indoor radon or for evaluating a potential threshold exposure below which radon-progeny exposure would not be harmful.

TABLE G-23    Summary of continuing studies of residential radon and lung cancer risk

| Country | Location | Cases | Cntrls | Estimated completion date | Comments |
|---|---|---|---|---|---|
| *European studies:* | | | | | |
| Belgium-France | Ardennes-Eiffel | 1,200 | — | 1996 | — |
| France | Brittany | 600 | 1,200 | 1996 | — |
| Germany | Western | 2,500 | | 1996 | — |
| | Eastern | 1,500 | | 1997 | — |
| | Tyrol | 250 | | 1997 | — |
| Sweden | | 480 | | 1998 | — |
| United Kingdom | Cornwall and Devon | 986 | | 1997 | — |
| *North American studies:* | | | | | |
| Canada | — | 800 | | 1998 | Includes only never-smoking subjects |
| United States | Connecticut | 960 | | 1995 | Jointly conducted with Utah |
| | Iowa | 450 | | 1998 | Includes subjects with at least 20 yr in current house |
| | Missouri | 700 | | 1996 | Extension of previous study, but includes ever-smoking and never-smoking women |
| | New Jersey | 787 | | 1995 | |
| | Utah | 600 | | 1997 | Jointly conducted with Connecticut |
| *Other:* | | | | | |
| China | Gansu Province | 900 | 1,800 | 1997 | About 50% of population live in homes built below ground level |

## Case-Control Studies

## Qualitative Summary of Results

Results of epidemiologic studies of indoor radon concentration and lung cancer that used surrogate measures were generally consistent with increased risk at higher exposures. Their interpretation is complicated by the inability to link the surrogate measure directly to an estimate of exposure to radon progeny for the study participants. Thus, although the results of these studies are an important step in establishing a link between residential radon and lung-cancer risk, their direct relevance in assessing either the risk posed by indoor exposure or the validity of miner-based risk extrapolations is limited. The most relevant epidemiologic studies of lung cancer are those which used in-home measurements of radon to estimate exposure, in that direct measurements provide the most accurate estimates of exposure. Eight major case-control studies have been reported that included direct radon measurements, along with a pooled analysis of 3 of the studies and a meta-analysis of the major studies.

Table G-24 summarizes the sizes of the various studies, the radon concentrations and overall results. The highest radon concentrations were found in the Finland's-I study [mean, 210.9 $Bqm^{-3}$ (5.7 $pCiL^{-1}$)], and the next highest in the Stockholm study [mean, 129.5 $Bqm^{-3}$ (3.5 $pCiL^{-1}$)] and Winnipeg study (mean, 118.4 and 199.8 $Bqm^{-3}$ (3.2 and 5.4 $pCiL^{-1}$) in the living room and basement, respectively). Intermeditate radon concentrations were measured in the Swedish national study [mean, 107.3 $Bqm^{-3}$ (2.9 $pCiL^{-1}$)], the Finland-II study [mean, 99.9 $Bqm^{-3}$ (2.7 $pCiL^{-1}$)], and the Shenyang study [median, 85.1 $Bqm^{-3}$ (2.3 $pCiL^{-1}$)]; and the lowest concentrations were measured in the New Jersey study [median, 22.2 $Bqm^{-3}$ (0.6 $pCiL^{-1}$)]. The relationship of the measurement information for both the Swedish national study and the Finland-I study relative to the other studies is uncertain, inasmuch as radon was measured in winter, with detectors placed for 3 and 2 mo, respectively.

Comparisons of results from subgroup analyses provide an additional framework for evaluating consistency among studies. Variations of risk patterns within subgroups and inconsistencies between studies compel a cautious interpretation of results. Three studies—Shenyang, Winnipeg, and Finland-II—found no association with exposure overall and after intense subgroup analysis. Results of the other studies offer mixed support for a positive association. In Finland-I, RRs exceeded 1.0 for all radon categories, but there was no significant trend with increasing radon concentration and the highest category had a low RR. In New Jersey, there was a significant linear trend, but RRs for radon categories, of less than 1.0, 1.0-1.9, 2.0-3.9, and at least 4.0 $pCiL^{-1}$ were 1.0, 1.2, 1.3, and 8.7, indicating that the trend was strongly influenced by the highest category, which included 5 cases and 1 controls. In Stockholm, there was a significant trend with radon concentration; however, the trend was affected by occupancy or when

TABLE G-24   Summary of results from case-control studies of residential radon exposure

| Study | Cases | Controls | Rn level- pCiL$^{-1}$ (med/mean) | Comment |
|---|---|---|---|---|
| Finland-I | 238 | 415 | 40% > 4.7<br>20% > 7.4 | Results show only a modest suggestion of an overall trend with increasing radon level, but all RRs exceeded 1. |
| Finland-II | 517 | 517 | cases, 103 Bqm$^{-3}$ (mean); controls, 96 Bqm$^{-3}$ (mean) | Results show no overall trend. Residential occupancy less than 12 h/d. |
| Israel | 35 | 35 | 1.0 (mean) | Study has few cases; radon concentrations are very low; no conclusions can be drawn. |
| Missouri | 538 | 1,183 | cases, 1.8 (mean); controls, 1.8 (mean) | Results show no overall trend with increasing radon level; suggestive trends were found when analyses restricted to adenocarcinoma cases or in-person interviews. |
| New Jersey | 480 | 442 | cases, 0.5 (med) controls, 0.5 (med) | Significant exposure-response trend, but mean exposures very low and results influenced strongly by highest exposure category with 5 cases and 1 control. |
| Port Hope | 27 | 49 | cases, 2.7[a] (mean); controls, 0.5[a] (mean) | Nonsignificant excess relative risk with or without adjustment for smoking. |
| Shenyang | 308 | 356 | cases, 2.8 (med); controls, 2.9 (med) | Results show no increasing RR with increasing radon level, overall and within categories of indoor air pollution. |
| Stockholm | 201 | 378 | cases, 3.1 (med); controls, 2.9 (med) | Results suggest positive increase, but cautious interpretation indicated because trend depends on cut-points and disappears after adjustment for occupancy or with BEIR IV weighting. |

TABLE G-24  Continued

| Study | Cases | Controls | Rn level- $pCiL^{-1}$ (med/mean) | Comment |
|---|---|---|---|---|
| Sweden | 1,281 | 2,576 | 1.5 (med) | RRs increase significantly with increasing radon level; RR patterns similar by histologic type and homogeneous across categories for never-smoker, ex-smoker, and number cigarettes per day. |
| Winnipeg | 738 | 738 | cases, 3.1 (mean); controls, 3.4 (mean) | Results show no increasing RR with increasing radon level, as measured in living area or in basement. |

[a]Estimated cumulative radon-progeny exposure in WLM.

exposures more than 15 yr before were given half the weight. Furthermore, it was found that the p value for the test of trend differed when continuous $Bqm^{-3}$ ($pCiL^{-1}$) was used as the quantitative value in the test statistic, as opposed to category-specific means. The Swedish national study offered the clearest pattern of increasing RR trend with radon concentration.

Subgroup analyses revealed inconsistencies within and between studies. In New Jersey, there was no trend in the RRs with radon concentration among never-smokers, a positive trend in light smokers (under 25 cigarettes/d), and a negative trend in heavy smokers (at least 25 cigarettes/d); in Stockholm the trend was observed only in never-smokers and in heavy smokers (at least 20 cigarettes/ d); and in the Swedish national study trends, were the same for never-smokers, ex-smokers, and current smokers. In New Jersey, the RR trend was steepest when the case group was restricted to large cell carcinomas (a relatively rare histologic type); in Stockholm, trends were most apparent with small and squamous cell carcinoma; and in Sweden, there was no difference by histologic type. By way of comparison, in miners, there is suggestive evidence that radon-progeny exposure might be more closely associated with small cell carcinoma (Land and others 1993; Yao and others 1994) and adenocarcinoma (Yao and others 1994).

## Quantitative Summary Based on Pooled Analysis of Pooled Data from 3 Studies

Results from studies of indoor radon and lung cancer are quantitatively summarized by either pooling data (Chekoway 1991; Friedenreich 1993) or conducting meta-analysis (Greenland 1987; Thacker 1988). In the former approach, original, primary data from multiple studies are combined and analyzed jointly. In the latter approach, only data from published papers are used (Glass 1976); that is, the study is the unit of analysis (Greenland 1987). Both approaches have well-known limitations due to differences among the studies in design, type and method of data collection, source population, quality-control procedures, information on important confounding variables, and time (Friedenreich 1993; Thacker 1988). Meta-analyses have added burdens associated with the need to rely on information that is available only in the published papers; that limits flexibility to assess the exposure of interest, adjust for important confounders, and evaluate subtle effects (Greenland 1984; Oakes 1990; Petitti 1994; Shapiro 1994).

An analysis of pooled primary data from residential case-control studies in New Jersey, Shenyang, and Stockholm—including almost 1,000 cases—concluded that these 3 studies were consistent with each other and that any differences among them could have arisen by chance (Lubin and others 1994b). The study-specific estimates of RR and 95% CIs at 150 per $Bqm^{-3}$ based on fitted linear excess-RR models were 1.7 (0.8-3.8), 0.9 (0.0-1.2), and 1.2 (0.8-2.4), respectively. The combined exposure-response relationship showed no trend, with a pooled RR estimate of 1.0 with 95% CI (0.8-1.3) at 150 per $Bqm^{-3}$. Results suggest that RRs were consistent with no effect of exposure; however, results were also consistent with extrapolations from miners.

## Quantitative Summary Based on Meta-Analysis of 8 Studies

A recent meta-analysis involved the 8 studies that had enrolled 200 or more lung-cancer cases is listed in Table G-24 (and shown in figure G-1) (Lubin and Boice 1997). Overall, 4,263 lung-cancer cases and 6,612 controls contributed to the meta-analysis. Figure G-1 suggests that RRs from indoor studies are consistent with the extrapolation based on miner studies, but also that RRs are quite variable. The CIs for the individual RRs are large, suggesting that results are also consistent with no effect of radon concentration. However, more of the RRs exceeded 1.0 than were less than 1.0, and there was a general tendency for higher RRs with higher radon concentrations.

Lubin and Boice (1997) obtained RR estimates and 95% CIs for categories of radon concentration in $Bqm^{-3}$ from published results and carried out weighted linear-regression analyses of the natural logarithm of the RR estimates using inverse variances as weights (Draper and Smith 1966). For each study, a loglinear RR model that passed through the quantitative value for the baseline category was fitted. For exposure at concentration x, the regression model was

$$\log[RR(x;x_0)] = \beta (x - x_0), \tag{1}$$

where $x_0$ was the exposure for the referent category and $\beta$ the unknown exposure-response parameter. Model (1) was fitted to each study, and an estimate of $\beta_i$, denoted $\hat{\beta}_i$, was obtained. A summary estimate for $\beta_1, \ldots, \beta_8$ was obtained with the same 2-step approach used by the committee in its analysis of miners.

Except for the Finland-I study, loglinear models provided good fits to the RR from the individual studies (Figure G-2), and there were no significant deviations from linearity. The study-specific values for the exponential of the estimates in units of 150 $Bqm^{-3}$, that is, $\exp(\hat{\beta}_i \times 150)$, are shown in Table G-25. The fitted RRs at 150 $Bqm^{-3}$ ranged from 0.8 to 1.8. A test of homogeneity of the estimates was rejected ($p < 0.001$). The fitted RR at 150 $Bqm^{-3}$ was $\exp(0.0009 \times 150) = 1.14$ with 95% CI of (1.0-1.3).

The baseline categories for the RRs differed for the various studies. RRs for each category and for each study were adjusted to a baseline concentration of "zero" radon (that is, ambient concentrations) by multiplying each RR by $\exp(\hat{\beta}_i x_{0i})$, where $x_{0i}$ was the concentration for the baseline category and $\hat{\beta}_i$ the estimate for the $i^{th}$ study. With the adjusted RRs, 5 categories of radon concentration were created on the basis of quintiles, less than 55.4, 55.5-88.7, 88.8-142.2, 142.3-250.8, and at least 250.9 $Bqm^{-3}$. Estimates of RR and 95% CIs for the 5 categories were 1.0, 1.05 (0.9-1.2), 1.05 (0.9-1.2), 1.25 (1.0-1.5), and 1.20 (1.0-1.4). Those RRs were in turn adjusted to a zero baseline by multiplying by $\exp(\beta x_0)$, where $x_0$ was the mean concentration for the lowest radon category, 34.2 $Bqm^{-3}$, and $\beta$ the parameter estimate. Figure 3-2 presented earlier in chapter 3 shows the adjusted RRs (solid squares). The figure also shows that the summary loglinear model, $\log[RR(x)] = 0.009x$, provided a good fit to the data.

Mean cumulative exposure in the miner studies was over 20 times greater than living 30 yr in an average US house at the mean concentration of 46 $Bqm^{-3}$. With data from the $< 0.175$ $Jhm^{-3}$ ($< 50$ WLM) restricted analysis of miners (see page 3-15), RRs in miners were compared with RRs from the residential studies. A correspondence between exposures for miners in $Jhm^{-3}$ (WLM) and radon concentrations in homes in $Bqm^{-3}$ was made assuming 30 yr of exposure, standard residential occupancy assumptions [living for 1 yr in a house at 37 $Bqm^{-3}$ and assuming 70% occupancy, and 0.4 equilibrium factor is approximately equal to 0.00014 $Jhm^{-3}$ (0.4 WLM) of exposure], and a 1.0 K factor. For example, a miner exposed to 0.0875 $Jhm^{-3}$ (25 WLM) was assumed to have about the same exposure as a person living 30 yr in a house with a radon concentration of 220 $Bqm^{-3}$ [$= 37 \times 25$ WLM/(30 yr $\times 0.14 \times 1.0$)]. For miners, RRs were calculated for 0, 1-9, 10-19, 20-29, 30-39, and 40-49 WLM. Figure 3-1 shows RRs and 95% CIs from the miner data (open squares).

The estimate for RR from a loglinear model fitted to the miner RRs under 0.175 $Jhm^{-3}$ (50 WLM) was 1.13 at 150 $Bqm^{-3}$ with a 95% CI of 1.0-1.2, essentially the same as the 1.14 (1.0, 1.3) estimate from the meta-analysis of residential studies. Thus, RRs for miner exposures under 0.175 $Jhm^{-3}$ (50 WLM) were similar to extrapolations with the miner-based risk model (Figure G-1), devel-

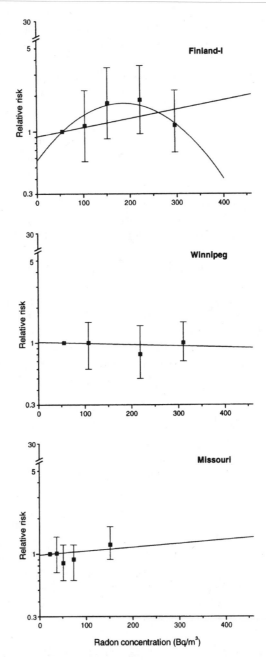

FIGURE G-2   Relative risks for radon-concentration categories and fitted exposure-response models for each case-control study.  Fitted lines adjusted to pass through quantitative value for baseline category.

FIGURE G-2 *Continued*

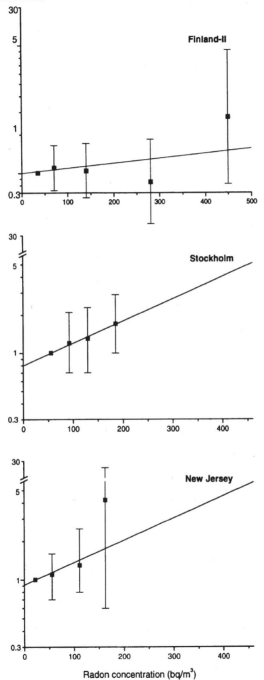

Radon concentration (bq/m³)

FIGURE G-2 *Continued*

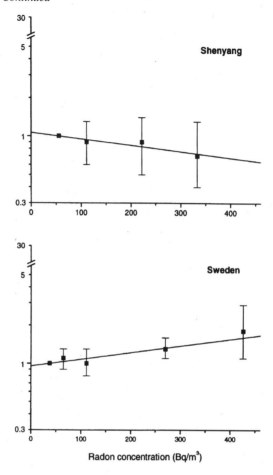

Radon concentration (Bq/m³)

oped from data with generally higher exposures, and similar to the RRs from the indoor studies.

Results from the indoor case-control studies do not provide direct information on lifetime risks posed by radon exposure. The excess risk of 14% at 150 Bqm$^{-3}$ corresponds to only 30 years of exposure in a house at a constant radon concentration and hence does not reflect the risk of lung cancer following lifetime exposure, where the estimated excess lifetime relative risk at 150 Bqm$^{-3}$ based on the miner models is 40 to 50% (Table 3-6). Estimated relative risks from indoor studies and from miner-based models reflect a 30-year exposure period at 148 Bqm$^{-3}$ and not lifetime exposures at this level. Thus, if exposures outside this 30-year period influence lung-cancer risk, as suggested by the miner data, then the 14% excess relative risk at 148 Bqm$^{-3}$ from indoor studies is a

TABLE G-25    Estimates of relative risk (RR) at 150 Bqm$^{-3}$ and the 95% confidence interval (CI) each study and for all studies combined

| Study | RR[a] | 95% CI | Reported in original paper[b] |
|---|---|---|---|
| Finland-I[c] | 1.30 | (1.09,1.55) | N.A. |
| Finland-II | 1.01 | (0.94,1.09) | 1.02 |
| New Jersey | 1.83 | (1.15,2.90) | 1.77 |
| Shenyang | 0.84 | (0.78,0.91) | 0.92[d] |
| Winnipeg | 0.96 | (0.86,1.08) | 0.97 |
| Stockholm | 1.83 | (1.34,2.50) | 1.79 |
| Sweden | 1.20 | (1.13,1.27) | 1.15 |
| Missouri | 1.12 | (0.92,1.36) | N.A. |
| Combined[e] | 1.14 | (1.01,1.30) | — |

[a]Values shown are estimated RR at 150 Bqm$^{-3}$ that is, $\exp(\beta \times 150)$, where $\beta$ was obtained from a weighted linear regression fitting the model $\log(RR) = \beta(x - x_0)$, where $x_0$ is the quantitative value for the lowest radon category and x is the category-specific radon level.
[b]RR at 150 Bqm$^{-3}$, on the basis of computed from exposure-response relationship provided in original reports. Exposure-response data not available (N.A.) in Finland-I and Missouri studies.
[c]For Finland-I, there was a significant departure from linearity (P = 0.03). Estimated RR for 150 Bqm$^{-3}$ with linear-quadratic model was 1.71.
[d]Taken from results in pooled analysis (Lubin and others 1994a).
[e]Combined estimate and confidence interval based on random-effects model. Fixed effects estimate was 1.11 with 95% CI (1.07,1.15).

biased estimate of the lifetime relative risk at this concentration and therefore cannot be used to estimate attributable risks for a population.

In the meta-analysis, study-specific exposure-response estimates differed significantly. In an attempt to explain the differences, values for overall mean radon level, percentage of exposure interval covered by radon-measurement data, mean number of homes per subject, mean number of measured homes per subject, percent of cases who smoked, percentage of eligible subjects included in the radon analysis, percentage of homes with year-long radon measurements, percentage living subjects, and percentage female subjects were obtained for each study. None of those variables, individually or jointly, explained the heterogeneity in the study-specific exposure-response estimates.

For the meta-analysis an influence analysis, in which summary estimates were computed on the basis of 7 of 8 studies, indicated that the overall estimates change very little when any single study is omitted.

In summary, there was a significant exposure-response relationship in the meta-analysis by Lubin and Boice (1997) with an estimated RR at 150 Bqm$^{-3}$ of 1.14, and results were generally confirmatory of miner-based extrapolations of risk and with RR among the least-exposed miners. However, meta-analyses are known to have numerous limitations, including an inability to explore adequately the consistency of results within and between studies and to control for potentially important confounding factors. Nonetheless, the results are consistent with a small effect on lung cancer associated with exposure to indoor radon progeny.

Finally, the results of the ecologic analysis by Cohen (1995) can be compared with the results of the meta-analysis. In his analysis, Cohen fitted a linear model that resulted in a declining linear excess-RR trend of 0.002/Bqm$^{-3}$. Figure 3-2 compares the ecologic regression line with RRs from residential studies and from miner studies. It is clear that the negative exposure-response relationship is contradicted by both the miner data and the data from the indoor radon studies.

## DESIGN LIMITATIONS OF INDOOR-RADON STUDIES

If miner studies are so unequivocal in showing the carcinogenicity of radon, why are results from current studies of indoor radon, particularly those which include measurements of indoor radon, variable and relatively inconclusive? Case-control studies of lung cancer and indoor radon are, of course, limited by factors that affect any epidemiologic study as summarized in Table G-26 (Lubin and others 1990a). Inadequate design elements might result in reduced power for a study to detect a significant effect and in biased or confounded estimates. However, studies of lung cancer and indoor radon have unique features that place additional burdens on the accurate assessment of the effects of exposure and the attainment of sufficient study power (Table G-27):

• The use of miner-based extrapolations provides uncertain estimates of the size of the RR in homes, although expected RRs are very small—an RR in the range of 1.1-1.3 for a 25-yr exposure at 148 Bqm$^{-3}$ (4 pCiL$^{-1}$). For a case-control study, this implies that the distribution of exposures for cases is very similar to the distribution of exposures for controls. As a consequence, substantial numbers of subjects are needed to establish a significant difference in the distributions and

TABLE G-26   Potential limiting factors of case-control studies of indoor radon and lung cancer

---

Error in estimation of radon exposure
Errors in estimation of tobacco use and other potential confounders
Data missing because of nonresponse
    To interview
    To radon measurement
    From use of surrogate responders
Misclassification of disease
Identification of appropriate target population for selection of controls
Inappropriate design assumptions for accurate assessment of sample size and power:
    Incorrect specification of dose-response
    Incorrect specification of true exposure distribution
    Failure to consider consequences of residential mobility
    Failure to consider effects of random error in exposure assessment

---

TABLE G-27    Sources of error in estimation of cumulative indoor radon progeny exposure

Errors related to measurement of radon:
  Counting error for α-track device
  Possible effects of airborne contaminants
  Measurement at fixed location in room
  Measurements limited to 1or 2 rooms only
  Diurnal and seasonal variation
  Use of contemporary measurements to characterize past levels
Sources of errors in duration of exposure:
  Variation in occupancy over time
  Imprecision of estimate of occupancy time
Radon exposures occurring outside home
Measurement gaps for homes within exposure period
Exposure as duration times mean exposure rate as an approximation of time-integrated
    exposure rate
Conversion of radon concentration to WL[a]

[a]WL denotes working levels, the unit of radon progeny measured in studies of underground mines. Conversion of radon to radon progeny is needed to estimate risk based on miner models.

to estimate effects precisely. In addition, because the distribution of radon concentrations is skewed, few homes exceed 148 Bqm$^{-3}$ (4 pCiL$^{-1}$); in the United States, only 5-7% of homes are estimated to exceed this level (EPA 1991; Marcinowski and others 1994).

• Subjects usually live in many homes during their lifetimes, thereby narrowing the range of exposures in the target population and reducing study power. The consequences of residential mobility can be demonstrated with an extreme example: If every member of the population moved every day and radon levels in homes were statistically independent, the total exposure of each subject after, say, 25 yr would be about 25 times the mean exposure rate; thus, there would be little or no exposure variation in the population, and this would complicate the detection of any risk.

• The use of contemporary measurements in current and past homes results in exposures estimated with great imprecision. Unless those formidable limitations can be addressed—with new measurement technologies, with studies in low-mobility or high-exposure populations, or with the pooling of data—it is uncertain whether definitive results can ever be achieved.

## EFFECTS OF ERROR, MOBILITY, AND MISSING DATA ON INDOOR-RADON STUDIES

The pattern of lung-cancer risk in miners suggests that exposures in the preceding 5-30 yr are the most relevant for estimating radon-associated lung-cancer risk (NRC 1988; Lubin and others 1994). That permits exposure-assessment efforts in residential studies to focus on more-recent years, which is fortunate because it is often impossible to locate residences and measure radon concentrations for homes in which subjects lived previously. However, the extent to which exposures before to the defined exposure-time window contribute to lung-cancer risk, then omitting them by design adds imprecision to the exposure estimates (Lubin and others 1990).

There is an important distinction between error in a measurement device and error in assessment of individual exposure. The most-common area dosimeter used in epidemiologic studies is the $\alpha$-track detector; radon concentration is determined by counting the number of etched tracks made on plastic film by alpha particles, and the number of tracks is proportional to concentration (Alter and Fleischer 1981; Lovett 1969). Both the counting and measurement processes are subject to error, which has been estimated to be about 15-25% (Létourneau and others 1994; Yeager and others 1991). The error in the measurement process defines the absolute lower bound of the accuracy of any exposure assessment based on $\alpha$-track devices.

Many other factors contribute to error in the estimation of personal exposure (Table G-27). Total exposure to radon progeny is the sum of exposures received in all environments, including the home, the workplace, and outdoors. $\alpha$-track detectors are usually left in place for several months to a year. In some studies, concentration measurements from short-term devices (3-7 days) supplemented those from long-term devices that might have been lost or unusable. Residential radon concentrations vary daily and seasonally (Swedjemark 1985), and short-term measurements or single-season measurements might not provide an accurate characterization of year-long radon levels.

For analysis, radon exposure in a defined time window is often computed as the time-weighted average concentration (TWA) or cumulative exposure in $Bqm^{-3}$-yr (or $pCiL^{-1}$-yr). For TWA, gaps in the measurement data for previous homes are often ignored when there is more than 1 residence. That can induce bias. Suppose that 1 subject lived for 30 yr in a home measured at 150 $Bqm^{-3}$ and a second subject lived for 15 yr in a home measured at the same level and 15 yr in an unmeasured home. If one ignores the missing data from the unmeasured home, each subject would have a computed TWA of 150 $Bqm^{-3}$. However, because of regression toward the mean, the TWA for the latter subject is likely an overestimate. If coverage of the exposure-time window is related to case status, ignoring measurement gaps is potentially biasing. To minimize such biases, an imputation procedure for missing data with adjustment for the variance estimates of parameters would be the preferred approach (Weinberg and others 1996).

In indoor-radon studies, only current radon levels can be measured, and they might not accurately reflect those of 15-30 yr earlier because of tightening of homes for energy conservation or other modifications (Kendall and others 1994). In addition, measurements are typically made only in a few rooms of a house.

## Simulation Studies

The effects of errors in exposure, residential mobility, and the inability to measure radon concentrations in all homes in the exposure-time window were illustrated in a series of simulation studies by Lubin and others (1995), expanding earlier analyses of Lubin and others (1990).

Steps in the simulation of data are shown in Table G-28. Case-control studies with M = 700 cases and N = 700 controls were selected from independently generated populations of 10,000 persons, with an overall lung cancer rate of 10%. Initially, it was assumed that each person lived in only 1 house. The lognormal distribution of US radon concentrations was used, with a geometric mean (GM) of 24.8 Bqm$^{-3}$ and a geometric standard deviation (GSD) of 3.11 (Marcinowski and others 1994).

A multiplicative error (U) was assumed; ln(U) was assumed to be normally distributed with a mean of 0 and a variance of $\tau^2$. Measurement error was specified by exp($\tau$) as 1.0 (no error), 1.50, 2.0, and 3.0, which roughly correspond to exposure errors of zero, ±50%, ±100%, and ±200%, respectively. For example, exp($\tau$) = 2.0 implies that a true exposure of 0.07 Jhm$^{-3}$ (20 WLM) will be estimated as 0.035-0.14 Jhm$^{-3}$ (10-40 WLM). For comparison with current case-control studies, radon progeny exposures were rescaled on the basis 25 yr of exposure to radon concentration, and RRs were computed by categories of Bqm$^{-3}$.

Simulations were also conducted to illustrate the effects of error, mobility, and missing radon-measurement data from past residences (Lubin and others 1995). Table G-29 provides an empirical calculation of study power for case-control studies with 700 cases and 700 controls and with 2,000 cases and 2,000

TABLE G-28   Steps in simulation study conducted by Lubin and others (1995)

1. For each individual, generate a true radon concentration by randomly sampling from a lognormal distribution with geometric mean 24.8 Bqm$^{-3}$ and GSD 3.11. Convert to radon progeny exposure in WLM on the basis of 25 yr of exposure[a] (X), multiplying by (0.18/37)25yr, where the first factor represents the conversion under standard assumptions of 1 yr residence in a house at 1 Bqm$^{-3}$ to exposure in WLM/yr.[a]
2. Compute the probability of lung cancer, on the basis of P(D = 1|X) = e$^a$(1+bX){1+e$^a$(1+bX)}$^{-1}$, with a and b specified. Randomly sample from a uniform 0-1 distribution to determine disease status D.
3. Include a multiplicative error, U, by randomly sampling from a lognormal distribution, where ln(U) is normal with mean 0 and variance $\tau^2$, and create an observed exposure Z = XU.
4. Repeat steps 1-3 10,000 times to generate a population.
5. Select M disease cases and N controls, categorize radon concentration, and compute RRs and test statistics, using continuous radon concentration as the quantitative variable.
6. Repeat steps 1-5 to generate each simulated case-control data set.

[a]Initial assumption was 25 yr of occupancy in a single house. This assumption was later relaxed.

TABLE G-29    The percentage of times P value for score test of no linear trend in relative risk with exposure is less than 0.05, based on 1,000 simulated case-control studies[a]

| | Number of homes occupied in 5 to 30 yr exposure window[b] | | | | | |
|---|---|---|---|---|---|---|
| | 1 | 2 | | 3 | | |
| Error distribution[c]: | Percent coverage of exposure-time window | | | | | |
| exp(τ) | 100% | 100% | 50% | 100% | 67% | 33% |
| | Study size: 700 cases and 700 controls | | | | | |
| 1.0 | 45.0 | 30.6 | 17.1 | 28.2 | 24.2 | 11.5 |
| 1.5 | 41.4 | 26.6 | 16.6 | 24.9 | 16.4 | 10.8 |
| 2.0 | 29.4 | 18.8 | 12.7 | 9.4 | 9.2 | 6.8 |
| 3.0 | 17.1 | 11.8 | 7.2 | 6.4 | 8.6 | 5.8 |
| | Study size: 2,000 cases and 2,000 controls | | | | | |
| 1.0 | 89.8 | 73.6 | 46.0 | 60.8 | 52.6 | 32.1 |
| 1.5 | 84.8 | 66.8 | 44.1 | 55.1 | 42.1 | 29.4 |
| 2.0 | 71.0 | 54.0 | 35.6 | 34.4 | 26.6 | 23.5 |
| 3.0 | 40.8 | 32.9 | 22.0 | 23.6 | 14.6 | 13.0 |

[a]Risk is based on a 0.10 background rate of lung cancer and an excess relative risk of 0.015 per working level month. Exposure is based on 25 yr of residence and a lognormal radon concentration distribution with geometric mean 24.8 Bqm$^{-3}$ and geometric standard deviation 3.11.

[b]For multiple homes, it is assumed that equal numbers of years are spent in each home. Thus, for 2 homes, 50% indicates that 12.5 yr of the exposure-time window was covered by radon-measurement data, for 3 homes, 33% and 67% indicate that 8.3 and 16.7 yr of the exposure window were covered by measurement data, respectively.

[c]The multiplicative error distribution is assumed to be lognormal, with the logarithm of the error having mean 0 and variance $\tau^2$. The row with exp(τ) = 1 shows results when exposure is measured without error.

controls, that is, the percentage of 1,000 simulated studies that rejected the null hypothesis of a radon effect. The table shows that a study with 700 cases and 700 controls, in which all subjects lived in a single residence for 30 yr and exposure is measured without error, has a power of 0.45 of rejecting a hypothesis test of no exposure effect when the true trend is 0.015/WLM. With 2,000 cases and 2,000 controls, the study has a power of 0.90. The table also illustrates the marked decline in power with increasing exposure error and mobility and with decreasing coverage of the exposure-time window.

## Sample Sizes for Case-Control Studies

Table G-30 shows the number of cases required for a study designed to have 90% power to reject a null hypothesis based on a 2-sided 0.05-level test if the alternative, b = 0.015 WLM, were true (Lubin and others 1995). This table updates the sample sizes provided in Lubin and others (1990a). With typical

TABLE G-30    The effects of measurement error, exp($\tau$), and mobility on sample size[a] (entries are number of lung-cancer cases required)

| Exp($\tau$)[b] | Mobility pattern | | |
|---|---|---|---|
| | 1 × 25 yr | 2 × 12.5 yr | 3 × 8.3 yr |
| | Control-to-case ratio = 1:1 | | |
| 1.0 | 2,033 | 2,447 | 3,408 |
| 1.5 | 2,521 | 3,292 | 4,879 |
| 2.0 | 3,716 | 5,365 | 8,484 |
| 3.0 | 8,429 | 13,530 | 22,694 |
| | Control-to-case ratio = 2:1 | | |
| 1.0 | 1,488 | 1,810 | 2,530 |
| 1.5 | 1,846 | 2,437 | 3,626 |
| 2.0 | 2,724 | 3,974 | 6,311 |
| 3.0 | 6,183 | . 10,034 | 16,895 |

[a]Study required to have 90% power to reject a trend in radon exposure, b = 0, when the true trend is b = 0.015, using a 2-sided 0.05-level test. Exposure based on 25 yr of exposure and occupancy in 1, 2, or 3 houses.
[b]It is assumed that error is multiplicative and the logarithm of the error is normally distributed with mean 0 and variance $\tau^2$.

mobility and with exp($\tau$) about 1.5-3.0, about 5,000-18,000 lung-cancer cases and an equal number of controls would be needed, or about 4,000-13,000 cases and twice the number of controls. Those numbers should be interpreted cautiously, perhaps as a lower bound, because calculations do not account for unmeasured houses and adjustment of other risk factors.

## CONCLUSIONS

Accurate exposure estimation is essential for any study of lung cancer and indoor radon. Estimating past exposures is a formidable task, and a present-day measurement, even if made for an entire year, might not accurately reflect radon concentrations of 30 yr ago or earlier. Exposure assessment is further burdened by subject mobility, which decreases the range of exposures in a population and thereby decreases study power. Mobility also creates the potential for gaps in the reconstruction of exposure histories because of an inability to measure all previous houses. When reasonable assumptions are made about measurement errors, mobility and gaps in the exposure-time window, any calculated dose-response relationship will probably be consistent with no exposure effect unless there are substantial numbers of cases and controls. On the basis of those observations, the committee concludes that the seeming inconsistency among case-control studies to date is in large part an inherent consequence of errors in dosimetry and residential mobility.

In recent years, new statistical techniques have been developed for analyses of case-control data that attempt to take errors in exposure assessment into account. The validity of the techniques and a resulting "corrected" risk estimate require direct information on the form of the error distribution. Thus, the estimation of the true effects of residential radon exposure could be enhanced by the collection of data that allows an evaluation of exposure errors.

The ability to estimate lung-cancer risk from indoor-radon studies is much more complex than the simple computer simulations, which were based only on measurement errors, considered only simple residential mobility patterns, and were defined by an ideal situation: occupancy known exactly, no variation in the equilibrium of radon with its decay products, and no misspecification of disease status. In reality, radon studies suffer from further uncertainties arising from a variable relationship between exposure and dose and from the potential confounding and potentiating effect of tobacco smoke, both active and passive, and the possible presence of other factors, such as indoor air quality or occupational exposures, which may also be measured only imprecisely.

Computer simulations document that small predicted levels of risk and misspecification of radon exposures contribute to the mixed results of current radon case-control studies. The public is perplexed by the seemingly conflicting results and uncertain as to the existence of an adverse effect. Although many of the newer studies are larger than published studies, the marked reduction in study power from (random) errors in exposure suggests that results from the newer studies might also be mixed (Samet 1994).

In the long term, several steps can help to address problems caused by extensive exposure-assessment errors. The most obvious is improvement in estimating exposures, which can be accomplished by selection of a stable target population so that the potential for gaps in exposure measurements is minimized or by using of improved technology for the measurement of radon concentrations.

The power of an indoor-radon study to detect an excess risk could also be enhanced by targeting special populations, such as a population with high exposures, a broad range of exposures, and low residential mobility.

Case-control studies of residential radon are limited by the generally low dose of alpha energy delivered to the lung, which reflects the low radon concentrations to which most of us are exposed every day. The anticipated excess risk is small and not readily measured, because of errors that affect estimation of exposure. On the basis of results of the simulations, the committee concludes the following:

• Because of error intrinsic in the measurement and estimation of prior radon exposures, a single study cannot be expected to provide sufficiently precise lung-cancer risk estimates.

- The inconsistencies in existing indoor radon concentrations are not surprising and are probably the consequence of mobility and errors in dosimetry related to missing radon measurements and poor exposure assessment.
- With increased residential mobility and errors in dosimetry, it is virtually impossible practically to distinguish between studies for which an underlying radon effect is assumed and studies for which no radon risk exists.
- Investigators are encouraged to include procedures to estimate the distribution of exposure errors in the design of indoor-radon studies, and analyses of studies should adjust for exposure errors.
- Additional studies of the consequences of missing exposure data within the exposure-time window, of ignoring the roughly 30% of time exposed to nonhome sources of radon progeny, and of incomplete coverage of radon concentrations in prior homes are needed.
- Improved technologies are needed for the accurate estimation of prior radon-progeny exposure.

Combining data from prior and current studies should be encouraged. However, even with a large sample size a clear picture of lung-cancer risk posed by residential radon exposure might not arise, because the substantial influence of errors in radon-exposure assessment.

# References

Aaltonen LA, Peltomaki P, Leach FS, Sistonen P, Pylkkanen L, Mecklin JP, Jarvinen H, Powell SM, Jen J, Hamilton SR, Petersen GM, Kinzler KW, Vogelstein B, de la Chapelle A. 1991. Clues to the pathogenesis of familial colorectal cancer. Science 260:812-816.

Abelson PH. 1990. Uncertainties about health effects of radon (editorial). Science 250:353.

Abelson PH. 1991. Mineral dusts and radon in uranium mines (editorial). Science 254:777.

Adamson IYR. 1985. Cellular kinetics of the lung. Pp. 289-317 in Toxicology of Inhaled Materials, H.P. Witschi and J.P. Brain, eds. Berlin: Springer-Verlag.

Agnew JE. 1984. Physical properties and mechanisms of deposition of aerosol. P. 49 in Aerosols and the Lung: Clinical and Experimental Aspects, S.W. Clarke and D. Pavia, eds. Boston: Butterworths.

Ahlberg J, Ahlbom A, Lipping H, Norell S, Osterblom L. 1981. Cancer among professional drivers: A problem-oriented register-based study. [Swed.] Lakartidningen 78:1545-1546.

Ahlman K, Koskela RS, Kuikka P, Koponen M, Annanmaki M. 1991. Mortality among sulfide ore miners. Am J Ind Med 19:603-617.

Alavanja MCR, Brownson RC, Lubin JH, Brown C, Berger C, Boice JD, Jr. 1994. Residential radon exposure and lung cancer among nonsmoking women. J Natl Cancer Inst. 86:1829-1837.

Albering HJ, Engelen JJ, Koulischer L, Welle IJ, Kleinjans JC. 1994. Indoor radon, an extrapulmonary genetic risk? Lancet 344:750-751.

Albrecht E and Kaul A. 1967. Continuous registration of 222Rn in air varying with time. In Assessment of Airborne Radioactivity in Nuclear Operations, Vienna: International Atomic Energy Agency.

Alexandrie AK, Sundberg MI, Seidegard J, Tornling G, Rannug A. 1994. Genetic susceptibility to lung cancer with special emphasis on CYP1A1 and GSTM1: A study on host factors in relation to age at onset, gender and histological cancer types. Carcinogenesis 15(9):1785-1790.

Alter HW and Fleischer RL. 1981. Passive integrating radon monitor for environmental monitoring. Health Phys 40:693-700.

Alter HW and Oswald RA. 1987. Nationwide distribution of indoor radon measurements: A preliminary data base. J Air Pollution Control Assoc 37:227-231.

Amandus H and Costello J. 1991. Silicosis and lung cancer in U.S. metal miners. Arch Environ Health 46:82-89.

Amstad P, Reddel RR, Pfeifer R, Malan-Shibley L, Mark GE, Harris CC. 1988. Neoplastic transformation of a human bronchial epithelial cell line by a recombinant retrovirus encoding viral Harvey ras. Mol Carcinog 1:151-60.

Amundson SA, Chen JD, Okinaka RT. 1996. Alpha particle mutagenesis of human lymphoblastoid cell lines. Int J Radiat Biol 70(2):219-226.

Anderson RE, Hill RB, Key CR. 1989. The sensitivity and specificity of clinical diagnosis during five decades. JAMA 261:1610-1617.

Anonymous. 1993. Cigarette smoking-attributable mortality and years of potential life lost in United States, 1990. MMWR 42(33):645-649.

Anttila SA, Hirvonen K, Husgafvel-Pursiainen A, Karjalainen AT, Nurminen H, Vainio H. 1994. Combined effect of CYP1A1 inducibility and GSTM1 polymorphism on histological type of lung cancer. Carcinogenesis 15:1133-1135.

Archer VE. 1987. Association of lung cancer mortality with precambrian granite. Arch Environ Health 42:87-91.

Archer VE. 1996. Radon, silicosis, and lung cancer. Health Phys 70:268.

Armitage P and Doll R. 1961. Stochastic models for carcinogenesis. Pp. 19-38 in Vol. 4 of the Proceedings of the Fourth Berkeley Symposium on Mathematical Statistics and Probability, J. Neyman, ed. Berkeley and Los Angeles: Univ Calif Press.

Armstrong B and Doll R. 1975. Environmental factors and cancer incidence and mortality in different countries, with special reference to dietary practices. Int J Cancer 15:617-631.

Arnold A, Cossman J, Bakshi A, Jaffe ES, Waldmann TA, Korsmeyer SJ. 1993. Immunoglobulin gene rearrangements as unique colonal markers in human lymphoid neoplasms. N Engl J Med 309:1593-1599.

Arnstein A. 1913. On the so-called "Schneeberg Lung Cancer." Verhandl di deutsch path Gesellsch 16:332-342.

Atencio EM. 1994. Epithelial cell kinetics of the upper respiratory tract of Wistar rats following radon exposure. Thesis, Masters of Science in Environmental Science, Washington State University.

Auvinen A, Mäkeläinen I, Hakama M, Castrén O, Pukkala E, Reisbacka H, Rytömaa T. In press. Indoor radon and the risk of lung cancer: A nested case-control study in Finland. J Natl Cancer Inst

Axelson O, Edling C, Kling H. 1979. Lung cancer and residency - a case-referent study on the possible impact of exposure to radon and its daughters in dwellings. Scand J Work Environ Health 5:10-15.

Axelson O, Andersson K, Desai G, Fagerlund I, Jansson B, Karlsson C, Wingren G. 1988. Indoor radon exposure and active and passive smoking in relation to the occurrence of lung cancer. Scand J Work Environ Health 14:286-292.

Baird SJS, Cohen JT, Graham JD, Shlyakhter AI, Evans JS. 1996. Noncancer risk assessment: A probabilistic alternative to current practice. Human and Ecological Risk Assessment 2:79-102.

Bale WF. 1980. Memorandum to the files, March 14, 1951: Hazards associated with radon and thoron. Health Phys 38:1062-1066.

Bao CY, Ma AH, Evans HH, Horng MF, Mencl J, Hui TE, Sedwick WD. 1995. Molecular analysis of hypoxanthine phosphoribosyltransferase gene deletions induced by a- and X-radiation in human lymphoblastoid cells. Mutat Res 326:1-15.

Bao S, Harwood PW, Chrisler WB, Groch KM, Brooks AL. 1997. Comparative clastogenic sensitivity of respiratory tract cells to gamma rays. Radiat Res 148:90-97.

Barendsen GW. 1985. Do fast neutrons at low dose rate enhance cell transformation in vitro? A basic problem of microdosimetry and interpretation. Int J Radiat Biol 47:731-734.

Barendsen GW, Coot CJ, Van Kersen GR, Bewley DK, Field SB, Parnell CJ. 1966. The effect of oxygen on impairment of the proliferative capacity of human cells in culture by ionising radiations of different LET. Int J Radiat Biol 10:317-327.

Bartlett S, Krewski D, Wang Y, Zielinski JM. 1993. Evaluation of error rates in large scale computerized record linkage studies. Survey Methodology 19:3-12.

Bartlett S, Richardson GM, Krewski D, Rai SN, Fyfe M. 1996. Characterizing uncertainty in risk assessment - Conclusions drawn from a workshop. Human and Ecological Risk Assessment 2:217-227.

Bartsch H, Hollstein M, Mustonen R, Schmidt J, Spiethoff A, Wesch H, Wiethege T, Muller KM. 1995. Screening for putative radon-specific p53 mutation hotspot in German uranium miners. Lancet 346(8967):121.

Bates MN, Smith AH, Hopenhayn-Rich C. 1992. Arsenic ingestion and internal cancers: A review. Am J Epid 135:462-476.

Battista G, Belli A, Carboncini F, Comba P, Levante G, Sartorelli P, Strambi F, Valentini F, Axelson O. 1988. Mortality among pyrite miners with low-level exposure to radon daughters. Scand J Work Environ Health 14:280-285.

Bauchinger M, Schmid E, Braselmann H, Kalka U. 1994. Chromosome aberrations in peripheral lymphocytes from occupants of houses with elevated indoor radon concentrations. Mutat Res 310:135-142.

Bauchinger M, Braselmann H, Kulka U, Iluber R, Georgiadau-Schumacher V. 1996. Quantification of FISH-painted chromosome aberrations after domestic random exposure. Int J Radiat Biol 70:657-663.

Bauer FW and Robbins SL. 1972. An autopsy study of cancer patients. I. Accuracy of the clinical diagnosis (1955 to 1965) Boston City Hospital. JAMA 221:1471-4.

Baverstock, KF. 1990. Radon and leukaemia (Letter). Lancet 335:1337-1338.

Bean JA, Isacson P, Hausler, Jr. WJ, Kohler J. 1982a. Drinking water and cancer incidence in Iowa. I. Trends and incidence by source of drinking water and size municipality. Am J Epidemiol 116(6):912-23.

Bean JA, Isacson P, Hahne RMA, Kohler J. 1982b. Drinking water and cancer incidence in Iowa. II. Radioactivity in drinking water. Am J Epidemiol 116(6):924-32.

Becker, KH, Reineking A, Scheibel HG, Porstendörfer J. 1984. Radon daughter activity size distributions. Radiat Prot Dosimet 7:147-150.

Bedford JS and Goodhead DT. 1989. Breakage of human interphase chromosomes by alpha-particles and X-rays. Int J Radiat Biol 55:211-216.

Beebe GW, Ishida I, Jablon S. 1962. Studies of the mortality of A-bomb survivors. I. Plan of study and mortality in the medical subsample (Selection 1), 1950-1958. Radiat Res 16:253-280.

Bender MA, Awa AA, Brooks AL, Evans HJ, Groer PG, Littlefield LG, Pereira C, Preston RJ, Wachholz BW. 1988. Current status of cytogenetic procedures to detect and quantify previous exposures to radiation. Mutat Res 196:103-159.

Bernhard S, Pineau JF, Rannou A, Zettwoog P. 1984. 1983: One year of individual dosimetry in French miners. Occupational Radiation Safety in Mining, Vol.2 Proceedings of the International Conference, Oct 14-18, Toronto. H. Stocker, Ed. 526. Canadian Nuclear Association.

Biberman R, Lusky A, Schlesinger T, Margaloit M, Neeman E, Modan B. 1993. Increased risk for small call lung cancer following residential exposure to low-dose radon: A pilot study. Arch Environ Health 48:209-12.

Bigu J. 1985. Theoretical models for determining 222Rn and 220Rn progeny levels in Canadian underground U mines - a comparison with experimental data. Health Phys 48:371-399.

Birchall A and James AC. 1994. Uncertainty analysis of the effective dose per unit exposure from radon progeny and implications for ICRP risk-weighting factors. Radiat Prot Dosimet 53:133-140.

Bishop JM. 1983. Cellular oncogene retroviruses. Ann Rev Biochem 52:301-354.

Bishop JM and Varmus HE. 1984. Functions and origins of retroviral transforming genes. Pp. 990-1108 in RNA Tumor Viruses: Molecular Biology of Tumor Viruses, 2nd ed., R. Weiss, N. Teich, H. Varmus, J. Coffin, eds. Cold Spring Harbor: Cold Spring Harbor Laboratory.

Bishop VJ. 1971. Control of radon daughters in the Colorado Plateau mines. Presented at the Mining Environmental Conference. University of Missouri-Rolla, October.

Bisson M, Collier CG, Poncy JL, Taya A, Morlier JP, Strong J, Baker S, Monchaux G, Fitsch P. 1994. Biological dosimetry in different compartments of the respiratory tract after inhalation of radon and its daughters. Radiat Prot Dosimet 56:89-92.

Blocher D. 1988. DNA double strand break repair determines the RBE of alpha particles. Int J Radiat Biol 54:761-771.

Blot WJ and Fraumeni JF, Jr. 1994. Arsenic and lung cancer. Pp. 207-218 in The Epidemiology of Lung Cancer, J. Samet, ed. New York: Marcell Dekker.

Blot WJ, Xu Z-Y, Boice, JD Jr., Zhao D-Z, Stone BJ, Sun J, Jing L-B, Fraumeni JF, Jr. 1990. Indoor radon and lung cancer in China. J Natl Cancer Inst 82:1025-30.

Bodmer WF, Bishop T, Karran P. 1994. Genetic steps in colorectal cancer. Nat Genet 6:217-219.

Boffetta P, Stellman SD, Garfinkel L. 1988. Diesel exhaust exposure and mortality among males in the american cancer society prospective study. Am J Ind Med 14:403-415.

Boffetta P, Harris RE, Wynder. 1990. Case-control study of occupational exposure to diesel exhaust and lung cancer risk. Am J Ind Med 17:577-591.

Bogen KT. 1994. A note on compounded conservatism. Risk Anal 14:379-381.

Bogen KT. 1995. Methods to approximate joint uncertainty and variabitity in risk. Risk Anal 15:411-419.

Boivin J. 1995. Smoking, treatment for Hodgkin's Disease, and subsèquent lung cancer risk. J Natl Cancer Inst 87:1502-3.

Boothman DA, Davis TW, Sahijdak WM. 1994 Enhanced expression of thymidine kinase in human cells following ionizing radiation. Int J Radiat Oncol Biol Phys 30:391-398.

Borek C, Ong A, Mason H. 1987. Distinctive transforming genes in X-ray - Transformed mammalian cells. Proc Natl Acad Sci USA 84:794-798.

Boring CC, Squires TS, Tong T. 1993. Cancer statistics. CA Cancer J Clin 43:7-26.

Bos JL. 1990. Ras oncogenes in human cancer: A review [published erratum appears in Cancer Res 50:1352. 1990] Cancer Res 49:4682-4689.

Bowie SHU. 1990. Radon and leukaemia (Letter). Lancet 335:1336.

Braby LA. 1992. Microbeam studies of the sensitivity of structures within living cells. Scanning Microscopy 6(1):167-174.

Brain JD and Valberg PA. 1974. Models of lung retention based on ICRP task group report. Arch Environ Health 28(1):1-11.

Brain JD and Valberg PA. 1979. Deposition of aerosols in the respiratory tract. Am Rev Respir Dis 120:1325-1373.

Brenner DJ. 1989. The effectiveness of single alpha particles. Pp. 477-480 in Low Dose Radiation: Biological Bases of Risk Assessment, J. Lancashire, ed. London and New York: Taylor and Francis.

Brenner DJ. 1992. Radon: current challenges in cellular radiobiology. Int J Radiat Biol 61:3-13.

Brenner DJ. 1994. The significance of dose rate in assessing the hazards of domestic radon exposure. Health Phys 67:76-79.

Brenner DJ and Hall EJ. 1990. The inverse dose-rate effect for oncogenic transformation by neutrons and charged particles. A plausible interpretation consistent with published data. Int J Radiat Biol 58: 745-758.

Brenner DJ and Hall EJ. 1992. Radiation induced oncogenic transformation: The interplay between dose, dose protraction, and radiation quality. Adv Radiat Biol 16: 859-885.

Brenner DJ and Sachs RK. 1994. Chromosomal "fingerprints" of prior exposure to densely ionizing radiation. Radiat Res 140:123-142.

Brenner DJ and Sachs RK. 1996. Comments on "Comment on the ratio of chromosome-type dicentric interchanges to centric rings for track-clustered compared with random breaks" by Savage and Papworth. Radiat Res 146(2):241-242.

Brenner DJ and Ward JF. 1992. Constraints on energy deposition and target size of multiply damaged sites associated with DNA double-strand breaks. Int J Radiat Biol 61:737-748.

Brenner DJ, Hall EJ, Randers-Pehrson G, Miller RC. 1993. Mechanistic considerations on the dose-rate/LET dependence of oncogenic transformation by ionizing radiations. Radiat Res 133:365-369.

Brenner DJ, Miller RC, Huang Y, Hall EJ. 1995. The biological effectiveness of radon-progeny alpha particles III Quality factors. Radiat Res 142:61-69.

Brenner DJ, Hahnfeldt P, Amundson SA, Sachs RK. 1996. Interpretation of inverse dose-rate effects for mutagenesis by sparsely ionizing radiation. Int J Radiat Biol 70(4):447-458.

Brenner H, Savitz DA, Jockel KH, Greenland S. 1992. The effects of nondifferential exposure misclassification in ecological studies. Am J Epidemiol 135(1):85-95.

Breslin AJ, George AC, Weinstein MS. 1969. Investigation of the radiological characteristics of uranium mine atmospheres. HASL-220. Health and Safety Laboratory. New York, NY: U.S. Atomic Energy Commission.

Breslow NE and Clayton DG. 1993. Approximate Inference in Generalized Liner Mixed Models. J Am Stat Assoc 88:9-25.

Breslow NE and Day NE. 1987. Statistical Methods in Cancer Research. Volume II. The Design and Analysis of Cohort Studies. Lyon: International Agency for Research on Cancer.

Breslow NE and Storer BE. 1985. General relative risk functions for case-control studies. Am J Epidemiol 122(1):149-62.

Breslow NE, Lubin JH, Marek P, Langholz B. 1983. Multiplicative models and the analysis of cohort studies. J Am Stat Assoc 78:1-12.

Bridges BA, Cole J, Arlett CF, Green MHL, Waugh APW, Beare D, Henshaw DL, Last RD. 1991. Possible association between mutant frequency in peripheral lymphocytes and domestic radon concentrations. Lancet 337:1187-1189.

Brightwell J, Fouillet X, Cassano-Zoppi AL, Bernstein D, Crawley F, Duchosal F, Gatz R, Percel S, Pfeifer H. 1989. Tumours of the respiratory tract in rats and hamsters following chronic inhalation of engine exhaust emissions. J Appl Toxicol 9:23-31.

Brodeur GM, Seeger RC, Schwab M, Varmus HE, Bishop JM. 1984. Amplification of N-myc in untreated human neuroblastomas correlates with advanced disease stage. Science 224:1121-1124.

Brooks AL. 1975. Chromosome damage in liver cells from low dose rate alpha, beta and gamma irradiation: derivation of RBE. Science 190:1090-1092.

Brooks AL. 1980. Low dose and low dose-rate effects on cytogenetics. Pp. 263-276 in Radiation Biology in Cancer Research, R.E. Meyn and H.R. Withers, eds. New York: Raven Press.

Brooks AL. 1996. Biodosimetry and molecular signatures (Cellular Changes) In: Radiation Risk Assessment, Statistical methodology and mechanisms, American Statistical Association 1996 Conference on Radiation and Health, June 23-27. Vail, Colorado.

Brooks AL, Benjamin SA, Jones RK, McClellan RO. 1982. Interaction of [144]CE and partial-hepatectomy in the production of liver neoplasms in the Chinese Hamster. Rad Res 91:573-588.

Brooks AL, Newton GJ, Shyr L-J, Seiler FA, Scott BR. 1990a. The combined effects of alpha-particles and X-rays on cell killing and micronuclei induction in lung epithelial cells. Int J Radiat Biol 58:799-811.

Brooks AL, Rithidech K, Johnson NF, Thomassen DG, Newton GJ. 1990b. Evaluating chromosome damage to estimate dose to tracheal epithelial cells. Pp. 601-614 in Part 2 of the Twenty-Ninth Hanford Life Sciences Symposium, Indoor Radon and Lung Cancer: Reality or Myth? Richland: Hanford Press.

Brooks AL, Khan MA, Duncan A, Buschbom RL, Jostes RF, Cross FT. 1994. Effectiveness of radon relative to acute $^{60}$Co gamma-rays for induction of micronuclei *in vitro* and *in vivo*. Int J Radiat Biol 66:801-808.

Brooks AL, Miick R, Buschbom RL, Murphy MK, Khan MA. 1995. The role of dose rate on the induction of micronuclei in deep-lung fibroblasts in vivo after exposure to cobalt-60 gamma rays. Radiat Res 144:114-118.

Brooks AL, McDonald KE, Mitchell C, Culp DS, Lloyd A, Johnson NF, Kitchin RM. 1996. The combined genotoxic effects of radiation and occupational pollutants. Appl Occup Environ Hyg 11(4):410-416.

Brooks AL, Bao S, Harwood PW, Wood BH, Chrisler WB, Khan MA, Gies RA, Cross FT. 1997. Induction of micronuclei in respiratory tract following radon inhalation. Int J Radiat Biol 72(5):485-495.

Brown CC and Chu KC. 1983. Implications of the multistage theory of carcinogenesis applied to occupational arsenic exposure. J Natl Cancer Inst 70:455-463.

Burch PRJ and Chesters MS. 1986. Neoplastic transformation of cells in vitro at low and high dose rates of fission neutrons: an interpretation. Int J Radiat Biol 49:495-500.

Burchall A and James AC. 1994. Uncertainty analysis of the effective dose per unit exposure from radon progeny and implications for ICRP risk-weighting factors. Radiat Prot Dosimet 53(1-4):133-40.

Burmaster DE and Anderson PD. 1994. Principles of good practice for the use of Monte Carlo techniques in human health and ecological risk assessments. Risk Anal 14:477-481.

Burmaster DE and Thompson KM. 1998. Fitting second-order parametric distributions to data using maximum likelihood estimation. Human and Ecological Risk Assessment 4(2):319-339.

Burmaster DE and Wilson AM. 1996. An introduction to second-order random variables in human health risk assessment. Human and Ecological Risk Assessment 2(4):892-919.

Burnett RT, Ross WH, Krewski D. 1995. Nonlinear random effects regression models. Environmentrics 6:85-99.

Burns DM. 1994. Tobacco smoking. In Epidemiology of Lung Cancer, JM Samet, ed. New York: Marcel Dekker, Inc.

Burns PB and Swanson GM. 1991. The occupational cancer incidence surveillance study (ociss): risk of lung cancer by usual occupation and industry in the Detroit metropolitan area. Am J Ind Med 19:655-671.

Busigin A, Van der Vooren AW, Babcock JC, Phillips CR. 1981. The nature of unattached 218Po (RaA) particles. Health Phys 40:333-343.

Butland BK, Muirhead CR, Draper GJ. 1990. Radon and leukaemia (letter). Lancet 335:1338-1339.

Butterworth BE and Goldsworthy TL. 1991. The role of cell proliferation in multistage carcinogenesis. Proc Soc Exp Biol Med 198:683-7.

Canadian Task Force on the Periodic Health Examination. 1990. Periodic health examination, 1990 update: 3. Interventions to prevent lung cancer other than smoking cessation. Can Med Assoc J 143:269-272.

Cao JM, Wells RL, Elkind MM. 1992. Enhanced sensitivity to neoplastic transformation by 137Cs gamma-rays of cells in the G2-/M-phase age interval. Int J Radiat Biol 62:191-199.

Cao JM, Wells RL, Elkind MM. 1993. Neoplastic transformation of C3H mouse embryo cells, 10T1/2: cell cycle dependence for 50 kV X-rays and UV-B light. Int J Radiat Biol 64:83-92.

Caporaso N, DeBaun MR, Rothman N. 1995. Lung cancer and CYP2D6 (the debrisoquine polymorphism): sources of heterogeneity in the proposed association. Pharmacogenetics 5:S129-34.

Cardis E, Gilbert ES, Carpenter L, Howe G, Kato I, Armstrong BK, Beral V, Cowper G, Douglas A, Fix J, Fry SA, Kaldor J, Lave C, Salmon L, Smith PG, Voelz GL, Wiggs LD. 1995. Effects of low doses and low dose rates of external ionizing radiation: Cancer mortality among nuclear industry workers in three countries. Radiat Res 142:117-32.

Carrano AV and Heddle JA. 1973. The fate of chromosome aberrations. J Theor Biol 38:289-304.

Carta P, Cocco PL, Casula D. 1991. Mortality from lung cancer among Sardinian patients with silicosis. Br J Ind Med 48:122-129.

Carta P, Cocco P, Picchiri, G. 1994. Lung cancer mortality and airways obstruction among metal miners exposed to silica and low levels of radon daughters. Am J Ind Med 25:489-506.

Cartwright BG and Shirk EK. 1978. A nuclear-track-recording polymer of unique sensitivity and resolution. Nucl Instrum Meth 153:457-460.

Cavanee WK. 1989. Tumor progression stage; specific losses of heterozygosity. Int Symp Princess Takamatsu Cancer Res Fund 20:33-42.

Cavenee WK, White RL. 1995. The genetic basis of cancer. Sci Am 272:72-79.

Cavenee WK, Hansen MF, Nordenskjold M, Kock E, Maumenee I, Squire JA, Phillips RA, Gallie BL. 1985. Genetic origin of mutations predisposing to retinoblastoma. Science 228:501-503.

CDC (Centers for Disease Control). 1995. Morbidity and Mortality Weekly Report, Atlanta, GA - Cigarette smoking among adults - United States 1993. JAMA 273(5):369-370.

Chameaud J, Perraud R, Chretien J, Masse R, Lafuma J. 1982. Lung carcinogenesis during in vivo cigarette smoking and radon daughter exposure in rats. Recent Results Cancer Res 82:11-20.

Chameaud J, Masse R, Lafuma L. 1984. Influence of radon daughter exposure at low doses on occurrence of lung cancer in rats. Radiat Prot Dosimet 7:385-391.

Chang WP and Little JB. 1992. Persistently elevated frequency of spontaneous mutations in progeny of CHO clones surviving X-irradiation; association with delayed reproductive death phenotype. Mutat Res 270:191-199.

Charles M, Cox R, Goodhead DT, Wilson A. 1990. CEIR forum on the effects of high-LET radiation at low doses/dose rates. Int J Radiat Biol 58:859-885.

Charlton DC, Nikjoo H, Humm JL. 1989. Calculation of initial yields of single and double-strand breaks in cell nuclei from electrons, protons and alpha particles. Int J Radiat Biol 56:1-19.

Chaudhry MA, Jiang Q, Ricanati M, Horng MF, Evans HH. 1996. Characterization of multilocus lesions in human cells exposed to X radiation and radon. Radiat Res 145:31-38.

Checkoway H. 1991. Data pooling in occupational studies. J Occup Med 33:1257-1260.

Chen DJ, Strinste GF, Tokita N. 1984. The genotoxicity of a-particles in human embryonic skin fibroblasts. Radiat Res 100:321-327.

Chen DJ, Carpenter S, Hanks T. 1990. Mutagenic effects of alpha particles in normal human skin fibroblasts. Pp. 569-581 in Twenty-Ninth Hanford Life Sciences Symposium, Indoor Radon and Lung Cancer: Reality or Myth? Richland: Battelle Press.

Chen JQ, McLaughlin JK, Zhang J, Stone BJ, Jiamo L, Rongan C, Dosemeci M, Rexing SH, Wu Z, Hearl FJ, McCawley MA, Blot WJ. 1991. Mortality among dust-exposed Chinese mine and pottery workers in Tongji Medical University China and National Cancer Institute USA (eds): "Study of Silicosis and Lung Cancer for Dust Exposed Workers (Silica, Silicosis and Lung Cancer)." 1:1-12.

Cheng, YS and Yeh HC. 1980. Theory of a screen-type diffusion battery. J Aerosol Sci 11:313-320.

Cheng YS, Keating JA, Kanapilly GM. 1980. Theory and calibration of a screen-type diffusion battery. J Aerosol Sci 11:549-556.

Cheng YS, Yeh HC, Mauderly JL, Mokler BV. 1984. Characterization of diesel exhaust in a chronic inhalation study. Am Ind Hyg Assoc J 45:547-555.

Chia SE, Chia KS, Phoon WH, Lee HP. 1991. Silicosis and lung cancer among Chinese granite workers. Scand J Work Environ Health 17:170-174.

Chiyotani K, Saito K, Okubo T, Takahashi K. 1990. Lung cancer risk among pneumoconiosis patients in Japan, with special reference to silicotics. In: Simonato L, Fletcher AC, Saracci R, Thomas TL (eds): "Occupational Exposure to Silica and Cancer Risk," IARC Sci Pub 97:95-104. Lyon, France: IARC.

Chmelevsky D, Kellerer AM, Lafuma J, Morin M, Masse R. 1984. Comparison of the induction of pulmonary neoplasms in Sprague-Dawley rats by fission neutrons and radon daughters. Radiat Res 98:519-535.

Churg A. 1994. Lung Cancer Cell Type and Occupational Exposure. In Epidemiology of Lung Cancer, JM Samet, ed. New York: Marcel Dekker, Inc.

Clayson DB, Nera EA, Lok E. 1989. The potential for the use of cell proliferation studies in carcinogen risk assessment. Regulat Toxicol Pharmacol 9:284-295.

Clifton KH, Groch KM, Domann FE, Jr. 1991. Thyroid clonogen biology and carcinogenesis. Prog Clin Biol Res 369:173-183.

Cohen AJ and Higgins MWP. 1995. Health effects and diesel exhaust: Epidemiology. Pp. 251-292 in Diesel Exhaust: A Critical Analysis of Emissions, Exposure, and Health Effects. A Special Report of the Institute's Diesel Working Group. Cambridge, MA: Health Effects Institute.

Cohen BL. 1990. A test of the linear-no threshold theory of radiation carcinogenesis. Environ Res 53:193-220.

Cohen BL. 1993. Relationship between exposure to radon and various types of cancer. Health Phys 65:529-537.

Cohen BL. 1994. Invited Commentary: In defense of ecologic studies for testing a linear-no threshold theory. Am J Epidemiol 139:756-768.

Cohen BL. 1995. Test of the linear-no threshold theory of radiation carcinogenesis for inhaled radon decay products. Health Phys 68:157-174.

Cohen BL and Colditz GA. 1990. Tests of the linear no-threshold theory of radon induced lung cancer. Environ Res 53:193-220.

Cohen BL and Colditz GA. 1995. Lung cancer mortality and radon exposure: A test of the linear-no-threshold model of radiation carcinogenesis. In Radiation and Public Perception. Benefits and Risks. Washington, DC: American Chemical Society.

Cohen BL and Gromicko N. 1988. Variation of radon levels in U.S. homes with various factors. JAPCA 38(2):129-134.

Cohen BL, Kulwicki DR, Warner KR Jr, Grassi CL. 1984. Radon concentrations inside public and commercial buildings in the Pittsburgh area. Health Phys 47(3):399-405.

Cohen SM and Ellwein LB. 1990. Cell proliferation in carcinogenesis. Science 249:1007-1011.

Cohen SM, Garland EM, Ellwein LB. 1992. Cancer enhancement by cell proliferation. Prog Clin Biol Res 374:213-229.

Cole J, Green MHL, Bridges BA, Waugh APW, Beare DM, Henshaw D, Last R, Liu Y, Cortopassi G. 1996. Lack of evidence for an association between the frequency of mutants or translocations in circulating lymphocytes and exposure to radon gas in the home. Radiat Res 145:61-69.

Cole LA. 1993. Elements of Risk: The Politics of Radon. Washington, DC: AAAS Press.

COMARE (Committee on Medical Aspects of Radiation in the Environment). 1996. Fourth Report: The incidence of cancer and leukaemia in young people in the vicinity of the Sellafield site. West Cumbria: Further studies and an update of the situation since the publication of the report of the Black Advisory Group in 1984. Conference on Occupational Radiation Safety in Mining. (Stocker, H., ed.). Toronto, Canada: Canadian Nuclear Association, pp. 344-349.

Cooper JA, Jackson PO, Langford JC, Petersen MR, Stuart BO. 1973. Characteristics of attached radon-222 daughters under both laboratory and filed conditions with particular emphasis upon underground uranium mine envirnments. US Bureau of Mines Report H0220029. Battelle: Pacific Northwest Laboratories (Report No. BN-SA-299).

Coquerelle TM, Weibezahn KF, Lücke-Huhle C. 1987. Rejoining of double-strand breaks in normal human and ataxia-telangiectasia fibroblasts after exposure to $^{60}$Co γ-rays, $^{241}$Am α-particles or bleomycin. Int J Radiat Biol 51:209-218.

Corkill DA, Dory AB. 1984. A retrospective study of radon daughter concentrations in the workplace in the fluorspar mines of St. Lawrence, NFLD. Report No. INFO-0127. Ottawa: Atomic Energy Control Board.

Cornforth MN and Goodwin EH. 1991. The dose-dependent fragmentation of chromatin in human fibroblasts by 3.5-MeV alpha particles from 238Pu: Experimental and theoretical considerations pertaining to single-track effects. Radiat Res 127:64-74.

Court-Brown WM and Doll R. 1958. Expectation of life and mortality from cancer among British radiologists. Br Med J 2:181-190.

Cowell JK and Hogg A. 1992. Genetics and cytogenetics of retinoblastoma. Cancer Genet Cytogenet 64:1-11.

Cox R. 1982. A cellular description of the repair detect in ataxia-telangiectasia. In: Bridges BA, Harnden DG, eds. Ataxia Telangiectasia: A Cellular and Molecular Link Between Cancer, Neuropathology, and Immune Deficiency. New York, NY: Wiley Medical Publications.

Cox R. 1994a. Human cancer predisposition and the implications for radiological protection. Int J Radiat Biol 66:643-647.

Cox R. 1994b. Molecular mechanisms of radiation oncogenesis. Int J Radiat Biol 65:57-64.

Cox R and Masson WK. 1979. Mutation and inactivation of cultured mammalian cells exposed to beams of accelerated heavy ions. III. Human diploid fibroblasts. Int J Radiat Biol 36:149-160.

Cox R, Thacker J, Goodhead DT. 1977a. Inactivation and mutation of cultured mammalian cells by aluminium characteristic ultrasoft X-rays. II. Dose-responses of Chinese hamster and human diploid cells to aluminium X-rays and radiations of different LET. Int J Radiat Biol Relat Stud Phys Chem Med (6):561-576.

Cox R, Thacker J, Goodhead DT, Munson RJ. 1977b. Mutation and inactivation of mammalian cells by various ionizing radiations. Nature 267:425-427.

Cross FT. 1981. Experimental studies on lung carcinogenesis and their relationship to future research on radiation-induced lung cancer in humans. Pp. 27-35 in The Future of Human Radiation Research, Gerber GB, Taylor DM, Cardis E, Thiessen JW, eds. (Section 2 Lung Cancer), Report 22, British Institute of Radiology, London. Madison: Medical Physics Publishing.

Cross FT. 1992. A review of experimental animal radon health effects data. Pp. 476-481 in Radiation Research: A Twentieth-Century Perspective, Vol. II. JD Chapman, WC Dewey, GF Whitmore, eds. San Diego: Academic Press.

Cross FT. 1994a. Invited commentary: residential radon risks from the perspective of experimental animal studies. Am J Epidemiol 140:333-339.

Cross FT. 1994b. Evidence of cancer risk from experimental animal radon studies. Pp. 79-87 in Radiation and Public Perception. American Chemical Society.

Cross FT, Palmer RF, Busch RH, Filipy RE, Stuart BO. 1981. Development of lesions in Syrian golden hamsters following exposure to radon daughters and uranium ore dust. Health Phys 41:135-153.

Cross FT, Palmer RF, Dagle GE, Busch RH, Buschbom RL. 1984. Influence of radon daughter exposure rate, unattachement fraction, and disequilibrium on occurrence of lung tumors. Radiat Prot Dosimet 7:381-384.

Cross FT, Palmer RF, Busch RH, Dagle GE, Filipy RE. 1986. An overview of PNL radon experiments with reference to epidemiological data. Pp. 608-623 in Life-span Radiation Effects Studies in Animals: What Can They Tell Us? CONF-830951, Office of Scientific and Technical Information, United States Department of Energy.

Cross FT, Dagle GE, Gies RA, Smith LG, Buschbom RL. 1995. Experimental animal studies of radon and cigarette smoke. Pp. 821-844 in Indoor Radon and Lung Cancer: Reality or Myth? Part 2, Cross FT, ed.

Crump KS. 1994a. Limitations of biological models of carcinogenesis for low-dose extrapolation. Risk Anal 14:883-6.

Crump KS. 1994b. Use of mechanistic models to estimate low-dose cancer risks. Risk Anal 14:1033-8.

Crump KS, Lambert T, Chen C. 1991. Assessment of Risk from Exposure to Diesel Engine Emissions: Report to the U.S. Environmental Protection Agency for Contract No. 68-02-4601 (Work Assignment No. 182, July). Office of Health Assessment, U.S. Environmental Protection Agency, Washington, DC.

Curtis SB. 1989. A possible role of the inverse dose-rate effect in the radon exposure problem. Pp. 547-553 in Low Dose Radiation, Biological Bases of Risk Assessment, Baverstock KF, Stather JW, eds. London: Taylor and Francis.

Dakins ME, Toll JE, Small MJ. 1994. Risk-based environmental remediation: Decision framework and roles of uncertainty. Environ Toxicol Chem 13:1907-1915.

Dalla-Favera RS, Martinotti S, Gallo R, Erikson J, Croce C. 1983. Translocation and rearrangement of the c-myc oncogene locus in human undifferentiated B-cell lymphomas. Science 219:963-997.

Damber L and Larsson LG. 1987. Lung cancer in males and type of dwelling. An epidemiologic pilot study. Acta Oncol 26:211-215.

Darby SC and Doll R. 1990. Radiation and exposure rate. Nature 344:824.

Darby S, Whitley E, Howe GR, Hutchings SJ, Kusiak RA, Lubin JH, Morrison HI, Timarche M, Tomášek L, Radford EP, Roscoe RJ, Samet JM, Yao SX. 1995. Radon exposure and cancers other than lung cancer in underground miners: a collaborative analysis of 11 studies. J Natl Cancer Inst 87:378-384.

Dare WL, Lindblom RA, Soule JH. 1953. Uranium mining on the Colorado Plateau. Bureau of Mines Information Circular 7726, September.

Davidian M and Giltinan DM. 1995. Nonlinear Models for Repeated Measurement Data. New York: Chapman and Hall.

Davies CN, ed. 1967. Aerosol Science. New York: Academic Press.

Davies CN. 1980. An algebraical model for the deposition of aerosols in the human respiratory tract during steady breathing-Addendum. J Aerosol Sci 11:213-224.

Davis AL. 1976. Bronchogenic carcinoma in chronic obstructive pulmonary disease. JAMA 235:621-622.

de Lara CM, Jenner TJ, Townsend KMS, Marsden SJ, O'Neill P. 1995. The effect of dimethyl sulfoxide on the production of DNA double-strand breaks in V79-4 mammalian cells by alpha particles. Radiat Res 144:43-49.

Dennis JA and Dennis A. 1988. Neutron dose effect relationships at low doses. Radiat Environ Biophys 27:91-101.

Deshpande A, Goodwin EH, Bailey SM, Marrone BL, Lehnert BE. 1996. Alpha-particle-induced sister chromatid exchange in normal human lung fibroblasts: Evidence for an extranuclear target. Radiat Res 145:260-267.

Doll R and Peto R. 1978. Cigarette smoking and bronchial carcinoma: Dose and time relationships among regular smokers and life-long non-smokers. J Epidemiol Community Health 32:303-313.

Doll R and Peto R. 1981. The Causes of Cancer. Oxford: Oxford University Press.

Doll R, Gray R, Hafner B, Peto R. 1980. Mortality in relation to smoking: 22 years' observations on female British doctors. Br Med J 2:967-71.

Dosemeci M, Wacholder S, Lubin JH. 1990. Does nondifferential miscassification of exposure always bias a true effect towards the null value? Am J Epidemiol 132:746-8.

Dousset M and Jammet H. 1985. Comparison of cancer mortality in Limousin and Poitou-Charentes. Radioprotection 20:61-7.

Draper NR and Smith H. 1966. Applied Regression Analysis. New York: Wiley.

Driscoll KE, Maurer JK, Hassenbein D, Carter J, Janssen YMW, Mossman BT, Osier M,. Oberdörster G. 1994. Contribution of macrophage-derived cytokines and cytokine networks to mineral dust-induced lung inflammation. Pp. 177-189 in Toxic and Carcinogenic Effects of Solid Particles in the Respiratory Tract, U. Mohn, D. L. Dungworth, J. L. Mauderly and G. Oberdörster, eds. Washington: International Life Sciences Institute Press.

Driscoll KE, Carter JM, Howard BW, Hassenbein DG, Pepelko W, Baggs RB, Oberdörster G. 1996. Pulmonary, inflammatory, chemokine, and mutagenic responses in rats after subchronic inhalation of carbon black. Toxicol Appl Pharmacol 136:372-380.

DSMA Atcon Ltd. 1985. Elliot Lake Study: Factors affecting the uranium mine working environment prior to the introduction of current ventilation practices. A research report prepared for the Atomic Energy Control Board, Ottawa, Canada. March 4, 1985.

Dua SK, Hopke PK, Raunemaa T. 1995. Hygroscopic Growth of Indoor Aerosols. Aerosol Sci Technol 23:331-340.

Eatough JP and Henshaw DL. 1992. Radon and thoron associated dose to the basal layer of the skin. Phys Med Biol 37:955-967.

Eatough JP and Henshaw DL. 1995. The theoretical risk of non melanoma skin cancer from domestic radon exposure. J Rad Protect 15(1)45-51.

Economou P, Lechner JF, Samet JM. 1994. Familial and genetic factors in the pathogenesis of lung cancer. In Epidemiology of Lung Cancer, JM Samet, ed. New York: Marcel Dekker, Inc.

Edelmann KG and Burmaster DE. 1997. Are all distributions of risk with the same 95th percentile equally acceptable? Human and Ecological Risk Assessment 3:223-234.

Edling C, Comba P, Axelson O, Flodin U. 1982. Effects of low-dose radiation — a correlation study. Scand J Work Environ Health 8:59-64.

Edling C, Kling H, Axelson O. 1984. Radon in homes - a possible cause of lung cancer. Scand J Work Environ Health 10:25-34.

Edling C, Wingren G, Axelson O. 1984. Radon daughter exposure in dwellings and lung cancer. Pp. 29-34 In Indoor Air. Radon, Passive Smoking, Particulates and Housing Epidemiology, Vol. 2, Berglund B, Lindvall T, Sundell J, eds. Stockholm Council for Building Res.

Elenitoba-Johnson K, Medeiros LJ, Khorsand J, King TC. 1995. Lymphoma of the mucosa-associated lymphoid tissue of the lung A multifocal case of common clonal origin. Am J Clin Pathol 103:341-345.

Elkind MM. 1991. Enhanced neoplastic transformation due to protracted exposures of fission-spectrum neutrons: a biophysical model (Letter). Int J Radiat Biol 59:467-1475.

Elkind MM. 1994. Radon-induced cancer: A cell-based model of tumorigenesis due to protracted exposures. Int J Radiat Biol 66(5):649-653.

Emerit I, Oganesian N, Sarkisian T, Arutykuknykan R, Pogosian A, Asrian K, Levy A, Cernjavski L. 1995. Clastogenic Factors in the plasma of chernobyl accident recovery workers: Anti-clastogenic effect of Ginko Biloba extract. Radiat Res 144:198-205.

EML (Environmental Measurements Laboratory). 1990. EML Procedures Manual, 27th edition. US Department of Energy Report No. HASL-300. Chapter 2. New York, NY: Environmental Measurements Laboratory.

Enderle GJ, Friedrich K. 1995. East German uranium miners (Wismut) - exposure conditions and health consequences. Stem Cells 13:78-89.

Ennemoser O, Ambach W, Brunner P, Schneider P. 1994. Unusual high radon exposure in homes and lung cancer. Lancet 344:127.

Enstrom JE. 1979. Rising lung cancer mortality among nonsmokers. J Natl Cancer Inst 62:755-760.

Enterline PE, Marsh GM. 1983. Mortality among workers exposed to arsenic and other substances in a copper smelter. Pp. 226-244 in Arsenic: Industrial, Biomedical Environmental Perspectives. New York: Van Nostrand Reinhold Co.

Enterline PE, Marsh GM, Esmen NA, Henderson VL, Callahan CM, Paik M. 1987. Exposure to arsenic and respiratory cancer: A reanalysis. Am J Epidemiol 125:929-938.

Enterline PE, Day R., Marsh GM. 1995. Cancers related to exposure to arsenic at a copper smelter. Occup Environ. Med. 52:28-32.

Evans HH. 1991. Cellular and molecular effects of radon and other alpha particle emitters. Adv Mutat Res 3:28-52.

Evans HH. 1992. Relationship of the cellular and molecular effects of alpha-particle irradiation to radon-induced lung cancer. Pp. 537-555 in Twenty-Ninth Hanford Life Sciences Symposium, Indoor Radon and Lung Cancer: Reality or Myth? Richland: Battelle Press.

Evans HH, Mencl J, Bakale G, Rao PS, Jostes RF, Hui TE, Cross FT, Schwartz JL. 1993. Interlaboratory comparison of the effects of radon on L5178Y cells: Dose contribution of radon daughter association with cells. Radiat Res 136:49-56.

Evans HJ. 1993. Molecular genetic aspects of human cancers: The 1993 Frank Rose Lecture. Br J Cancer 68:10951-1060.

Fearon ER, Cho KR, Nigro JM, Kern SE, Simons JW, Ruppert JM, Hamilton SR, Preisinger AC, Thomas G, Kinzler KW, Vogelstein B. 1990. Identification of a chromosome 189 gene that is altered in colorrectal cancers. Science 247:49-56.

Fialkow PJ. 1976. Colonal origin of human tumors. Biochem Biophys Acta 458:283-321.

Fidler IJ and Talmadge JE. 1986. Evidence that intravenously derived murine pulmonary melanoma metastases can originate from the expansion of a single tumor cell. Cancer Res 46(10):5167-5171.

Finkelstein MM. 1995. Silicosis, radon, and lung cancer risk in Ontario miners. Health Phys 69:396-399.

Finkelstein MM and Kusiak RA. 1995. Clinical Measures, Smoking, Radon Exposure and Lung Cancer Risk Among Elliot Lake Uranium Miners. Ontario: Ontario Ministry of Labour.

Finley B and Paustenbach D. 1994. The benefits of probabilistic exposure assessment: 3 case studies involving contaminated air, water, and soil. Risk Anal 14:53-74.

Finley B, Proctor D, Scott P, Harrington N, Paustenbach D, Price P. 1994. Recommended distributions for exposure factors frequently used in health risk assessment. Risk Anal 14:533-554.

Fleischer RL. 1981. A possible association between lung cancer and phosphate mining and processing. Health Phys 41:171-175.

Fleischer RL. 1986. A possible association between lung cancer and a geological outcrop. Health Phys 50:823-827.

Folkard M, Prise KM, Vojnovic B, Newman HC, Roper MJ, Hollis KJ, Michael BD. 1995. Radiat Prot Dosimet 61:215-218.

Forastiere F, Valesini S, Arca' M, Magliola ME; Michelozzi P; Tasco C. 1985. Lung cancer and natural radiation in an Italian province. Sci Total Environ 45:519-526.

Ford JR, Terzaghi-Howe M. 1992a. Basal cells are the progenitors of primary tracheal epithelial cell cultures. Exp Cell Res 198:69-77.

Ford JR, Terzaghi-Howe M. 1992b. Characteristics of magnetically separated rat tracheal epithelial cell populations. Am J Physiol 263 (Lung Cell. Mol. Physiol. 7): L568-L574.

Fox JC and McNally NJ. 1990. The rejoining of DNA double-strand breaks following irradiation with $^{238}$Pu α-particles: evidence for a fast component of repair as measured by neutral filter elution. Int J Radiat Biol 57:513-521.

Franko AJ, Sharplin J, Ward WF, Taylor J. 1996. Evidence for two patterns of inheritance of sensitivity to lung fibrosis in mice by radiation, one of which involves two genes. Radiat Res 146:68-74.

Fried BM. 1931. Primary carcinoma of the lung: Bronchiogenic cancer—a clinical and pathological study. Medicine X:373-188.

Friendenreich CM. 1993. Methods for pooled analyses of epidemiologic studies. Epidemiology 4:295-302.

Frost SE. 1983. Beaverlodge working level calculations, draft number 4. Eldorado Resources, Ltd., Ottawa. August 10.

Fry RJM, Powers-Risius PE, Alpen L, Ainsworth EJ. 1985. High-LET radiation carcinogenesis. Radiat Res 104:S188-195.

Fuchs, NA. 1964. The Mechanics of Aerosols. Elmsford: Pergamon Press.

Furst A. 1983. A new look at arsenic carcinogenesis. Pp. 151-164 in Arsenic—Industrial, Biomedical Environmental Perspectives. New York: Van Nostrand Reinhold Co.

Furuse T, Otsu H, Noda Y, Kobayashi S, Ohara H. 1992. Induction of liver tumors and lung tumors in C57BL/6J male mice irradiated with low doses of high LET radiations. In Proceedings of the International Conference on Low Dose Irradiation and Biological Defense Mechanisms, Kyoto, Japan, Sugahara T, Sagan LA, Aoyama T, eds. Amsterdam: Elsevier Science Publishers.

Garfinkel L. 1981. Time trends in lung cancer mortality among nonsmokers and a note on passive smokers. J Natl Cancer Inst 66:1061-1066.

Garshick E, Schenker MB, Muñoz A, Segal M, Smith TJ, Woskie SR, Hammond SK, Speizer FE. 1987. A case-control study of lung cancer and diesel exhaust exposure in railroad workers. Am Rev Respir Dis 135:1242-1248.

Garshick E, Schenker MB, Muñoz A, Segal M, Smith TJ, Woskie SR, Hammond SK, Speizer FE. 1988. A retrospective cohort study of lung cancer and diesel exhaust exposure in railroad workers. Am Rev Respir Dis 137:820-825.

Geard CR. 1985. Charged particle cytogenetics: Effects of LET, fluence, and particle separation on chromosome aberrations. Radiat Res Suppl 8:S112-S121.

Geard CR, Brenner DJ, Randers-Pehrson G, Marino S. 1991. Single-particle irradiation of mammalian cells at the Radiological Research Accelerator Facility: Induction of chromosomal changes. Nucl Instr Meth B54:411-416.

George AC. 1972. Measurement of the uncombined fraction of radon daughters with wire screens. Health Phys 23:390-392.

George AC. 1976. Scintillation flasks for the determination of low level concentrations of radon. In Proceedings of Ninth Midyear Health Physics Symp. Denver.

George AC. 1977. A passive environmental radon monitor. In: Breslin, AJ, ed., Radon Workshop-February 1977, New York: Health and Safety Laboratory; HASL-325, 25-30.

George AC. 1984. Passive integrated measurement of indoor radon using activated carbon. Health Phys 46:867-872.

George AC. 1993. Measurement of airborne 222Rn daughters by filter collection and alpha radioactivity collection. IARC Sci Publ 109:173-179.

George AC and Breslin AJ. 1980. The Distribution of Ambient Radon and Radon Daughters in Residential Buildings in the New Jersey-New York Area, National Radiation Environment III, Vol. 2, CONF-780422. Technical Information Center, U.S. Department of Energy, p. 1272.

George AC and Weber T. 1990. An improved passive activated C collector for measuring environmental 222Rn in indoor air. Health Phys 58:583-589.

George AC, Hinchliffe L, Sladowski R. 1975. Size distribution of radon daughter particles in uranium mine atmospheres. Am Ind Hyg Assoc J 36:484-490.

George AC, Hinchliffe L, Sladowski R. 1977. Size distribution of radon daughter particles in uranium mine atmposheres. Report No. HASL-326.

George AC, Wilkening MH, Knutson EO, Sinclair D, Andrews L. 1984. Measurements of Radon and Radon Daughter Aerosols in Socorro, New Mexico, Aerosol Sci Technol. 3:277-281.

Gilbert ES. 1994. Smoking as an explanation for the negative relationship between exposure to radon and certain types of cancer. Health Phys 67:192.

Gilbert ES, Cross FT, Dagle GE. 1996. Analysis of lung tumor risks in rats exposed to radon. Radiat Res 145(3):350-360.

Gilliland FD and Samet JM. 1994. Lung Cancer. London, England: Imperial Cancer Research Fund. 175 19. Trends in Cancer Incidence and Mortality.

Gilliland DG, Blanchard KL, Levy J, Perrin S, Bunn F. 1991. Clonality in myeloproliferative disorders: analysis by means of the polymerase chain reaction. Proc Natl Acad Sci USA 88:6848.

Glass GV. 1976. Primary, secondary, and meta-analysis of research. Educ Researcher 5:3-8.

Goddard MJ and Krewski D. 1995. The future of mechanistic research in risk assessment: where are we going and can we get there from here? Toxicology 102:53-70.

Goddard MJ, Krewski D, Zhu Y. 1994. Measuring carinogenic potency. Pp. 193-208 in Environmental Statistics, Assessment and Forecasting, C.R. Cothern and N.P. Ross, eds. Boca Raton: Lewis Publishers.

Goldsmith DF and Samet JM, editors. 1994. Silica exposure and pulmonary cancer. Epidemiology of Lung Cancer. New York, New York: Marcel Dekker, Inc. 11:248-298.

Goldsmith DF, Beaumont JJ, Morrin LA, Schenker MB. 1995. Respiratory cancer and other chronic disease mortality among silicotics in California. Am J Ind Med 28:459-467.

Goldstein SD and Hopke PK. 1985. Environmental neutralization of polonium-218. Environ Sci Technol 19:146-150.

Goldsworthy TL, Morgan KT, Popp JA, Butterworth BE. 1991. Guidelines for measuring chemically induced cell proliferation in specific rodent target organs. Pp. 253-284 in Chemically Induced Cell Proliferation: Implications for Risk Assessment. Wiley-Liss, Inc.

Goodhead DT. 1988. Spatial and temporal distribution of energy. Health Phys 55:231-240.

Goodhead DT. 1994. Initial events in the cellular effects of ionizing radiations: clustered damage in DNA. Int J Radiat Biol 65:7-17.

Goodhead DT and Nikjoo H. 1989. Track structure analysis of ultrasolf X-rays compared to high- and low-LET radiations. Int J Radiat Biol 55:513-529.

Goodhead DT, Munson RJ, Thacker J, Cox R. 1980. Mutation and inactivation of cultured mammalian cells exposed to beams of accelerated heavy ions: IV. Biophysical interpretation. Int J Radiat Biol 37:135-167.

Goodhead DT, Bance DA, Stretch A, Wilkinson RE. 1991. A versatile plutonium-238 irradiator for radiobiological studies with alpha-particles. Int J Radiat Biol 59:195-210.

Goodwin EH and Cornforth MN. 1994. RBE: Mechanisms inferred from cytogenetics. Adv Space Res 14:249-255.

Gopalakrishnan S, Lui-X, Patel DJ. 1992. Solution structure of the covalent sterigmatocystin-DNA adduct. Biochemistry 31:10790-10801.

Gorgojo L and Little JB. 1989. Expression of lethal mutations in progeny of irradiated mammalian cells. Int J Radiat Biol 55:619-630.

Gray DJ and Windham ST. 1987. EERF standard operating procedures for radon-222 measurement using charcoal cannisters. U.S. Environmental Protection Agency Report EPA 520/5-87-005, Washington, D.C.

Gray RG, Lakuma J, Parish SE, Peto R. 1986. Lung tumors and radon inhalation in over 2,000 rats: approximate linearity across a wide range of doses and potentiation by tobacco smoke. In: Lifespan radiation effects studies in animals: what can they tell us? Pp. 592-607 in Proceedings of the 22nd annual Hanford Life Sciences Symposium (CONF-830951). Springfield: National Technical Information Service.

Greenland S. 1983. Tests for interaction in epidemiologic studies: a review and s study of power. Stat Med 2:243-251.

Greenland S. 1984. Bias methods for deriving standardized morbidity ratio and attributable fraction estimates. Stat Med 3:131-141.

Greenland S. 1987. Quantitative methods in the review of epidemiologic literature. Epidemiol Reviews 9:1-30.

Greenland S. 1993. Basic problems in interaction assessment. Environ Health Perspect 101:59-66.

Greenland S. 1994. Can meta-analysis be salvaged? Am J Epidemiol 140:783-787.

Greenland S and Morgenstern H. 1989. Ecological bias, confounding, and effect modification. Int J Epidemiol 18:269-274.

Greenland S and Robins J. 1994a. Accepting the limits of ecologic studies: Drs. Greenland and Robins reply to Drs. Piantadosi and Cohen. Am J Epidemiol 139:769-771.

Greenland S and Robins J. 1994b. Invited Commentary: Ecologic studies - biases, misconceptions, and counterexamples. Am J Epidemiol 139:747-760.

Griffin CS, Harvey AN, Savage JRKS. 1994. Chromatid damage induced by $^{238}$Pu alpha particles in G2 and S-phase Chinese hamster cells. Int J Radiat Biol 66:85-98.

Griffin CS, Marsden SJ, Stevens PL, Simpson P, Savage JRK. 1995. Frequencies of complex chromosome exchange aberrations induced by 238Pu alpha-particles and detected by fluorescence in situ hybridization using single chromosome-specific probes. Int J Radiat Biol 67:431-439.

Griffiths SD, Marsden SJ, Wright EG, Greaves MF, Goodhead DT. 1994. Lethality and mutagenesis of B lymphocyte progenitor cells following exposure to α-particles and X-rays. Int J Radiat Biol 66:197-205.

Groch KM, Khan MA, Brooks AL, Saffer JD. 1997. Differential lung cancer response following inhaled radon by A/J and C57BL/6J mice. Int J Radiat Biol 71(3):301-308.

Guerrero I, Villasante A, Corces V, Pellicer A. 1984. Activation of a c-K-ras oncogene by somatic mutation in mouse lymphomas induced by gamma radiation. Science 225:1159-1162.

Gunby JA, Darby SC, Miles JC, Green BM, Cox DR. 1993. Factors affecting indoor radon concentration in the United Kingdom. Health Phys 64:2-12.

Haenszel W, Shimkin MB, Mantel N. 1958. A retrospective study of lung cancer in women. J Natl Cancer Inst 21:825-842.

Hall EJ. 1987. Radiobiology for the Radiologist, 3rd Edition. Philadelphia: Lippincott.

Hall EJ and Freyer GA. 1991. The molecular biology of radiation carcinogenesis. Basic Life Sci 58:3-19; discussion 19-25.

Hall EJ and Hei TK. 1990. Modulating factors in the expression of radiation-induced oncogenic transformation. Environ Hlth Perspect 88:149-155.

Hall EJ, Geard CR, Brenner DJ. 1992. Risk of breast cancer in ataxia telangiectasia. N Engl J Med 326:1358-9.

Hall JM, Lee MK, Newman B, Morrow JE, Anderson LA, Huey B, King MC. 1990. Linkage of early-onset familial breast cancer to chromosome 17q21. Science 250:1684-1689.

Hall NEL and Wynder EL. 1984. Diesel exhaust exposure and lung cancer: A case-control study. Environ Res 34:77-86.

Hamilton LD, Swent LW, Chambers DB. 1990. Visit to the Centre of Radiaion Hygiene, Institute of Hygiene and Epidemiology, Prague, Czechoslovakia. Trip report to Division of Environmental Health, World Health Organization, Geneva, Switzerland. December, SENES Consultants Limited, Ontario, Canada.

Hammon EC, Garfinkel L. 1968. Changes in cigarette smoking 1959-1965. Am J Public Health Nations Health 68(1):30-45.

Han A and Elkind MM. 1979. Transformation of mouse C3H/10T1/2 cells by single and fractionated doses of x-rays and fission-spectrum neutrons. Cancer Res 39:123-30.

Han A, Hill CK, Elkind MM. 1980. Repair of cell killing and neoplastic transformation at reduced dose rates of $^{60}$Co gamma rays. Cancer Res 40:3328-3332.

Harber P, Oren A, Moshenifar Z, Lew M. 1986: Obstructive airway disease as risk factor for asbestos-associated malignancy. J Occup Med 28:82-86.

Harley JH. 1952. Sampling and measurement of airborne daughter products of radon. Doctoral dissertation. Rensselaer Polytechnic Institute.

Harley JH. 1953. Sampling and measurement of airborne daughter products of radon. Nucleonics 11:12-15.

Harley JH. 1980. Sampling and measurement of airborne daughter products of radon. Reprinted in Health Phys 38:1067.

Harley NH. 1988. Radon daughter dosimetry in the rat tracheobronchial tree. Radiat Prot Dosimet 24:457-461.

Harley NH and Pasternack BS. 1972. Alpha absorption measurements applied to lung dose from radon daughters. Health Phys 23:771-782.

Harley NH and Pasternack BS. 1981. A model for predicting lung cancer risks induced by environmental levels of radon daughters. Health Phys 40:307-316.

Harley NH and Pasternack BS. 1982. Environmental radon daughter alpha dose factors in a five-lobed human lung. Health Phys 42:789-799.

Harley NH and Robbins ES. 1992. $^{222}$Rn alpha dose to organs other than the lung. Radiat Prot Dosimet 45: 617-622.

Harley NH, Cohen BS, Robbins ES. 1996. The variability in radon decay product bronchial dose. Environmental International 22:S959-S964.

Harris CC and Hollestein M. 1993. Clinical implications of the p53 tumor-suppressor gene. N Engl J Med 329:1318-1327.

Harris CC, Reddel R, Pfeifer A, Iman D, McMenamin M, Trump BF, Weston A. 1991. Role of oncogenes and tumour suppressor genes in human lung carcinogenesis. IARC Sci Publ 105:294-304.

Harting FH and Hesse W. 1879. Der Lungenkrebs, die bergkrankheit in den schneeberger gruben. Viertel Gerichtl Med Oeff Sanitaetswes 31:102-132, 313-337.

Harvey GJ, Jr. 1977. Trackless mining at Union Carbide's operations in the Uravan District in southwestern Colorado and southeastern Utah. Presented at First Conference. Uranium Mining Technology. University of Nevada, Reno, April 24-29, 1977.

Hattis D and Barlow K. 1996. Human interindividual variability in cancer risks - Technical and management challenges. Human and Ecological Risk Assessment 2:194-220.

Hattis D and Burmaster DE. 1994. Assessment of variability and uncertainty distributions for practical risk analysis. Risk Anal 14:713-730.

Hattis D and Silver K. 1994. Human interindividual variability - A major source of uncertainty in assessing risks for noncancer health effects. Risk Anal 14:421-432.

Hayes RB, Thomas T, Silverman DT, Vineis P, Blot WJ, Mason TJ, Pickle LW, Correa P, Fontham TH, Schoenberg JB. 1989. Lung cancer in motor exhaust-related occupations. Am J Ind Med 16:685-895.

Haynes RM. 1988. The distribution of domestic radon concentrations and lung cancer mortality in England and Wales. Radiat Prot Dosimet 25:93-96.

Heady JA, Kennaway EL. 1949. The increase in deaths attributed to cancer of the lung. Br J Cancer III:311-321.

HASL (Health and Safety Laboratory). 1960. Experimental Environmental Survey of AEC Leased Uranium Mines. Report No. HASL-91. New York: U.S. Atomic Energy Agency, Health and Safety Laboratory.

HASL (Health and Safety Laboratory). 1969. Investigation of the radiological characteristics of uranium mine atmospheres. New York: U.S. Atomic Energy Commission.

HEI (Health Effects Institute). 1995. Diesel Exhaust: A Critical Analysis of Emissions, Exposure, and Health Effects. A Special Report of the Institute's Diesel Working Group, Health Effects Institute, Cambridge, MA, April 1995.

Hei TK, Bedford J, Waldren CA. 1994a. p53 mutation hotspot in radon-associated lung cancer. Lancet 343:1158-1159

Hei TK, Piao CQ, Willey JC, Thomas S, Hall EJ. 1994b. Malignant transformation of human bronchial epithelial cells by radon-simulated alpha particles. Carcinogenesis 15:431-437.

Hei TK, Zhu LX, Waldren CA. 1994c. Molecular mechanisms of mutagenesis by radiation of different qualities. Pp 171-176 in Molecular Mechanisms in Radiation Mutagenesis and Carcinogenesis, Chadwick KH, Cox R, Leehouts HP, Thacker J, eds. Brussels: European Commission.

Heinrich U, Muhle H, Takenaka S, Ernst E, Furst R, Mohr U, Pott F, Stöber W. 1986. Chronic effects on the respiratory tract of hamsters, mice and rats after long-term inhalation of high concentrations of filtered and unfiltered diesel engine emissions. J Appl Toxicol 6:383-395.

Heinrich U, Furst R, Rittinghausen S, Creutzenberg O, Bellmann B, Koch W, Levsen K. 1995. Chronic inhalation exposure of wistar rats and two different strains of mice to diesel engine exhaust, carbon black, and titanium dioxide. Inhal Toxicol 7:533-556.

Henshaw DL, Eatough JP, Richardson RB. 1990. Radon as a causative factor in induction of myeloid leukemia and other cancers. Lancet 335(8696):1008-1012.

Henshaw DL, Eatough JP, Richardson RB. 1990. Radon and leukaemia (Letter). Lancet 335:1339.

Hertz-Picciotto I, Smith AH. 1993. Observations on the dose-response curve for arsenic exposure and lung cancer. Scand J Environ Health 19(4):217-226.

Hertz-Picciotto I, Smith AH, Holtzman D, Lipsett M, Alexeeff G. 1992. Synergism between occupational arsenic exposure and smoking in the induction of lung cancer. Epidemiology 3:23-31.

Hess CT, Weiffenbach CV, Norton SA. 1983. Environmental radon and cancer correlations in Maine. Health Phys 45:339-348.

Hessel PA, Sluis-Cremer GK, Hnizdo E. 1990. Silica exposure, silicosis, and lung cancer: A necropsy study. Br J Ind Med 47:4-9.

Heyder J and Schuech G. 1983. Diffusional transport of nonspherical aerosol particles. Aerosol Sci Technol 2:41.

Heyder J, Gebhart J, Stahlhofen W, Stuck B. 1982. Biological variability of particle deposition in the human respiratory tract during controlled and spontaneous mouth-breathing. Ann Occup Hyg 26:137.

Heyder J, Gebhart J, Rudolf G, Schiller CF, Stahlhofen W. 1986. Deposition of particles in the human respiratory tract in the size range 0.0015-15 m. J Aerosol Sci 17:811.

Hickman AW, Jaramillo RJ, Lechner JF, Johnson NF. 1994. Alpha particle-induced p53 protein expression in a rat lung epithelial cell strain. Cancer Res 54:5797-5800.

Hill AB and Faning EL. 1948. Studies in the incidence of cancer in a factory handling inorganic compounds of arsenic. I. Mortality experience in the factor. Br J Ind Med 5:1-6.

Hill CK, Han A, Elkind MM. 1984. Fission-spectrum neutrons at a low dose rate enhance neoplastic transformation in the linear, low dose region (0-10 cGy). Int J Radiat Biol 46:11-15.

Hill CK, Carnes BA, Han A, Elkind MM. 1985. Neoplastic transformation is enhanced by multiple low doses of fission-spectrum neutrons. Radiat Res 102:404-410.

Hill CK, Renan M, Buess E. 1991. Is neoplastic transformation by high-LET radiations dose rate dependent or cell cycle dependent? P. 344 in Proceedings of the International Congress on Radiation Research, JD Chapman, WC Dewey, and GF Whitmore, eds. San Diego: Academic Press.

Hinds WC. 1982. Aerosol Technology-Properties, Behavior, and Measurement of Airborne Particles. New York: Wiley-Interscience.

Hlatky L, Sachs RK, Hahnfeldt P. 1992. The ratio of dicentrics to centric rings produced in human lymphocytes by acute low-LET radiation. Radiat Res 129:304-308.

Hnizdo E and Sluis-Cremer GK. 1991. Silica exposure, silicosis, and lung cancer: A mortality study of South African gold miners. Br J Ind Med 48:53-60.

Hodgkins PS, O'Neill P, Stevens D, Fairman MP. 1996. The severity of a a-particle induced DNA damage is revealed by exposure to cell-free extracts. Radiat Res 146:660-667.

Hodgson RD and Jones RD. 1990. Mortality of a cohort of tin miners 1941-86. Br J Ind Med 47:665-676.

Hoffman FO and Hammonds JS. 1994. Propagation of uncertainty in risk assessments: The need to distinguish between uncertainty due to lack of knowledge and uncertainty due to variability. Risk Anal 14:707-712.

Hofmann W. 1982. Dose calculations for the respiratory tract from inhaled natural radionuclides as a function of age, II. Basal cell dose distributions and associated lung cancer risk. Health Phys 43:31-44.

Hofmann W, Katz R, Zhang CX. 1985. Lung cancer in a Chinese high background area - epidemiological results and theoretical interpretation. Sci Total Environ 45:527-534.

Holaday, DA. 1969. History of the exposure of miners to radon. Health Phys 16(5):547-552.

Holaday DA, Rushing DE, Coleman RD, Woolrich PF, Kusnetz HL. 1957. Control of radon and daughters in uranium mines and calculations on biological effects. PHS Publication No. 494. Washington, D.C. U.S. Government Printing Office.

Holley WR and Chatterjee A. 1996. Clusters of DNA damage induced by ionizing radiation: Formation of short DNA fragments. I. Theoretical modelling. Radiat Res 145:188-199.

Hollstein M, Sidransky D, Vogelsetein B, Harris CC. 1991. p53 mutations in human cancers. Science 253:49-53.

Hollstein M, Bartsch H, Wesch H, Kure EH, Mustonen R, Muhlbauer KR, Spiethoff A, Wegener K, Wiethege T, Muller KM. 1997. P53 gene mutation analysis in tumors of patients exposed to alpha-particles. Carcinogenesis 18(3):511-516.

Holub RF and Knutson EO. 1987. Measuring polonium-218 diffusion-coefficient spectra using multiple wire screens. Pp. 340-356 in Radon and Its Decay Products: Occurrence, Properties and Health Effects, Symposium Series 331, Hopke, P.K., ed. Washington, DC: American Chemical Society.

Hoover HC and Hoover LH. 1950. Georgius Agricola De Re Metalallica. New York: Dover Publications, Inc.

Hopke PK. 1992. Some thoughts on the "unattached" fraction of radon decay products. Health Phys 63:209-212.

Hopke PK, Ramamurthi, Knutson EO, Tu KW, Scofield P, Holub RF, Cheng YS, Su YF, Winklmayr W, Strong JC, Solomon S, Reineking A. 1992. The measurement of activity-weighted size distribution of radon progeny: methods and laboratory intercomparison studies. Health Phys 63(5):560-570.

Hopke PK, Montassier N, Wasiolek P. 1993. Evaluation of the effectiveness of several air cleaners for reducing the hazard from indoor radon progeny. Aerosol Sci Technol 19:268-278.

Hopke PK, Jensen B, Montassier N. 1994. Evaluation of several air cleaners for reducing indoor radon progeny. J Aerosol Sci 25:395-405.

Hopke PK, Jensen B, Li CS, Montassier N, Wasiolek P. 1995. Assessment of the exposure to and dose from radon decay products in normally occupied homes. Environ Sci Technol 29:1359-1364.

Hornung RW and Meinhardt TJ. 1987. Quantitative risk assessment of lung cancer in U.S. uranium miners. Health Phys 52:417-430.

Hornung RW, Deddens J, Roscoe R. 1995. Modifiers of exposure-response estimates for lung cancer among miners exposed to radon progeny. Environ Health Perspect 103(Supp 2):49-53.

Howe GR and RH Stager. 1996. Risk of lung cancer mortality after exposure to radon decay products in the Beaverlodge cohort based on revised exposure estimates. Radiat Res 146:37-42.

Howe GR, Nair RC, Newcombe HB, Miller AB, Abbatt JD. 1986. Lung cancer mortality (1950-80) in relation to radon daughter exposure in a cohort of workers at the Eldorado Beaverlodge Uranium Mine. J Natl Cancer Inst 77:357-362.

Howe GR, Nair RC, Newcombe HB, Miller AB, Burch JD, Abbat JD. 1987. Lung cancer mortality (1950-1980) in relation to radon daughter exposure in a cohort of workers at the Eldorado Port Radium uranium mine: Possible modification by exposure rate. J Natl Cancer Inst 79:1255-1260.

Hui TE, James AC, Jostes RF, Schwartz JL, Swinth KL, Cross FT. 1993. Evaluation of an alpha probe detector for in vitro cellular dosimetry. Health Phys 64(6):647-652.

Hunter T. 1991. Cooperation between oncogenes. Cell 64:249-270.

IARC (International Agency for Research on Cancer). 1986. Monographs on the Evaluation of the Carcinogenic Risk of Chemicals to Humans. Tobacco smoking.

IARC (International Agency for Research on Cancer). 1987. International Agency for Research Monography on the Evaluation of Carcinogenic Risks to Humans. Overall Evaluations of Carcinogenicity: An Updating of IARC Monographs Volumes 1 to 42, Supplement 7. Lyon, IARC.

IARC (International Agency for Research on Cancer). 1988. Man-made Mineral Fibres and Radon. Lyon, France: International Agency for Research on Cancer. p. 143. IARC Monographs on the Evaulation of Carcinogenic Risks to Humans.

ICRP (International Commission on Radiological Protection). 1987. Lung cancer risk from indoor exposures to radon daughters. Oxford, England: Pergamon Press 50.

ICRP (International Commission on Radiological Protection). 1991. 1990 Recommendations of the International Commission on Radiation Protection, ICRP Publication 60, Ann. of ICRP 21.

ICRP (International Commission on Radiological Protection). 1993. Protection Against Radon-222 at Home and at Work. London, England: Pergamon Press 23(2):1-65. ICRP Publication 65.

ICRP (International Commission on Radiological Protection). 1994. Human Respiratory Tract Model for Radiological Protection. A Report of Committee 2 of the ICRP. ICRP Publication 66. Ann ICRP 24(1/4) (Oxford: Pergamon Press).

ICRU (International Commission on Radiation Units and Measurements). 1986. The Quality Factor in Radiation Protection. ICRU Report 40. Washington, DC: ICRU Publications.

Iliakis G. 1984. The mutagenicity of a-particles in Ehrlich ascites tumor cells. Radiat Res 99:52-58.

Ishinishi N, Kuwabara N, Nigase S, Suzuki T, Ishiwata S, Kohno T. 1986. Long-term inhalation studies on effects of exhaust from heavy and light duty diesel engines on F344 rats. In: Carcinogenic and Mutagenic Effects of Diesel Engine Exhaust. Eds. N Ishinishi, A Koizumi, RO McClellan, and W Stöber, pp. 329-348. New York: Elsevier.

Ives JC, Buffler PA, Selwyn BJ, Hardy RJ, Decker M. 1988. Lung cancer mortality among women employed in high-risk industries and occupations in Harris County, Texas, 1977-1980. Am J Epidemiol 127:65-74.

Iwai K, Udagawa T, Yamagishi M, Yamada H. 1986. Long-term inhalation studies of diesel exhaust on F344 SPF rats. Incidence of lung cancer and lymphoma. In: Carcinogenic and Mutagenic Effects of Diesel Engine Exhaust. Eds. N Ishinishi, A Koizumi, RO McClellan, W Stöber, pp. 349-360. Amsterdam: Elsevier.

Jaberaboansari A, Dunn WC, Preston RJ, Mitra S, Waters LC. 1991. Mutations induced by ionizing radiation in a plasmid replicated in human cells. II. Sequence analysis of alpha-particle-induced point mutations. Radiat Res 127(2):202-210.

Jablon S, Tachikawa K, Belsky JL, Steer A. 1971. Cancer in Japanese exposed as children to atomic bombs. Lancet 1(7706):927-932.

Jacobi W. 1994. The History of the Radon Problem in Mines and Homes.Protection Against Radon-222 at Home and at Work. Pergamon Press; London, England.

Jacobi W and Eisfeld K. 1980. Dose to Tissues and Effective Dose Equivalent by Inhalation of Radon-222, Radon-220 and Their Short-lived Daughters, GSF Report S-626. Gesellschaft für Strahlen-und Umweltforschung, Munich-Neuherberg, West Germany.

Jacobi W, Henrichs K, Barclay D. 1992. Verura schungswarhscheinlichkeit von Lungenkrebs durch die berufliche Strahlenexposition von Uran-Bergarbeitern der WISMUT AG. GSF Research Center for Environment and Health, Neugerberg, Germany. GSF-Report S-14/92.

James AC. 1988. Lung dosimetry. Pp. 259-309 in Radon and Its Decay Products in Indoor Air, Nazaroff WW and Nero AV, eds. New York: Wiley Interscience.

James AC. 1992. Dosimetry of radon and thoron exposures: Implications for risks from indoor exposure. Pp. 167-198 in Indoor Radon and Lung Cancer: Reality or Myth, Cross FT, ed. Columbus: Battelle Press.

James AC. 1994. Dosimetry of inhaled radon and thoron progeny. Pp. 161-180 in Integrally Radiation Dosimetry, Health Physics Society 1994 Summer School, Raabe OG, ed. Madison: Medical Physics Publishing.

James AC, Fisher DR, Hui TE, Cross FT, Durham JS, Gehr P, Egan MJ, Nixon W, Swift DL, Hopke PK. 1991. Dosimetry of Radon Progeny. In: Pacific Northwest Laboratory Annual Report for 1990 to the DOE Office of Energy Research, Pt. 1, pp 55-63. PNL-7600, Pacific Northwest Laboratory, Richland, Washington.

Jarvis NS, Birchall A, James AC, Bailey MR, Dorrian MD. 1993. LUDEP 1.0 Personal Computer Program for Calculating Internal Doses Using the New ICRP Respiratory Tract Model, NPRB-SR264, National Radiological Protection Board, Chilton, Didcot Oxon, England.

Jenner TJ, de Lara CM, O'Neill P, Stevens DL. 1993. Induction and rejoining of DNA double strand breaks in V79-4 mammalian cells following gamma-and alpha-irradiation. Int J Radiat Biol 64:265-273.

Jin Y, Yie TA, Carothers AM. 1995. Non-random deletions at the dihydrofolate reductase locus of Chinese hamster ovary cells induced by a-particles simulating radon. Carcinogenesis 16(8):1981-1991.

Johnson JH. 1988. Automotive emissions. In Air Pollution, The Automobile, and Public Health, Watson AY, Bates RR, Kennedy D, eds. Washington, DC: National Academy Press. pp. 39-75.

Johnson NF. 1995. Radiobiology of lung target cells. Radiat Prot Dosimet 60:327-330.

Johnson NF and Hubbs AF. 1990. Epithelial progenitor cells in the rat trachea. Am J Respir Cell Molec Biol 3:579-585

Johnson NF and Newton GJ. 1994. Estimation of the dose of radon progeny to the peripheral lung and the effect of exposure to radon progeny on the alveolar macrophage. Radiat Res 139:163-169.

Jones GDD, Milligan JR, Ward JF, Calabro-Jones PM, Aguilera JA. 1993. Yield of strand breaks as a function of scavenger concentration and LET for SV40 irradiated with He ions. Radiat Res 136:196-196.

Jostes RF, Hui TE, Cross FT. 1993. Use of the single-cell gel technique to support hit probability calculations in mammalian cells exposed to radon and radon progeny. Health Phys 64:675-679.

Jostes RF, Fleck EW, Morgan TL, Stiegler GL Cross FT. 1994. Southern blot and polymerase chain reaction exon analyses of HPRT- mutations induced by radon and radon progeny. Radiat Res 137:371-379.

Jostes RF. 1996. Genetic, cytogenetic and carcinogenic effects of radon: A review. Mutat Res 340:125-139.

Kadhim MA, Macdonald DA, Goodhead DT, Lorimore SA, Marsden SJ, Wright EG. 1992. Transmission of chromosomal instability after plutonium a-particle irradiation. Nature 355:738-740.

Kadhim MA, Lorimore SA, Hepburn MD, Goodhead DT, Buckle VJ, Wright EG. 1994. Alpha-particle-induced chromosomal instability in human bone marrow cells. Lancet 344:987-988.

Kadhim MA, Lorimore SA, Townsend KM, Goodhead DT, Buckle VJ, Wright EG. 1995. Radiation-induced genomic instability: delayed cytogenetic aberrations and apoptosis in primary human bone marrow cells. Int J Radiat Biol 67:287-293.

Kaldor JM, Day NE, Bell J, Clarke EA, Langmark F, Karjalainen S, Band P, Pedersen D, Choi W, Blair V, Henryamar M, Prior P, Assouline D, Pompekirn V, Cartwright RA, Koch M, Arslan A, Fraser P, Sutcliffe SB, Host H, Hakama M, Stovall M. 1992. Lung cancer following Hodgkin's Disease: A case-control study. Int J Cancer 52:677-681.

Kasten MB, Onyuuekwere O, Sidransky D, Vogelstein B, Craig R. 1991. Participation of p53 protein in the cellular response to DNA damage. Cancer Res 51:6304-6311.

Kaur GP and Athwal RS. 1989. Complementation of a DNA repair defect in xeroderma pigmentosum cells by transfer of human chromosome 9. Proc Natl Acad Sci USA 86:8872-8876.

Keenan KP, Wilson TS, McDowell EM. 1983. Regeneration of hamster tracheal epithelium after mechanical injury. V. Histochemical immunocytochemical and ultrastructural studies. Vircros Arch [B] 43:213-240.

Kelly G, Stegelmeir BL, Hahn FF, 1995. p53 alterations in plutonium-induced F344 rat lung tumors. Radiat Res 142:263-269.

Kemp CJ, Wheldon T, Balmain A. 1994. p53 deficient mice are extremely susceptible to radiation-induced tumorigenesis. Nat Genet 8:66-69.

Kendall GM, Miles JCH, Cliff KD, Green BMR, Muirhead CR, Dixon DW, Lomas PR, Goodridge SM. 1994. Exposure to radon in UK dwellings. Publication No. NRPB-R272, National Radiological Protection Board, Chilton, Didcot, Oxfordshire, UK.

Kennedy AR and Little JB. 1984. Evidence that a second event in X-ray induced oncogenic transformation in vitro occurs during cellular proliferation. Radiat Res 99:228-248.

Khan MA, Cross FT, Jostes RF, Hui E, Morris JE, Brooks AL. 1994. Micronuclei induced by radon and its progeny in deep-lung fibroblasts of rats in vivo and in vitro. Radiat Res 139:53-59.

Khan MA, Cross FT, Buschbom RL, Brooks AL. 1995. Inhaled radon-induced genotoxicity in Wistar rat, Syrian hamster, and Chinese hamster deep-lung fibroblasts in vivo. Mutat Res 334:131-137.

Kihara M, Kihara M, Noda K. 1995. Risk of smoking for squamous and small cell carcinomas of the lung modulated by combinations of CYP1A1 and GSTM1 gene polymorphisms in a Japanese population. Carcinogenesis 16:2331-2336.

Kilburn KH. 1977. Clearance mechanisms in the respiratory tract. In Handbook of Physiology, Section 9, Reactions to Environmental Agents, Lee DHK, Falk HL, and Murphy SD, eds. Bethesda: American Physiological Society.

Kimbell JS, Gross EA, Joyner DR, Godo MN, Morgan KT. 1993. Application of computational fluid dynamics to regional dosimetry of inhaled chemicals in the upper respiratory tract of the rat. Tox Appl Pharm 121:253-263.

Klotz JB, Schoenberg JB, Wilcox HB. 1993. Relationshop among short- and long-term radon measurements within dwellings: Influence of radon concentrations. Health Phys 65(4):367-374.

Knudson AG. 1971. Mutation and cancer: Statistical study of retinoblastoma. Proc Natl Acad Sci USA 68:820-823.

Knutson EO and George AC. 1992. Reanalysis of data on particle size dsitributions of radon progeny in uranium mines. Pp. 149-164 in Indoor Radon and Lung Cancer: Reality or Myth, Cross FT, ed. Columbus: Battelle Press.

Knutson EO, George AC, Knuth RH, Koh BR. 1984. Measurements of radon daughter particle size. Radiat Prot Dosimet 7:121-125.

Kodell RL, Krewski D, Zielinski JM. 1991. Additive and multiplicative relative risks in the two-stage clonal expansion model of carcinogenesis. Risk Anal 11:483-490.

Kopecky KJ, Nakashima E, Yamamato T, Kato H. 1988. Lung cancer, radiation, and smoking among A-bomb survivors, Hiroshima and Nagasaki. Radiation Effects Research Foundation. 13-86. p.1 RERF Technical Report Series.

Koskela RS, Klockars M, Jarvinen E, Rossi A, Kolari PJ. 1990. Cancer mortality of granite workers 1940-1985. In Simonato L, Fletcher AC, Saracci R, Thomas TL (eds): "Occupational Exposure to Silica and Cancer Risk." IARC Sci Publ no. 97:105-111. Lyon, France: IARC.

Kotin P, Falk HL, Thomas M. 1955. Aromatic hydrocarbons. III. Presence in the particulate phase of diesel-engine exhaust extracts. Arch Ind Health 11:113-120.

Krewski D, Goddard MJ, Zielinski JM. 1992. Dose-response realtionships in carcinogenesis. In: Mechanisms of Carcinogenesis in Risk Identification (H.Vainio, P.N. Magee, D.B. McGregor & A.J. McMichael, eds). IARC Scientific Publications No. 116, International Agency for Research on Cancer, Lyon, pp. 579-599.

Krolewski B, Little JB. 1989. Molecular analysis of DNA isolated from the different stages of X-ray-induced transformation. Vitro Mol Carcinog 2:27-33.

Kronenberg A and Little JB. 1989. Molecular characterization of thymidine kinase mutants of human cells induced by densely ionising radiation. Mutat Res 211:215-224.

Kronenberg A, Gauny S, Criddle K, VannaisD, Ueno A, Kraemer S, Waldren CA. 1995. Heavy ion mutagenesis: linear energy transfer effects and genetic linkage. Radiat Environ Biophys 34:73-78.

Kunz E, Sevc J, Placek V. 1978. Lung cancer mortality in uranium miners. Health Phys 35:379-580.

Kunz E, Sevc J, Placek V, Horacek. 1979. Lung cancer in man in relation to different time distributions of radiation exposure. Health Phys 36:699-706.

Kurosu K, Yumoto N, Mikata A, Taniguchi M, Kuriyama T. 1996. Monoclonality of B-cell lineage in primary pulmonary lymphoma demonstrated by immunoglobulin heavy chain gene sequence analysis of histologically non-definitive transbronchial biopsy specimens. J Pathol 178:316-22.

Kusiak RA, Springer J, Ritchie AC, Muller J. 1991. Carcinoma of the lung in Ontario gold miners: possible aetiological factors. Br J Ind Med 48:808-817.

Kusiak RA, Ritchie AC, Muller J, Springer J. 1993. Mortality for lung cancer in Ontario uranium miners. Br J Ind Med 50:920-928.

Kysela BP, Arrand JE, Michael BD. 1993. Relative contributions of levels of initial damage and repair of double-strand breaks to the ionizing radiation-sensitive phenotype of the Chinese hamster cell mutant, XR-V15B. Part II. Neutrons. Int J Radiat Biol 64:531-538.

L'Abbe KAL, Howe GR, Burch JF, Miller AB, Abbatt J, Band P, Choi W, Du J, Feather J, Gallagher R, Hill G, Matthews B. 1991. Radon exposure, cigarette smoking, and other mining experience in the Beaverlodge uranium miners cohort. Health Phys 60:489-495.

Laird NM and Mosteller F. 1990. Some statistical methods for combining experimental results. International J Tech Asses in Hlth Care 6:5-30.

Land CE, Shimosato Y, Saccomanno G, Tokuoka S, Auerbach O, Tateishi R, Greenberg SD, Nambu S, Carter D, Akiba S, Keehn R, Madigan P, Mason TJ, Tokunaga M. 1993. Radiation-associated lung cancer: A comparison of the histology of lung cancers in uranium miners and survivors of the atomic bombings of Hiroshima and Nagasaki. Radiat Res 134:234-243.

Land H, Parada LF, Weinberg RA. 1983. Tumorigenic conversion of primary embryo fibroblasts requires at least two cooperating oncogenes. Nature 304:596-602.

Last JM. 1983. A Dictionary of Epidemiology. Second Edition. New York: Oxford University Press.

Lauer GR, Gang QT, Lubin JH, Jun-Yao L, Kan CS, Xiang YS, Jian CZ, Yi H, WanDe G, Blot WJ. 1993. Skeletal lead $^{210}$Pb levels in lung cancer among radon-exposed tin miners in southern China. Health Phys 64:253-259.

Lavin MF, Bennett I, Ramsay J, Gardiner RA, Seymour GJ, Farrell A, Walsh M. 1994. Identification of a potentially radiosensitive subgroup among patients with breast cancer. J Natl Cancer Inst 86:1627-1634.

Lee JM, Abramson JLA, Kandel R, Donehower A, Bernstein A. 1994. Susceptibility to radiation-carcinogenesis and accumulation of chromosomal breakage in p53 deficient mice. Oncogene 9:3731-3736.

Lee WH, Bookstein R, Hong, Young LH, Shew, Lee EY-HP. 1987. Human retinoblastoma susceptibility gene: Cloning identification and sequence. Science 235:1394-1399.

Lee-Feldstein A. 1983: Arsenic and respiratory cancer in man: Follow-up of an occupational study. In Arsenic: Industrial, Biomedical Environmental Perspectives. Van Nostrand Reinhold Co., New York. pp 245-254.

Leenhouts HP, Chadwick KH. 1994. A two-mutation model of radiation carcinogenesis: application to lung tumors in rodents and implications for risk evaluation. J Radiol Prot 14:115-130.

Lees REM, Steele R, Roberts JH. 1987. A case-control study of lung cancer relative to domestic radon exposure. Int J Epidemiol 16:7-12.

Létourneau EG, Mao Y, McGregor RG, Semenciw R, Smith MH, Wigle DT. 1983. Lung cancer mortality and indoor radon concentrations in 18 Canadian cities. Epidemiology applied to health physics. Proceedings of a conference; 1983 Jan 10; Albuquerque, NM. Pp 470-483.

Létourneau EG, Krewski D, Choi NW, Goddard MJ, McGregor RG, Zielinski JM, Du J. 1994. Case-control study of residential radon and lung cancer in Winnipeg, Manitoba, Canada. Am J Epidemiol 140:310-322.

Leupker RV and, Smith ML. 1978. Mortality in unionized truck drivers. JOM 20:677-682.

Levin ML. 1953. The occurrence of lung cancer in man. Acta Unio International Contra Cancrum 9:531-541.

Li CS and Hopke PK. 1991. Characterization of radon decay products in a domestic environment. Indoor Air 1:539-561.

Li CS and Hopke PK. 1991. The efficacy of air cleaners in controlling indoor radon decay products. Health Phys 61:785-797.

Li CS and Hopke PK. 1993. Initial size and distributions and hygroscopicity of indoor combustion aerosol particles. Aerosol Sci Tech 19:305-316.

Li CS and Hopke PK. 1994. Hygroscopic growth of consumer spray products. Aerosol Sci 25:1342-1351.

Lippmann, M., D.B. Yeates, and R.E. Albert. 1980. Deposition, retention, and clearance of inhaled particles. Br J Ind Med 37:337.

Little JB. 1968. Delayed initiation of DNA synthesis in irradiated human diploid cells. Nature 218:1064-1065.

Little MP, Hawkins MM, Charles MW, Hildreth NG. 1992. Fitting the Armitage-Doll model to radiation-exposed cohorts and implications for population cancer risks. Radiat Res 132:207-221 (see also erratum in Radiat Res 137:124-128, 1994).

Little MP. 1995. Are two mutations sufficient to cause cancer? Some generalizations of the two-mutation model of carcinogenesis of Moolgavkar, Venzon, and Knudson, and of the multistage model of Armitage and Doll. Biometrics 51:1278-91.

Liu BYH and Pui DYH. 1975. On the performance of the electrical aerosol analyzer. J Aerosol Sci 6:249-264.

Lloyd DC, Edwards AA, Prosser JS, Bolton D, Corp MJ. 1984. Chromosome aberrations induced in human lymphocytes by D-T neutrons. Radiat Res 98:561-578.

Lloyd EL and Henning CB. 1981. Morphology of cells malignantly transformed by alpha particle irradiation. Scanning Electron Microscopy IV:87-92.

Lloyd EL, Gemmell MA, Henning CB, Zabransky BJ. 1979. Transformation of mammalian cells by alpha particles. Int J Radiat Biol 36:467-478.

Lo YM, Darby S, Noakes L, Whitley E, Silcocks PBS, Fleming KA, Bell JI. 1995. Screening for codon p53 mutation in lung cancer associated with domestic radon exposure. Lancet 345:60.

Löbrich M, Rydberg B, Cooper PK. 1994. DNA double-strand breaks induced by high-energy neon and iron ions in human fibroblasts. II. Probing individual *NotI* fragments by hybridization. Radiat Res 139:143-151.

Löbrich M, Cooper PK, Rydberg B. 1996. Non-random distribution of DNA double strand breaks induced by particle irradiation. (submitted for publication).

Lorenz E. 1944. Radioactivity and lung cancer: A critical review of lung cancer in the miners of Schneeberg and Joachimsthal. J Natl Cancer Inst 5:1-15.

Lorimore SA, Goodhead DT, Wright EG. 1993. Inactivation of haemopoietic stem cells by slow a-particles. Int J Radiat Biol 63:655-660.

Lorimore SA, Goodhead DT, Wright EG. 1995. The effect of p53 status on the radiosensitivity of haemopoietic stem cells. Cell Death and Differentiation 2:233-234.

Loucas BD and Geard CR. 1994. Initial damage in human interphase chromosomes from alpha particles with linear energy transfers relevant to radon exposure. Radiat Res 139:9-14.

Lovett DB. 1969. Track etch detectors for alpha exposure estimates. Health Phys 16:623-628.

Lubin JH. 1994. Invited commentary: Lung cancer and exposure to radon. Am J Epidemiol 140:323-332.

Lubin JH and Boice JD, Jr. 1989. Estimating radon-induced lung cancer in the U.S. Health Phys 7:417-427.

Lubin JH and Boice JD, Jr. 1997. Lung cancer risk from residential radon: meta-analysis of eight epidemiologic studies. J Natl Cancer Inst 89:49-57.

Lubin JH and Gaffey W. 1988. Relative risk model for assessing the joint effects of multiple factors. Am J Ind Med 13:149-167.

Lubin JH and Steindorf K. 1995. Cigarette use and the estimation of radon-attributable lung cancer in the U.S. Radiat Res 141:79-85.

Lubin JH, Blot WJ, Berrino F, Gillis CR, Kunze M, Shamahl D, Visco G. 1984. Patterns of lung cancer risk according to type of cigarette smoked. Int J Cancer 33:569-576.

Lubin JH, Samet JM, Weinberg C. 1990a. Design issues in epidemiologic studies of indoor exposure to Rn and risk of lung cancer. Health Phys 59:807-817.

Lubin JH, You-Lin Qiao, Taylor PR, Schatzkin A, Bao-Lin Mayo, Jian-Yu Rao, Xiang-Zhen Xuan, Jun-Yao Li. 1990b. Quantitative evaluation of the radon and lung cancer association in case control study of Chinese tin miners. Cancer Res 50:174-180.

Lubin JH, Boice Jr. JD, Edling C, Hornung RW, Howe G, Kunz E, Kusiak RA, Morrison HI, Radford EP, Samet JM, Tirmarche M, Woodward A, Yao SX, Pierce DA. 1994a. Radon and Lung Cancer Risk: A Joint Analysis of 11 Underground Miners Studies. National Institutes of Health, National Cancer Institute. NIH Publication No. 94-3644. Washington, D.C. U.S. Department of Health and Human Services.

Lubin JH, Liang Z, Hrubec Z, Pershagen G, Schoenberg JB, Blot WJ, Klotz JB, Xu Z-Y, Boice JD Jr. 1994b. Radon exposure in residences and lung cancer among women: Combined analysis of three studies. Cancer Causes Control 5:114-128.

Lubin JH, Boice JD, Jr., Edling C, Hornung RW, Howe G, Kunz E, Kusiak RA, Morrison HI, Radford EP, Samet JM, Tirmarche M, Woodward A, Yao SX. 1995a. Radon-exposed underground miners and inverse exposure-rate (protraction enhancement) effects. Health Phys 69:494-500.

Lubin JH, Boice JD, Jr., Edling C, Hornung RW, Howe GR, Kunz E, Kusiak RA, Morrison HI, Radford EP, Samet JM, Tirmarche M, Woodward A, Yao SX, Pierce DA. 1995b. Lung cancer in radon-exposed miners and estimation of risk from indoor exposure. J Natl Cancer Inst 87:817-27.

Lubin JH, Boice JD Jr, Samet JM. 1995c. Errors in exposure assessment, statistical power, and the interpretation of residential radon studies. Radiat Res 144:329-341.

Lubin JH, Tomášek L, Edling C, Hornung RW, Howe G, Kunz E, Kusiak RA, Morrison HI, Radford EP, Samet JM, Timarche M, Woodward A, Yao, SX. 1997. Estimating lung cancer mortality from residential radon using data for low exposures in miners. Radiat Res 147(2):126-134.

Lucas JN, Awa A, Straume T, Poggensee M, Kodama Y, Nakano M, Ohtaki K, Weier HU, Pinkel D, Gray D, Littlefield G. 1992. Rapid translocation frequency analysis in humans decades after exposure to ionizing radiation. Int J Radiat Biol 62:53-63.

Lucie NP. 1989. Radon exposure and leukemia. Lancet 2(8654):99-100.

Lucke-Huhle C, Comper W, Hieber L, Pech M. 1982. Comparative study of G2 delay and survival after 241Americium-alpha and 60cobalt-gamma irradiation. Radiat Environ Biophys 20:171-185.

Ludwig P and Lorenser S. 1924. Untersuchungen der grubenluft in den Schneeberger gruben auf den gehalt an radiumemanation. Z Physik 22:178-185.

Luebeck EG, Curtis SB, Cross FT, Moolgavkar SH. 1996. Two-stage model of radon-induced malignant lung tumors in rats: Effects of cell killing. Radiat Res 145(2):163-173.

Lundin FE, Wagoner JK, Archer VE. 1971. Radon Daughter Exposure and Respiratory Cancer Quantitative and Temporal Aspects. NIOSH-NIEHS Joint Monograph No. 1. U.S. Department of Health, Education and Welfare.

Lutze LH, Winegar RA, Jostes R, Cross FT, Cleaver JE. 1992. Radon-induced deletions in human cells: Role of nonhomologous strand rejoining. Cancer Res 52:5126-5129.

Lutze LH, Cleaver JE, Winegar RA. 1994. Factors affecting the frequency, size, and location of ionizing-radiation-induced deletions in human cells. In: Molecular Mechanisms in Radiation Mutagenesis and Carcinogenesis (Eds. K.H. Chadwick, R. Cox, H.P. Leenhouts and J. Thacker). European Commission EUR 15294:41-46.

Lyon JL, Gardner JW, West DW. 1980. Cancer in Utah: Risk by religion and place of residence. J Natl Cancer Inst 65:1063-1071.

Mabuchi K, Lilienfeld AM, Snell LM. 1985. Cancer and occupational exposure to arsenic: A study of pesticide workers. Prev Med 9:51-77.

Mabuchi K, Land CE, Akiba S. 1992. Radiation, smoking and lung cancer. RERF Update 7-8.

Macklin MT. 1942. Has a real increase in lung cancer been proved? Am J Ind Med 17:308-324.

Magnus K, Engeland A, Green BM, Haldorsen T, Muirhead CR, Strand T. 1994. Residential radon exposure and lung cancer - An epidemiological study of Norwegian municipalities. Int J Cancer 58:1-7.

Mahaffey JA, Parkhurst MA, James AC, Cross FT, Alavanja MCR, Boice JD Jr., Ezrine S, Henderson P, Brownson RC. 1993. Estimating past exposure to indoor radon from household glass. Health Phys 64:381-391.

Maher EF and Laird NM. 1985. EM algorithm reconstruction of particle size distributions from diffusion battery data. J Aerosol Sci 16:557-570.

Maillie HD, Simon W, Greenspan BS, Watts RJ, Quinn BR. 1994. The influence of life table corrections for smokers and nonsmokers on the health effects of Radon using the BEIR IV method. Health Phys 686:615-620.

Maity A, McKenna WG, Muschel RJ. 1994 The molecular basis for cell cycle delays following ionizing radiation: A review. Radiother Oncol 31:1-13.

Mäkeläinen I, Voutilainen A, Castrén O. 1987. Uppskattning av radonhalten i småhus på basen av lokation och byggnadsdata. Nordiska Sällskapet för Strålskydd, mötet 26-28 Aug., Mariehamn.

Mapel DW, Samet JM, Coultas DB. 1996. Corticosteroids and the treatment of idiopathic pulmonary fibrosis. Past, present, and future. Chest 110(4):1058-1067.

Marchaux G, Morlier J-P, Morin M, Chameaud J, Lafuma J, Masse R. 1994. Carcinogenic and cocarcinogenic effects of radon and radon daughters in rats. Environ Health Perspect 102:64-73.

Marcinowski F, Lucas RM, Yeager WM. 1994. National and regional distributions of airborne radon concentrations in U.S. homes. Health Phys 66:699-706.

Marder BA, Morgan WF. 1993. Delayed chromosomal instability induced by DNA damage. Mol Cell Biol 13:6667-6677.

Martins MB, Sabatier L, Ricoul M, Pinton A, Dutrillaux B. 1993. Specific chromosome instability induced by heavy ions: a step towards transformation of human fibroblasts. Mutat Res 285:229-237.

Martland HS. 1931. Occurrence of malignancy in radioactive persons. Am J Cancer 15:2435-2516.

Martz DE, Falco RJ, Langner Jr. GH. 1990. Time-averaged exposures to 220Rn and 222Rn progeny in Colorado homes. Health Phys 58:705-713.

Mason TJ. 1994. The descriptive epidemiology of lung cancer. Samet JM, Ed. Epidemiology of Lung Cancer. Marcel Dekker, Inc. New York, New York.

Masse R and Cross FT. 1989. Risk considerations related to lung modeling. Health Phys 57(Suppl 1):283-9.

Mauderly JL. 1992. Diesel exhaust. Pp. 119-162 in Environmental Toxicants–Human Exposures and Their Health Effects, Lippmann M, ed. New York: Van Nostrand Reinhold.

Mauderly JL. 1993. Toxicological approaches to complex mixtures. Environ Health Perspect 101:155-165.

Mauderly JL. 1994. Contribution of inhalation bioassays to the assessment of human health risks from solid airborne particles. Pp. 355-365 in Toxic and Carcinogenic Effects of Solid Particles in the Respiratory Tract, Mohr U, Dungworth DL, Mauderly JL, and Oberdörster G, eds. Washington: International Life Sciences Institute Press.

Mauderly JL, Jones RK, Griffith WC, Henderson RF, McClellan RO. 1987. Diesel exhaust is a pulmonary carcinogen in rats exposed chronically by inhalation. Fundam Appl Toxicol 9:208-221.

Mauderly JL, Snipes MB, Barr EB, Belinsky SA, Bond JA, Brooks AL, Chang IY, Cheng YS, Gillett NA, Griffith WC, Henderson RG, Mitchell CF, Nikula KJ, Thomassen DG. 1994. Pulmonary toxicity of inhaled diesel exhaust and carbon black in chronically exposed rats. Part I: Neoplastic and nonneoplastic lung lesions. Res Rep Health Eff Inst 68:1-75.

Mauderly JL, Banas DA, Griffith WC, Hahn FF, Henderson RF, McClellan RO. 1996. Diesel exhaust is not a pulmonary carcinogen in CD-1 mice exposed under conditions carcinogenic to F344 rats. Fundam Appl Toxicol (in press).

May KR. 1945. The cascade impactor: An instrument for sampling coarse aerosols. J Sci Instr 22:187-195.

Mayer SA. 1973. Present regulations on control of radiation in uranium mines. U.S. Atomic Energy Commission.

Mazumdar S, Redmond CK, Enterline PE, Marsh GM, Costantino JP, Zhou SYJ, Patwardham RN. 1989. Multistage modeling of lung cancer mortality among arsenic-exposed copper-smelter workers. Risk Anal 9:551-563.

McClellan RO. 1987. Health effects of exposure to diesel exhaust particles. Am Rev Pharmacol Toxicol 27:279-300.

McClellan RO. 1990. Particle overload in the lung: Approaches to improving our knowledge. J Aerosol Med (Suppl.) 3:S197-S207.

McClellan RO. 1996. Lung cancer in rats from prolonged exposure to high concentrations of carbonaceous particles: Implications for human risk assessment. Inhal Toxicol 8(Suppl):193-226.

McDonald JW, Taylor JA, Watson MA, Saccomanno G, Devereux TR. 1995. p53 and K-ras in radon-associated lung adenocarcinoma. Cancer Epidemiol Biomarkers Prev 4(7):791-793.

McDowell EM and Trump BF. 1984. Histogenesis of preneoplastic lesions in tracheobronchial epithelium. Surv Synth Pathol Res 2:235-279.

McLaughlin JK, Chen J-Q, Dosemeci M, Chen R-A, Rexing SH, Wu Z, Hearl FJ, McCawley MA, Blot WJ. 1992. A nested case-control study of lung cancer among silica exposed workers in China. Br J Ind Med 49:167-171.

Meijers JMM, Swaen GMH, Volovics A, Slangen JJM, Van Vliet K. 1990. Silica exposure and lung cancer in ceramic workers: A case-control study. Int J Epidemiol 19:19-25.

Melhorn J, Selig R, Pabst R. 1992. Zum silikosegeschehen i3m Uranerzbergbau der DDR. In Kreutz R, Piekarski C (eds): Arbeitsmedizinische Aspekte...Stuttgart: Gentner Verlag 415-419.

Menck HR and Henderson BE. 1976. Occupational differences in rates of lung cancer. JOM 18:797-801.

Mercer TT. 1981. Production of therapeutic aerosols; principles and techniques. Chest 80(Suppl.):181-189.

Merlo F, Doria M, Fontana L, Ceppi M, Chesi E, Santi L. 1990. Mortality from specific causes among silicotic subjects: A historical prospective study. In Simonato L, Fletcher AC, Saracci R, Thomas TL (eds): "Occupational Exposure to Silica and Cancer Risk." IAARC Sci Publ no. 97:105-111. Lyon, France: IARC.

Metting NF and Little JB. 1995. Transient failure to dephosphorylate the cdc2-cyclin B1 complex accompanies radiation-induced G2-phase arrest in HeLa cells. Radiat Res 143:286-292.

Metting NF, Palayoor ST, Macklis RM, Atcher RW, Liber HL, Little JB. 1992. Induction of mutations by bismuth-212 a-particles at two genetic loci in human B-lymphoblasts. Radiat Res 132:339-345.

Mifune M, Sobue T, Arimoto HG, Komoto Y, Kundo S, Tanooka H. 1992. Cancer mortality survey in a spa area (Misasa, Japan) with a high radon background. Japan J of Cancer Res 83:1-5.

Miki Y, Swensen J, Shattuck-Eidens D, Futreal PA, Larshman KI, Tavtigian S, Liu Q, Cochran C, Bennet LM, Ding W, Bell R, Rosenthral J, Hussey C, Tran T, McClure M, Frye C, Hattier T, Phelps R, Haugen-Strano A, Katcher H, Yakumo K, Gholami Z, Shaffer D, Stone S, Bayer S, Wray C, Bogden R, Dayananth P, Ward J, Tonin P, Narod S, Bristow PK, Norris FH, Helvering L, Morrison P, Rosteck P, Lai M, Barrett JC, Lewis C, Neuhausen S, Cannon-Albright L Goldgar D, Wiseman R, Kamb A, Skolnick MH. 1994. A strong candidate for the breast and ovarian cancer susceptibility gene BRCA1. Science 266:66-71.

Miller RC, Brenner DJ, Randers-Person G, Marino SA, Hall EJ. 1990. The effects of temporal distribution of dose on oncogenic transformation by neutrons and charged particles of intermediate LET. Radiat Res 124:S62-S68.

Miller RC, Geard CR, Geard MJ, Hall EH. 1992. Cell-cycle dependent radiation-induced transformation of C3H10T$^{1}/_{2}$ cells. Radiat Res 130:129-133.

Miller RC, Randers-Pehrson G, Hieber L, Marino SA, Richards M, Hall EJ. 1993. The inverse dose rate effect for oncogenic transformation by charged particles is LET dependent. Radiat Res 133:259-263.

Miller RC, Marino SA, Brenner DJ, Martin SG, Richards M, Randers-Pehrson G, Hall EJ. 1995. The biological effectiveness of radon-progeny alpha particles II Neoplastic transformation as a function of LET. Radiat Res 142:54-60.

Mole RH. 1990. Radon and leukaemia (Letter). Lancet 335:1336.

Moolgavkar SH. 1993. Cell proliferation and carcinogenesis models: general principles with illustrations from the rodent liver system. Environ Health Perspect 101(Suppl 5):91-94.

Moolgavkar SH. 1994. Biological models of carcinogenesis and quantitative cancer risk assessment. Risk Anal 14:879-882.

Moolgavkar SH and Luebeck EG. 1990. Two-event model for carcinogenesis: Biological, mathematical and statistical considerations. Risk Anal 10:323-341.

Moolgavkar SH and Leubeck EG. 1993. Two-mutation model for radiation carcinogenesis in humans and rodents. In New Frontiers in Cancer Causation. Proceedings of the Second International Conference on Theories of Carcinogenesis, Iversen OH, ed. Taylor & Francis.

Moolgavkar SH, Cross FT, Luebeck G, Dagle GE. 1990. A two-mutation model for radon-induced lung tumors in rats. Radiat Res 121:28-37.

Moolgavkar SH, Luebeck EG, Krewski D, Zielinski JM. 1993. Radon, cigarette smoke, and lung cancer: A re-analysis of the Colorado plateau uranium miners' data. Epidemiology 4:204-217.

Morgan M, Henrion M, Small M. 1990. Uncertainty. New York: Cambridge University Press.

Morgan WF, Day JP, Kaplan MI, McGhee FM, Limoli CL. 1996. Genomic instability induced by ionizing radiation. Radiat Res 146:247-258.

Morgan WKC, Clague HW, Vinitski S. 1983. On paradigms, paradoxes, and particles. Lung 161:195.

Morgenstern H. 1995. Ecologic studies in epidemiology: Concepts, principles, and methods. Ann Rev Public Health 16:61-81.

Morken DA. 1973. The biological effects of radon on the lung. In: Nobel Gases, Edt. by R.E. Stanley, A.A. Moghissi, CONF-730915, U.S. Energy Development and Research Agency. National Environmental Research Center, Washington D.C., pp. 501-506.

Morlier J-P, Morin M, Monchaux G, Fritsch P, Chameaud J, Lafuma J, Masse R. 1994. Lung cancer incidence after exposure of rats to low doses of radon: Influence of dose rate. Radiat Prot Dosimet 56:93-97.

Morrison HI, Semenciw RM, Mao Y, Wigle DT. 1988. Cancer mortality among a group of fluorspar miners exposed to radon progeny. Am J Epidemiol 128:1266-1275.

Morrow PE. 1988. Possible mechanisms to explain dust overloading of the lungs. Fundam Appl Toxicol 10:369-384.

Muirhead CR, Butland BK, Green BM, Draper GJ. 1991. Childhood leukemia and a natural radiation. Lancet 337:503-504.

Muller J, Wheeler WC, Gentleman JF, Suranji JF Kusick. R. 1983. Study of mortality of Ontario miners 1955-1957 Part 1. Toronto: Ontario Ministry of Labor.

Muller J, Wheeler WC, Gentleman JP, Suranyi G, Kusiak R. 1984. Study of mortality of Ontario miners. In: Proceedings of the International Conference on Occupational Radiation Safety in Mining, (Stocker, H., ed). Toronto, Canada: Canadian Nuclear Association, pp 335-343.

Murnane JP. 1995. Cell cycle regulation in response to DNA damage in mammalian cells: A historical perspective. Cancer and Metastasis Reviews 14:17-29.

Muscat JE, Wynder EL. 1995. Diesel exhaust, diesel fumes, and laryngeal cancer. Otolaryngol-Head Neck Surg 112:437-440.

Nagarkatti M, Nagarkatti PS, Brooks AL. 1996. Effect of radon on the immune system: Alterations in the cellularity and functions of T cells in lymphoid organs of mouse. J Toxicol Environ Health 47:535-545.

Nagasawa H and Little JB. 1992. Induction of sister chromatid exchanges by extremely low doses of alpha-particles. Cancer Res 52:6394-6396.

Nagasawa H, Robertson J, Little JB. 1990. Induction of chromosomal aberrations and sister chromatid exchanges by alpha particles in density-inhibited cultures of mouse 10T1/2 and 3T3 cells. Int J Radiat Biol 57:35-44.

NCRP (National Council of Radiation Protection and Measurements). 1984a. Exposures from the Uranium Series with Emphasis on Radon and Its Daughters. NCRP Report 77. Washington, DC: National Council on Radiation Protection and Measurements.

NCRP (National Council on Radiation Protection and Measurements). 1984b. Evaluation of occupational and environmental exposures to radon and radon daughters in the United States. NCRP Report 78. Washington, DC: National Council on Radiation Protection and Measurements.

NCRP (National Council on Radiation Protection and Measurements). 1988. Epidemiological studies of lung cancer in underground miners. Proceedings of the Twenty-Fourth Annual Meeting of the National Council on Radiation Protection and Measurements. Bethesda, MD. Pp. 30-50.

NCRP (National Council on Radiation Protection and Measurements). 1990. The Relative Biological Effectiveness of Radiations of Different Quality. NCRP Report 104. Issued December 15, 1990. Bethesda, Maryland. Nationa Councel on Radiaiton Protection and Measurements.

NCRP (National Council on Radiation Protection and Measurements). 1991. Radon Exposure of the U.S. Population — Status of the Problem. Bethesda, Maryland: National Council on Radiation Protection and Measurements. p. 16.

NCRP (National Council on Radiation Protection and Measurements). 1993. Research Needs for Radiation Protection. NCRP Report 117. Washington, DC: National Council on Radiation Protection and Measurements.

NEA (Nuclear Energy Agency Group of Experts). 1983. Dosimetry Aspects of Exposure to Radon and Thoron Daughter Products. Paris: Organization for Economic Cooperation and Development.

NIOSH (National Institute for Occupational Safety and Health). 1987. Criteria for a Recommended Standard: Occupational Exposure to Radon Progeny in Underground Mines. Washington, DC. U.S. Government Printing Office.

NRC (National Research Council). 1980. Committee on the Biological Effects of Ionizing Radiations (BEIR III). The Effects on Populations of Exposure to Low Levels of Ionizing Radiation; Washington, D.C. National Academy Press.

NRC (National Research Council). 1988. Committee on the Biological Effects of Ionizing Radiations. Health risks of radon and other internally deposited alpha-emitters: BEIR IV; Washington, D.C. National Academy Press.

NRC (National Research Council). 1990. Committee on the Biological Effects of Ionozing Radiations. Health Effects of Exposure to Low Levels of Ionizing Radiation: BEIR V; Washington, D.C. National Academy Press.

NRC (National Research Council). 1991. Panel on Dosimetric Assumptions Affecting the Application of Radon Risk Estimates. Comparative Dosimetry of Radon in Mines and Homes. Washington, D.C. National Academy Press.

NRC (National Research Council). 1994a. Committee on Health Effects of Exposure to Radon (BEIR VI), and Commission on Life Sciences. Health Effects of Exposure to Radon: Time for Reassessment? Washington, D.C. National Academy Press.

NRC (National Research Council). 1994b. Committee on Risk Assessment of Hazardous Air Pollutants. Science and Judgment in Risk Assessment. National Academy Press, Washington DC.

Nauss KM, The HEI Diesel Working Group. 1995. Critical issues in assessing the carcinogenicity of diesel exhaust: A synthesis of current knowledge. Pp. 11-61 in Diesel Exhaust: A Critical Analysis of Emissions, Exposure, and Health Effects. Cambridge: Health Effects Institute.

Nelson JM, Brooks AL, Metting NF, Khan MA, Buschbom RL, Duncan A, Miick R, Braby LA. 1996. Clastogenic effects of defined numbers of 3.2 MeV alpha particles on individual CHO-K1 cells. Radiat Res 145:568-574.

Nero AV, Schwehr MB, Nazaroff WW, Revzan KL. 1986. Distribution of airborne radon-222 concentrations in U.S. homes. Science 234:992-997.

Nero AV, Nazaroff WW, Nero Jr AV, eds. 1988. Radon and its decay products. Pp. 1-53 in Indoor Air: An Overview. New York: John Wiley & Sons, Inc.

Nero AV, Gadgil AJ, Nazaroff WW, Revzan KL. 1990. Indoor radon and decay products: concentrations, causes, and control strategies. Report prepared for the U.S. Department of Energy, Office of Health and Environmental Research, Washington, D.C., November 1990. DOE/ER-0480P, NTIS, Springfield, VA.

Neuberger JS, Lynch CF, Kross BC, Field RW, Woolson RF. 1994. Residential radon exposure and lung cancer: Evidence of an urban factor in Iowa. Health Phys 66:263-269.

Neuberger O. 1947. Arsenical cancer: A review. Br J Cancer 1:192-251.

Neugat AI and Murray T. 1994. Increased risk of lung cancer after breast cancer radiation therapy in cigarette smokers. Cancer 73:1615-20.

Newcomb EW, Steinberg JJ, Pellicer A. 1988. Ras oncogenes and phenotypic staging in N-methyl nitrosourea and g-irradiation-induced mymic lymphomas in C57BL/6J mice. Cancer Res 48:5514-5521.

Newman B, Austin MA, Lee M, King MC. 1988. Inheritance of human breast cancer: evidence for autosomal dominant transmission in high-risk families. Proc Natl Acad Sci USA 85(9):3044-3048.

Newman JA, Archer VE, Saccomanno G, Kuschner M, Auerbach O, Grondahl D, Wilson JC. 1976. Histologic types of bronchogenic carcinoma among members of copper mining and smelting communities. Ann NY Acad Sci 271:260-268.

Ng TP, Chan SL, Lee J. 1990. Mortality of a cohort of men in a silicosis register: Further evidence of an association with lung cancer. Am J Ind Med 17:163-171.

Nikula KJ, Snipes MB, Barr EB, Griffith WC, Henderson RF, Mauderly JL. 1995. Comparative pulmonary toxicities and carcino-genicities of chronically inhaled exhaust and carbon black in F344 rats. Fundam Appl Toxicol 25:80-94.

Oakes M. 1990. On meta-analysis. Pp. 157-163 in Statistical Inference. Chestnut Hill, MA: Epidemiology Resources Inc.

Oberdörster G. 1996. Significance of particle parameters in the evaluation of exposure-dose-response relationships of inhaled particles. Inhal Toxicol 8(Suppl):73-89.

Ott MG, Holder BB, Gordon HL. 1974. Lung cancer among pesticide workers exposed to inorganic arsenicals. Arch Environ Health 29:250-255.

Page S. 1993. EPA's strategy to reduce risk of radon. J Environ Health 56:27-36.

Palmer RF, Stuart BO, Filipy RE. 1973. Biological effects of daily inhalation of radon and its short-lived daughters in experimental animals. In: Nobel gases, Edt. by R.E. Stanley, A.A. Moghissi, CONF-730915. U.S. Energy Development and Research Agency, National Environmental Research Center, Washington D.C. Pp. 507-519.

Park MS, Hanks T, Jaberaboansari A, Chen DJ. 1995. Molecular analysis of gamma-ray-induced mutations at the hprt locus in primary human skin fibroblasts by multiplex polymerase chain reaction. Radiat Res 141(1):11-18.

Pathak S. 1990. Cytogenetic abnormalities in cancer: With special emphasis on tumor heterogeneity. Cancer and Metastases Reviews 8:299-318.

Pavia D. 1984. Lung mucociliary clearance. In Aerosols and the Lung: Clinical and Experimental Aspects, Clarke SW, and Pavia D, eds. London: Butterworths.

Peltomaki P, Aaltonen LA, Sistonen P, Pylkkanen L, Mecklin J-P, Jarvinen H, Green JS, Jass JR, Weber JL, Leach FS, Petersen GM, Hamilton SR, de la Chapelle A, Vogelstein B. 1993. Genetic mapping of a locus predisposing to human colorectal cancer. Science 260:810-812.

Percy C, Stanek E 3d, Gloeckler L. 1981. Accuracy of cancer death certificates and its effect on cancer mortality statistics. Am J Public Health 71(3):242-250.

Pershagen G, Damber L, Falk R. 1984. Exposure to radon in dwellings and lung cancer: A pilot study. Pp. 29-34 in Indoor Air. Radon, Passive Smoking, Particulates and Housing Epidemiology, vol 2, Berglund B, Lindvall T, and Sundell J, eds. Stockholm Council for Building Res.

Pershagen G, Bergman F, Klominek J, Damber L, Wall S. 1987. Histological types of lung cancer among smelter workers exposed to arsenic. Br J Ind Med 44:454-458.

Pershagen G, Liang Z-H, Hrubec Z, Svensson C, Boice JD Jr. 1992. Residential radon exposure and lung cancer in Swedish women. Health Phys 63:179-186.

Pershagen G, Åkerblom G, Axelson O, Clavensjö B, Damber L, Desai G, Enflo A, Lagarde F, Mellander H, Svartengren M, Swedjemark GA. 1994. Residential radon exposure and lung cancer in Sweden. N Engl J Med 330:159-164.

Petersen DD, Gonzalez FJ, Rapic V, Kozak CA, Lee JY, Jones JE, Nevert DW. 1989. Marked increases in hepatic NAD(P)H:oxidoreductase gene transcription and mRNA levels correlated with a mouse chromosome 7 deletion. Proc Natl Acad Sci USA 86(17):6699-6703.

Petersen GR, Gilbert ES, Buchanan JA, Stevens RG. 1990. A case-cohort study of lung cancer, ionizing radiation, and tobacco smoking among males at the Hanford Site. Health Phys 58:3-11.

Petitti DB. 1994. Of babies and bathwater. Am J Epidemiol 140:779-782.

Peto J. 1990. Radon and the risks of cancer. Nature 345:389-390.

Peto R, Lopez AD, Boreham J, Thun M, Heath C Jr. 1992. Mortality from tobacco in developed countries: Indirect estimation from national vital statistics. Lancet 339:1268-1278.

Pfeifer AMA, Jones RT, Bowden PE, Mann D, Spillare E, Klien-Szanto AJP, Trump BF, Harris CC. 1991. Human bronchial epithelial cells transformed by c-raf-1 and c-myc protooncogenes induce multidifferentiated carcinomas in nude mice: A model for lung carcinogenesis. Cancer Res 51:3793-3801.

Piantadosi S. 1994. Invited commentary: Ecologic biases. Am J Epidemiol 139:761-764.

Piantadosi S, Byar D, Green S. 1988. The ecological fallacy. Am J Epidemiol 127:893-904.

Piao CQ, Hei TK. 1993. The biological effectiveness of radon daughter alpha particles I. Radon, cigarette smoke and oncogenic transformation. Carcinogenesis 14:497-501.

Piechowski JW, LeGac J, Brenot J, Nenot JC, Zettwoog P. 1981. Exposure to short-lived radon daughters: Comparison of individual and ambient monitoring in a French uranium mine. Radiation Hazards in Mining: control, measurement, and medical aspects. Golden Conference, Oct 4-9. M. Gomez, Editor. American Institute of Mining, Metallurgical and Petroleum Engineering, Inc. New York City, NY.

Pierce DA, Stram DO, Vaeth M. 1990. Allowing for random errors in radiation dose estimates for the atomic bomb survivor data. Rad Res 123:275-284.

Pirchan A, Sikl H. 1932. Cancer of the lung in the miners of Jachymov. Am J Cancer 16:681-722.

Porstendörfer J, Röbig G, Ahmed A. 1979. Experimental determination of the attachment coefficients of atoms and ions on monodisperse particles. J Aerosol Sci 10:21-28.

Prentice AG, Copplestone, JA. 1990. Radon and leukaemia (Letter). Lancet 335:1337.

Prentice RL, Yoshimoto Y, Mason MW. 1983. Relationship of cigarette smoking and radiation exposure to cancer mortality in Hiroshima and Nagasaki. J Natl Cancer Inst 70:611-622.

Preston DL, Lubin JH, Pierce DA, McConney ME. 1991. EPICURE: User's Guide, HiroSoft International. Corporation. 1463 E. Republican Ave. Suite 103, Seattle, WA 98112, USA.

Prise K. 1994. Use of radiation quality as a probe for DNA lesion complexity. Int J Radiat Biol 65:43-48.

Prise KM, Davies S, Michael BD. 1987. The relationship between radiation-induced DNA double-strand breaks and cell kill in hamster V79 fibroblasts irradiated with 250 kVp X-rays, 2.3 MeV neutrons or $^{238}$Pu a-particles. Int J Radiat Biol 52:893-902.

Probart CK. 1989. Issues related to radon in schools. J Sch Health 59(10):441-443.

Proctor RN. 1995. Cancer Wars. How Politics Shapes What We Know and Don't Know About Cancer. New York: Basic Books.

Puskin JS. 1992. An analysis of the uncertainties in estimates of radon-induced lung cancer. Risk Anal 12:277-285.

Puskin JS, Nelson CB. 1989. EPA's perspective on risks from residential radon exposure. JAPCA 39:915-920.

Qiao YL, Taylor RP, Yao SX, Schatzkin A, Mao BL, Lubin JH, Rao JY, Li JY. 1989. The relation of radon exposure and tobacco use to lung cancer among tin miners in Yunnan Province, China. Am J Ind Med 16:511-521.

Raabe OG. 1982. Deposition and clearance of inhaled aerosols. In Mechanisms in Respiratory Toxicology, Vol. I, Witschi H and Nettesheim P, eds. Boca Raton: CRC Press.

Radford EP and St. Clair Renard KG. 1984. Lung cancer in Swedish iron miners exposed to low doses of radon daughters. N Engl J Med 310(23):1485-1494.

Rai SN and Krewski D. 1998. Uncertainty and variability analysis in multiplicatie risk models. Risk Anal 18:37-45.

Rai SN, Krewski D, Bartlett S. 1996. A general framework for the analysis of uncertainty and variability in risk assessment. Human and Ecological Risk Assessment 2(4):972-989.

Raju MR, Eisen Y, Carpenter S, Jarret K, Harvey WF. 1993. Radiobiology of alpha particles IV. Cell inactivation by alpha particles of energies 0.4-3.5 MeV. Radiat Res 133:289-96.

Ramamurthi M and Hopke PK. 1989. On improving the validity of wire screen unattached fraction daughter measurements. Health Phys 56:189-194.

Ramamurthi M and Hopke PK. 1991. An automated, semi-continuous system for measuring indoor radon progeny activity-weighted size distributions, dp: 0.5-500 nm. Aerosol Sci Technol 14:82-92.

Ramamurthi M., Strydom R, Hopke PK. 1990. Assessment of wire and tube penetration theories using a $218P_oO_x$ cluster aerosol. J Aerosol Sci 21:203-211.

Rannou A. 1987. Contribution a l'étude du risque lie a la presence du radon 220 et du radon 222 dans l'atmosphere das habitations. Rapport CEA-R-5378. Commissariat a l'Energie Atomic. Sarclay, France.

Rannou A, Mouden A, Renouard H, Kerlau G, Tymen G. 1988. An assessment of natural radiation exposure in granitic areas in the west of France. Radiat Prot Dosimet 24(1/4):327-331.

Raunio H, Husgafvel-Pursiainen K, Anttila S, Hietanen E, Hirvonen A, Pelkonen O. 1995. Diagnosis of polymorphisms in carcinogen-activating and inactivating enzymes and cancer susceptibility—a review. Gene 159:113-121.

Redpath JL and Sun C. 1990. Sensitivity of a human hybrid cell line (HeLa x skin fibroblast), to radiation-induced neoplastic transformation in G2, M, and mid-G1 phases of the cell cycle. Radiat Res 121:206-211.

Reineking A and Porstendörfer J. 1986. High-volume screen diffusion batteries and a-spectroscopy for measurement of the radon daughter activity size distributions in the environment. J Aerosol Sci 17:873-879.

Reineking A and Porstendörfer J. 1990. "Unattached" fraction of short-lived Rn decay products in indoor and outdoor environments: An improved single-screen method and results. Health Phys 58(6):715-727.

Reineking A, Becker KH, Porstendörfer J. 1985. Measurements of the unattached fractions of radon daughters in houses. Sci Total Environ 45:261-270.

Reineking A, Becker KH, Porstendörfer J. 1988. Measurement of activity size distributions of the short-lived radon daughters in the indoor and outdoor environment. Radiat Prot Dosimet 24:245-250.

Reist PC. 1984. Introduction to Aerosol Science. New York: McMillan.

Report of the Royal Commission regarding radiation, compensation, and safety at the Fluorospar mines of St. Laurence, Newfoundland. 1969.

Rice AJ. 1956. Radiation concentrations in uranium mines and methods of control on the Colorado Plateau. Address to the Colorado Plateau Section, AIME, Grand Junction, Colorado, March 3, 1956.

Richardson RB, Eatough JP, Henshaw DL. 1991. Dose to red bone marrow from natural radon and thoron exposure. Br J Radiol 64:608-824.

Risch HA, Howe GR, Jain M, Burch JD, Holowaty EJ, Miller AB. 1993. Are female smokers at higher risk for lung cancer than male smokers? A case control analysis by histologic type [see comments]. Am J Epidemiol 93:281-293.

Ritter MA, Cleaver JE, Tobias CA. 1977. High-LET radiations induce a large proportion of non-rejoining DNA strand breaks. Nature 266:653-655.

Robbins ES and Meyers OA. 1995. Cycling cells of human and dog tracheobronchial mucosa: Normal and repairing epithelia. Technology: Journal of the Franklin Institute 332A:35-42.

Roberts CJ and Goodhead DT. 1987. The effect of 238Pu a-particles on the mouse fibroblast cell line C3H 10T$^{1}$/$_{2}$: Characterization of source and RBE for cell survival. Int J Radiat Biol 52:871-882.

Rogers VC and Nielson KK. 1991. Correlations for predicting air permeabilities and $^{222}$Rn diffusion coefficients in soils. Health Phys 61(2):225-230.

Rogot E and Murray JL. 1980. Smoking and causes of death among U.S. veterans: 16 years of observation. Public Health Rep 95:213-222.

Rossi HH. 1991. Point mutations and radiation carcinogenesis. Radiat Res 128(1):115.

Rossi HH and Kellerer AM. 1986. The dose rate dependence of oncogenic transformation by neutrons may be due to variation of response during the cell cycle. Int J Radiat Biol 50:353-361.

Roth R. 1957. The sequelae of chronic arsenic poisoning in Moselle vinters. Germ Med Month 2:172-176.

Rothman KJ. 1986. Modern Epidemiology. Boston: Little, Brown and Company.

Rothman KJ, Greenland S, Walker A. 1980. Concepts of interaction. Am J Epidemiol 112:467-470.

Rowley R. 1998. Mammalian cell cycle responses to DNA damaging agents. In: DNA Damage and Repair: Vol. 2. DNA Repair in Higher Eucaryotes, Nickoloff, JA, Hoekstra, MF, eds. Totowa, NJ: Humana Press.

Ruffle B, Burmaster DE, Anderson PD, Gordon HD. 1994. Lognormal distributions for fish consumption by the general U.S. population. Risk Anal 14:395-404.

Rundo J, Markun F, Plondke NJ. 1979. Observation of high concentrations of radon in certain houses. Health Phys 36(6):729-730.

Ruosteenoja E. 1991. Indoor radon and risk of lung cancer: An epidemiologic study in Finland. Doctoral Disseration, Department of Public Health, University of Tampere, Finnish Government Printing Centre, Helsinki.

Ruosteenoja E, Makelainen I, Rytomaa T, Hakulinen T, Hakama M. 1996. Radon and lung cancer in Finland. Health Phys 71(2):185-189.

Rutter CM and Elashoff RM. 1994. Analysis of longitudinal data: Random coefficient regression modeling. Stat Med 13:1211-1231.

Ruzer LS, Nero AV, Harley NH. 1995. Assessment of lung deposition and breathing rate of underground miners in Tadjikistan. Radiat Prot Dosimet 58:261-268.

Rydberg B. 1966. Clusters of DNA damage induced by ionizing radiation: Formation of short DNA fragments. II. Experimental detection. Radiat Res 145:200-209.

Sabitier L, Dutrillaux B, Martins MB. 1992. Chromosomal instability. Nature 357:548.

Sabatier L, Lebeau J, Dutrillaux B. 1994. Chromosomal instability and alterations of telomeric repeats in irradiated human fibroblasts. Int J Radiat Biol 66(5):611-613.

Saccomanno GS, Yale C, Dixon W, Auerback O, Huth GC. 1986. An epidemiological analysis of the relationship between exposure to Rn progeny, smoking and bronchogenic carcinoma in the U-mining population of the Colorado Plateau 1960-1980. Health Phys 50(5):605-618.

Saccomanno GS, Huth GC, Auerbach O, Kuschner M. 1988. Relationship of radioactive radon daughters and cigarette smoking in the genesis of lung cancer in uranium miners. Cancer 62:1402-1408.

Saccomanno G, Auerbach O, Kuschner M, Harley NH, Michels RY, Anderson MW, Bechtel JJ. 1996. A comparison between the localization of lung tumors in uranium miners and in nonminers from 1947 to 1991. Cancer 77:1278-1283.

Sachs RK, Awa A, Kodama Y, Nakano M, Ohtaki K, Lucas JN. 1993. Ratios of radiation-produced chromosome aberrations as indicators of large-scale DNA geometry during interphase. Radiat Res 133:345-350.

Samet JM. 1988. Involuntary exposure to tobacco smoke. Ann Sports Med 4:1-15.

Samet JM. 1989. Radon and lung cancer. J Natl Cancer Inst 81:745-757.

Samet JM. 1992. Diseases of uranium miners and other underground miners exposed to radon. In Environmental and Occupational Medicine, 2nd ed, Rom WN, ed. Boston: Little, Brown and Company.

Samet JM. 1994. Indoor radon and lung cancer: Risky or not? (Editorial). J Natl Cancer Inst 86(24):1813-1814.

Samet JM. 1995. Lung cancer. In: Cancer Prevention and Control, Greenwald P, Kramer BS, Weed DL, eds. New York: Marcel Dekker, Inc.

Samet JM, ed. (In Press). Changes in Cigarette Related Disease Risks and Their Implication for Prevention and Control. Bethesda, Maryland: U.S. Government Printing Office.

Samet JM and Spengler JD. 1991. Introduction. In: Indoor Air Pollution. A Health Perspective, Samet JM and Spengler JD, eds. Baltimore: Johns Hopkins University Press.

Samet JM, Humble CG, Pathak DR. 1986a. Personal and family history of respiratory disease and lung cancer risk. Am Rev Respir Dis 134:466-470.

Samet JM, Morgan MV, Key MV, Pathak DR, Valdivia AA. 1986b. Studies of uranium miners in New Mexico. Proceedings of the International Conference in Health of Miners, Cincinnati, American Conference of Governmental Hygienists, RW Wheeler, ed. Pp. 351-355.

Samet JM, Wiggins CL, Humble CG, Pathak DR. 1988. Cigarette smoking and lung cancer in New Mexico. Am Rev Resp Dis 88(5):1110-1113.

Samet JM, Pathak DR, Morgan MV, Marbury MC, Key CR, Valdivia AA. 1989. Radon progeny exposure and lung cancer risk in New Mexico uranium miners: A case-control study. Health Phys 56:415-421.

Samet JM, Pathak DR, Morgan MV, Key CR, Valdivia AA, Lubin. JJ. 1991. Lung cancer mortality and exposure to radon progeny in a cohort of New Mexico uranium miners. Health Phys 61:745-752.

Samet JM, Pathak DR, Morgan MV, Coultas DB, James DS, Hunt WC. 1994. Silicosis and lung cancer risk in underground uranium miners. Health Phys 66:450-453.

(SAMMEC) Smoking-Attributable Mortality, Morbidity, and Economic Cost [computer program]. 1992. Shultz JM, Novotny TE, Rice DP. Vers. 2.1. Atlanta, GA: Centers for Disease Control and Prevention.

Sanders CL. and Lundgren DL. 1995. Pulmonary carcinogenesis in F 344 and Wistar rat after inhalation of plutonium dioxide. Radiat Res 144:206-214.

Sanders CL, Lauhala KE, McDonald KE. 1989. Tritiated thymidine-labeled bronchioloalveolar cells and radiation dose following inhalation of plutonium in rats. Exp Lung Res 15:755-769.

Sanford KK, Parshad R, Gantt R, Tarone RE, Jones GM, Price FM. 1989. Factors affecting and significance of G chromatid radiosensitivity in predisposition to cancer. Int J Radiat Biol 55:963-981.

Sankaranarayanan K and Chakraborty R. 1995. Cancer predisposition, radiosensitivity and the risk of radiation-induced cancers. I. Background. Radiat Res 143:121-143.

Savage JR. 1996. Insight into sites. Mutat Res 366(2):81-95.

Saxon PJ, Srivatsan ES, Stanbridge EJ. 1986. Introduction of normal human chromosome 11 via microcell transfer controls tumorigenic expression of HeLa cells. EMBO J:5:3461-3466.

Scheibel HG and Porstendörfer J. 1984. Penetration measurements in the ultrafine particle size range. J Aerosol Sci 15:549-556.

Schenker MB, Smith T, Muñoz A, Woskie S, Speizer FE. 1984. Diesel exposure and mortality among railway workers: result of a pilot study. Br J Ind Med 41:320-327.

Schery SD. 1990. Thoron in the environment. J AWWA 40:493-497.

Schery SD. 1992. Thoron and its progeny in the atmospheric environment. Chapter 10 of Gaseous Pollutants: Characterization and Cycling. J Wiley & Sons.

Scheuch G. 1991. Die Dispersion, Deposition, und Clearance von Aerosolpartikeln in den menschlkichen Atemwegen (Ph.D. thesis), J.W. Goethe-Universität. Frankfurt am Main, Germany.

Scheuch G, Kreyling W, Haas F, Stahlhofen W. 1993. Effect of settling velocity on particle recovery from human conducting airways after breath holding. J Aerosol Med 6(Suppl.) 47.

Scheutzle D and Jensen TE. 1985. Analysis of nitrated polycyclic aromatic hydrocarbons (nitro-PAH) by mass spectrometry. Pp. 121-167 in Nitrated Polycyclic Aromatic Hydrocarbons, White C, ed. Heidelberg: Huthig Verlag.

Scheutzle D and Lewtas J. 1986. Bioassay-directed chemical analysis in environmental research. Anal Chem 58:1060A-1075A.

Schlesinger RB. 1985. Clearance from the respiratory tract. Fund Appl Toxicol 5:435.

Schoenberg JB, Wilcox HB, Mason TJ, Bill J, Stemhagen A, Benhamou E, Benhamou S, Auquier A, Flamant R. 1989. Variation in smoking-related lung cancer risk among New Jersey women. Changes in patterns of cigarette smoking and lung cancer risk: Results of a case-control study. Am J Epidemiol 89(4):601-604.

Schoenberg JB, Klotz JB, Wilcox GP, Gil-del-Real MT, Stemhagen A, Mason TJ. 1990. Case-control study of residential radon and lung cancer among New Jersey women. Cancer Res 50:6520-6524.

Schwab M, Alitalo K, Klempnauer KH, Varmus HE, Bishop JM, Gilbert F, Brodeur G, Goldstein M, Trent J. 1983. Amplified DNA with limited homology to myc cellular oncogene is shared by human neuroblastoma cell-lines and a neuroblastoma tumor. Nature 305:245-248.

Schwartz JL, Rotmensch J, Atcher RW, Jostes RF, Cross FT, Hui TE, Chen D, Carpenter S, Evans HH, Mencl J, Bakale G, Roe PS. 1992. Interlaboratory comparison of different alpha-particle and radon sources: cell survival and relative biological effectiveness. Health Phys 62:458-461.

Schwartz JL, Rotmensch J, Sun J, An J, Xu Z, Yu Y, Hsie A. 1994. Multiplex PCR reaction based deletion analysis of spontaneous, gamma-rays and alpha induced mutants of CHO-K1 cells. Mutagenesis 9:537-540.

Scott AG. 1992. Site characterization for radon supply potential: A progress review. Health Phys 62:422-428.

Scott AG. 1993. Comparison of criteria to define radon-prone areas (letter). Health Phys 64(4):435-436.

Scott D, Spreadborough AR, Jones LA, Roberts SA, Moore CJ. 1996. Chromosomal radiosensitivity in G2-phase lymphocytes as an indicator of cancer predisposition. Radiat Res 145:3-16.

Searle AG, Beechey CV, Green D, Humphreys ER. 1976. Cytogenetic effects of protracted exposures to alpha-particles from plutonium-239 and to gamma-rays from cobalt-60 compared in male mice. Mutat Res 41:297-310.

Selvanayagam CS, Davis CM, Cornforth MN, Ullrich RL. 1995. Latent expression of *p53* mutations and radiation-induced mammary cancer. Cancer Res 55:3310-3317.

SENES Consultants, Limited 1989. Uncertainty in exposure of underground miners to radon daughters and the effect of uncertainty on risk estimates. Report to the Atomic Energy Control Board. Ottawa.

SENES Consultants, Limited 1991. Detailed reconstruction of radon daughter exposures of Eldorado Beaverlodge uranium mine employees. Report to the Atomic Energy Control Board. Ottawa.

SENES Consultants Limited. 1995. Preliminary feasibility study into the re-evaluation of exposure data for the Colorado Plateau uranium miner cohort study. A report prepared for the National Mining Association, Washington, D.C. Ontario Canada, November.

SENES Consultants Limited. 1996a. An algorithm for estimating radon decay product exposures from underground employment at the Eldorado Beaverlodge mine. A report prepared for Atomic Energy Control Board, Ottawa , Canada. Ontario, Canada, March.

SENES Consultants Limited. 1996b. A re-evaluation of radon decay product exposures to underground workers at the Port Radium mine. Report to Atomic Energy Control Board of Canada. Ontario, Canada.

Ševc JE and Placek V. 1976. Lung cancer in uranium miners and long-term exposure to radon daughter products. Health Phys 30:433-437.

Ševc J, Kunz E, Tomášek L, Placek V, Horacek J. 1988. Cancer in man after exposure to Rn daughters. Health Phys 54:27-46.

Ševc J, Tomášek L, Kunz E, Placek V, Chemelevsky D, Barclay D, Kellerer AM. 1993. A survey of the Czechoslovak follow-up of lung cancer mortality in uranium mines. Health Phys 64:355-369.

Seymour CB, Mothersill C, Alper T. 1986. High yields of lethal mutations in somatic mammalian cells that survive ionizing radiation. Int J Radiat Biol 50:167-179.

Shapiro S. 1994. Meta-analysis/shmeta-analysis. Am J Epidemiol 140:771-778.

Shiloh Y. 1995. Ataxia-telangiectasia: closer to unraveling the mystery. Eur J Human Genet 3:116-138.

Shimizu H, Morishita M, Mizuno K, Masuda T, Ogura Y, Santo M, Nishimura M, Kunishima K, Karasawa K, Nishiwaki K, Yamamoto M, Hisamichi S, Tominaga S. 1988. A case-control study of lung cancer in nonsmoking women. Tohoku J Exp Med 154:389-397.

Short SR and Ptesonk EL. 1993. Respiratory health risks among nonmetal miners. Occup Med 8:57-70.

Shuin J, Billings PC, Lillehaug JR, Patierno SR, Roy-Burman P, Landolph. JR. 1986. Enhanced expression of c-myc and decreased expression of c-fos protooncogenes in chemically and radiation transformed C3H/10T1/2 mouse embryo cell line. Cancer Res 46:5302-5311.

Siemiatycki J, ed. 1991. Risk Factors for Cancer in the Workplace. Boca Raton: CRC Press.

Simmonds JR, Robinson CA, Phipps AW, Muirhead CR, Fry FA. 1995. Risks of leukaemia and other cancers in Seascale from all sources of ionizing radiation exposure. Report NRPB-R276 HMSO, London.

Simmons JA, Cohn P, Min T. 1996. Survival and yields of chromosome aberrations in hamster and human lung cells irradiated by alpha particles. Radiat Res 145:174-180.

Simonato L, Moulin JJ, Javelaud B, Ferro G, Wild P, Winkelmann R, Saracci R. 1994. A retrospective mortality study of workers exposed to arsenic in a gold mine and refinery in France. Am J Ind Med 25:625-633

Simpson SD, Crosby EH, Yourt GR. 1954. A survey of radioactivity, dust, and ventilation at Eldorado Beaverlodge, Unpublished report, Eldorado Mining & Refining, Ltd., Ottawa, Ontario.

Simpson SG and Comstock GW. 1983. Lung cancer and housing characteristics. Arch Environ Health 38:248-251.

Sinclair D, George AC, Knutson EO. 1977. Application of diffusion batteries to measurement of submicron radioactive aerosols. In: Airborne Radioactivity. La Grange Park, IL: American Nuclear Society. Pp. 103-114.

Skillrud DM, Offord KP, Miller RD. 1986. Higher risk of lung cancer in chronic obstructive pulmonary disease. Ann Intern Med 105:503-507.

Small MJ. 1994. Invariably Uncertain about Variability? Try the Normal-Gamma Conjugate!, Presented at the 87th Annual Meeting & Exhibition of the Air & Waste Management Association, Cincinnati, Ohio, June 19-24.

Smith ML, Zhan Q, Bae I, Fornace AJ, Jr. 1994. Role of retinoblastoma gene product in p53-mediated DNA damage response. Exp Cell Res 215:386-389.

Solli HM, Andersen A, Stranden E, Langard S. 1985. Cancer incidence among workers exposed to radon and thoron daughters at a niobium mine. Scand J Work Env Health 11(1):7-13.

Solomon S. 1989. Personal communication.

Sram RJ, Binkova B, Dobias L, Rossner P, Topinka J, Vesela D, Vesely D, Stejskalova J, Bavorova H, Rericha V. 1993. Monitoring genotoxic exposure in uranium miners. Environ Health Perspect 99:303-305.

Stahlhofen W. 1989. Human lung clearance following bolus inhalation of radioaerosols. Pp. 153-166 in Extrapolation of Dosimetric Relationships for Inhaled Particles and Gases. Washington: Academic Press.

Stahlhofen W, Gebhart J, Heyder J. 1980. Experimental determination of the regional deposition of aerosol particles in the human respiratory tract. Am Ind Hyg Assoc J 41:385.

Stahlhofen W, Gebhart J, Rudolf G, Scheuch G. 1986a. Measurement of lung clearance with pulses of radioactively-labelled aerosols. J Aerosol Sci 17:333-336.

Stahlhofen W, Gebhart J, Rudolf G, Scheuch G, Philipson K. 1986b. Clearance from the human airways of particles of different sizes deposited from inhaled aerosol boli. In: Aerosols: Formation and Reactivity, Second International Aerosol Conference, West Berlin, Germany, September 22-26, 1986, pp. 192-196, Pergamon Press, Oxford, U.K.

Stahlhofen W, Gebhart J, Rudolf G, Scheuch G. 1987a. Retention of radiolabelled $Fe_2O_3$-particles in human lungs. In: Deposition and Clearance of Aerosols in the Human Respiratory Tract, Second International Symposium, Salzburg, Austria, September 18-20, 1986, pp 123-128, (ed. Hofmann, W.) Facultas Universitätsverlag Ges.m.b.H. Vienna Austria.

Stahlhofen W, Gebhart J, Rudolf G, Scheuch G, Bailey MR. 1987b. Human lung clearance of inhaled radioactively labelled particles in horizontal and vertical position of the inhaling person. J Aerosol Sci 18:741-744.

Stahlhofen W, Koebrich R, Rudolf G, Scheuch G. 1990. Short-term and long-term clearance of particles from the upper human respiratory tract as function of particle size. J. Aerosol Sci. 21(Suppl. 1):S407-S410.

Stahlhofen W, Scheuch G Bailey MR. 1994. Measurement of the tracheobronchial clearance of particles after aerosol bolus inhalation. In: Inhaled Particles VII, Proceedings of an International Symposium on Inhaled Particles Organized by the British Occupational Hygiene Society, 16-22 September 1991. Dodgson J, McCallum RI, eds. Ann Occup Hyg 189.

Stanbridge EJ. 1976. Suppression of malignancy in human cells. Nature 260:17-20.

Stather JW, Dionian J, Brown J, Fell TP, Muirhead CR. 1986. The risks of leukaemia and other cancers in Seascale from radiation exposure. Addendum to Report R171. Report NRPB-R171. Addendum HMSO, London.

Stayner LT and Wegman DH. 1983. Smoking, occupation, and histopathology of lung cancer: A case-control study with the use of the Third National Cancer Survey. J Natl Cancer Inst 70:421-426.

Steenland K and Goldsmith DF. 1995. Silica exposure and autoimmune disease. Am J Ind Med 28:603-608.

Steenland K, Silverman D, Zaebst D. 1992. Exposures to diesel exhaust in the trucking industry and possible relationships with lung cancer. Am J Ind Med 21:887-890.

Steinhäusler F. 1996. Environmental 220Rn: A review. Environ Int 22(Suppl 1):S1111-1123.

Stenerlow B, Blomquist E, Grusell E, Hartman T, Carlsson J. 1996. Rejoining of DNA double-strand breaks induced by accelerated nitrogen ions. Int J Radiat Biol 70:413-420.

Stewart BW. 1994. Mechanisms of apoptosis: integration of genetic, biochemical, and cellular indicators. J Natl Cancer Inst 86(17):1286-1296.

Stidley CA and Samet JM. 1993. A review of ecological studies of lung cancer and indoor radon. Health Phys 65:234-251.

Stidley CA and Samet JM. 1994. Assessment of ecologic regression in the study of lung cancer and indoor radon. Am J Epidemiol 139:312-322.

Stranden E. 1987. Radon-222 in Norwegian dwellings. In Radon and Its Decay Products: Occurrence, Properties, and Health Effects. Washington: American Chemical Society.

Strong JC. 1988. The size of attached and unattached radon daughters in room air. J Aerosol Sci 19:1327-1330.

Strong JC. 1989. Design of the NRPB activity size measurement system and results, presented at the Workshop on "Unattached" Fraction Measurements, University of Illinois, Urbana, IL, April 1989.

Stuart BO. 1984. Deposition and clearance of inhaled particles. Environ Health Perspect 55:369.

Stuart BO, Palmer RF, Fillip RE, Gaven J. 1978. Inhaled radon daughters and uranium ore dust in rodents, pp. 3.70-3.72. In: Pacific Northwest Laboratory Annual Report for 1977 to the DOE assistant secretary for environment, PNL-2500, Pt. 1, JNTIS, Springfield, VA.

Svensson C, Eklund G, Pershagen G. 1987. Indoor exposure to radon from the ground and bronchial cancer in women. Int Arch Occup Environ Health 59:123-131.

Svensson C, Pershagen G, Hrubec Z. 1988. A comparative study on different methods of measuring Rn concentrations in homes. Health Phys 55(6):895-902.

Svensson C, Pershagen G, Klominek J. 1989. Lung cancer in women and type of dwelling in relation to radon exposure. Cancer Res 49:1861-1865.

Swartout HO and Webster RG. 1940. To what degree are mortality statistics dependable? Am J Public Health 30:811-815.

Swedjemark GA. 1985. Radon and its decay products in housing. Estimation of the radon daughter exposure to the Swedish population and methods for evaluation of the uncertainties in annual averages. Doctoral Dissertation, Department of Radiation Physics, University of Stockholm.

Swift M, Morrell D, Cromartie E, Chamberlin AR, Skolnick MH, Bishop, DT. 1986. The incidence and gene frequency of ataxia-telangiectasia in the United States. Am J Hum Genet 39:573-583.

Swift M, Morrell D, Massey RB, Chase CL. 1991. Incidence of cancer in 161 families affected by ataxia-telangiectasia. N Engl J Med 325:1831-1836.

Swift DL, Montassier N, Hopke PK, Karpen-Hayes K, Cheng YS, Su YF, Yeh HC, Strong JC. 1992. Inspiratory deposition of ultrafine particles in human nasal replicate casts. J Aerosol Sci 23:65-72.

Takeshima Y, Seyama T, Bennett WP, Akiyama M, Tokuoka S, Inai K, Mabuchi K, Land CE, Harris CC. 1993. p53 mutations in lung cancers from non-smoking atomic-bomb survivors. Lancet 342:1520-1521.

Taya A., Morgan A, Baker ST, Humphreys JA, Bisson M, Collier CG. 1994. Changes in the rat lung after exposure to radon and its progeny: effects on incorporation of bromodeoxyuridine in epithelial cells and on the incidence of nuclear aberrations in alveolar macrophages. Radiat Res 139:170-177.

Taylor JA, Watson MA, Devereux TR, Michels RY, Saccomanno G, Anderson M. 1994. p53 muitation hotspot in radon-associated lung cancer. Lancet 343:86-87.

Taylor PR, Qiao YL, Schatzkin A, Yao SX, Lubin JH, Mao BL, Rao JY, McAdams M, Xuan XZ, Li JY. 1989. The relation of arsenic exposure to lung cancer among tin miners in Yunnan Province, China. Br J Ind Med 46:881-886.

Thacker J. 1986. The nature of mutants induced by ionising radiation in cultured hamster cells. III. Molecular characterization of HPRT-deficient mutants induced by gamma-rays or alpha-particles showing that the majority have deletions of all or part of the hprt gene. Mutat Res 160:267-275.

Thacker J. 1988. Meta-analysis. A quantitative approach to research integration. JAMA 88(11):1685-1689.

Thacker J. 1994. The study of responses the model DNA breaks induced by restriction endonucleases in cells and cell-free systems: Achievements and difficulties. Int J Radiat Biol 66:591-596.

Thacker J. 1995. Molecular mechanisms of radiation mutagenesis in normal and radiosensitive human cells. In: Proceedings of the 10th International Congress on Radiation Research, Würzburg, 27 August - 1 September 1995.

Thacker J, Stretch A, Stephens MA. 1979. Mutation and inactivation of cultured mammalian cells exposed to beams of accelerated heavy ions. II. Chinese hamster cells. Int J Radiat Biol 36:37-148.

Thacker J, Stretch A, Goodhead DT. 1982. The mutagenicity of α-particles from plutonium-238. Radiat Res 92:343-352.

Thacker SB. 1988. Meta-analysis: A quantitative approach to research integration. JAMA 259:1685-1689.

Thomas DC. 1981. General relative risk models for survival time and matched case-control studies. Biometrics 37:673-686.

Thomas DC. 1988. Models for exposure-time-response relationships with applications to cancer epidemiology. Annu Rev Public Health 9:451-482.

Thomas DC and McNeill KG. 1982. Risk Estimates for the Health Effects of Alpha Radiation. Info-0081. Ottawa: Atomic Energy Control Board.

Thomas DC, McNeill KG, Dougherty C. 1985. Estimates of lifetime lung cancer risks resulting from Rn progeny exposures. Health Phys 49:825-846.

Thomas D, Stram D, Dwyer J. 1993. Exposure measurement error: Influence on exposure-disease relationships and methods of correction. Ann Rev Public Health 14:69-93.

Thomas D, Pogoda J, Langholz B, Mack W. 1994. Temporal modifiers of the radon-smoking interaction. Health Phys 66:257-262.

Thomassen DG, Seiler FA, Shyr LJ, Griffith WC. 1990. Alpha-particles induce preneoplastic transformation of rat tracheal epithelial cells in culture. Int J Radiat Biol 57:395-405.

Thomassen DG, Newton GJ, Guilmette RA, Johnson, NF. 1992. A biodosimetric approach for estimating radiation dose to the respiratory epithelium from inhaled radon progeny. Radiat Prot Dosimet 38:65-71.

Thompson CB. 1995. Apoptosis in the pathogenesis and treatment of disease. Science 267(5203):1456-1462.

Thun MJ, Day-Lally C, Myers DG, Calle EE, Flanders WD, Zhu B, Namboodiri MM, Health CW. 1997. Trends in tobacco smoking and mortality from cigarette use in Cancer Prevention Studies I (1959 through 1965) and II (1982 through 1988). In Changes in Cigarette-Related Disease Risks and Their Implication for Prevention and Control, Shopland DR, Burns DM, Garfinkel L, Samet JM, eds. National Cancer Institute. Monograph 8: Smoking and Tobacco Control. NIH Publication no. 97-4213.

Tirmarche M, Brenot J, Piechowski J, Chameaud J, Paradel J. 1984. The Present State of an Epidemiological Study of Uranium Miners in France. Pp. 344-349 in Proceedings of the International Conference on Occupational Radiation Safety in Mining, Vol 1, Stocker H, ed. Toronto: Canadian Nuclear Association.

Tirmarche M, Raphalen A, Allin F, Chameaud J, Bredon P. 1993. Mortality of a cohort of French uranium miners exposure to relatively low radon concentrations. Br J Cancer 67:1090-1097.

Tockman MS and Samet JM, eds. 1994. Other host factors and lung cancer susceptibility. Pp. 397-412 in Epidemiology of Lung Cancer. New York: Marcel Dekker, Inc.

Tockman MS, Anthonisen NR, Wright EC, Donithan MG. 1987. Airway obstruction and risk for lung cancer. Ann Intern Med 106:512-518.

Tomášek L and Darby SC. 1995. Recent results from the study of West Bohemian uranium miners exposed to radon and its progeny. Environ Health Perspect 103(Suppl 2):55-57.

Tomášek L, Darby SC, Swerdlow AJ, Placek V, Kunz E. 1993. Radon exposure and cancers other than lung cancer among uranium miners in West Bohemia. Lancet 341:919-923.

Tomášek L, Darby SC, Fearn T, Swerdlow AJ, Placek V, Kunz E. 1994a. Patterns of lung cancer mortality among uranium miners in West Bohemia with varying rates of exposure to radon and its progeny. Radiat Res 137:251-261.

Tomášek L, Swerdlow AJ, Darby SC, Plcek V, Kunz E. 1994b. Mortality in uranium miners in West Bohemia: A long term cohort study. Occup Environ Med 41:308-315.

Trump BF, McDowell EM, Harris CC. 1984. Chemical carcinogenesis in the tracheobronchial epithelium. Environ Health Perspect 55:77-84.

Tu KW and Knutson EO. 1988a. Indoor radon progeny particle size distribution measurements made with two different methods. Radiat Prot Dosimet 24:251-255.

Tu KW and Knutson EO. 1988b. Indoor outdoor aerosol measurements for two residential buildings in New Jersey. Aerosol Sci Technol 9:71-82.

Twomey S. 1975. Comparison of constrained linear inversion and an iterative nonlinear algorithm applied to the indirect estimation of particle size distributions. J Comp Phys 18:188-200.

Ullrich RL. 1983. Tumor induction in BALB/c female mice after fission neutron or $\gamma$ irradiation. Radiat Res 93:506-515.

Ullrich RL, Jernigan MC, Cosgrove GE, Satterfield LC, Bowles ND, Storer JB. 1976. The influence of dose and dose rate on the incidence of neoplastic disease in RFM mice after neutron irradiation. Radiat Res 68:115-131.

UNSCEAR (United Nations Scientific Committee on the Effects of Atomic Radiation). 1988. Sources, Effects and Risks of Ionizing Radiation. Report of the United Nations Scientific Committee on the Effects of Atomic Radiation. New York: United Nations.

UNSCEAR (United Nations Scientific Committee on the Effects of Atomic Radiation). 1994. Sources and Effects of Ionizing Radiation. United Nations, New York.

U.S. Department of Health Education and Welfare. 1964. Smoking and Health. Report of the Advisory Committee to the Surgeon General. Washington, DC: U.S. Government Printing Office.

USDHHS (U.S.Department of Health and Human Services). 1985. A report of the Surgeon General: The health consequences of smoking — Cancer and chronic lung disease in the workplace. Washington, DC: U.S. Government Printing Office.

USDHHS (U.S.Department of Health and Human Services). 1988. A report of the Surgeon General: The health consequences of smoking: Nicotine addiction. Washington, DC: U.S. Government Printing Office.

USDHHS (U.S. Department of Health and Human Services). 1989. A report of the Surgeon General: Reducing the Health Consequences of Smoking. 25 Years of Progress. Washington, DC: U.S. Government Printing Office.

USDHHS (U.S. Department of Health and Human Services). 1990. A report of the Surgeon General: The health benefits of smoking cessation. Washington, DC: U.S. Government Printing Office.

USDHHS (U.S. Department of Health and Human Services). 1991. Strategies to control tobacco use in the United States: A blueprint for public health action in the 1990's. Washington, DC: U.S. Government Printing Office.

USDHHS (U.S. Department of Health and Human Services). 1995. SEER Cancer Statistics Review 1973-1990. National Institutes of Health 93-2789.

USDHHS (U.S. Department of Health and Human Services). 1996. Public Health Service, and National Cancer Institute (NCI). Burns, D.M., Garfinkel, L., and Samet, J.M. editors. Changes in Cigarette-Related Disease Risks and Their Implication for Prevention and Control. Bethesda, Maryland: U.S. Government Printing Office. 1996; 6 (In Press). Smoking and Tobacco Control Monograph.

U.S. Department of Labor, Bureau of Labor Statistics. 1972. Railroad Technology and Manpower in the 1970's. USGPO, Washington, D.C.

USEPA (U.S. Environmental Protection Agency). 1991. EPA's National Residential Radon Survey Preliminary Results. U.S. Government Printing Office, Washington, D.C.

USEPA (U.S. Environmental Protection Agency). 1992a. Respiratory health effects of passive smoking: Lung cancer and other disorders. EPA/600/006F. U.S. Government Printing Office, Washington, D.C.

USEPA (U.S. Environmental Protection Agency). 1992b. Technical Support Document for the 1992 Citizen's Guide to Radon. 1992b; EPA-400-R-92-011. U.S. Government Printing Office, Washington, D.C.

USEPA (U.S. Environmental Protection Agency). 1992c. A Citizen's Guide to Radon. The Guide to Protecting Yourself and Your Family from Radon. 2nd edition. U.S. Government Printing Office, Washington, DC.

USEPA (U.S. Environmental Protection Agency). 1993. National School Radon Survey: Report to Congress. Washington, DC: Environmental Protection Agency.

U.S. Surgeon General. 1989. Reducing the Heath Consequences of Smoking. 25 Years of Progress. U.S. Government Printing Office, Washington, D.C.

Vahakangas KH, Samet JM, Metcalf RA, Welsh JA, Bennett WP, Lane DP, Harris CC. 1992. Mutations of p53 and ras genes in radon-associated lung cancer from uranium miners. Lancet 339:576-580.

Van Leeuwen FE, Klokman WJ, Stovall M, Hagenbeek A, van-den-Belt-Dusebout AW, Noyon R, Boice JD Jr, Burgers JM, Somers R. 1995. Roles of radiotherapy and smoking in lung cancer following Hodgkin's disease. J Natl Cancer Inst 87:1530-1537.

Venitt S and Biggs PJ. 1994. Radon, mycotoxins, p53, and uranium mining. Lancet 343:795.

Vogel F. 1979. Genetics of retinoblastoma. Hum Genet 52:1-54.

Vogelstein B. 1990. A deadly inheritance. Nature 348:681.

Vonstille WT and Sacarello HLA. 1990. Radon and cancer: Florida study finds no evidence of increased risk. J Environ Health 53:25-28.

Vostal JJ. 1986. Factors limiting the evidence for chemical carcinogenicity of diesel emissions in long-term inhalation experiments. In Carcinogenic and Mutagenic Effects of Diesel Engine Exhaust, eds. N. Ishinishi, A. Koizumi, R. O. McClellan and W. Stöber, pp. 381-396, Proceedings of the International Satellite Symposium on Toxicological Effects of Emissions from Diesel Engines held in Tsukuba Science City, Japan, July 26-28, 1986. Amsterdam: Elsevier.

Wainscoat JS and Fey MF. 1990. Assessment of clonality in human tumors: A review. Cancer Res 50:1355-1360.

Wang Y, Krewski D, Lubin JH, and Zielinski JM. 1995. Meta-analysis of multiple cohorts of underground miners exposed to radon. Pp. 21-28 in Proceeding of Statistics Canada Symposium 95: From Data to Information—Methods and Systems. Ottawa: Statistics Canada.

Ward JF. 1985. Biochemistry of DNA lesions. Radiat Res 104:S103-S111.

Ward JF. 1988. DNA damage produced by ionizing radiation in mammalian cells: identities, mechanisms of formation and repairability. Progress in Nucleic Acids and Molecular Biology 35:95-125.

Ward JF. 1994. The complexity of DNA damage: relevance to biological consequences. Int J Radiat Biol 66:427-432.

Wasiolek PT, Hopke PK, James AC. 1992. Assessment of exposure to radon decay products in realistic living conditions. J Exposure Anal Environ Epidemiol 2:309-322.

Waxweiler RJ, Stringer W, Wagoner JK, Jones J, Falk H, Carter C. 1976. Neoplastic risk among workers exposed to vinyl chloride. Ann NY Acad Sci (5NM) 271:40-48.

Wei L, Sha Y, Tao Z, He W, Chen D, Yuan Y. 1990. Epidemiological investigation in high background radiation areas of Yangjiang, China. Sohrabi M, Ahmed U, Durrani SA, eds. Proc Int Conf On High Levels of Natural Radioactivity.

Weinberg CR, Moledor ES, Umbach DM, Sandler D: Submitted. Imputation for exposure histories with gaps.

Westin JB, Cramer Z, Richter ED, Shani J, Ne'eman E, Elyakim O, Tal Y. 1991. Radon in a self-selected sample of Isreali homes, schools, and workplaces. Public Health Rev 19(1-4):199-203.

White E. 1994. p53, guardian of Rb. Nature 371:21-22.

White SB, Bergston JW, Alexander BV, Rodman NF, Phillips JL. 1992. Indoor radon $^{222}$Rn concentrations in a probability sample of 43,000 houses across 30 states. Health Phys 62:41-50.

Whitehead A and Whitehead J. 1991. A general parametric approach to the meta-analysis of randomized clinical trials. Stat Med 10:1665-1677.

Whittemore AS and McMillan A. 1983. Lung cancer mortality among U.S. uranium miners: A reappraisal. J Natl Cancer Inst 71:489-499.

Wicks M, Archer V, Auerbach O, Kuschner M. 1981. Arsenic exposure in a copper smelter as related to histological type of lung cancer. Am J Ind Med 2:25-31.

Wiggins CL, Becker TM. 1993. Racial and ethnic patterns of mortality in New Mexico. Albuquerque: University of New Mexico Press.

Williams RR, Stegens NL, Goldsmith JR. 1977. Associations of cancer site and type with occupation and industry from the third national cancer survey interview. J Natl Cancer Inst 59:1147-1185.

Wolff SP. 1991. Leukaemia risks and radon. Nature 352:288.

Wolff S, Afzal V, Wiencke JK, Olivieri G, Michaeli A. 1988. Human lymphocytes exposed to low doses of ionizing radiations become refractory to high doses of radiation as well as to chemical mutagens that induce double-strand breaks in DNA. Int J Radiat Biol 53:39-48.

Wolff S, Jostes RF, Cross FT, Hui TE, Afzal V, Wiencke JK. 1991. Adaptive response of human lymphocytes for the repair of radon-induced chromosomal damage. Mutat Res 250:299-306.

Woodward AD, Roder AJ, McMichael P, Crouch, Mylvaganam A. 1991. Radon daughter exposures at the Radium Hill Uranium Mine and lung cancer rates among former workers, 1952-87. Cancer Causes and Control 2:213-220.

Wooster R, Heuhausen SL, Mangion J, Quirk Y, Ford D, Collins N, Nguyen K, Seal S, Tran T, Averill D, Fields P, Marshall G, Narod S, Lenoir GM, Lynch H, Feunteun J, Devilee P, Cornelisse CJ, Menko FH, Daly PA, Ormiston W, McManus R, Pye C, Lewis CM, Cannon-Albright LA, Peto J, Ponder BAJ, Skolnick MH, Easton DF, Goldgar DE, Stratton MR. 1994. Localization of a breast cancer susceptibility gene, BRCA2, to Chromosome 13q12-13. Science 265:2088-2090.

Wright HA, Magee JL, Hamm RN, Chatterjee A, Turner JE, Klots CE. 1985. Calculations of physical and chemical reactions produced in irradiated water containing DNA. Radiat Prot Dosimet 13:133-136.

Wu AH, Henderson BE, Pike MC, Yu MC. 1985. Smoking and other risk factors for lung cancer in women. J Natl Cancer Inst 85(4):747-751.

Wu-Williams AH and Samet JM. 1994. Lung cancer and cigarette smoking. In Epidemiology of Lung Cancer, Samet JM, ed. New York: Marcel Dekker, Inc.

Wynder E and Graham EA. 1950. Tobacco smoking as a possible etiologic factor in bronchiogenic carcinoma. A study of six hundred and eighty-four proved cases. JAMA 143:329-346.

Wynder EL, Bross IJ, Cornfield J. 1956. Lung cancer in women - a study of environmental factors. N Engl J Med 255:1111-1121.

Wynder EL, Mabuchi K, Beattie EJ. 1970. The epidemiology of lung cancer: Recent trends. J Am Med Assoc 213:2221-2228.

Xiang-Zhen X, Lubin JH, Jun-Yao L, Li-Fen Y, Qing-Sheng L, Lan Y, Jian-Zhang W, Blot WJ. 1993. A cohort study in southern China of tin miners exposed to Radon and radon decay products. Health Phys 64:120-131.

Xu Z-Y, Blot WJ, Xiao H-P, Wu A, Feng Y-P, Stone BJ, Sun J, Ershow AG, Henderson BE, Fraumeni JF Jr. 1989. Smoking, air pollution and the high rates of lung cancer in Shenyang, China. J Natl Cancer Inst 81:1800-1806.

Xuan XZ, Lubin JH, Li JY, Blot WJ. 1993. A cohort study in southern China of workers exposed to radon and radon decay products. Health Phys 64:120-131.

Yalow RS. 1995. Radiation and Public Perception. In Radiation and Public Perception. Benefits and Risks. Washington: American Chemical Society.

Yao SX, Lubin JH, Qiao YL, Boice JD, Jr., Li JY, Cai SK, Zhang FM, Blot WJ. 1994. Exposure to radon progeny, tobacco use and lung cancer in a case-control study in southern China. Radiat Res 138:326-336.

Yeager WM, Lucas RM, Daum KA, Sensintaffar E, Poppell S, Feldt L, Clarkin M. 1991. A performance evaluation study of three types of $\alpha$-track detector radon monitors. Health Phys 60:507-515.

Yeh HC, Cheng YS, Orman MM. 1982. Evaluation of various types of wire screens as diffusion battery cells, J Colloid Interface Sci 86:12-16.

You M, Wang Y, Stoner G, You L, Maronpot R, Reynolds SH, Anderson M. 1992. Parental bias of Ki-ras oncogenes detected in lung tumors from mouse hybrids. Proc Natl Acad Sci USA 89:5804-5808.

Zeger S, Liang KY, Albert PS. 1988. Models for longitudinal data: A general estimating equation approach. Biometrics 44:1049-1060.

Zhu LX, Waldren CA, Vannias D, Hei TK. 1996. Cellular and molecular analysis of mutagenesis induced by charged particles of defined linear energy transfer. Radiat Res 145(3):251-259.

Zielinski JM and Krewski D. 1990. Application of the two-stage model clonal expansion model in characterizing the joint effect of exposure to two carcinogens. Pp. 846-880 in Indoor Radon and Lung Cancer: Reality or Myth? 29th Hanford Symposium on Health and Environment, Columbus, Ohio. Health and Environment. Columbus: Battle Press.

# Glossary

*Absolute risk.* An expression of excess risk based on the assumption that the excess risk from radiation exposure *adds* to the underlying (baseline) risk by an increment dependent on dose but independent of the underlying natural risk.

*Absorbed dose.* The mean energy imparted by ionizing radiation to an irradiated medium per unit mass. Units: gray (Gy), rad. 1 Gy = 100 rad.

*Action level.* A concentration of radon in air that the Environmental Protection Agency recommends should be the maximum concentration in homes; i.e., concentrations above this level should be reduced or mitigated to this level. In 1998, EPA recommends an action level of 148 Bqm$^{-3}$ or 4 pCiL$^{-1}$.

*Activity.* The amount of radionuclide radioactivity defined as the mean number of decays per unit time. Units: becquerel (Bq), curie (Ci). 1 Bq = 2.7 x 10$^{-11}$ Ci.

*Activity Median Aerodynamic Diameter (AMAD).* The diameter of a unit-density sphere with the same terminal settling velocity in air as that of the aerosol particle whose activity is the median for the environment.

*Additive effects.* Equal to the sum of effects from two agents when acting alone.

*Aerosol.* Solid or liquid particles that are dispersed in a gaseous medium, are able to remain suspended for a relatively long time, and that have a high surface area to volume ratio.

*Alpha particle.* Two neutrons and two protons bound as a single particle that is emitted from the nucleus of certain radioactive isotopes in the process of decay or disintegration. Is positively charged and indistinguishable from a helium atom nucleus.

*Apoptosis.* Programmed cell death. The cell death is characterized by a distinctive fragmentation of DNA which is regulated by cellular functions.

*Attributable risk (AR).* The estimated burden of a disease (such as lung cancer) that could, in theory, be prevented if all exposures to a particular causative agent (such as radon) were eliminated.

*Background radiation.* The amount of radiation to which a member of the population is exposed from natural sources, such as terrestrial radiation due to naturally occurring radionuclides in the soil, cosmic radiation originating in outer space, and naturally occurring radionuclides deposited in the human body.

*Baseline rate.* The cancer incidence observed in a population in the absence of the specific agent being studied; the baseline rate includes cancers from a number of other causes, such as smoking, background radiation, etc.

*Becquerel (Bq).* SI unit of activity. (see Units). 1 Bq = 1 disintegration per second.

*BEIR IV.* Refers to the report of the fourth National Research Council Committee on *B*iological *E*ffects of *I*onizing *R*adiation; the report was published in 1988.

*Bias.* Factors that influence the outcome of data collection such as causing certain measurements to have a greater chance of being included than others.

*Bronchial morphometry.* Characterization of the cellular and anatomical structure of the bronchial region of the lung.

*Cancer.* A malignant tumor of potentially unlimited growth, capable of invading surrounding tissue or spreading to other parts of the body by metastasis.

*Carcinogen.* An agent that is believed to be able to cause cancer. Ionizing radiations are physical carcinogens; there are also chemical and biologic carcinogens and biologic carcinogens may be external (e.g., viruses) or internal (genetic defects).

*Carcinoma.* A malignant tumor (cancer) of epithelial origin.

*Case-control study.* An epidemiologic study in which people with disease and a similarly composed control group are compared in terms of exposures to a putative causative agent.

*Cell Culture.* The growing of cells in vitro (in a glass or plastic container, or in suspension) in such a manner that the cells are no longer organized into tissues.

*Cohort study.* An epidemiologic study in which groups of people (the cohort) are identified with respect to the presence of, or absence of, exposure to a disease-causing agent and in which the outcomes in terms of disease rates are compared; also called a follow-up study.

*Competing risks.* Other causes of death which affect the outcome of the risk being studied. Persons dying from other causes are not recorded at risk of dying from the factor in question.

*Confidence limits or intervals.* A measure of the reliability of a risk estimate. A 90% confidence interval means that 9 times out of 10 the estimated risk would be within the specified interval.

*Constant relative risk (CRR).* A risk model which assumes that, after a certain time, the ratio of the risk at a specific dose to the risk in the absence of the dose does not change with time.

*Curie (Ci).* A unit of activity equal to 3.7 x $10^{10}$ disintegrations/s or 3.7 x $10^{10}$ becquerels. (see Units).

*DNA.* Deoxyribonucleic acid; the genetic material of cells.

*Deletions.* Type of mutation in which sections of DNA are removed; can refer to the removal of a single base or several bases.

*Dose.* The quantity of energy or chemical agent delivered to a specific tissue following exposure. (see Absorbed dose).

*Dose-distribution factor.* A factor which accounts for modification of the dose effectiveness in cases in which the radionuclide distribution and the resultant dose are non-uniform.

*Dose-effect (dose-response) model.* A mathematical formulation and description of the way the effect (or biological response) depends on the dose.

*Dose equivalent.* A quantity that expresses, for the purposes of radiation protection and control, an assumed equal biological effectiveness of a given absorbed dose on a common scale for all kinds of ionizing radiation. SI unit is the Sievert (see Units). 1 Sv = 100 rem.

*Dose rate.* The quantity of absorbed dose delivered per unit time.

*Dose Rate Effectiveness Factor (DREF).* A factor by which the effect caused by a specific type of radiation changes at low (protracted or fractionated delivery of dose) as compared to high (or acute) dose rates.

*Dosimetric model.* A method for estimating risk based on the use of physical models for doses to target cells and the use of results from epidemiologic studies of exposures to humans from other types of radiations.

*Ecologic study.* A method of epidemiologic analysis in which, for radon, regional rates of lung cancer are related to the measure of regional radon concentrations.

*Effective attributable risk (EAR).* The reduced attributable risk such as the fraction of total lung-cancer deaths that would be eliminated by implementing a radon-mitigation scenario.

*Electron volt (eV).* A unit of energy = 1.6 x $10^{-12}$ ergs or 1.6 x $10^{-19}$ J; 1 eV is equivalent to the energy gained by an electron in passing through a potential difference of 1 V; 1 keV-1,000 eV; 1 MeV - 1,000,000 eV.

*Empirical model.* A model that is derived from measurements in populations as opposed to a theoretical model.

*Epidemiology.* The study of the determinants of the frequency of disease in humans. The two main types of epidemiological studies of chronic disease are cohort (or follow-up) studies and case-control (or retrospective) studies.

*Equilibrium fraction.* In equilibrium, the radioactivity from the parent nuclides and progeny nuclides are equal.

*Etiology.* The science or description of cause(s) of disease.

*Ever-smokers.* People who smoked cigarettes for a period of time, regardless of whether they have stopped smoking.

*Excess relative risk (ERR).* A model that describes the risk imposed by exposures as a multiplicative increment to the excess disease risk above the background rate of disease.

*Exposure.* The condition of having contact with a physical or chemical agent.

*Exposure-age-concentration model.* Risk model based on the average radon concentration.

*Exposure-age-duration model.* Risk model based on the duration of exposure to radon.

*Fibrosis.* Damage to normal tissue which results in a modification of tissue structure but which is not cancer.

*Fractionation.* The delivery of a given dose of radiation as several smaller doses, separated by intervals of time.

*Gamma radiation.* Also gamma rays; short wavelength electromagnetic radiation of nuclear origin, similar to x rays but usually of higher energy (100 keV to several MeV).

*Geometric mean.* The geometric mean of a set of positive numbers is the exponential of the arithmetic mean of their logarithms. The geometric mean of a lognormal distribution is the exponential of the mean of the associated normal distribution.

*Geometric standard deviation (GSD).* The geometric standard deviation of a lognormal distribution is the exponential of the standard deviation of the associated normal distribution.

*Gray (Gy).* SI unit of absorbed dose (see Units). 1 Gy = 100 rad or deposition of 1 joule per Kg.

*Half-life, biologic.* Time required for the body to eliminate half of an administered dose of any substance by regular processes of elimination; it is approximately the same for both stable and radioactive isotopes of a particular element.

*Half-life, radioactive.* Time required for a radioactive substance to lose 50% of its activity by decay.

*Histologic types.* Pathologists have identified 4 principal lung cancers based upon microscopic analysis of cellular characteristics: squamous cell carcinoma, adenocarcinoma, small cell carcinoma, and large cell carcinoma.

*Incidence.* Or incidence rate; the rate of occurrence of a disease within a specified period of time, often expressed as a number of cases per 100,000 individuals per year.

*Inhalation.* To draw air into the lungs by breathing; considered an exposure route for radon and radon progeny.

*In utero.* In the womb, i.e., before birth.

*Inverse dose-rate effect.* An effect in which, for a given exposure, the probability of effect increases as the dose rate is lowered.

*In vitro.* Refers to cell culture conditions in glass or plastic containers as opposed to *in vivo*, in the living individual.

*In vivo.* In the living organism.

*Ionizing radiation.* Radiation sufficiently energetic to dislodge electrons from an atom thereby causing an ion pair. Ionizing radiation includes x and gamma radiation, electrons (beta radiation), alpha particles (helium nuclei), and heavier charged atomic nuclei. Neutrons ionize indirectly by first colliding with components of atomic nuclei.

*Isotopes.* Nuclides that have the same number of protons in their nuclei, and hence the same atomic number, but that differ in the number of neutrons, and therefore in the mass number; chemical properties of isotopes of a particular element are almost identical.

*K factor.* A dimensionless parameter in the risk model that characterizes the comparative doses to lung cells for exposures in homes compared to similar exposures in mines.

*Latent period.* The period of time between exposure and expression of the disease. After exposure to a dose of radiation, there is a delay of typically several years (the latent period) before any cancers are observed.

*Life table.* A table showing the number of persons who, of a given number born or living at a specified age, live to attain successive higher ages, together with the numbers who die in each interval.

*Lifetime relative risk (LRR).* The relative increment in lung cancer risk resulting from exposure to an agent such as radon.

*Linear energy transfer (LET).* Average amount of energy lost per unit track length.

*Low LET radiations.* Light, charged particles such as electrons or x rays and gamma rays that produce sparse ionizing events far apart on the scale of a cellular nucleus.

*High LET radiations.* Heavy, charged particles such as protons and alpha particles that produce dense ionizing events close together on the scale of a cellular nucleus.

*Linear (L) model or relationship.* Also, linear dose-effect relationship; expresses the effect (e.g., mutation or cancer) as a proportional (linear) function of the dose.

*Linear-quadratic (LQ) model.* Also, linear-quadratic dose-effect relationship; expresses the effect (e.g., mutation or cancer) as a function of two components, one directly proportional to the dose (linear term) and one proportional to the square of the dose (quadratic term). The linear term will predominate at lower doses, the quadratic term at higher doses.

*Lognormal distribution.* When the logarithms of a randomly distributed quantity have a normal (Gaussian) distribution.

*Mechanistic basis.* An explanation derived from a knowledge of the individual stages leading to an effect.

*Meta-analysis.* A method for analyzing epidemiologic data based on grouping or pooling information obtained from several studies.

*Mitigation.* The act of reducing radon concentrations in homes.

*Model.* A schematic description of a system, theory, or phenomenon that accounts for its known or inferred properties and may be used for further study of its characteristics.

*Monte Carlo Calculation.* The method for evaluation of a probability distribution by means of random sampling.

*Mortality (rate).* The rate to which people die from a disease, e.g., a specific type of cancer, often expressed as deaths per 100,000 per year.

*Multiplicative interaction model (MIM).* The assumption that the relative risk (the relative excess risk plus one) resulting from the exposure to two risk factors is the product of the relative risks from the two factors taken separately.

*Neoplasm.* Any new and abnormal growth, such as a tumor; neoplastic disease refers to any disease that forms tumors, whether malignant or benign.

*Never-smokers.* People who have not smoked cigarettes.

*Nonstochastic.* Describes effects whose severity is a function of dose; for these, a threshold may occur; some examples of somatic effects believed to be nonstochastic are cataract induction, nonmalignant damage to skin, hematological deficiencies, and impairment of fertility.

*Normal distribution.* Referring to the so-called "bell-shaped curve" of randomly distributed quantities; also referred to as a "Gaussian distribution."

*Nuclide.* A species of atom characterized by the constitution of its nucleus, which is specified by its atomic mass and atomic number (Z), or by its number of protons (Z), number of neutrons (N), and energy content.

*Oncogenes.* Genes which encode the potential for cancer.

*Person-gray.* Unit of population exposure obtained by summing individual dose-equivalent values for all people in the exposed population. Thus, the number of person-grays by 1 person exposed to 1 Gy is equal to that contributed by 100,000 people each exposed to 10 µGy.

*Person-years-at-risk (PYAR).* The number of persons exposed times the numbers of years after exposure minus some lag period during which the dose is assumed to be unexpressed (minimum latent period).

*Pooled analysis.* When data from more than one study is combined for evaluation.

*Potential Alpha Energy Concentration (PAEC).* The concentration of potential alpha energy from radon progeny suspended in a volume of air. PAEC is measured in quantities of $Jm^{-3}$ or Working Levels (1 WL = 2.08 × $10^{-5}$ $Jm^{-3}$)

*Prevalence.* The number of cases of a disease in existence at a given time per unit of population, usually 100,000 persons.

*Probability of causation.* A number that expresses the probability that a given cancer, in a specific tissue, has been caused by a previous exposure to a carcinogenic agent, such as radiation.

*Progeny.* The decay products resulting after a series of radioactive decay. Progeny can also be radioactive, and the chain continues until a stable nuclide is formed.

*Projection model.* A mathematical model that simultaneously describes the excess cancer risk at different levels of some factor such as dose, time after exposure, or baseline level of risk, in terms of a parametric function of that factor. It becomes a projection model when data in a particular range of observations is used to assign values to the parameters in order to estimate (or project) excess risk for factor values outside that range.

*Promoter.* An agent which is not by itself carcinogenic, but which can amplify the effect of a true carcinogen by increasing the probability of late-stage cellular changes needed to complete the carcinogenic process.

*Protraction.* The spreading out of a radiation dose over time by continuous delivery at a lower dose rate.

*Pulmonary interstitium.* The spacing between the linings in the structure of the lung.

*Quadratic-dose model.* A model which assumes that the excess risk is proportional to the square of the dose.

*Quality factor (Q).* An LET-dependent factor by which the absorbed doses are multiplied to obtain (for radiation protection purposes) a quantity which corresponds more closely to the degree of the biological effect produced by x or low-energy gamma rays. Dose in $Gy \times Q$ = Dose equivalent in Sv.

*Rad.* A unit of absorbed dose equal to 100 ergs of energy absorbed per gram of tissue. Replaced by the gray in SI units. 100 rad = 1 Gy. (see Units).

*Radiation.* Energy emitted in the form of waves or particles by radioactive atoms as a result of radioactive decay.

*Radioactivity.* The property of nuclide decay in which particles or gamma radiations are usually emitted.

*Artificial radioactivity.* Man-made radioactivity produced by fission, fusion, particle bombardment, or electromagnetic irradiation.

*Natural radioactivity.* The property of radioactivity exhibited by more than 50 naturally occurring radionuclides.

*Radiogenic.* Caused by radiation.

*Radioisotope.* A radioactive atomic species of an element with the same atomic number and usually identical chemical properties.

*Radionuclide.* A radioactive species of an atom characterized by the constitution of its nucleus.

*Radon.* A naturally occurring radioactive gas produced from uranium; decays to form radon progeny.

*Radon progeny.* The radioactive products formed following the radioactive decay of radon; radionuclides which when inhaled can expose living cells to their emitted alpha particles.

*Relative risk.* An expression of excess risk relative to the underlying (baseline) risk; if the excess equals the baseline risk the relative risk is 2.

*Relative biological effectiveness (RBE).* An adjustment factor used to qualify an absorbed dose to account for its relative potential to do damage in biologic tissues. RBE is standardized to effects caused by x rays of a standard energy.

*Rem.* (*rad* equivalent *mammal*); unit of dose equivalent. The unit of dose equivalent or "rem" is numerically equal to the absorbed dose in "rad" multiplied by the "quality factor" (see Quality factor), "relative biological effectiveness" (see Relative biological effectiveness), the distribution factor, and any other necessary modifying factor. 1 Rem = 0.01 Sievert.

*Respiratory epithelium.* The cells lining the lung surfaces.

*Retinoblastoma.* An eye tumor that is an example of an inherited malignant tumor with a dominant autosomal gene inheritance pattern.

*Risk.* A chance of injury, loss, or detriment. A measure of the deleterious effects that may be expected as the result of an action or inaction.

*Risk assessment.* The process by which the risks associated with an action or inaction are identified and quantified.

*Risk coefficient.* The increase in the annual incidence or mortality rate per unit dose: (1) absolute risk coefficient is the observed minus the expected number of cases per person year at risk for a unit dose; (2) relative-risk coefficient is the fractional increase in the baseline incidence or mortality rate for a unit dose.

*Risk estimate.* The number of cases (or deaths) that are projected to occur in a specified exposed population per unit dose for a specified exposure regime and expression period: number of cases per person-gray or, for radon, the number of cases per person cumulative working-level-month.

*SI units.* The International Systems of Units as defined by the General Conference of Weights and Measures in 1960. These units are generally based on meter/kilogram/second units, with special quantities for radiation including the becquerel, gray, and sievert.

*Sievert.* The SI unit of radiation dose equivalent. It is equal to dose in gray times a quality factor or times other modifying factors; for example, a distribution factor; 1 sievert (Sv) equals 100 rem. (see Units).

*Specific Activity.* Total activity of a given nuclide per gram of a compound, element, or radioactive nuclide.

*Specific energy.* The actual energy per unit mass deposited per unit volume in a given event. This is a stochastic quantity as opposed to the average value over a larger number of instances (i.e., the absorbed dose).

*Squamous cell carcinoma.* A cancer composed of cells that are scaly or platelike.

*Standard mortality ratio (SMR).* Standard mortality ratio is the ratio of the disease or accident mortality rate in a certain specific population compared with that in a standard population. The ratio is based on 100 for the standard so that an SMR of 200 means that the test population has twice the mortality from that particular cause of death.

*Stochastic.* Random events leading to effects whose probability of occurrence in an exposed population (rather than severity in an affected individual) is a direct function of dose; these effects are commonly regarded as having no threshold; hereditary effects are regarded as being stochastic; some somatic effects, especially carcinogenesis, are regarded as being stochastic.

*Submultiplicative effects.* Effects less than the anticipated effect if the joint effect were the product of the risks from the two agents individually (but more than if the joint effect were the sum of the individual risks).

*Suppressor gene.* A gene which can suppress another gene such as an oncogene. Changes in suppressor genes can lead to expression by genes such as oncogenes.

*Synergistic.* An increased effectiveness resulting from an interaction between two agents so that the total effect is greater than the sum of the effects of the two agents acting alone.

*Target cells.* Cells in a tissue that have been determined to be the key cells in which changes occur in order to produce an endpoint such as cancer.

*Threshold hypothesis.* The assumption that no radiation injury occurs below a specified dose.

*Time-since-exposure (TSE) model.* A model in which the risk is not constant but varies with the time after exposure.

*Transfection.* The introduction of DNA into a host cell.

*Transformed cells.* Tissue-culture cells changed from growing in an orderly pattern exhibiting contact inhibition to growing in a pattern more like that of cancer cells.

*Uncertainty.* The range of values within which the true value is estimated to lie. It is a best estimate of possible inaccuracy due to both random and systemic errors.

    *Random Errors.* Errors that vary in a nonreproducible way around a limiting mean. These errors can be treated statistically by use of the laws of probability.

    *Systemic Errors.* Errors that are reproducible and tend to bias a result in one direction. Their causes can be assigned, at least in principle, and they can have constant and variable components. Generally, these errors cannot be treated statistically.

*Units*

| Units[a] | Conversion Factors |
|---|---|
| Becquerel (SI) | 1 disintegration/s = $2.7 \times 10^{-11}$ Ci |
| Curie | $3.7 \times 10^{10}$ disintegrations/s = $3.7 \times 10^{10}$ Bq |
| Gray (SI) | 1 J/kg – 100 rad |
| Rad | 100 erg/g; also 0.01 Gy |
| Rem | 0.01 Sievert |
| Sievert (SI) | 100 Rem |

[a]International Units are designated SI.

*UNSCEAR.* *U*nited *N*ations *S*cientific *C*ommittee on the *E*ffects of *A*tomic *R*adiation publishes periodic reports on sources and effects of ionizing radiation.

*Variability.* The variation of a property or a quantity among members of a population. Such variation is inherent in nature and is often assumed to be random. However, it can sometimes be represented by a frequency distribution.

*Working levels (WL).* Any combination of the short-lived progeny of radon in 1 liter of air, under ambient temperature and pressure, that results in the ultimate emission of $1.3 \times 10^5$ MeV of alpha particle energy. This is approximately the total amount of energy released over a long period of time by the short-lived progeny in equilibrium with 100 pCi of radon. 1 WL = $2.08 \times 10^{-5}$ Jm$^{-3}$

*Working level months (WLM).* A cumulative exposure equivalent to one working level for a working month (170 hours). 1 WLM = $2 \times 10^{-5}$ Jhm$^{-3} \times 170$ h = $3.5 \times 10^{-3}$ Jhm$^{-3}$.

*x radiation.* Also x rays; penetrating electromagnetic radiation, usually produced by bombarding a metallic target with fast electrons in a high vacuum.

*Xeroderma pigmentosum (XP).* An inherited disease in which skin cells are highly susceptible to cancer; XP cells have a defect in DNA repair which results in the accumulation of DNA damage after ultraviolet irradiation which apparently accounts for the development of cancer.

# Committee Biographies

**Jonathan M. Samet, M.D., M.S., (Chairman)**, is presently professor and chairman of the Department of Epidemiology of the Johns Hopkins University School of Hygiene and Public Health. Dr. Samet received a Bachelor's degree in Chemistry and Physics from Harvard College, an M.D. degree from the University of Rochester School of Medicine and Dentistry, and a Master of Science in epidemiology from the Harvard School of Public Health. He is trained as a clinician in the specialty of internal medicine and in the subspecialty of pulmonary diseases. From 1978 through 1994, he was a member of the Department of Medicine at The University of New Mexico School of Medicine and he is presently co-director of the Risk Sciences and Public Policy Institute at the Johns Hopkins University School of Hygiene and Public Health. His research has addressed the effects of environmental agents on human health, including risks of lung cancer and other diseases in uranium miners. He was a member of the BEIR IV Committee and chair of the Panel on Dosimetric Assumptions Affecting the Application of Radon Risk Estimates of the National Research Council. He has served on the Science Advisory Board for the U.S. Environmental Protection Agency. He has been a member of the Commission on Life Sciences and the Board on Radiation Effects Research and was elected to the Institute of Medicine in 1997.

**David Brenner, Ph.D.,** is a professor of Radiation Oncology and Public Health at Columbia University, New York. He earned a M.Sc. from St. Bartholomews' Hospital, University of London, in Radiation Physics, a Ph.D. from the University of Surrey in Physics, and has been awarded an honorary D.Sc. degree by Oxford University. He is a radiation biophysicist focusing on mechanisms of induction of DNA and chromosomal damage by different radiations. His research also involves modeling and analyzing low dose epidemiological studies, as well as the analysis and design of fractionation schedules for radiotherapy. He has authored books on *Radiation, Risk and Remedy* and *Making the Radiation Therapy Decision* as well as publishing around 120 papers in the peer-reviewed literature. He has served on various EPA, NAS, IAEA, and NCRP committees. He is a past recipient of the Radiation Research Society Annual Research Award, and the Robert D. Moseley Annual Award from the National Council on Radiation Protection and Measurements.

**Antone L. Brooks, Ph.D.**, is currently working as a senior research scientist at Washington State University Tri-Cities. Formerly he was the Cell and Molecular Biology Group head at Lovelace Inhalation Toxicology Research Institute and then was the manager of the Cellular and Mammalian Biology section at Battelle Pacific Northwest Laboratories. He has conducted extensive research on the health effects of internally deposited radioactive materials with special emphasis on the induction of chromosome aberrations and cancer from low-dose and dose-rate exposures. His research has used cellular and molecular changes to relate exposure to dose and to determine the role of early cellular changes during the process of cancer induction. He has conducted research using radon in experimental animals and cellular systems and using biodosimetry, has defined the relationships that exist between exposure, dose, and alpha-particle traversals. A major goal of this research is to link mechanistic cellular and molecular biology studies to human-health risk.

**William H. Ellett, Ph.D.**, is retired and lives in Crofton, Maryland. He was formerly a senior staff officer in the Board on Radiation Effects Research at the National Academy of Sciences. Dr. Ellett was study director for the BEIR IV and V Studies of the National Research Council.

**Philip K. Hopke, Ph.D.**, is the R.A Plane Professor of Chemistry at Clarkson University. He holds a joint appointment in the Department of Civil and Environmental Engineering and was recently appointed as the Dean of the Graduate School. Dr. Hopke has been studying radon-progeny behavior since 1975 and has made major contributions to the understanding of the neutralization, deposition, and radiolytic formation of particles. His group has made assessments of radon decay-product exposure in normally occupied homes and the incremental effects of showering with radon-laden water. He has been a member of several National Research Council committees assessing the risk of exposure to hazardous air pollutants. He previously served as one of the principal scientists in the U.S. Department of Energy's Radon Research Program and was the General Chair of the Sixth International Symposium on the Natural Radiation Environment.

**Ethel S. Gilbert, Ph.D.**, spent several years as a biostatistician and senior staff scientist at Battelle, Pacific Northwest Laboratories, but recently joined the Radiation Epidemiology Branch of the National Cancer Institute as a special expert. Her research has focused on epidemiologic studies of nuclear workers, which has included combined analyses of national and international data and the development of statistical methods for examining the relationship of health effects and low-level chronic radiation exposures. She has also analyzed data on experimental animals exposed to radon and to inhaled plutonium. Dr. Gilbert was a member of the working group responsible for revising the health effects model in the U.S. Nuclear Regulatory Commission's Reactor Safety Study, where she provided and updated a model for estimating cancer risks. Dr. Gilbert is a fellow of the American Statistical Association and a member of the National Council on Radiation Protection and Measurements.

**Dudley T. Goodhead**, Ph.D., heads the Radiation and Genome Stability Unit of the Medical Research Council (MRC) in the United Kingdom and he is an honorary professor of Brunel University. He was formerly deputy director of the MRC Radiobiology Unit. Dr. Goodhead earned his D.Phil. from the University of Oxford in high-energy physics and subsequently held academic appointments at the Universities of California, London, and Natal. His own research has been predominately in the field of the biophysics of radiation effects. Dr. Goodhead has served on the Task Group on the Biological Effectiveness of Neutrons of the Committee on Interagency Radiation Research and Policy Coordination, the Committee on Medical Aspects of Radiation in the Environment (UK), and other national and international committees involved with the biological effects of radiation. He has served as consultant to the United Nations Scientific Committee on the Effects of Atomic Radiation, the International Atomic Energy Agency, and the General Accounting Office of the U.S. Congress. He is Councillor for Physics of the International Association of Radiation Research and has served on the editorial boards of the *Radiation Research Journal* and the *International Journal of Radiation Biology*.

**Eric J. Hall, Ph.D.**, is the Higgins Professor of Radiation Biophysics, professor of Radiology and Radiation Oncology, and director of the Center for Radiological Research. Dr. Hall received his M.A., D. Phil., and D.Sc. degrees in radiation biology from Oxford University. He is past-president of the Radiation Research Society and currently president-elect of the International Association of Radiation Research and Secretary of the American Radium Society. He is Senior editor for biology of the *International Journal of Radiation Oncology Biology Physics*. Dr. Hall has received the Gold Medal of both the Radiological Society of North America (1992) and the American Society of Therapeutic Radiology and Oncology (1993), the Janeway Medal of the American Radium Society as well as the Failla Medal of the Radiation Research Society (1991). Dr. Hall is a member of the National Council for Radiation Protection and served on the BEIR V Committee.

**Daniel Krewski, Ph.D., M.H.A.**, is the director of Risk Management, Health Canada, professor of epidemiology and biostatistics in the Department of Epidemiology and Community Medicine at the University of Ottawa, and adjunct research professor of statistics from Carleton University. Dr. Krewski earned his M.S. and Ph.D. in mathematics and statistics from Carleton University, and his M.H.A. from the University of Ottawa. He is associate editor of *Risk Analysis*, the *Journal of Epidemiology and Biostatistics*, and *Regulatory Toxicology and Pharmacology*. He is currently a member of the NRC Board on Environmental Studies and Toxicology and its Committee on Toxicology, and he recently chaired the NRC's Colloquium on Scientific Advances and the Future in Toxicologic Risk Assessment.

**Jay H. Lubin, Ph.D.**, is a mathematical statistician in the Division of Cancer Epidemiology and Genetics at the U.S. National Cancer Institute. He holds a B.A. and a M.A. in mathematics from the University of California, Los Angeles, and a Ph.D. in Biostatistics from the University of Washington, Seattle. He has been a member of the Regional Advisory Board of the Biometrics Society and a recipient of the U.S. Public Health Service Special Recognition Award. He is currently associate editor of *Biometrics*. In 1994, Dr. Lubin was elected a Fellow of the American Statistical Association. His statistical research interests are the design and analysis of epidemiology studies. Dr. Lubin has conducted extensive epidemiology research into the etiology of cancer, particularly in the areas of occupational exposures and radiation, and of lung cancer and exposure to radon and radon progeny.

**Roger O. McClellan, D.V.M.**, serves as president of the Chemical Industry Institute of Toxicology located in Research Triangle Park, NC, a position he has held since 1988. He is well known for his work in the related fields of toxicology and risk assessment, especially concerning the potential human risks of airborne materials. McClellan has previously served as president of the Society of Toxicology and American Association for Aerosol Research. He is a Fellow of the Society for Risk Analysis and an elected member of the Institute of Medicine of the National Academy of Sciences. He has served in an advisory role to many public and private organizations including Chairmanship of the U.S. Environmental Protection Agency Clean Air Scientific Advisory Committee and the National Research Council Committee on Toxicology. McClellan is a strong advocate of the need to integrate data from epidemiological, controlled clinical, laboratory animal, and cell studies to assess human health risks of occupational or environmental exposures to chemicals.

**Paul L. Ziemer, Ph.D.**, is professor and head of the School of Health Sciences, Purdue University, a position he has held since 1983. From August, 1990, to January, 1993, Dr. Ziemer took a leave of absence from Purdue to serve as the Assistant Secretary of Energy for Environment, Safety, and Health at the U.S. Department of Energy during the Bush Administration. He has held several positions at Purdue relating to health sciences, health physics, and bionucleonics. As head of the School of Health Sciences, Dr. Ziemer is administratively responsible for teaching and research programs in industrial hygiene, health physics, medical physics, environmental health, and medical technology. Dr. Ziemer has served as a health physicist at Oak Ridge National Laboratory, as a Radiological Physics Fellow at Vanderbilt University, and as a physicist at the U.S. Naval Research Laboratory in Washington, D.C. He is a Certified Health Physicist and has been national president of the Health Physics Society, president of the American Academy of Health Physics, and a member of the American Board of Health Physics.

# Index

## A

Absolute risk, 43-44, 64, 74, 75, 90, 134-135, 137, 164, 201, 226-228, 301

Action levels

   EPA, 3, 13, 19, 24, 31, 94, 356, 357

   foreign authorities, 31, 358

Adenocarcinomas, 33, 239, 249, 250, 251, 346, 355, 386, 390, 392, 396, 398, 399, 402-410 (passim), 414, 415

Aerosol characteristics, general, 199, 203, 208, 214, 223

   mine measurements, 219, 194

   size distribution, 31, 180-181, 183-184, 311

   smoking-generated, 252

Age factors, 31, 71, 82, 83, 88, 89, 102, 103, 104, 110, 137, 140, 141, 144, 167, 210, 212, 213, 255, 274

   attained age, 8, 74, 75-76, 80-84 (passim), 110, 114, 134-139 (passim), 144, 147, 149, 151, 257-272 (passim), 275, 280, 281, 285, 286, 287, 376

   diesel fumes as carcinogen, 352

   dosimetry, 9, 119

   ecologic studies, 360, 361, 366-360, 376

   exposure-age-concentration, 8, 11, 12, 14, 15, 18, 25-26, 80-82, 84-87, 92-99 (passim), 105-109, 113, 115, 147-149, 151-154, 173, 175

   exposure-age-duration, 8, 11, 12, 14, 15, 18, 25-26, 80-87, 92-99 (passim), 105, 106-109, 113, 147-149, 151-154, 174, 175

   first exposure, age at, 134, 136, 138, 139, 140, 165, 254, 255, 256-275, (passim), 280, 282, 284, 285-286, 290, 292, 315, 317, 319, 322, 325

   gender factors and, 12, 232, 233

   inverse dose rate effect, 60, 61

   lifetime relative risk, 110-112, 134-135, 167, 171

   relative risk/excess relative risk, 8, 25, 81, 84, 87, 88, 90, 93, 110, 114, 134-135, 136, 138, 147, 151-154, 164-165, 280, 284

   smoking, 95-96, 143, 226, 232, 233-234, 238, 239, 240, 371

   *see also* Children; Infants

Airways and airway diseases, 47, 50, 205, 208

American Cancer Society, 238, 239

Animal studies, 5, 22, 27, 37

   diesel fumes as carcinogen, 351-354

   extrapolation to human effects, 63, 243

   genetic susceptibility, 38, 42, 43-44, 68, 200-201

   mycotoxins as carcinogens, 355

   nonmalignant respiratory diseases, 117